T0230152

Lecture Notes in Computer Science 1155

Edited by G. Goos, J. Hartmanis and J. van Leeuwen

Advisory Board: W. Brauer D. Gries J. Stoer

Springer

Berlin
Heidelberg
New York
Barcelona
Budapest
Hong Kong
London
Milan
Paris
Santa Clara
Singapore
Tokyo

James Roberts   Ugo Mocci
Jorma Virtamo   (Eds.)

Broadband
Network Traffic

Performance Evaluation and Design
of Broadband Multiservice Networks

Final Report of Action COST 242

 Springer

Series Editors

Gerhard Goos, Karlsruhe University, Germany

Juris Hartmanis, Cornell University, NY, USA

Jan van Leeuwen, Utrecht University, The Netherlands

Volume Editors

James Roberts
France Telecom CNET/PAA/ATR
38-40 rue du Général-Leclerc, F-92131 Issy-les-Moulineaux Cedex, France
E-mail: roberts@issy.cnet.fr

Ugo Mocci
Fondazione Ugo Bordini
Via B. Castiglione 59, I-00142 Rome, Italy

Jorma Virtamo
Technical Research Centre of Finland, VTT Information Technology
P.O. Box 1202, 02044 Espoo, Finland

Cataloging-in-Publication data applied for

Die Deutsche Bibliothek - CIP-Einheitsaufnahme

Broadband network teletraffic : performance evaluation and
design of broadband multiservice networks ; final report of
action COST 242 / James Roberts ... (ed.). - Berlin ;
Heidelberg ; New York ; Barcelona ; Budapest ; Hong Kong ;
London ; Milan ; Paris ; Santa Clara ; Singapore ; Tokyo :
Springer, 1996
  (Lecture notes in computer science ; Vol. 1155)
  ISBN 3-540-61815-5
NE: Roberts, James [Hrsg.]; GT

CR Subject Classification (1991): C.2, H.4.3, H.5.1, D.4.4, J.2, K.6

ISSN 0302-9743
ISBN 3-540-61815-5 Springer-Verlag Berlin Heidelberg New York

© Springer-Verlag Berlin Heidelberg 1996
Printed in Germany

Typesetting: Camera-ready by author
SPIN 10525565   06/3142 - 5 4 3 2 1 0   Printed on acid-free paper

Preface

This is the second COST book on the performance evaluation and design of multiservice networks. The first book was the final report of Action COST 224, published in 1992 by the Commission of the European Communities [Rob92a]. This book was very well received by the telecommunications networking research community and its success has encouraged the members of Action COST 242, which has been built on the nucleus of COST 224, to again draft its final report in the form of a monograph. This book is thus the record of a collaboration over four years of a group of telecommunications industry and university researchers participating in Action COST 242.

COST stands for COoperation in Science and Technology and represents a form of cooperation between European countries established since 1970. COST actions in a wide range of disciplines proceed on the basis of a Memorandum of Understanding (MoU) setting out the objectives and scope of the activity to be undertaken by signatory nations over a period of several years. The organisation of each action is determined by a Management Committee containing up to two members per signatory nation. The type of organisation can be quite different from one action to another, flexibility being a hallmark of cooperation in the COST framework. There are currently 25 COST member countries: the fifteen European Union countries plus Croatia, the Czech Republic, Hungary, Iceland, Norway, Poland, Slovakia, Slovenia, Switzerland, and Turkey. Telecommunications is one of the most active of the 15 areas of research covered by COST with some 25 actions currently in progress.

COST 242 began in 1992 and ended in 1996. It was built on the basis of the previous Action COST 224 which itself grew out of COST 214. The origin of this collaboration on multiservice networks thus dates back to 1985. Working methods have been based on two-day Management Committee meetings held at three to four-monthly intervals. The majority of meeting time is devoted to the presentation and discussion of research performed by participants. The number of participating organisations has grown quite sharply in the latest action and a typical meeting has had an attendance of around 40 with the presentation of some 20 research reports. The scope of the action covers a wide range of broadband network design issues and is accurately reflected in the table of contents of the present report.

Collaboration between COST 242 participants has been spontaneous rather than planned, proceeding by a mixture of emulation and constructive criticism. The creation of a critical mass of research effort in the field of multiservice network performance evaluation and traffic engineering has been a major advantage of COST 242 and its predecessors. Interesting new topics introduced by one member naturally lead to further research by others while blind alleys are rapidly identified. COST 242, more than the previous actions, has benefitted from a synergy between the telecommunications operators and university researchers. This form of collaboration is complementary to that of more structured projects in RACE, the European Framework programme on broadband communications. COST 242 has, moreover, maintained useful contacts with related RACE projects, notably COMBINE, BAF, EXPLOIT, and BRAVE.

The countries participating in COST 242 are shown in the table below.

Country	Members	Organisation
Belgium	H. Bruneel	University of Ghent
	G. Latouche	Free University of Brussels
Croatia	B. Zovko-Cihlar	University of Electrotechnic and Computing (Zagreb)
Denmark	B. Veirø	Tele Denmark
	S. Blaabjerg	Technical University of Denmark
Finland	K. Kilkki	Telecom Research Centre
	J. T. Virtamo	VTT
France	A. Gravey	France Télécom - CNET
	J. Roberts	France Télécom - CNET
Germany	R. Kleinewilling-höfer-Kopp	Deutsche Telekom AG
	P. Tran-Gia	University of Würzburg
Hungary	C. Szabo	Technical University of Budapest
Ireland	T. Curran	Dublin City University
Italy	A. Tonietti	CSELT
	U. Mocci	Fondazione Ugo Bordoni
Netherlands	A. H. Roosma	KPN Research
	J. L. van den Berg	KPN Research
Norway	P. T. Huth	Telenor Research
	O. Østerbø	
Poland	W. Burakowski	Technical University of Warsaw
	A. Pach	University of Mining and Metallurgy (Krakow)
Slovakia	P. Podhradsky	Technical University of Bratislava
	M. Konvit	University of Transport and Communications, Zilina
Spain	M. Villen	Telefonica I&D
	O. Casals	Polytechnic University of Catalonia
Sweden	K. Lindberger	Telia Research
	K. Nivert	Telia Research

	S-E. Tidblom	Telia Research
Switzerland	J-Y. Le Boudec	EPFL
	R. Sawaaf	PTT
United Kingdom	P. B. Key	BT Labs
	R. J. Gibbens	University of Cambridge

Participating organisations also included the University of Antwerp, the Polytechnic of Turin, Eindhoven University of Technology, and IBM Zurich. In addition to officially designated members, many "experts" attended Management Committee meetings on a regular basis. In fact, the distinction between member and expert was never overtly recognised in the working of COST 242. The following is a list of persons having attended at least one of the 15 COST 242 meetings, classified according to the number of meetings attended:

15 : K. Lindberger, J. Roberts

14 : A. Tonietti, P. Tran Gia, B. Veirø

13 : W. Burakowski, P. T. Huth, S-E. Tidblom

12 : J. Andrade, P. B. Key, O. Østerbø, C. Scoglio

11 : J. Garcia Vidal, U. Mocci

10 : A. Gravey, C. Szabo

 9 : J. T. Virtamo

 8 : S. Blaabjerg, H. Christiansen, S. Manthorpe, S. Molnár, K. Nivert, I. Norros, M. Ritter, A. Simonian

 7 : J.L. van den Berg, G. Dallos, R. Slosiar

 6 : L. Jereb, O. Rose, F. Brichet, F. Hübner, K. Kilkki

 5 : O. Casals, R. Dittmann, T. V. Do, R. J. Gibbens, K. Laevens, R. Lo Cigno, B. Zovko-Cihlar

 4 : C. Blondia, D. Botvich, H. Bruneel, S. F. Møller, M. Pioro, A. H. Roosma, B. Steyaert

 3 : H. Azmoodeh, S. Fuhrman, A. Dupuis, A. Farago, R. Kleinewillinghöfer-Kopp, M. Konvit, G. Latouche, J. Lubacz, A-M. Møller, S. Murphy, B. F. Nielsen, A. Pach, S. Robert, T. R. Griffiths

 2 : S. Aalto, A. Bak, M. Collier, H. Hennenberg, T. Kalkowski, F. P. Kelly, M. Klimo, T. Kovacikova, C. Lavrijsen, J-Y. Le Boudec, A. Marcek, J. Resing, R. Sawaaf, N. Vicari, M. Villen, K. Wajda

 1 : E. Aarstad, M. Ajmone Marsan, M. D'Ambrosio, A. Andersson, D. Bursztynowski, T. Cinkler, T. Curran, V. Elek, P. Farkas, G. Fodor, P. Droz, S. Garcia, P. Geise, M. de Graaf, R. Grünenfelder, J. Guibert, L. Gyalog, H. Hemmer, I. Iliadis, V. B. Iversen, V. Kalashnikov, P. Kukura, M. Kotuliak, E. Martin, J. McGibney, Z. Papir, H. Pettersen, A. Pfening, P. Podhradsky, A. Skliros, T. Smit, I. Svinnset, K. van der Wal, S. Wittevrongel, M. ten Wolde, F. Wyler, I. Svinnset, K. van der Wal, S. Wittevrongel, M. ten Wolde, F. Wyler

The chairman of the action was J. Roberts, the vice-chairman was B. Veirø, and the secretary was first A. Simonian and then F. Brichet.

The members of the action were deeply distressed by the untimely death in 1994 of Krister Nivert, Management Committee member for Sweden. We hope that this report may be seen as a tribute to his exceptional contribution to successive COST actions on telecommunications network design (COST actions 201, 214, 224, and 242) as well as to his work in traffic engineering in general.

While the main results of Action COST 242 have been to advance European research in the field of multiservice network design and performance evaluation and to encourage cooperation between participating organisations and nations, this final report may be seen as its tangible output. It is a collective work written by the action Management Committee members and experts. We have not wished to identify authors of specific sections because this is frequently not possible due to the collaborative nature of the action. It is appropriate, however, to recognise the considerable effort put into the drafting of this report by a number of people.

Part I was coordinated by J. Roberts with contributions from the following authors: J. Andrade, F. Brichet, A. Gravey, R. Gibbens, F. Kelly, P. Key, K. Lindberger, S. Manthorpe, I. Norros, M. Ritter, J. Roberts, O. Rose, S-E. Tidblom.

Part II was coordinated by U. Mocci with contributions from the following: S. Blaabjerg, V. Elek, A. Farago, J. Garcia-Vidal, L. Jereb, Z. Kovacs, J-Y. Le Boudec, K. Lindberger, J. Lubacz, U. Mocci, F. Panken, M. Pioro, C. Scoglio, R. Slosiar, C. Szabo, A. Tomaszewski, A. Tonietti, J. Virtamo.

Part III was coordinated by J. Virtamo with contributions from the following: S. Aalto, A. Andersen, J. Andrade, H. Azmoodeh, N. Bean, J.L. van den Berg, S. Blaabjerg, C. Blondia, F. Brichet, H. Bruneel, R. Gibbens, A. Gravey, K. Laevens, G. Latouche, K. Lindberger, I. Norros, O. Østerbø, J. Resing, M. Ritter, S. Robert, J. Roberts, O. Rose, A. Simonian, R. Slosiar, B. Steyaert, P. Tran-Gia, J. Virtamo, S. Zachary.

R. J. Gibbens assembled the extensive bibliography file containing some 535 entries.

The content of the report is largely based on some 240 research reports (COST 242 Temporary Documents) presented and discussed during the COST 242 meetings. The authors of these reports are listed below:

S. Aalto, E. Aarstad, J. Adams, M. Ajmone Marsan, A. T. Andersen, J. Andrade, E. Andries, J. Arnold, L. Ast, H. Azmoodeh, A. Bak, J. M. Barcelo, S. Bauer, K. Begain, B. Bensaou, J. L. van den Berg, A. Bianco, S. Blaabjerg, C. Blondia, A. Bohn Nielsen, D. D. Botvich, J. Boyer, P. Boyer, B. Bremer, F. Brichet, N. Brook, H. Bruneel, C. Bruni, K. Budai, W. Burakowski, D. Bursztynowski, Y. Canetti, O. Casals, P. Castelli, E. Cavallero, F. Cerdan, J. Cherbonnier, I. Chlamtac, M. Christiansen, T. Curran, M. D'Ambrosio,

P. D'Andrea, G. Dallos, R. Dittmann, T. V. Do, D. Down, A. Dupuis,
V. Elek, A. Farago, P. Farkas, G. Fodor, A. Forcina, M. Frater, B. Friis
Nielsen, T. Fritsch, P. Gajowniczek, Z. Gal, J. Gamo, S. Garcia, J. Garcia
Vidal, R. J. Gibbens, M. Grasse, A. Gravey, T. R. Griffiths, G. Gripenberg,
J. Guibert, F. Guillemin, L. Gyalog, H. Haapasalo, M. Hamdi, G. Hébuterne,
H. Hemmer, T. Henk, S. Herrera, W. Holender, F. Hubner, J. Ivansson,
V. B. Iversen, A. Jensen, L. Jereb, T. Johnsson, V. Kalashnikov, R. Kaloc,
F. P. Kelly, P. B. Key, K. Kilkki, M. Klimo, M. Koivula, M. Konvit, Z. Kop-
ertowski, S. Kornprobst, M. Kotuliak, T. Kovacikova, K. Kvols, K. Laevens,
C. Lavrijsen, J-Y. Le Boudec, K. Lindberger, R. Lo Cigno, J. Lubacz, A. Mac
Fhearraigh, M. Mandjes, S. Manthorpe, A. Marek, E. Martin-Rebello, M. Mar-
tinez Pasque, L. Massoulié, R. Melen, M. Menozzi, J. Mignault, M. Mit-
tler, U. Mocci, A-M. Møller, S. Molnár, W. Monin, J. Mulawka, M. Mu-
nafo, S. Murphy, R. N. Macfadyen, M. F. Neuts, B. F. Nielsen, I. Norros,
O. Østerbø, F. Panken, P. Pannunzi, Z. Papir, R. Pasquali, P. Perfetti, H. Pet-
tersen, A. Pfening, H. L. Phan, M. Pioro, J. Resing, M. Ritter, S. Robert,
J. W. Roberts, A. H. Roosma, O. Rose, C. Scoglio, M. Servel, K. Sevilla,
A. Simonian, A. Skliros, R. Slosiar, T. Smit, B. Steyaert, R. Syski, P. Sz-
ablowski, C. Szabo, A. Szentesi, M. Telek, W. M. ten Wolde, S-E. Tidblom,
A. Tomaszewski, A. Tonietti, P. Tran Gia, M. Tvarozek, D. Veitch, M. Villen,
B. Vinck, J. T. Virtamo, K. van der Wal, S. Wittevrongel, G. Wolfner,
T. Zhang, B. Zovko-Cihlar, Z. Zuaha

While the report is much more than a compilation of research papers, it
certainly lacks the homogeneity of a monograph written by a single author.
On some of the more controversial issues, the reader may also discern conflicting points of view expressed in different sections. It has certainly never
been an ambition of the action to arrive at a consensus on, for example,
what constitutes the "best" traffic control solution. This is the role of standards organisations. The content has also been determined by the interests
of the COST 242 participants and the external reader should not wonder
why some aspect of network design or a particular approach to traffic modelling has not been included: the objective has never been to produce an
encyclopaedia. Nevertheless, the authors are confident that the report will
be of considerable value as a reference work not only to themselves and their
colleagues but also to the networking community at large. It is the product
of an amicable and fruitful collaboration and we hope that it will be received
in the same spirit of generosity that motivated its numerous co-authors.

July 1996   Jim Roberts

Contents

List of abbreviations

A-M-S	Anick-Mitra-Sondhi [model] [AMS82]
AAL	ATM Adaption Layer
ABR	Available Bit Rate
ABT	ATM Block Transfer
ACR	Allowed Cell Rate
ADSL	Asynchronous Digital Subscriber Loop
AIR	Additive Increase Rate
ANSI	American National Standards Institute
ARIMA	Auto-Regressive Integrated Moving Average
ARMA	Auto-Regressive Moving Average
ATC	ATM Transfer Capabilities
ATM	Asynchronous Transfer Mode
B-ISDN	Broadband Integrated Services Digital Network
BAF	Broadband Access Facilities
BECN	Backward Explicit Congestion Notification
BPP	Belt Permit Programmer
BT	Burst Tolerance
BTI	Basic Time Interval
CAC	Connection Admission Control
CAT	Capacity Assignment Tables
CATV	Community Antenna TeleVision
CBR	Constant Bit Rate
CDV	Cell Delay Variation
CDVT	Cell Delay Variation Tolerance
CI	Congestion Indication
CLP	Cell Loss Priority
CLR	Cell Loss Ratio
COST	COoperation in Science and Technology
CP	Complete Partitioning
CPP	Circular Permit Programmer
CS	Complete Sharing
CSMA/CD	Carrier Sense Multiple Access/Collision Detection
CTD	Cell Transfer Delay
D-BMAP	Discrete-time Batch Markovian Arrival Process
DBR	Deterministic Bit Rate
DES	Destination End System
DQDB	Dual Queue - Dual Bus
EFCI	Explicit Forward Congestion Indication
ER	Explicit Rate

ERF	Explicit Reduction Factor
FARIMA	Fractional Auto-Regressive Integrated Moving-Average
FBM	Fractional Brownian Motion
FCFS	First Come First Served
FDDI	Fiber Distributed Data Interface
FDM	Frequency Division Multiplexing
FECN	Forward Explicit Congestion Notification
FIFO	First In First Out
FTTH	Fibre To The Home
FTTC	Fibre To The Curb
GCRA	Generic Cell Rate Algorithm
GF	Global Fifo
GFC	Generic Flow Control
GI	General Independent
GMLP	General Multi-Level Process
GOP	Group of Pictures
GoS	Grade of Service
GPS	Generalized Processor Sharing
HOL	Head-Of-Line
iid	independent and identically distributed
IBT	Intrinsic Burst Tolerance
ICI	Inter-Carrier Interface
ICR	Initial Cell Rate
IDC	Index of Dispersion for Counts
IEEE	Institute of Electrical and Electronics Engineers
IETF	Internet Engineering Task Force
IP	Internet Protocol
IPP	Interupted Poisson Process
ISDN	Integrated Services Digital Network
ISO	International Standards Organization
ITU	International Telecommunication Union
IWUs	InterWorking Units
JPEG	Joint Photographic Experts Group
LAN	Local Area Network
LT	Line Termination
MAC	Medium Access Control
MAN	Metropotital Area Network
MAP	Markovian Arrival Process
MBS	Maximum Burst Size
MC	Markov Chain
McFred	MAC FRame basED
MCR	Minimum Cell Rate
MCRT	Multi-Class Real Time
MLE	Maximum Likelihood Estimator

MMBP	Markov Modulated Bernoulli Process
MMPP	Markov Modulated Poisson Process
MMRP	Markov Modulated Rate Process
MPEG	Moving Picture Experts Group
NPC	Network Parameter Control
NT	Network Termination
O-D	Origin – Destination
ODN	Optical Distribution Network
OLT	Optical Line Termination
ONU	Optical Network Unit
OSI	Open Systems Interconnection
PACS	Priority Access Control Scheme
PBS	Partial Buffer Sharing
PCR	Peak Cell Rate
PCV	Permanent Virtual Channel
PDA	Permit Distribution Algorithm
PEI	Peak Emission Interval
PGF	Probability Generating Function
PGPS	Packet by packet Generalized Processor Sharing
PMF	Probability Mass Function
PON	Passive Optical Network
PS	Partial Sharing
QoS	Quality of Service
RACE	Research on Advanced Communications in Europe
RAU	Request Access Unit
RB	Request Blocks
RDF	Rate Decrease Factor
REM	Rate Envelope Multiplexing
RM	Resource Management
RSVP	ReSerVation Protocol
RTT	Round-Trip Time
SBR	Statistical Bit Rate
SCFQ	Self Clocked Fair Queueing
SCR	Sustainable Cell Rate
SDH	Synchronous Digital Hierarchy
SDM	Space Division Multiplexing
SDU	Service Data Unit
SES	Source End System
SMDS	Switched Multi-megabit Data Service
STM	Synchronous Transfer Mode
TCM	Time Compression Multiplexing
TCP	Transmission Control Protocol
TDM	Time Division Multiplexing
TDMA	Time Division Multiple Access

TE	Terminal Equipment
TR	Trunk Reservation
TTRT	Target Token Rotation Time
UBR	Unspecified Bit Rate
UDP	User Datagram Protocol
UNI	User – Network Interface
UPC	Usage Parameter Control
VBR	Variable Bit Rate
VC	Virtual Channel
VCC	Virtual Channel Connection
VCI	Virtual Channel Identifier
VCR	Video Cassette Recorder
VP	Virtual Path
VPC	Virtual Path Connection
VPI	Virtual Path Identifier
VPN	Virtual Private Network
WAN	Wide Area Network
WCT	Worst Case Traffic
WDM	Wavelength Division Multiplexing
WFQ	Weighted Fair Queueing
WWW	World Wide Web

Part I

Traffic Control in
Broadband Networks

Introduction to Part I

This, the first part of the COST 242 report, is concerned with the mechanisms and protocols necessary to ensure that a broadband network can offer the quality of service required by a wide range of audio, video and data communications applications. The term traffic control is employed in ITU Recommendation I.371 [ITU96b] to cover a variety of functions acting over a range of time scales from the priorities afforded to individual ATM cells to the network management decisions which attribute sets of network resources to given traffic relations. In this report we adopt a more restricted meaning for "traffic control", considering just the actions which have an impact on the quality of service of an established connection. Quality of service at connection level, mainly manifested by the probability of network blocking at connection set up, is considered in Part II on network design.

To meet the quality of service requirements of connections of a variety of imagined types, a number of transfer capabilities have been specified by the ITU and the ATM Forum for ATM based broadband networks. Similarly, the Internet Engineering Task Force (IETF) is in the process of defining new service classes with guaranteed performance to augment the current ubiquitous "best effort" IP datagram service class. The definition of transfer capabilities or service classes is based on the principle that the network must satisfy a number of performance criteria concerning information loss or transmission delays, on condition that the traffic offered by users abides by a set of rules. While the rules and the nature of the performance guarantees are specified in considerable detail, the standards do not state how the network should be operated in order to fulfill its quality of service committments. Nor do they give guidelines about which transfer capabilities are most appropriate for given applications. Such concerns are, of course, of great importance to network operators and considerable research effort has been employed over the last ten years or so to provide satisfactory solutions. Despite this intense activity, however, many important traffic control issues remain unresolved. Indeed, even the most fundamental networking options may now be considered to be in question with the increasingly popular Internet, and its World Wide Web, challenging the ATM based B-ISDN as the foundation of the future information superhighway.

Designing effective traffic controls is difficult essentially because of the requirement to guarantee QoS criteria when multiplexing performance depends on traffic characteristics which are often both difficult to know *a priori* and impossible to enforce at network ingress. This leads to the need for a compromise between accurate performance prediction and the characterisation of traffic through crude rule-based parameters. A further complication comes from the adaptability of communication applications to the networking environment which makes it difficult to specify their characteristics in absolute terms. For instance, telephony has traditionally relied on the transparent transmission quality of circuit switching but it is clear from recent developments that even an unreliable datagram service like that of the Internet can be used for voice communication between appropriately adaptive terminal equipments.

The COST 242 group has discussed a variety of traffic control issues but it has never been an objective to arrive at a concensus or to issue recommendations. The present chapter should therefore be seen as a record of some of the approaches developed by members of COST 242. We have not sought to eliminate contradictions, preferring the expression of all points of view. Furthermore, the whole range of traffic control issues is not dealt with here. The bias within the project has been towards the preventive control strategies considered for the ATM based B-ISDN and this is reflected in the chapters of this part of the report.

In Chapter 1 we present some considerations on the nature of the applications to be handled in a broadband network and on the requirements they are likely to place on traffic control. We suggest a classification of applications according to their timeliness requirements as real time, playback and "elastic" and identify information retrieval and virtual private networking as particularly important applications to test the efficiency of proposed controls. The question of long range dependence in traffic streams is discussed qualitatively in anticipation of a more mathematical treatment in Part III. Video traffic will clearly be a major component in any broadband network. The second section of this chapter provides a succinct description of MPEG video coding. Traffic characteristics are presented based on an extensive statistical analysis and mathematical traffic models are proposed.

Chapter 2 is devoted to a discussion on the transfer capabilities and service classes defined by the ITU, the ATM Forum and the IETF. As the definitions all rely on the ability to characterise and control the parameters of traffic streams using rule-based parameters, this chapter begins with a description of the rule in question. This is the "leaky bucket", also known as the virtual scheduling algorithm, the generic cell rate algorithm or the token bucket filter. The description of ATM transfer capabilities is based on ITU Recommendation I.371 [ITU96b] and the ATM Forum traffic management specification [For96]. The description of Internet service classes is derived from drafts published electronically by the IETF "Int-serv" working group.

ATM traffic controls, notably connection admission control, which are based on knowledge of connection rates must take account of cell delay variation. This phenomenon is discussed at some length in Chapter 3. We first consider the impact of the so-called CDV clumping and dispersion effects on the control of connection rates and on the realisation of a circuit emulation service, respectively. The difficult question of fixing the rate and CDV tolerance parameters for a given connection after its cells have been delayed in one or several multiplexing stages is then discussed. This question applies both to fixing the parameters of the generic cell rate algorithm for traffic control and to setting the play-out parameter in the case of circuit emulation. The final section of the chapter deals with the impact on CDV of various networking factors, providing a wide range of quantitative comparisons.

Chapter 4 discusses aspects of preventive and reactive traffic controls and their impact on service quality and network element dimensioning. After presenting the notion of traffic contract, we distinguish two types of statistical multiplexing using preventive traffic control: "rate envelope multiplexing" aims to preserve the bit rate of a connection throughout the network by limiting admission to ensure that the sum of rates is greater than link capacity with negligible probability; "rate sharing" allows an input rate excess using large buffers to absorb excess traffic at the cost of longer network delays. Statistical multiplexing using reactive traffic control is considered in the particular case of the ABR transfer capability recently defined by the ATM Forum and the ITU. The performance evaluation proceeds by a deterministic analysis in both steady state and transient phases leading to simple formulae for required buffer capacities. Both binary congestion indication and explicit rate options of the ABR capability are considered.

Chapter 5 considers the key issue for preventive, open loop control of designing efficient connection admission control (CAC) strategies. The objective of CAC is to accept new connection requests only if their traffic, added to that of existing connections, would not lead to infringement of quality of service criteria. The problem is considered successively in three cases of resource allocation: peak rate allocation, rate envelope multiplexing and rate sharing. For peak rate allocation the major problem is to take account of CDV. We first consider a certain notion of "negligible CDV", allowing a simple CAC procedure, before considering an approach accounting directly for the declared CDV tolerance. Rate envelope multiplexing relies on the combined input rate of multiplexed connections being less than the link rate with very high probability. CAC thus depends on an estimation of the probability distribution of this rate. We distinguish three broad approaches: assuming the rate distribution is known (leading to the useful concept of effective bandwidth); making worst case assumptions based on given rule-based parameters; estimating the rate distribution by means of real time measurements. CAC in the case of rate sharing is rather more complicated due to the fact that performance depends on many complex traffic parameters. A notion of effective bandwidth applicable to certain types of traffic source is outlined. We then discuss a possible CAC framework for high speed data connections where the main performance objective is to guarantee long term throughput with negligible loss. Lastly, we consider the possibility of directly policing significant traffic parameters describing the burst structure of a traffic source. The final section of the chapter is devoted to a discussion on the relationship between effective bandwidth and pricing. Introducing the generalised definition of an effective bandwidth function, it is demonstrated how a tariffing policy can be designed to give incentive to users to declare appropriate traffic parameter values to be used for admission control.

Chapter 6 considers the weighted fair queueing (WFQ) scheduling algorithm applied in the context of an ATM network. The chapter discusses different implementations of WFQ concluding with a preference for the so-called "self-clocked fair queueing" or "virtual spacing" algorithm. We then present some performance and fairness results for this algorithm before discussing realisation issues. The way WFQ parameters can be chosen to fulfill the requirements of diverse services is considered in the final section.

1 Broadband traffic characteristics

Both the performance evaluation and the design of broadband multiservice networks obviously depend on the traffic characteristics and performance requirements of the telecommunications services to be provided. A prior characterisation is, however, very difficult since most communications services have few intrinsic properties but adapt to the facilities provided by the available network. For example, most user applications can be handled by either packet switched networks with very limited performance guarantees such as the Internet or by an essentially circuit switched network such as the narrowband ISDN. In this section we nevertheless attempt to identify a number of types of application likely to have a significant impact on the future broadband network and examine their traffic characteristics.

In Section 1.1 we discuss various aspects of the nature of broadband traffic and the requirements they are likely to place on the network. We highlight the importance of information retrieval systems like the World Wide Web and discuss a number of difficult problems posed by the creation of virtual private multiservice networks. An introduction to the long range dependent nature of many traffic sources is provided in anticipation of a more mathematical description of traffic models in Part III, Section 13.

Section 1.2 is devoted to the important particular case of video traffic. The section presents a succinct description of MPEG coding and suggests mathematical modelling approaches based on the results of an extensive statistical analysis. The possibility of shaping variable rate video to conform to leaky bucket traffic parameters used for traffic control is considerd in the final subsection.

1.1 Nature of broadband traffic

In this section we discuss different types of user applications which need to be provided by a public broadband network and identify their particular traffic characteristics. The objective is to provide a qualitative description of broadband traffic necessary to identify the most appropriate types of traffic control and service categories to be implemented in the network. However, the present discussion is only loosely correlated with the definition of ATM transfer capabilities or Internet service classes (see Chapter 2): the most appropriate mapping of applications to service classes is still a matter of some controversy and it would be necessary to represent many diverging points of view to reflect the different opinions in the COST 242 group. We confine our considerations to the transport requirements of individual communications; network control issues, including the required network intelligence and signalling capabilities, are beyond the scope of the present discussion.

1.1.1   Classification by timeliness requirements

The type of transfer capability or service class to be provided by a broadband network depends largely on the timeliness requirements of user applications. We distinguish the following three main categories:

- interactive communications: either audio or video, bidirectional or multiparty, conversations, conferences, transmission of live events, ... ;

- transfer of stored information for temporary storage: text and pictures obtained from data bases, updating or safeguarding data bases, off line transfer of movies, transfer of files or large documents, ... ;

- transfer of stored audio or video information for immediate playback: consultation of multimedia data bases, Video on Demand, mail, ...

These services can be used conjointly in a variety of multimedia applications. Telecommunications services may also combine more than one category successively, e.g., voice messaging is first an interactive communication for the sender and then a stored audio transmission for the receiver. A network may not always see the above applications directly but be required to establish user to user virtual path connections or virtual private networks in which they are handled in an integrated manner.

Some more specialized applications such as distributed computing would perhaps require an additional category. For the present discussion we choose to suppose that these applications are marginal for the type of network we consider.

Interactive audio and video communications

Speech can be considered as a variable bit rate signal consisting of a succession of talkspurts and silences as depicted in Figure 1.1.1. It is the archetypal on/off traffic source. It is not certain, however, that in a broadband ATM network the gain in bandwidth is worth the processing effort needed to eliminate silences. The bit rate of interactive audio signals can vary from a few kilobits/s to hundreds of kilobits/s depending on quality.

Low rate signals imply a relatively long packetisation delay which can be prohibitive for interactive services. This has led to propositions for a composite cell based telephone service where each ATM cell carries samples of several communications [SA95, MB95]. In this case, the traffic entity seen by the broadband network is not a call but rather a bandwidth requirement, varying according to the current number of conversations in progress as depicted in Figure 1.1.2. The "user" is not the telephone subscriber but the

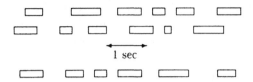

Figure 1.1.1: Audio traffic as an on/off process.

telephone service provider. This interpretation of telephone traffic may also be appropriate even when each cell is dedicated to only one call. The traffic is then characterized as a piecewise constant bit rate connection where the rate is typically counted in megabit/s or tens of megabits/s depending on the size of the traffic relation.

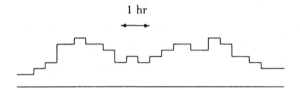

Figure 1.1.2: Audio traffic as a variable rate connection between specialized nodes.

Like the telephone, interactive video communications require real time responsiveness which implies constraints on the time necessary to perform video coding and the extent of network queueing delays. The bit rate of video signals may be constant (Figure 1.1.3a) or variable (Figure 1.1.3b). Constant rate is obtained by closed loop control which adjusts coder parameters (notably, quantization) to feed a transmit buffer with sufficient data to maintain the constant output rate. When this closed loop is removed, the rate is naturally variable due to varying picture contents (see Section 1.2). Constant rate coding introduces a systematic delay due to the smoothing buffer whereas open loop coding leads to output whose rate can vary widely depending on the filmed subject making it difficult to control network delays. A possible compromise solution is to shape the coder output by performing looser closed loop control than that required for CBR output, as discussed in Section 1.2.8 (Figure 1.1.3c). The output is shaped to satisfy the parameters of a leaky bucket: the fullness of the leaky bucket virtual queue is used to modulate the rate generated in successive frames to ensure an average rate equal to the leak rate (i.e., the virtual buffer is not allowed to empty or overflow). Such shaping avoids the buffering delay of CBR coding while facilitating network traffic control and allowing statistical multiplexing with small and predictable delays. The bit rate of the video signal determines perceived quality, e.g., around 1.5 Mbit/s for VCR-like quality and 20 or 30 Mbit/s for high definition TV.

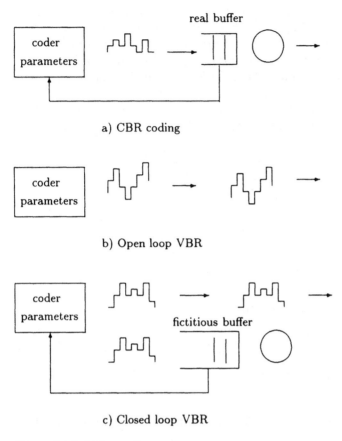

a) CBR coding

b) Open loop VBR

c) Closed loop VBR

Figure 1.1.3: Video coding options.

Transfer of information for temporary storage

In this category the temporary storage duration may vary from seconds, when short texts or pictures are displayed immediately on a screen, to hours, days or more in the case of bulk data transfer. The essential feature is that there are no strict time integrity requirements and bit rate (or throughput) is less an intrinsic characteristic than a (fairly loose) quality of service requirement. The bit rate is "elastic" and can be adjusted to fill available capacity without unduly affecting user appreciation of the service provided.

Transfers of texts and pictures arise in data base consultation services such as Teletel in France or the Internet's World-Wide Web (WWW). A page of unformatted text represents a very small amount of data while a photograph can be coded in around one megabit. More specialised applications, especially in the medical area, involve pictures with much higher resolution representing greater volumes of data (e.g., 50 Mbit for an X-Ray).

Bulk data transfer is required for data base management operations including updates and safeguards. The volume of data can be measured in giga-octets although the number of potential users is not likely to be great. Off-line transfers of films to update libraries, for example, may also constitute a non-negligible source of traffic. Many such bulk transfers can take place in off-peak periods.

Table 1.1.1 gives an indication of some typical data volumes which it might be required to transmit in a broadband network.

Table 1.1.1: Typical data unit sizes.

Data entity	Bytes
Page of ASCII text	10^3
Photograph	10^5
X-ray picture	10^7
Line drawing in Postscript	10^5
Ten page paper in Word	5×10^4
Same paper in Postscript	10^5
JPEG coded Star Wars movie	5×10^9
3.5" floppy disc	1.4×10^6
CD ROM	6×10^8

Transfer of audio and video information for immediate playback

The main difference with respect to the first category is the absence of a real time responsiveness requirement: signals should be capable of being played back without noticeable jitter but absolute transfer delay is not an issue. For low rate audio signals this means there is no strict constraint on the packet size and no need in an ATM Network to form composite cells; for video, more time can be taken to compress the signal, e.g., by using B-frames in MPEG coding (see Section 1.2).

Video on Demand is often seen as a service which should at least reproduce the functions of a video recorder: pause, slow motion, fast forward, ... From the networking point of view this could imply a connection whose traffic varies randomly between a certain number of types. Note that requirements on responsiveness to user commands place severe constraints on network performance.

Open loop coded feature films have been shown to produce variable rate output with so-called long term dependence which can have serious consequences on the performance of statistical multiplexing [GW94]. It may be prefered to use closed loop coding, as suggested above for interactive video communications, to eliminate low frequency rate variations by shaping to the parameters of a leaky bucket. There are obvious advantages in ensuring that the same networking techniques and the same decoders can be used for both video communications and video information retrieval.

Note that while the first category of applications require strict time integrity and the second are "elastic", the present category lies between these two extremes. A certain degree of elasticity in the rate provided by the network is acceptable if a playback buffer is inserted between the end user and the network termination. This buffer would need to be sufficiently large to absorb the fluctuations in the data arrival rate due to varying network delays; conversely, a network would have to be designed to limit delay variations to an amplitude comensurate with the size of the playback buffers provided.

1.1.2   Information retrieval paradigms

Consultation of a data base typically gives rise to a succession of information transfers from a server in response to user requests. If the same server supplies all the information as in the French Teletel service, for example, it is natural to envisage setting up a connection to be used for the entire session. The traffic on this connection is then particularly variable with indeterminate silent periods separating bursts whose traffic characteristics may correspond to different types of stored information (text, pictures, audio, video). In the WWW information retrieval paradigm currently implemented in the Internet the traffic must be seen in a different way.

In the WWW, each selection of a highlighted word or icon leads to the establishment of one or more new TCP connections. Indeed, as well as corresponding to a different type of transfer, each interrogation may involve a new information source located anywhere in the world. We thus identify the need to establish a succession of connections of various types. The traffic entity is not a multimedia session whose statistical characteristics are extremely difficult to characterise but a single connection whose characteristics are perfectly well known in advance: a given amount of text to be transfered, a video or audio sequence of given bit rate, ...

Statistical observations performed on WWW traffic are revealing interesting features about information retrieval traffic [CBC95, CB96, BC95]. A large majority of documents requested over the Web are small (a few kilobytes) but the distribution of document size is of power law or Pareto type indicating the presence of a significant number of very large documents. The use of cache memories to store frequently requested documents in local "proxy servers" can considerably reduce the amount of traffic handled in the wide area network.

1.1.3   Virtual private networking

Present demand for broadband communications is frequently expressed as a requirement for LAN interconnection or, more generally, for the constitution of virtual private networks (VPN). From a transport point of view these may consist simply of a collection of links between a set of user nodes. Alternatively, links may terminate in public nodes allowing the construction of more efficient networks with shared resources.

The relevant characteristics of traffic on the links of a VPN depend on the nature of the service provided by the broadband network. If the network simply reserves a fixed rate for each link, this rate is the essential parameter. VPN links may alternatively be described by peak (maximum) and sustainable (committed) information rates as in the frame relay service, for example. For efficient network operation it may, however, be necessary to know further traffic characteristics describing rate fluctuations. It is now well know that the characteristics of the composite output from a LAN are particularly difficult to parameterise due to the phenomenon of self-similarity or long term dependence.

Traffic on a VPN originates from a variety of different interactive communication, data transfer and information retrieval applications, as discussed above. A public VPN service needs to be able to provide the different qualities of service required by these applications and to allow the user to freely share his capacity between them. Special scheduling in network queues may be necessary to guarantee time integrity requirements for interactive or playback audio and video applications. Thus, from a traffic handling point of view, short of providing all traffic with the quality of service of the most demanding applications, it may prove impossible to consider VPN links as entities. Traffic management functions such as admission control would need to be exercised on individual connections or groups of connections with like QoS requirements.

1.1.4   Relative traffic volumes

It is important in deciding what traffic handling capabilities should be implemented by the network to have an idea of the relative volumes corresponding to the above traffic types. We first note the significant imbalance between the size of the telephone network and that of existing data networks. The volume in bits per second of telephone traffic is currently of the order of 20 times greater than the volume of data traffic generated by the same population of users. This is reflected in the investment represented by the respective networks. While the computer industry is roughly equivalent in value to the telecommunications industry, there is a completely opposite distribution of cost between terminals (computers and their software or telephone sets) and the network.

Traffic in data networks is growing rapidly but is also undergoing a mutation, notably with the appearance of multimedia terminals allowing voice and video communications. The future broadband network is likely to be largely dominated by audio and video traffic whether this be for interactive communications or for information retrieval. Bulk data transfers will develop with the need to move information around but it may be expected that tariffs will entice most of this traffic to the off-peak periods of audio and video services. The retrieval of text or photographs from data bases is likely to be a very widely used service, given current interest in the WWW and the "information super highway". However, although the number of transactions and an increase in their complexity are likely to be very significant for network signalling capabilities and intelligence, their impact on traffic volume will probably remain marginal. Humans simply cannot consume words and still pictures fast enough for this type of traffic to come remotely close in volume to that generated by audio or video applications.

1.1.5 Tariffing considerations

Charging is increasingly seen as a major factor in the design of broadband networks. However, these considerations have been largely outside the scope of the COST 242 project. We just briefly mention some of the more important issues.

Factors determining tariffs for telecommunications services include the following:

- subscription charge giving access to the network

- connection type and duration

- volume of transmitted information

- value of the information provided or service rendered.

Subscription is often the only basis for charging users of the Internet. This is satisfactory as long as the charge is light but questions of fairness arise since the financial contribution is then only loosely related to usage. The introduction of usage based charging is currently an important issue for the Internet [MMV95a, MMV95b].

Charging in the telephone network is currently based on the first two factors (although the second coincides with the third for constant rate connections). The division of charges between subscription and usage is currently changing with a bias to the former as increasing competition obliges operators to make charges more closely reflect costs (non-usage dependent costs include the very expensive distribution network between the user and the core network).

Variable rate connections might be charged on the basis of the volume of information transmitted or on the amount of network resources allocated to account for this variability. However, for VBR video it may be considered unreasonable that cost should be related to the activity of the participants in a video conference or to the proportion of action scenes in a movie, for example.

For information retrieval, the volume of information transmitted is clearly the most important factor determining network charges. Note, however, that the charge for a transaction will generally also include payment for the information provided (including royalties) or the service rendered. The network charge may only be marginal in the overall cost to the user of information retrieval services. The relative cost of communications services is particularly important for a video on demand service if this is to be competitive with the existing cassette rental service.

1.1.6 Long range dependence

One of the important recent findings of teletraffic research has been the presence of so called *long range dependence* in many common real traffic processes. This raises considerable challenges for statistical analysis and modelling of teletraffic. In this section, the concept of long range dependence is introduced in prose, and its various practically relevant aspects are discussed. The corresponding mathematical theory is presented in Part III, Chapter 13.

A weakly stationary time series is called long range dependent if the correlation between neighbouring exclusive blocks does not asymptotically vanish when the block size is increased. This correlation is always zero for the Poisson process and decays exponentially fast towards zero for finite state space Markov processes and autoregressive processes of finite order and, as a consequence, for all stochastic processes traditionally used as mathematical models for teletraffic.

A stochastic process is called self-similar if it behaves, up to a scaling factor, in exactly the same way at all time scales. A classical example of such a process is the Brownian motion. A self-similar process is long range dependent as soon as it has any positive correlations, since these remain the same through all time scales. On the other hand, by a general theorem by Lamperti, it is rather typical that a stochastic process is approximately self-similar when considered at large time scales. More precise statements can be found in Section 13.4.

Long range dependence became a topic in teletraffic research when statistical analysis of very extensive Ethernet traffic measurement data at Bellcore led to the consequence that the traffic was quite close to self-similar with strong positive correlations through at least 5–6 important time scales (from milliseconds to hours) [LTWW94]. Long range dependence has also been observed in more application specific packet traffic [MHWYH91], wide area traffic [PF95] and in VBR video traffic (see [BSTW95] and Section 1.2).

Impact of long range dependence on performance and its analysis

Long range dependence implies that traffic behaves worse than would be expected on the basis of short range dependent traffic models. The fluctuations are larger, and congestion builds up more easily. Some results on the queue-ing theory with long range dependent traffic are presented in Part III, Section 17.4.

Another consequence is that the estimation of traffic parameters is an entirely different problem compared with traditional modelling. For example, such an ordinary notion as the mean rate of traffic has a quite different nature in a case when the reliable estimation of the corresponding model parameter requires five minutes than when it takes five hours. Statistical aspects of long range dependence are considered in Part III, Section 13.6.

What causes long range dependence?

Discussion is going on about the physical causes of long range dependence and self-similarity in teletraffic streams. Essentially, there are two basic ex-planations.

The first explanation refers to the well known hierarchical nature of tele-traffic. As an illustration, Figure 1.1.4 shows the traffic of (both to and from) a multipurpose mailserver-workstation during a working day at 4 time scales. There are several layers of communication activity which cause traffic variation at several small time scales. The ladder continues further from the connection time scale upwards reflecting the organization of human work and other activities: single actions are usually parts of more extensive tasks and so forth. This explains to some extent long range dependence in data traffic, but it does not explain self-similarity.

The other explanation is more sophisticated and based on mathematical theory. It has been observed that many of the basic communication processes have strongly varying sojourn times at their different levels of activity, e.g., idleness periods, video scenes etc. More exactly, the empirical sojourn time distributions are often well described by so called heavy tailed distributions which have infinite variance. It is a mathematical fact (discovered by Man-delbrot; see Part III, Section 17.4.1) that heavy tails of sojourn time distri-butions are equivalent to long range dependence in this type of traffic model. Moreover, long range dependence itself means approximate self-similarity of the correlation function at large time scales. Thus, it is not even necessary to have a hierarchy of activities or anything having a complex structure in order to end up with a self-similar process: it is sufficient, for example, that all traffic sources simply send small bursts with heavy tailed idle periods in between, all sources and periods being independent. This explanation has been supported by source level analysis of the Bellcore data [WTSW95].

Long range dependence *vs.* non-stationarity

It has been argued that long range dependence might not exist in teletraffic or is at least not a very important notion in this context since the same em-pirical observations can be explained by non-stationarity of the processes at

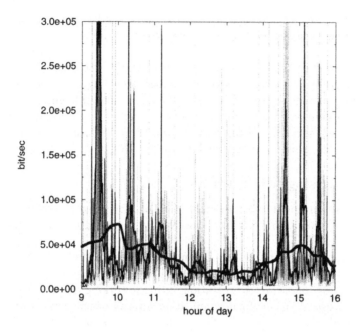

Figure 1.1.4: The traffic of a workstation during a working day. Integration periods 5 sec, 50 sec, 500 sec and 5000 sec.

larger time scales (see the argumentation by Duffield, Lewis *et al.* [DLO+94]). It is, indeed, difficult or impossible to distinguish with statistical methods between the two explanations for large scale variations in a given time series. Thus, part of the problem is methodological. However, it should be noted that often there are no deterministic patterns behind large scale variations (for example, the overall shape of the large scale traffic variation of the workstation of Figure 1.1.4 looked different every day of the week during which it was measured), and unpredictable large scale variation cannot be seen as a consequence of non-stationarity.

Time scales relevant for queueing and congestion management

A question often raised when long range dependence in teletraffic is discussed is: what relevance does the variation at a particular time scale have for observable network performance? For example, does it matter whether the hours are correlated or not, since all the queueing phenomena important for observable performance happen in fractions of a second. It turns out that even for long range dependent traffic, there is for each buffer size and utilization level a characteristic time scale at which the traffic variation is important for queueing performance and outside which it is less important, at least when only the stationary congestion probabilities are considered.

This problem is studied mathematically in Part III, Section 13.1.

1.1.7   Self regulation of data traffic

Traffic characterization in teletraffic theory has often meant the identification of a statistical law according to which the traffic sources "blindly" transmit bits. A distinguishing feature of data communication is, however, the presence of feedback: the well-designed control functions of transport layer protocols (in particular, TCP [Jac88]) and, recently, the proposed ABR service in ATM networks give the sources flexibility and intelligence in their utilization of the shared network resources.

TCP works end-to-end only, and the feedback obtained by the source is implicit: if the acknowledgement for a packet does not arrive before timeout, the source starts retransmission and reduces the packet transmission rate to a minimum. With ABR, network nodes can give explicit feedback about the load situation and even tell the cell rate the source should use next.

Through these feedback mechanisms, features of the network (e.g. transmission speeds, distances, traffic situations) strongly affect the output of the source. Therefore, it is sometimes misleading to apply empirically obtained data traffic characteristics in a context different from their origin.

In order to make the conceptual distinction between traffic that is or is not influenced by network feedback, we suggest the notion "free traffic" as an ideal notion for "what the traffic would be if the network resources were unlimited". Note that this does not mean infinite transmission speeds since it is assumed that the sources and destinations still have only their limited capabilities. For example, one could speak about dimensioning a network according to free traffic corresponding to the users' needs, meaning that the network is made so powerful that even the feedback-based rate reductions are rare events. This is overdimensioning for pure data traffic but not if one wants to run real-time applications like packet video with satisfactory quality over a network which cannot reserve bandwidth for connections.

1.2   VBR video traffic

In the ATM based B-ISDN, a major part of the traffic will be produced by multimedia sources like teleconferencing terminals and video-on-demand servers. Most of the video encoding will be done using the MPEG (ISO Moving Picture Expert Group) standard.

There remain a number of open issues concerning the transmission of MPEG video on high-speed networks including finding of the appropriate ATM adaption layer, dimensioning multiplexer buffers, shaping of video traffic and video cell stream monitoring. The solution of these problems relies on performance analysis using models for MPEG video traffic streams. Such models must be based on a thorough analysis of the statistical data sets corresponding to encoded videos.

In the next section, a brief outline of the MPEG coding technique is given. We then summarize the statistical characteristics of a number of coded sequences. Video traffic modeling is discussed with respect to a layered modeling scheme. Finally, we discuss the possibility of shaping video output to make it confirm to the parameters of a leaky bucket.

1.2.1   MPEG video encoding

Due to the high bandwidth needs of uncompressed video data streams, several coding algorithms for the compression of these streams have been developed.

At the moment, the MPEG coding scheme is widely used for all types of video applications. There are two schemes, MPEG-I [LeG91, ISO93b] and MPEG-II [ISO93a], where the MPEG-I functionalities are a subset of those of MPEG-II. The most significant difference for video transmission on ATM is that MPEG-II allows for layered coding. This means the video data stream consists of a base layer stream, which contains the most important video data together with one or more enhancement layers which can be used to improve video quality. MPEG-I and MPEG-II permit both CBR and VBR video encoding. CBR encoding leads to easy-to-handle transmission at the cost of varying video quality. In contrast, VBR encoding leads to constant quality at the cost of a more complex transmission.

In this section, we focus on one-layer video data streams of MPEG-I type. Most of the encoders will use this scheme and in case of multi-layer encoding the statistical properties of the base layer will be almost identical to this type of stream. We will only consider VBR encoded video sequences since CBR video is trivial from statistical analysis and modeling point of view.

The MPEG encoder input sequence consists of a series of frames, each containing a two-dimensional array of *picture elements*, called pels. The number of frames per second as well as the number of lines per frame and pels per line depend on national standards. For each pel, both luminance and chrominance information is stored. The compression algorithm is used to reduce the data rate before transmitting the video stream over communication networks.

This is done by both reducing the spatial and the temporal redundancy of the video data stream. The spatial redundancies are reduced by transforms and entropy coding and the temporal redundancies are reduced by prediction of future frames based on motion vectors. This is achieved using three types of frames (cf. Figure 1.2.1):

I-frames use only intra-frame coding, based on the discrete cosine transform and entropy coding;

P-frames use a similar coding algorithm to I-frames, but with the addition of motion compensation with respect to the previous I- or P-frame;

B-frames are similar to P-frames, except that the motion compensation can be with respect to the previous I- or P-frame, the next I- or P-frame, or an interpolation between them.

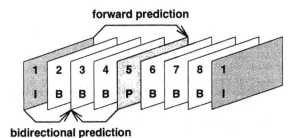

Figure 1.2.1: Group of Pictures of an MPEG stream.

Typically, I-frames require more bits than P-frames. B-frames have the lowest bandwidth requirement. After coding, the frames are arranged in a deterministic periodic sequence, e.g. "IBBPBB" or "IBBPBBPBBPBB", which is called *Group of Pictures* (GOP).

1.2.2   Statistical analysis of MPEG video sequences

In the following, we present some statistical measurements from several movies, TV sport events and TV shows, which were encoded at the Institute of Computer Science of the University of Würzburg using the UC Berkeley MPEG-I software encoder [Gon94]. Table 1.2.1 shows the sequences which we used to produce the data sets.

All sequences mentioned below were encoded using the following parameter set:

- Frame rate: 25 frames per second;

- Each frame consists of one slice;

- GOP pattern: IBBPBBPBBPBB (12 frames);

- Quantizer scales: 10 (I), 14 (P), 18 (B);

- Motion vector search: logarithmic/simple; window: half pel, 10; reference frame: original;

- Encoder input: 384 x 288 pels with 12 bit color information;

- Number of frames per sequence: 40000 (about half an hour of video)

Some parameters might not be optimal with respect to the quality of the MPEG video sequence because of some hardware limitations. We used a Sun Sparc 20 for the image processing and encoding and captured the sequence from a VCR with a SunVideo SBus board.

Table 1.2.1: Overview of encoded sequences.

Movies (cassettes)	
dino	Jurassic Park
lambs	The Silence of the Lambs
TV sports events (recorded from cable TV)	
soccer	Soccer World Cup 1994 Final: Brazil - Italy
race	Formula 1 car race at Hockenheim/Germany 1994
atp	ATP Tennis Final 1994: Becker - Sampras
Other TV sequences (recorded from cable TV)	
terminator	Terminator 2
talk1	German talk show
talk2	Political discussion
simpsons	Cartoon
asterix	Cartoon
mr.bean	Three slapstick episodes
news	German news show
mtv	Music clips
Set top camera	
settop	Student sitting in front of workstation

Overview

Table 1.2.2 shows the compression rates and the most important moments of the frame sizes, the GOP sizes, and the corresponding bit rates of the MPEG sequences.

For the sake of comparison we also present the statistical data of *Star Wars* as reported by Garrett and Willinger [GW94].

From Table 1.2.2 we conclude that typical TV sequences like sports, news, and music clips lead to MPEG sequences with a high peak bit rate and a high peak-to-mean ratio compared to movie sequences. These properties result from the rapid movements of a lot of small objects, which increase the amount of data necessary to encode the sequence.

Unfortunately, even the statistical properties of the sequences of the same category, like movies or cartoons, are not stable. For example, the measurements of *terminator* and *lambs* or of *simpsons* and *asterix* have no moments lying close together. This will lead to difficulties in finding traffic classes for MPEG video to be used for network admission control.

In the remainder of this section, we present a detailed analysis of the statistical data of the *dino*, *soccer*, and *starwars* sequences.

Table 1.2.2: Simple statistics of the encoded sequences.

Sequence	Compr. rate X : 1	Frames			GOPs		
		Mean [bits]	CoV	Peak/ Mean	Mean [bits]	CoV	Peak/ Mean
asterix	119	22,348	0.90	6.6	268,282	0.47	4.0
atp	121	21,890	0.93	8.7	262,648	0.37	3.0
dino	203	13,078	1.13	9.1	156,928	0.40	4.0
lambs	363	7,312	1.53	18.4	87,634	0.60	5.3
mr.bean	150	17,647	1.17	13.0	211,368	0.50	4.1
mtv	134	19,780	1.08	12.7	237,378	0.70	6.1
news	173	15,358	1.27	12.4	184,299	0.47	6.0
race	86	30,749	0.69	6.6	369,060	0.38	3.6
settop	305	6,031	1.92	7.7	72,379	0.18	2.0
simpsons	143	18,576	1.11	12.9	222,841	0.43	3.8
soccer	106	25,110	0.85	7.6	301,201	0.48	3.9
starwars	130	15,599	1.16	11.9	187,185	0.39	5.0
talk1	183	14,537	1.14	7.3	174,278	0.32	2.7
talk2	148	17,914	1.02	7.4	214,955	0.27	3.1
terminator	243	10,904	0.93	7.3	130,865	0.35	3.1

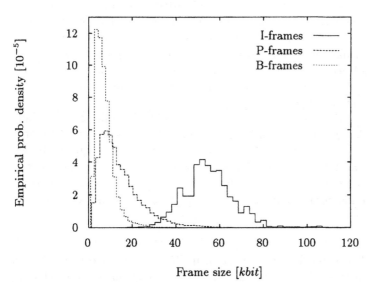

Figure 1.2.2: Frame size histograms of the **dino** sequence.

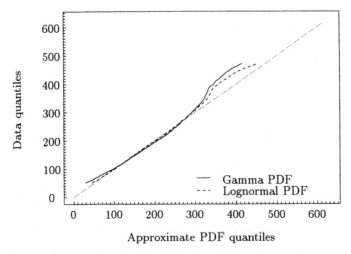

Figure 1.2.3: QQ-plot for the GOP sizes of the **dino** sequence.

Distributions

Figure 1.2.2 shows the frame size histograms of the I, P, and B frames of the *dino* sequence. The shapes of the curves indicate that the I-frames may be approximated by a Normal pdf, whereas the P and B-frames have a histogram resembling a Gamma or a lognormal pdf. The good agreement of the histogram and the Gamma curve becomes more obvious if we use a QQ-plot (quantile-quantile-plot), where Gamma or lognormal quantiles are plotted against the histogram quantiles.

The QQ-plots show that the frame size histograms of all three frame types of almost all encoded sequences can either be approximated by a Gamma or a Lognormal pdf. The differences between Gamma and Lognormal approximation performance are not too large in most cases. Perfect agreement of histogram and approximation cannot be achieved due to finite frame sizes. This leads to the conclusion, that for the modeling of the frame sizes, either Gamma or Lognormal pdfs can be used.

If we look at the GOP size distributions, we obtain similar results. Figure 1.2.3 shows the QQ-plot for the *dino* sequence. The Gamma and Lognormal quantiles are plotted against the histogram quantiles. An agreement with the dotted line indicates that histogram and model pdf are equal.For the sequence considered, the Lognormal distribution is a good approximation of the GOP size histogram, but the Gamma distribution is also adequate.

Correlations

Time-dependent statistics are important in the case of video traffic because correlations in the data streams can affect the performance of the ATM network.

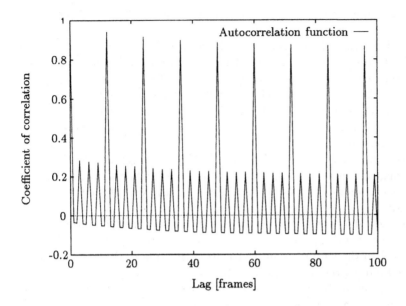

Figure 1.2.4: Autocorrelation function of the frame sizes of the **dino** sequence.

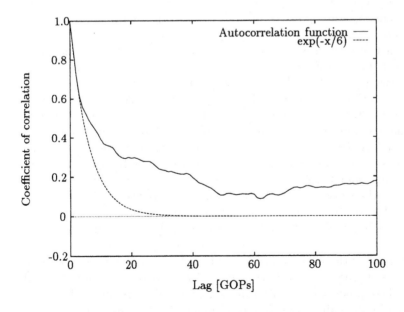

Figure 1.2.5: Autocorrelation function of the GOP sizes of the **dino** sequence.

First, autocorrelation functions of frame sizes and of GOP sizes are presented. The frame-by-frame correlations depend on the pattern of the GOP and, in principle, always look like Figure 1.2.4, assuming the same GOP pattern is used for the whole sequence. The autocorrelation function clearly reflects the 12 frame GOP structure. Thus, the pattern between two I frame peaks is repeated with slowly decaying amplitude of the peaks.

If a model is needed which reflects the frame-by-frame correlations of an MPEG video traffic stream, the GOP-pattern based shape of the autocorrelation function has to be considered. An approximation of the autocorrelations function is presented in [Ens94].

Based on the frame level correlations, it is difficult to get a clear picture of the long-range correlations of the video traffic stream. We therefore consider the autocorrelation function of the GOP sizes, i.e. the sum of the frame sizes of one GOP.

Figure 1.2.5 shows the autocorrelation functions of the GOP sizes of the *dino* sequences. For comparison, the dashed line shows the exponential function matched to the empirical autocorrelation function of the first few lags. Exponential autocorrelation appears if the GOP size process is memoryless. If the autocorrelation function of the statistical data is above the exponential function, this indicates dependences in the GOP size process. In Figure 1.2.5 this is clearly the case, whereas the autocorrelation curve and the exponential curve are matching well for other sequences such as *soccer*.

This result makes it difficult to find a GOP layer model which is appropriate for all types of video sequence. Some modeling approaches, such as Markov chains or autoregressive processes ignore the dependences at GOP level. In that case, the models can be kept simple at the cost of accuracy.

A succinct way of measuring long-term dependence is through the Hurst parameter H [Ber94, LWTW94]. We have estimated the H parameter of all sequences using the R/S analysis technique [MT79].

Table 1.2.3 gives the values of H for all the considered sequences. Note that time series without any long-range dependence have a Hurst parameter of 0.5, whereas time series of computer traffic can have H-values up to 1.0 [GW94]. It has been suggested that, in case of video traffic, a larger H-value reflects a larger amount of movement [Ber94]. The H-values, however, show that one cannot necessarily conclude from a high H-value to a lot of movement in the video. Even political discussion can have an H-value larger than that of a soccer match. Except for the *settop* sequence, all sequences have H-values higher than 0.7 and the existence of long-range dependence is obvious.

If the model of the video traffic should have long-range dependence properties, a class of processes called *fractional differencing processes* may be used [GW94]. These processes generate time series with given H-values. However, it may then be difficult to match a given marginal distribution for the generated samples.

Table 1.2.3: Hurst exponents of the encoded sequences.

Sequence	Hurst exponent H
mr.bean	0.95
lambs	0.92
simpsons	0.88
asterix	0.86
mtv	0.86
dino	0.85
starwars	0.85
talk2	0.85
talk1	0.84
news	0.84
terminator	0.80
atp	0.77
soccer	0.77
race	0.71
settop	0.53

1.2.3   A layered modeling scheme

There are several reasons to develop models for video traffic and to use them for the performance analysis of ATM networks.

The first reason is to extract the statistical properties of video traffic which have a significant impact on the network performance. We gain a lot of insight if we are able to reduce the statistical complexity of the empirical video data sets. It is true that only the frame size trace from the output of an MPEG encoder contains all statistical information about the encoded video; however, the large number of properties makes it difficult to decide which one may be causing performance problems.

The second reason is the computational complexity of simulations of ATM networks, particularly at cell level. It often takes long simulation runs to obtain results of high accuracy. In some cases numerical complexity can be considerably reduced using traffic models and standard analytical tools like matrix analysis or discrete time analysis.

The third reason is the need for connection traffic descriptors for video traffic. If the traffic model is simple, i.e., it has only a small number of parameters, these parameters can be used as traffic descriptors for CAC and UPC.

For the development of video traffic models we can both use the knowledge about the coding technique, MPEG-I or MPEG-II in our case, and the statistical analysis of the frame size sequence which we obtain from measurements. In this section, we present a layered modeling scheme for the development of MPEG video traffic models. The main information for model development of MPEG coding can be summarized as follows:

- There are three frame types: I, P, and B frames.

- A pattern of frame types, called GOP, is repeated continuously to create the encoded frame sequence.

- The frames of one single GOP depend strongly on each other.

Moreover, if we wish to create a model at cell level, both the AAL which is used for video transmission and the details of any shaping applied to the cell stream before it enters the network should also be taken into account.

Based on the information presented up to this point, we are already able to develop a scheme with three layers (cf. Figure 1.2.6): GOP layer, frame layer and cell layer. Higher layers, such as scenes, can be added if the statistical properties of the scene change process are available. Having decided on the layers, we have to define the statistical properties of each layer and of the way the layers interact.

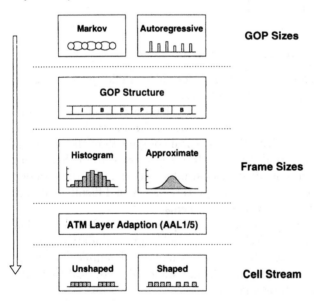

Figure 1.2.6: Layered video traffic modeling scheme.

Based on the results of Section 1.2.2, we are able to select an appropriate stochastic process for each layer. We then have to lay down the way the layers depend on each other. For example, if we want to generate a frame size sequence based on the GOP size process, we have to consider the structure of the GOP pattern giving the order of the frame types. The simplest way to find the frame sizes based on a GOP size sample is to use a scaling factor for each frame of the GOP, where the scaling factors are the mean sizes of the frames of one GOP divided by the mean GOP size of a given data set. More complex models may use frame size histograms or approximate pdfs to generate the frame size sequence (cf. Figure 1.2.6).

If we want to obtain a cell level model, we have to decide how frames are broken into cells. This will depend on the considered ATM Adaption Layer (AAL) and on the existence of shaping facilities between the video source and ATM network. If a statistical analysis of video cell stream measurements is available, it will be possible to base models directly on this material. This may lead to simpler models for the cell process.

The presented model development scheme is not a recipe for a perfect video traffic model. It is more the outline of a variety of stochastic modules and the description how they interact in the case of video traffic. The model developer will have to choose the modules which are appropriate for his analysis.

It should be noted that any model needs to be validated. Even quite complex models rely on simplifying assumptions and may ignore significant correlation effects. To obtain useful and reliable performance analysis results, it is important to know how these assumptions affect the results of the analysis.

1.2.4   Video modeling literature

A large variety of papers about video traffic modeling can be found in the current teletraffic literature. The modeling approaches can be divided into three main classes:

- Markov chains, e.g. [BC92, MAS+88, PElZ93],

- Autoregressive processes (including TES models), e.g. [HTL92, RS92, RMR94],

- Self-similar or fractal models, e.g. [BSTW95].

An overview on video models can be found in e.g. [FM94, RF94]. Most of the papers deal with Markov chain approaches since the estimation of the model parameters is straightforward and there is a large number of analysis techniques available to examine queuing systems with this type of input. The main disadvantage of these models is the implied assumption of an exponentially decaying autocorrelation function. This leads to inaccurate performance estimates if they depend on the long-range correlation properties of the video sequence. As shown in the following, considerable effort is necessary to obtain a Markov chain model whose generated sequence has a high coefficient of autocorrelation even for larger lags.

Autoregressive models also suffer from the difficulty to approximate the autocorrelation function of empirical data sets. For more complex models the estimation of model parameters and the queueing analysis techniques are often more difficult than in the Markov chain case.

Recently, a lot of attention has been given to fractal modeling of traffic streams in communication networks. Up to now, most of the papers deal mainly with the statistical analysis of data sets, i.e. in most cases the estimation of the Hurst parameter of an empirical sequence, and provide little information about traffic models and the analysis of queuing systems with fractal input sources.

To conclude this brief outline on video traffic models, we point out that all classes of models have their pros and cons, and that some care has to be taken in choosing a model for the analysis of a given communication system. In the following, we focus on Markov chain models of varying complexity in order to examine the advantages and limitations of this comparatively simple model class.

1.2.5   Markov chain models

Layered video traffic modeling

From Section 1.2.3 we conclude that we can identify several layers of an MPEG video traffic stream in ATM systems:

- Scene layer: intervals where the content of the pictures is almost the same, several seconds,

- GOP layer: period of one GOP, hundreds of milliseconds,

- Frame layer: period of a single frame, tens of milliseconds,

- Cell layer: period of a single ATM cell, microseconds.

First, we have to decide which layer should be modeled by the Markov chain because it will be difficult to find a model which is able to cover all the time scales mentioned above. For the cell loss simulations in Section 1.2.6 we need a sequence of frame sizes which we can either generate directly or indirectly by generating higher layer sequences which we divide into frames. We decided to model the GOP layer process for the following reasons.

When choosing the frame layer we have to model a periodic autocorrelation function as shown in Figure 1.2.4 to obtain reasonable cell loss estimates for a large range of buffer sizes [RF95]. This can only be achieved with a Markov chain with a sufficiently large number of states. On the other hand, if we choose the sequence layer model we first have to determine a GOP size sequence and then to generate the frame sizes from this sequence. This adds a considerable amount of complexity to the model but it will also improve its long-range dependence behavior.

Before we present the three GOP size process models, we give an outline of the procedure which generates the frame sizes from the GOP sizes. First, we use an empirical video frame size trace to estimate the mean size of each frame type (I, P, or B) of the GOP. If we divide the mean frame sizes by the mean GOP size, we receive a scaling factor for each frame of the GOP. To generate a sequence of frame sizes we use one of the models introduced later to generate the GOP sizes and compute the frame sizes by multiplying the GOP size with the scaling factors. As the results of this section show, this simple method leads to a good approximation of the frame process of the video trace. In particular, the periodic nature of the frame process is approximated with little effort. Due to the fact that both GOP and frame sizes of either type are often approximately Gamma distributed [Ros95], this method also leads to frame size distributions which are close to the original ones. The only frame layer information which is lost consists of the frame by frame correlation in addition to the correlation which is induced by the GOP pattern. However, from recent work we conclude that this un-modeled piece of information has almost no influence on cell loss results. The study shows that replacing the frame sizes of the P and B frames by their average values leads to essentially the same cell loss results as of the original sequence [RF95].

Next, we will present three Markov chain models which are considered in the following. The basic structures of these models are taken from [SSD93], [PEIZ93], and [BC92], where they are suggested for the frame size process of VBR video traffic. We point out that the intended use of these models differs from these studies in several aspects including using the GOP instead of frame process, choice of empirical data sets, studied performance measures. Therefore, some care has to be taken in comparing the results of those papers and the present study.

All models have in common, that their parameters are estimated from a descretized GOP size sequence. A GOP size sequence of length N is computed from the empirical MPEG trace by summing up the frames of each GOP. Next, the range of GOP size values is divided into fixed size, non-overlapping intervals. Each of these intervals relates to a state of the Markov chain. During GOP size generation, each time the Markov chain enters a state, a GOP size is generated equal to the mean value of the GOP sizes which fall into the interval related to this state. Thus, all models generate GOP size sequences in which the number of different GOP sizes is equal to the number of states of the Markov chain.

Histogram model

The *Histogram model* is the simplest model for the GOP size process. We compute a GOP size histogram from the empirical data set, where the number of histogram bins is equal to the number of states of the Markov chain. The probability P_i to enter state i is equal to n_i/N, where n_i is the number of

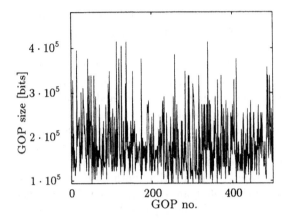

Figure 1.2.7: Model GOP trace.

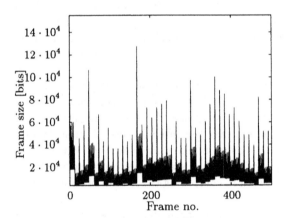

Figure 1.2.8: Model frame trace.

GOP sizes which fall into interval (or bin) i. The GOP process is generated by drawing samples according to the probability vector $\{P_i\}$. Strictly speaking, this is not an ordinary, i.e. first-order, Markov chain, where the current state depends on the last one, because in the *Histogram model* the current state is independent of any past states.

The *Histogram model* may be considered as a 0th-order Markov chain.

Figure 1.2.7 and Figure 1.2.8 show a GOP and frame size trace generated by the *Histogram model* with 100 states. These figures are shown for the sake of illustration and will not be used to compare the models. As expected, the Q-Q-plot (cf. Figure 1.2.9) shows that the model provides a good estimate of the empirical GOP size distribution. The disadvantage of this model is the lack of any GOP correlation information (cf. Figure 1.2.10).

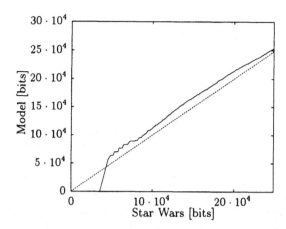

Figure 1.2.9: Q–Q-plot.

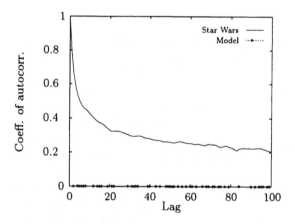

Figure 1.2.10: Autocorrelation function.

Simple MC model

The *Simple MC (Markov Chain) model* consist of an ordinary 1st-order Markov chain. The number of states M is G_{max}/σ_G, where G_{max} denotes the size of the largest GOP and σ_G the standard deviation of the GOP sizes. Thus, the size of the quantization interval is σ_G. It is also possible to choose other interval sizes but it turned out in our experiments that in most cases σ_G led to reasonable results. Only if the Q-Q-plots indicate that the model GOP size distribution does not have the desired degree of accuracy should the quantization interval be reduced. The entries of the transition probability matrix $\{P_{ij}\}$ are estimated by

$$P_{ij} = \frac{n_{ij}}{\sum\limits_{k=1}^{M} n_{ik}}, \qquad (1.2.1)$$

where n_{ij} denotes the number of transitions from interval i to interval j.

This model includes the correlations from one GOP to the next but no correlations over larger lags. This is also indicated by the exponentially decaying autocorrelation function.

Figure 1.2.11 and Figure 1.2.12 show a GOP and frame size trace generated by the *Simple MC model* with 13 states. The estimate of the empirical GOP sizes is adequate (cf. Figure 1.2.13), but the approximation of the autocorrelation function is bad (cf. Figure 1.2.14) although better than that of the *Histogram model*.

Scene-oriented model

The *Scene-oriented model* consists of a Markov chain which controls the scene change process and a number of Markov chains, such as the *Simple MC model* type, which generate the GOP sequence for each scene class.

For the Markov chain generating the scene process we need to divide the GOP process into scenes. From a statistical point of view, it is not necessary to determine the scene boundaries of the original video sequence by watching the movie. We prefer to use a method to find these boundaries which only depends on a few statistical parameters. These parameters should be available by simply scanning the GOP size sequence and without any knowledge of the content of the scenes. We suggest the following algorithm to find the scene boundaries. G_i denotes the size of GOP i and n the GOP number. The scene number is denoted by s and the coefficient of variation c of a sample $\{x_k : k = 1, \ldots, M\}$ is defined by

$$c = \frac{\frac{1}{M} \sum_{k=1}^{M} x_k}{\frac{1}{M-1} \left\{ \sum_{k=1}^{M} x_k^2 - \frac{1}{M} \left[\sum_{k=1}^{M} x_k \right]^2 \right\}}. \qquad (1.2.2)$$

(i) Set $n = 1$ and $s = 1$. Set current left scene boundary $b_{left}(s) = 1$.

(ii) Increment n by 1. Compute the coefficient of variation c_{new} of the sequence $G_{b_{left}}$ to G_n.

(iii) Increment n by 1. Set $c_{old} = c_{new}$. Compute the coefficient of variation c_{new} of the sequence $G_{b_{left}(s)}$ to G_n.

  (a) If $|c_{new} - c_{old}|(n - b_{left} + 1) > \epsilon$, set the right scene boundary $b_{right}(s) = n - 1$ and the left scene boundary of the new scene $b_{left}(s+1) = n$. Increment s by 1 and go to step (ii).

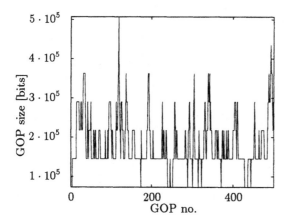

Figure 1.2.11: Model GOP trace.

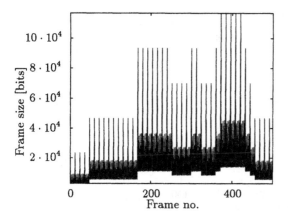

Figure 1.2.12: Model frame trace.

(b) If the above does not hold go to step (iii).

Iterating this algorithm over the whole GOP sequence provides a series of scene boundary pairs. The value ϵ limits the amount of variation which is tolerated for one scene. If adding a new GOP to the current scene increases or decreases the variation too much a new scene is assumed to start at this GOP number. Our experiments show that ϵ should be chosen so that the resulting average scene length is larger than ten frames to obtain a good approximation quality of the autocorrelation function.

Next, we compute the mean GOP size of each scene. After defining the number of scene classes which is equal to the number of states N_s of the scene change Markov chain, we have to classify the scenes by means of their mean GOP sizes and to compute the transition probability matrix for the scene

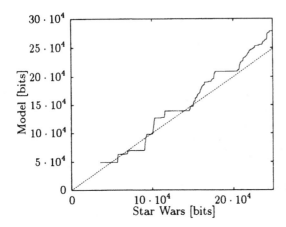

Figure 1.2.13: Q–Q–plot.

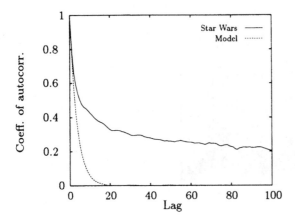

Figure 1.2.14: Autocorrelation function.

classes. This is done analogously to the procedure for the GOP sizes of the *Simple MC model*. It is important to note that the transition matrix is not simply based on the changes because of scene endings but on the GOP-by-GOP changes, i.e. if scene class i lasts five GOPs the four transitions from class i to i also have to be counted. This is done to obtain the same average scene lengths as the empirical sequence.

After creating the transition probability matrix of the scene change process, the matrices for the GOP process of each scene class have to be computed. We concatenate all GOP sizes of the scenes belonging to the same scene class in the order of their appearance and use the method of the *Simple MC model* to estimate the N_s transition matrices including the GOP sizes which are related to their states.

The GOP generation process of the *Scene-oriented model* for a simulation works as follows. First, the new state of the scene change Markov chain is determined. If the state or scene class does not change we keep on using the same GOP generation Markov chain as before and compute the current GOP size based on the last one. If the scene class changes we determine the adequate GOP Markov chain and select the starting state at random.

Fig. 1.2.15 shows a summary of the parameter estimation procedure and the GOP size generation process for the *Scene-oriented model*.

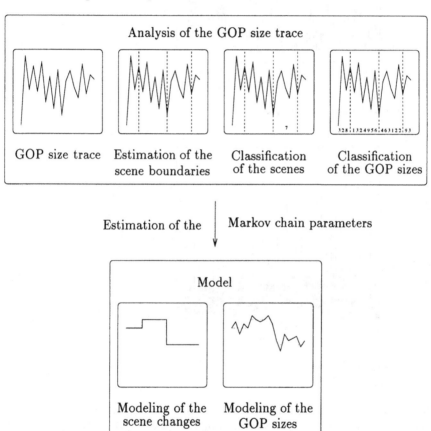

Figure 1.2.15: Parameter estimation for the Scene-oriented model.

For an analysis, the transition matrix of the nested Markov chains can be estimated by elementary matrix operations. Unfortunately, this already leads to rather large matrices, which only describe one MPEG traffic source. If a superposition of sources has to be modeled, simpler models should be considered.

Figure 1.2.16 and 1.2.17 show a GOP and frame size trace generated by the *Scene-oriented model*. We used $\epsilon = 0.9$ to compute the scene boundaries and obtained 10 scene classes with 30 states each, i.e. 300 states in total.

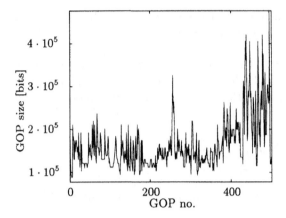

Figure 1.2.16: Model GOP trace.

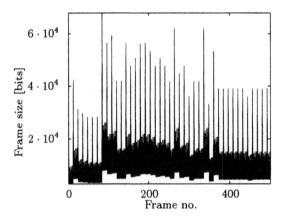

Figure 1.2.17: Model frame trace.

The estimate of the empirical GOP sizes is good (Figure 1.2.18), and the approximation of the autocorrelation function is very good for the range of lags shown in Figure 1.2.19.

Comparing the plots of the three models leads to the conclusion that the most complex model, the *Scene-oriented model*, shows very good agreement with the empirical data sets, whereas the the simpler models are unable to model the correlation behavior of the data sets over a larger period of time. In the next section, we will examine how these properties affect the capabilities of the models if they are used to predict cell losses at an ATM multiplexer buffer with VBR MPEG video input.

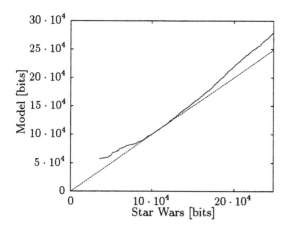

Figure 1.2.18: Q–Q–plot.

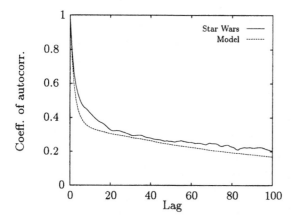

Figure 1.2.19: Autocorrelation function.

1.2.6   Comparison of the models

In this section, we present the cell loss simulation results which are used to compare the three models, where the model parameters are the same as in Section 1.2.5. For the simulation, we used a *fluid flow* approach, i.e. we simulated on frame level assuming a constant video source transmission rate for each frame. The frame level results are essentially equal to the cell level results if we assume that each video source spaces the cells of one frame over the frame duration.

The multiplexer has a link rate of 10 Mbps and buffer sizes of 100, 1000, 10 000, and 100 000 cells were considered and the multiplexer load is determined by the number of video input sources, ranging from 11 (load of about 0.4) to 33 (load of about 1.2). We scaled down the link rate to 10 Mbps

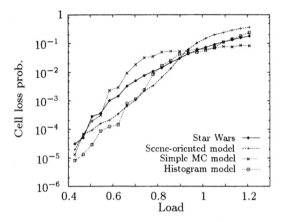

Figure 1.2.20: Loss for 100 cells buffered.

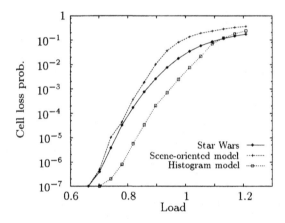

Figure 1.2.21: Loss for 1000 cells buffered.

because we wanted to obtain realistic cell loss results in spite of the fact that the empirical data set has a bit rate of only about 5 to 10 % of high quality full-screen MPEG video sequences.

Figure 1.2.20 shows the cell loss results for the three models and the *Star Wars* data set for a buffer size of 100 cells. The *Scene-oriented* and the *Simple MC model* show almost the same good approximation quality, but the *Histogram model* performs only slightly worse. For the sake of clarity, in Figure 1.2.21, 1.2.22, and 1.2.22 we do not show the curve for the *Simple MC model* because it is very close to the *Histogram model* curve.

Increasing the buffer size to 1000 cells (cf. Figure 1.2.21) and correspondingly increasing the influence of correlations on the cell loss estimates shows that the *Scene-oriented model* still shows the best approximation quality, whereas the estimates of the simpler model are becoming more and more op-

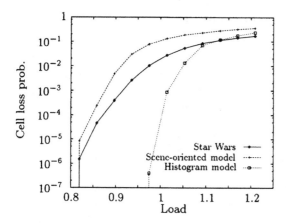

Figure 1.2.22: Loss for 10,000 cells buffered.

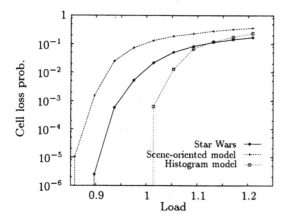

Figure 1.2.23: Loss for 100,000 cells buffered.

timistic. However, for dimensioning purposes the simpler *Histogram model* might still be useful up to buffer sizes of 1000 cells. Only for very large buffer sizes of 10 000 cells (cf. Figure 1.2.22) or 100 000 cells (cf. Figure 1.2.23)is the *Scene-oriented model* substantially better than the simple model. Moreover, for network dimensioning purposes it is more convenient to use a model which behaves worse than the real traffic. This is clearly the case for the *Scene-oriented model* which overestimates the cell losses for larger buffer sizes.

These results show that only for buffer sizes which are rather large is it necessary to use scene level correlation information to obtain good cell loss estimates. In addition, the simulation results prove that the simple procedure to obtain the frame sizes by multiplying scaling factors to the GOP sizes is a useful way to generate the frame sequence.

1.2.7   Higher-order Markov chains

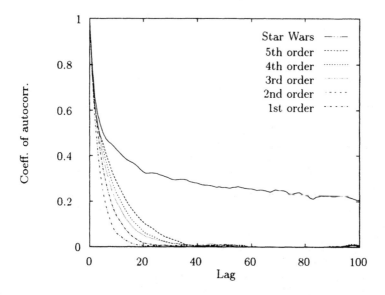

Figure 1.2.24: Autocorrelation function of kth-order Markov chains.

The results presented in the last two sections show that an improvement in the correlation behavior and, correspondingly, the capability to predict cell losses for rather large buffers requires the use of a complex model. It would be helpful to find a Markovian model which is more powerful than the *Histogram model* or the *Simple MC model* but still simpler than the Scene-oriented model. The straightforward approach is to avoid the use of scene-level information and to increase the Markovian order of the model until it shows an appropriate correlation behavior. Unfortunately, this approach has two important drawbacks.

First, we consider the number of model parameters and their estimation. In our case, the most important set of parameters of the Markov chain is the size of the transition probability matrix. The number of elements of this matrix for an M-state Markov chain is $M^{order+1}$, if we do not consider elements which are definitely zero. M, however, should not be chosen too small because it determines the approximation quality of the marginal distribution of the data set. In current systems, this will cause memory problems for orders larger than 10. Another problem is to obtain statistically significant estimates for the transition probabilities, because with an increasing order the number of samples for each estimate is very small even if we use very long empirical data sets.

After this more technical drawback, the properties of higher order Markov models both from autocorrelation function and coding theoretical point of view are evaluated. A first insight is derived from Figure 1.2.24 which shows

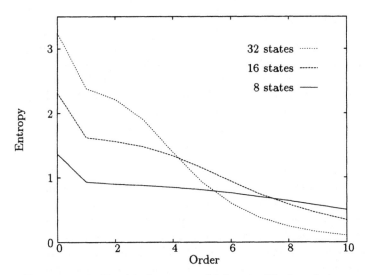

Figure 1.2.25: Empirical entropy of kth-order Markov chains.

the autocorrelation function of the *Star Wars* GOP sequence and some higher-order Markov chain models of it. As already mentioned, the decay of the empirical curve is clearly not exponential and increasing the order of the Markov chains leads only to marginal improvements in the approximation quality.

In coding theory, the empirical entropy of a time series is used to determine the Markovian order of its generation process. The method we used can be found in [MGZ89]. In the following we will only present the results of the algorithm and their interpretation. The presentation of the algorithm is beyond the scope of this section.

Figure 1.2.25 shows the empirical entropies for Markov chains up to the order of 10 with 8, 16, and 32 states. An entropy of 0 is equivalent to the fact that the model contains all the properties of the empirical data set. From the curves, we conclude that there is a gain in model quality if we increase the order of the model. However, to obtain good models for 16 or 32 states it seems that the order should be larger than 5, i.e., at least one million matrix elements in the 16 states case. On the other hand, if we consider 8 states, there is almost no improvement for Markov chains with an order larger than 1. Even for an order of 10 the entropy is still rather high. Of course, the empirical entropy will also decrease to zero but only for large orders.

To sum up, both from a technical and a statistical viewpoint it is almost impossible to improve the correlation behavior of the Markov chain models by simply increasing the Markovian order of the model.

1.2.8 Shaped VBR video

The MPEG standards currently allow two types of coding: open loop coding, producing VBR video as described in the previous sections, and closed loop coding, where coding parameters are adjusted to produce a constant bit rate output stream. In this section we consider a third variant where closed loop coding is used to produce a VBR output with controlled burstiness.

MPEG CBR

Codecs for video transmission have traditionally aimed to produce a constant bit rate stream suitable for transport over circuit switched telecommunications networks. The MPEG closed loop algorithm acts at the level of a subframe unit called the macroblock. To regulate the amount of data produced for a given picture the coder quantization parameter Q is modified from macroblock to macroblock. A fixed quantity of bits is allocated to each GOP and apportioned progressively to successive pictures and, within pictures, to successive macroblocks. Bit rate variability persists even at the GOP scale since the number of bits used may be different to the *a priori* assignment. The difference is taken into account in fixing the bit allocation of the next macroblock or GOP. Residual variability is smoothed in an emission buffer emptied at the contracted constant rate.

The drawback of CBR coding is that the same bit rate is generated independently of the scene contents thereby resulting in variable visual quality. Additionally, the need for a smoothing buffer in the coder and a compensating playback buffer in the receiver introduces coding delay which can be significant for real time interactive applications.

Controlled burstiness VBR

The variability of the signal produced by MPEG open loop coding has been described in the previous paragraphs. It is observed that the rate generated by a given coder varies widely depending on the scene filmed and that correlations between the rates of successive frames persist over very long time lags. These effects make it difficult to efficiently handle VBR video in the network: the rate of a given communication is difficult to predict and rate overloads, when they occur, can persist over significant periods whose length is determined by scene scale rate variations. To achieve an output with predictable bit rates without the severe constraints of CBR coding, it has been proposed to include in the MPEG standard a controlled burstiness coding algorithm as described below [HR95].

It is first argued that the full variability of open loop coding is not necessary to maintain the subjective quality of video sequences containing scenes of different types. Quality from the user point of view depends mainly on the visual capacity to capture the information displayed on the screen. In fast moving scenes with complex image structure, the human eye does not have enough time to discover all image details. It may be considered that the high bit rate generated for such scenes by an open loop coder is unnecessarily generous. On the other hand, scenes with little motion and simple structure are more sensitive to signal degradations. Their quality should be maintained at a satisfactory level. Subjective quality may actually be improved by restricting the scope for scene scale variations: for a given mean rate, higher resolution in low activity scenes more than compensates for poorly perceived detail in fast moving and complex sequences. It is for this reason that CBR coders generally produce acceptable visual quality.

The algorithm considered here uses a closed loop algorithm to ensure that the volume of data emitted satisfies the following burstiness constraint: in any sequence of k successive GOPs, the number of bits emitted $N(k)$ satisfies:

$$N(k) \leq rk + b. \tag{1.2.3}$$

This choice is motivated by the widespread use of leaky-bucket-like algorithms to control the traffic offered to packet and ATM networks. We consider the leaky bucket access controller of leak rate r and token pool size b where, for notational convenience, r is measured in bits/GOP and b in bits. Tokens are consumed as data enters the network and are replenished at rate r; input data will be discarded if it arrives to find no tokens available. By maintaining an image of the leaky bucket counter, the coder can implement a closed loop control to ensure that its output conforms and thus avoid data discard by the access controller.

We assume for present purposes that the counter is adjusted on a GOP by GOP basis and say the coder conforms to the leaky bucket when its output satisfies the burstiness constraint (1.2.3). (In practice, an access controller typically works on a packet or cell basis and it would be necessary to allow for cell and frame scale variations to ensure conformity). The task of the closed loop control is to maintain the leaky bucket counter within in the range (0, b). A zero value would mean the risk of packet discard while a counter value of b means that not all the available rate r is being used.

The shaping algorithm

In this section we describe a rate control algorithm to be implemented in the MPEG coder to ensure that its output conforms to a leaky bucket with parameters r and b. The algorithm operates in open loop to code the different frames and macroblocks of a GOP while the quantization parameter Q is adjusted from GOP to GOP to control the extent of bit rate variations. Note that since we allow variations over several GOPs, there is no need here to adjust Q on a finer scale as is done in the CBR algorithm. The shaping algorithm is thus considerably less complex than the CBR algorithm described in the MPEG standards.

The adjustments are derived from the value of a counter $X(k)$ which records the number of leaky bucket credits spent at the start of the kth GOP.

Let $R(k)$ be the number of bits generated in GOP k (i.e., $R(k)$ is the rate measured in bits/GOP). $X(k)$ then evolves as follows:

$$X(k+1) = \min\{b, \max\{0, X(k) - r\} + R(k)\}. \qquad (1.2.4)$$

The algorithm aims to adjust the GOP-k quantization parameter $Q(k)$ to ensure that $X(k)$ is neither too close to b nor too close to 0. In the former case, the coding tends to CBR coding at rate r; in the latter, the coder does not fully use the available average bit rate r. The adjustments to $Q(k)$ should allow flexible, open-loop-like control when $X(k)$ is in a middle range around $b/2$ while attracting it back to this range if it tends to approach either extreme, 0 or b. Note that the value of b, unlike the size of the CBR smoothing buffer, is largely unconstrained in the present algorithm since data is never actually delayed.

A detailed algorithm fulfilling the above objectives is described in [HR95]. An alternative coder rate control satisfying leaky bucket constraints is proposed by Reibman and Haskell [RH92]. Ortega et al. [OGV95] also discuss the merits of "constrained VBR coding" and suggest the utilization of rate adjustment algorithms adapted from optimal CBR schemes.

The best choice of coding algorithm remains an open question. Among the criteria for choice are, obviously, the resulting visual quality and how easy it is to predict network performance based on the induced traffic characteristics.

Network performance

Shaping the coder output simplifies the task of the network in verifying conformity to declared traffic parameters. It remains to determine how these parameters can be used by the network to guarantee quality of service by adequate resource provision.

The shaping algorithm ensures that the mean source rate is equal to the leak rate of the leaky bucket. Together with the peak rate, this parameter can be used to define a worst case on/off source behaviour: peak rate bursts, possibly at the start of each frame, separated by silence intervals and such that the source on-probability is equal to the mean to peak ratio.

The binomial rate distribution corresponding to this worst case traffic can be used to determine necessary resource provision, notably in the case of Rate Envelope Multiplexing (REM) as discussed in Section 4.1. With REM, the objective is to maintain the combined input rate of multiplexed sources within the multiplexer output capacity. This is feasible and efficient for the worst case traffic assumption if the source peak rate is a small fraction of the multiplexer rate.

More efficient REM can be achieved if less pessimistic assumptions can be made concerning the source stationary rate distribution. For example, if data emission is smoothed to a constant rate throughout a frame or GOP duration, the resulting rate distribution will be much less variant than that of the worst case traffic assumption. It remains to determine if the considered shaping algorithm or, indeed, any other algorithm can sufficiently characterize the source output rate distribution.

With REM, only a small cell scale multiplexer buffer is required and network delays are negligible. The use of greater buffering in the network allows greater link utilization. The fact that the source obeys the burstiness constraint (1.2.3) determined by the leaky bucket parameters allows multiplexer loss to be closely controlled [HR96]. However, it remains debatable whether the increased latency implied by the large network buffers might not be better exploited in further smoothing coder output at source and retaining the principle of REM.

2 Broadband service models

We define a service model to be the set of traffic control mechanisms and protocols enabling a network to offer a range of telecommunications services differentiated by their particular traffic characteristics and performance requirements. The service model of the B-ISDN is based on a set of ATM transfer capabilities (ATC) standardised by ITU-T Study Group 13 [ITU96b]. This model is closely paralleled by that of the ATM Forum where different "service categories" (broadly similar to the ATC) are provided to the user. The range of service options of the ATM networks and their relative complexity is contrasted with the service model of the present Internet where a single "best effort" class of service is provided based on IP, the Internet protocol. However, the Internet Engineering Task Force (IETF) is currently deliberating on the definition of additional service classes offering the performance guarantees which some see as necessary to provide audio and video services, notably for real time applications.

The definition of the service models by the ITU, the ATM Forum and the IETF all rely on the ability to characterise and control the parameters of traffic streams using rule-based parameters. The particular rule which is employed is basically the leaky bucket, known variously as the virtual scheduling algorithm, the generic cell rate algorithm or the token bucket filter. This algorithm and its different representations are described in Section 2.1 where the definition of so-called "worst case traffics" compatible with leaky bucket control are also defined. The ATC/service-categories defined for ATM networks are then outlined in Section 2.2 and the notion of traffic contract is presented in Section 2.3. The service model of a future integrated services Internet is succinctly described in Section 2.4 although it is to be noted that this model is still evolving and, at the time of writing, no firm proposals have yet emerged.

2.1 Rule based traffic parameters

Parameters used in the traffic descriptor of an ATM connection should ideally have the following properties:

- they should be easily understandable by the user;

- they should be significant for resource allocation;

- they should be verifiable by the network.

In fact, for variable rate connections, it proves difficult to meet all three requirements simultaneously. For example, the mean rate of a connection is certainly highly significant for resource allocation (it determines link loads) but may be difficult for a user to predict in advance; it is above all impossible to verify in real time at the network ingress. In view of the importance of being able to prevent users sending traffic in excess of their declaration and thus provoking congestion, standardized traffic parameters are rule-based, sacrificing somewhat the first two requirements in favour of the latter. Even the peak rate of a connection must be defined with respect to a rule allowing a certain CDV tolerance. In this section we describe the rule adopted for the B-ISDN, the Generic Cell Rate Algorithm (GCRA), and the related "token bucket filter" chosen to define parameters in an integrated services Internet.

2.1.1   Generic Cell Rate Algorithm

Traffic characteristics of an ATM connection are currently defined using the Generic Cell Rate Algorithm (GCRA). This was first defined by the ATM Forum [For96] by generalizing the Peak Cell Rate definition given by ITU-T in Recommendation I.371 [ITU96b]. The $GCRA(T, \tau)$ relates a time period T (reciprocal of a cell rate) to a tolerance τ quantifiyng the maximal deviation of the considered cell stream from purely periodic behaviour. It is necessary to introduce the tolerance τ for the peak rate since, due to multiplexing and buffering procedures in the network that cause cell delay variation, the rate of a connection observed at a given interface may be momentarily larger than the source original rate: for example, if some cells of the connection have been delayed in a buffer and then transmitted consecutively, the instantaneous rate observed at the output of the buffer is momentarily equal to the link rate. A tolerance for the sustainable cell rate, the intrinsic burst tolerance (IBT), is naturally introduced for variable bit rate connections. The IBT parameter limits the maximum size of a peak rate burst, as discussed below.

The $GCRA(T, \tau)$ determines if a cell stream conforms to a nominal rate $1/T$ subject to the tolerance τ. More precisely, cell k is "conforming" to $GCRA(T, \tau)$ at a given interface if and only if $y_k = c_k - a_k$ is smaller than τ, where a_k is the time at which cell k is observed at the interface and c_k is a "theoretical time" defined recursively as follows:

$$
\begin{aligned}
c_{k+1} &= c_k + T & \text{if } \tau \geq y_k \geq 0, \\
&= a_k + T & \text{if } y_k < 0, \\
&= c_k & \text{if } y_k > \tau.
\end{aligned} \tag{2.1.1}
$$

In the case of a peak cell rate, T is the Peak Emission Interval ($T = 1/\text{PCR}$), and τ is the CDV tolerance as defined in [ITU96b]. More generally, for the sustainable cell rate, T is the reciprocal of the Sustainable Cell Rate (SCR) and τ is the sum of the Intrinsic Burst Tolerance (IBT) and the CDV tolerance at sustainable cell rate level [For96].

The PCR and SCR parameters of a connection are meant to represent some traffic characteristics of the considered connection. For example, the PCR can be considered as the cell rate that is needed to accommodate the peak bit rate of the application using the connection and the SCR as a cell rate that allows transmission of the mean bit rate of the application. However, the peak bit rate of the application may be modified by several procedures (use of ATM Adaptation Layers, multiplexing in the upstream network) and the PCR of the connection may differ from the original peak bit rate of the application. Furthermore, any cell rate that allows transmision of the mean bit rate of the application can be considered as a possible candidate for the SCR. The exact choice, for both PCR and SCR then depends on the values chosen for the CDV tolerances and IBT.

While rule-based traffic descriptors are very attractive from a practical stand point, notably for conformance testing and traffic control purposes, the optimal selection of these parameters to meet user requirements remains an open problem. Strictly speaking, the selection of $GCRA(T, \tau)$ and $GCRA(T_{SCR}, \tau_{SCR})$ should be done simultaneously to ensure that the total probability of non-conforming cells is less than a prescribed ratio. Looking at the restricted problem of choosing a $GCRA(T_{SCR}, \tau_{SCR})$ assuming that a source is described by a set of statistical parameters with a fixed peak rate and neglecting jitter, it can be shown that an infinite number of parameter pairs (T_{SCR}, τ_{SCR}) can be considered as suitable for the source [RH94], [GRM95]. For instance, all couples (T_{SCR}, τ_{SCR}) rejecting (or delaying, in case of a buffered mechanism) cells with a probability less than $\varepsilon = 10^{-n}$ could be considered as admissible.

Note that the selection of these parameters is important from a network efficiency perspective since the parameters (T_{SCR}, τ_{SCR}) are directly related to the network resources allocated to the source.

2.1.2   Measurement Phasing

Due to the cell-by-cell definition of conformance, the set of cells which are identified as non-conforming depends on the initialization time of the conformance algorithm. This is the so-called "measurement phasing" problem : the time at which a conformance test is initialized influences the conformance decisions taken during the test.

An example of this phenomenon is given in Figure 2.1.1: the connection is characterized in the traffic contract by a pair $(T = 4, \tau = 4)$ (where the time unit is taken to be a cell emission time). The user transmits twice the negotiated rate, transmiting blocks of two back to back cells every 4 cell times. If the starting point of the conformance checking process is the first cell of a cell block, after a short transient period, all first cells of subsequent blocks are identified as conforming (y_k value equal to 4) and all second cells of subsequent blocks are identified as non-conforming (y_k value equal to 7). Conversely, if the starting point of the conformance checking process is the

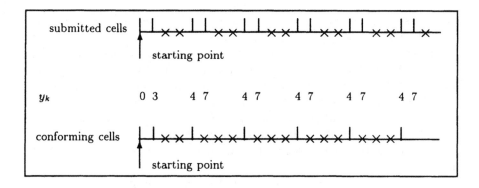

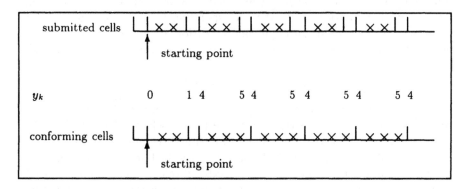

Figure 2.1.1: An example of the measurement phasing problem.

second cell of a cell block, opposite conformance choices are reached. Note that, in both cases, the proportion of non-conforming cells is 1/2.

More generally, it is shown in [GH94] that the long term ratio of conforming cells is a characteristic of a cell stream: when the number of observed cells tends to infinity, the limit of the ratio of conforming cells is independent of the starting time of the testing procedure. This property extends in a straightforward manner to the case where two cell rates are simultaneously defined for all cells of the same connection (as is the case for the SBR1 transfer capability), but not to the case where two cell rates are defined, one for the aggregate PCR and another of the CLP=0 cell flow. It is shown in [Gra94] that the limit of the ratio of CLP=0 conforming cells may then depend on the starting time of the testing procedure.

2.1.3   A queueing model representation of the GCRA

As discussed above, a cell rate is specified together with a corresponding tolerance. Let (T, τ) denote a pair of emission interval and CDV tolerance as defined in the GCRA formalism. The set of actual arrival times at the considered interface $\{a_k\}_{k \geq 0}$ defines a (general) arrival process to a (virtual) deterministic queue whose service duration is T. When a user enters the queue the virtual waiting time thus increases by T.

The set of theoretical times $\{c_k\}_{k \geq 0}$ computed by equation (2.1.1) then represent the (virtual) departure times from the virtual single server queue. More precisely, c_k represents the departure time of the previous cell to have entered the queue in the G/D/1 model.

If $y_k = c_k - a_k$ is negative, cell k initiates a busy period in the G/D/1 queue and $(a_k - c_k)$ is the duration of the preceding idle period. Conversely, if y_k is non-negative, it represents the "waiting time" of cell k in the G/D/1 model.

A given cell is conforming for the GCRA if and only if $y_k \leq \tau$, which is equivalent in the G/D/1 queueing model to a waiting time smaller than τ. If τ is a integer multiple of T, the corresponding G/D/1 queue is finite of capacity τ/T. This yields an upper bound B_c for the number of conforming cells accessing the conformance procedure at a given rate:

$$B_c = 1 + \lfloor \tau/(T - \Delta) \rfloor$$

where $\lfloor x \rfloor$ is the integer part of x and Δ is the minimum interval between two cell arrivals, i.e., $1/\Delta$ is the rate of the line on which cells are observed.

The G/D/1 model with limited waiting time directly corresponds to the "continuous-state leaky bucket" formalism described in [ITU96b]: in this formalism, the finite capacity bucket is emptied out at a continuous rate of 1 unit of content per time unit and its content is increased by T for cells finding the level of the bucket less than or equal to τ. It is thus seen that the level of the continuous-state leaky bucket is modelled by the value taken by the virtual waiting time in the G/D/1 queue.

Table 2.1.1 displays the relationships between the variables used in the G/D/1 model, in the GCRA formalism and in the continuous-state leaky bucket formalism (in this formalism, LCT denotes the Last Conformance Time and X the level of the leaky bucket immediately after incrementation).

2.1.4   The token bucket filter

Rates in the proposed Internet integrated services model are defined with respect to a "token bucket filter" [CSZ92]. This device is closely related to the GCRA although applied to the more general context of variable length packets. In fact, prior to the formal definition of the GCRA and its virtual scheduling algorithm and continuous state leaky bucket interpretations, the telecommunications community used the term leaky bucket for the ATM version of the token bucket filter described below (see, for example, [Rat91]).

The token bucket filter is characterised by two parameters, a rate r and a depth b. The token bucket may be considered to fill with tokens continuously at rate r up to a maximum depth b. Every time a packet is generated it takes p tokens from the bucket where p is the packet size. A traffic source

Table 2.1.1: Correspondence between the GCRA formalism, the G/D/1 model and the continuous-state LB formalism.

Cell Conformance Algorithm	G/D/1 queue with limited waiting time	Continuous state leaky bucket (LB)
emission interval	duration of service time	LB increment
a_k = arrival of cell k	arrival time of kth customer	potential LB incrementation time
c_k = theoretical time for cell k	departure time of previous entering customer	$X + LCT$ at time a_{k-1}^+
$y_k = c_k - a_k$	waiting time for k^{th} customer (if non-negative)	observed level in bucket at a_k^- (if non-negative)
$y_k < 0$ cell k late and conforming	k^{th} arrival initiates a busy period	empty bucket at a_k^- LB incremented
$0 \leq y_k \leq \tau$ cell k early and conforming	customer enters a non-empty queue	non-empty bucket at a_k^- LB incremented
$y_k > \tau$ cell k non-conforming	k^{th} customer rejected	bucket level larger than τ no LB increment
	virtual waiting time at a_k^+	X after LB increment
amount of clumping $= \lceil (c_k - a_k)/T \rceil$	number of customers in buffer at time a_k	

conforms to the token bucket filter (r, b) if there are always enough tokens in the bucket whenever a packet is generated.

More precisely, let a_k and p_k denote the arrival epoch and length, respectively, of the k-th packet of a traffic process. The process conforms to the token bucket (r, b) if the sequence n_k defined by:

$$n_0 = b \qquad (2.1.2)$$

$$n_k = \min\{b, n_{k-1}\} + (a_k - a_{k-1})r - p_k \text{ for } k > 0, \qquad (2.1.3)$$

is such that $n_k \geq 0$ for all k.

Packet lengths and corresponding bucket parameters may be measured in bits, bytes or any other convenient unit. Two cases are of particular relevance here:

- packets are of constant length (like ATM cells) and the token bucket parameters r and b are measured in packets/second and packets, respectively;

- the fluid approximation applies and the bucket depth can be a real number.

In the first case where b is an integer, the token bucket filter (r, b) corresponds exactly to the GCRA$(1/r, (b-1)/r)$. The second case corresponds to a useful modelling device when evaluating the effect of a token bucket/GCRA on a given traffic stream represented as a fluid process (see Section 5.3.2). Clearly, like the GCRA, the token bucket can be represented as a virtual queue. In this case, the queue has constant service rate r and maximum queue size b.

It is useful to note the bound on the amount of traffic input to a network when access is controlled by a token bucket filter. Let $\nu(s, t)$ be the amount of data admitted to the network in interval (s, t). The token bucket filter (r, b) then guarantees the inequality:

$$\nu(s, t) \le b + r(t - s), \tag{2.1.4}$$

as can be readily seen from the interpretation of the device as a queue. This allows the derivation of certain performance bounds as discussed in Chapter 4.1.2. It also constitutes the basis of the strict QoS guarantees envisaged for the "guaranteed service" category of the Internet service model (see Section 2.4).

2.1.5   Worst Case traffics based on rule based traffic parameters

A Worst Case Traffic (WCT) for a connection with given traffic contract is defined as a cell stream for which all cells are conforming to the negotiated traffic contract, and which requires the greatest amount of resources. This section describes WCTs for connections that have negotiated either a DBR or an SBR ATM transfer capability (see Section 2.2 for a definition of ATM tranfer capabilities). A similar definition clearly applies in the case of packet traffic controlled by one or two token bucket filters.

We assume the WCT for a DBR connection with a PCR characterized by (T, τ) to be a periodic pattern of bursts of length $B_c = 1 + \lfloor \tau/(T-1) \rfloor$, with an average cell rate given by $1/T$ (the cell transmission time on the link where the rate is being crontrolled is chosen as time unit, i.e., $\Delta = 1$). This WCT is illustrated in Figure 2.1.2. In certain cases, more complex arrival patterns have been shown to result in worse performance than that considered here [Dos94]. However, the present definition of WCT is commonly accepted as useful working assumption.

Clearly, if τ is smaller than $(T-1)$, B_c is equal to 2^1. However, B_c can be much larger than 2 depending on the values taken by T and τ. A lower

---

[1]Indeed, if $\tau < (T-1)$, two successive cells should never be observed back to back. However, since they may be spaced with a period smaller than T, it is a pessimistic assumption to assume that the MBS is 2.

Figure 2.1.2: WCT pattern for a DBR connection.

bound for τ (expressed in cell transmission times) is provided in [ITU96b]:

$$\tau_{\min} = \max[T, f(T)], \quad \text{where} \quad f(T) = 80 \times (1 - 1/T), \qquad (2.1.5)$$

i.e., it should not be expected that a source should have a CDV tolerance less than $\tau_{\min}$.

The linear bound $f(T)$ is obtained by fitting a linear curve to a set of points obtained as quantiles of the waiting time distribution for the cells of a deterministic cell stream with period T multiplexed with a Poisson background process in a FIFO queue loaded at 0.85. Function $f(T)$ thus corresponds to an upper bound on the CDV induced on a PCR spaced cell stream by a single multiplexing stage loaded at 0.85 and fed solely by PCR spaced traffic. Reference [ITU96b] proposes to set T as a lower bound for τ in order to ensure that a DBR cell stream with a slightly overestimated T value experiences a correspondingly small non-conformance ratio (see Appendix III to [ITU96b]).

Assume now that, in addition to the Peak Cell Rate traffic descriptor (T_{PCR}, τ_{PCR}), the connection is also characterized by a sustainable cell rate parameter set (SCR, IBT) and a burst level CDV tolerance. Let the sustainable cell rate parameter set first describe the aggregate CLP=0+1 traffic as for an SBR1 ATM transfer capability (see Section 2.2).

Let T_{SCR} and τ_{SCR} denote the reciprocal of SCR and the sum of IBT and the burst level CDV tolerance, respectively. The WCT corresponding to these traffic parameters at the considered interface is a periodic on/off source with bursts at two levels (see Figure 2.1.3):

- the on periods consist of emission periods at link rate, followed by silence periods; basically, the traffic pattern during an on period corresponds to the WCT of a DBR connection.

- during an on period, a fixed number of cells MBS (see [ITU96b]) is transmitted where

$$\text{MBS} = (1 + \lfloor \tau_{SCR}/(T_{SCR} - T_{PCR}) \rfloor)$$

As a particular case, we see that if PCR spacing is performed on user traffic, the WCT is a periodic on/off process, the traffic pattern during an on period being periodic with period T_{PCR}: the SBR WCT is then "smooth". This is illustrated in Figure 2.1.4.

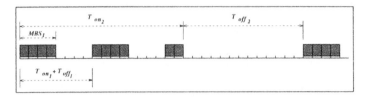

Figure 2.1.3: SBR WCT with bursts at two levels.

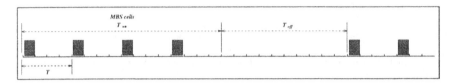

Figure 2.1.4: Smooth SBR WCT corresponding to PCR spaced traffic.

The above WCTs correspond to SBR1 connections; if either an SBR2 or an SBR3 connection is now considered (see Section 2.2), the sustainable cell rate parameter set then corresponds only to the CLP=0 sub-stream of the aggregate CLP=0+1 stream submitted by the user. WCTs in this case are obtained as described below:

- if the SBR connection is not PCR spaced, consider first a DBR WCT as described in Figure 2.1.2; CLP=0 cells in this WCT are located as described in Figure 2.1.3, the other cells being CLP=1 cells;

- if the SBR connection is PCR spaced, consider then a purely periodic cell pattern with period T_{PCR}; CLP=0 cells in this WCT are located as described in Figure 2.1.4, the other cells being CLP=1 cells.

2.2   ATM transfer capabilities

A number of different ATM transfer capabilities (ATC) are being defined by ITU-T [ITU96b] and by the ATM Forum [For96]. We briefly describe these different capabilities below indicating some differences which still exist between the ITU and the ATM Forum. An ATC is intended to translate a service model (i.e., a combination of traffic characterization and QoS requirements) into a set of traffic characterizations and traffic management procedures suitable for some types of applications and allowing efficient resource allocation. QoS requirements are only partially characterized by the ATC; indeed, a QoS class needs to be associated with the ATC in order to completely characterize the type of connection required by the user.

2.2.1 Deterministic Bit Rate (DBR) or Constant Bit Rate (CBR)

A user may be able to negotiate and control only its peak rate, in which case the DBR/CBR capability is negotiated at connection establishment. DBR is the ITU denomination and CBR that of the ATM Forum. This is the most straightforward ATM transfer capability where traffic characterization is reduced to the specification of the peak cell rate (PCR) and its associated CDV tolerance.

In the DBR capability, cells are handled independently of the value of the CLP bit. QoS requirements are expressed in terms of objectives set to the Cell Loss Ratio (CLR) parameter (for the aggregate CLP=0+1 cell flow), and possibly to delay related parameters: Cell Transfer Delay (CTD) and end-to-end Cell Delay Variation (CDV).

The DBR capability is not solely intended for connections carrying Continuous Bit Rate (CBR) applications since a user may or may not continuously transmit at the negotiated Peak Cell Rate; however, the network should be able to offer the user a QoS class compatible with the use of a DBR connection for carrying a constant rate application.

Note that the PCR parameter can be changed using signalling while a connection is in progress.

2.2.2 Statistical Bit Rate (SBR) or Variable Bit Rate (VBR)

The ITU uses the name SBR while the ATM Forum prefers VBR for the same ATC. SBR/VBR connections are described by three traffic parameters: PCR, sustainable cell rate (SCR) and intrinsic burst tolerance (IBT), the latter two being defined with respect to the generic cell rate algorithm (GCRA) (see Section 2.1). SCR and IBT are rule based parameters which provide some information on the characteristics of variable bit rate connections and are expected to lead to more efficient network operation by allowing controlled statistical multiplexing.

The ITU currently distinguishes three different SBR capabilities, which differ in their use of the CLP bit. In the SBR1 capability, cells are not discriminated on the basis of the CLP bit and CLR requirements are expressed for the aggregate CLP=0+1 cell flow. Objectives may or may not be set for CTD and CDV. In the SBR2 and SBR3 capabilities, cells are discriminated on the basis of the CLP bit. The traffic parameters negotiated for these capabilities are the PCR of the aggregate CLP=0+1 cell flow and the SCR parameter set for the CLP=0 cell flow. QoS requirements are expressed in terms of a CLR objective for the CLP=0 cell flow; there is no QoS commitment concerning either delay related parameters or the CLR for the aggregate cell flow. SBR3 differs from SBR2 only by the use of the tagging option: in SBR3, the network is allowed to tag non-conforming cells (i.e., downgrade the cells by setting their CLP bit to 1), whereas tagging is not allowed in SBR2.

The ATM Forum also distinguishes VBR1, VBR2 and VBR3 capabilitieswhich mirror the ITU definitions. An additional distinctioon between real time and non-real-time VBR services corresponds in the ITU to the choice of different QoS classes. Real time VBR requires tight delay guarantees suitable, for example, for interactive applications.

It is possible to view the SCR, IBT parameters in two ways:

- as limits defining an envelope into which traffic must be made to fit by shaping,

- as an approximate description of the statistical characteristics of a VBR traffic stream.

The first interpretation corresponds, for example, to the use of similar rule based parameters to define data connections in frame relay networks: it is up to the user to choose the best compromise between cost and performance, the latter including delays induced by the shaping function. The second interpretation is well known to be problematical since the statistical multiplexing performance of any random traffic depends on parameters which are very poorly represented by SCR and IBT: SCR is an upper bound on the mean rate and IBT determines a maximum burst length; however, these bounds must generally be very loose if the GCRA is not to constitute a severe filter on incoming random traffic.

2.2.3 ATM Block Transfer (ABT)

ABT ATM transfer capabilities are defined only at the ITU. In the ABT capabilities, the user is able to define and control a block structure in its data stream. The PCR for each block is negotiated between the user and the network by means of Resource Management (RM) cells. An ABT capability may be implemented using a Fast Reservation Protocol [BT92a].

Two variants are currently being standardized:

- ABT with Delayed Transmission (ABT/DT) in which the user waits for the network to either explicitly or implicitly accept a new block,

- ABT with Immediate Transmission (ABT/IT) in which the user does not wait for the network to accept a new block (blocks may then be discarded by the network).

Both ABT capabilities can be either *rigid* or *elastic*. In the *rigid* version, the rate required by the user is either accepted or not by the network. Repeated reservation requests are necessary in the event of blocking (when a requested bit rate is not available). In the *elastic* version, the network may offer a smaller rate than that requested by the user. Parameters in addition to the

PCR can be declared either to specify a minimum guaranteed rate or to supply additional information on expected connection traffic allowing more efficient connection admission control.

The QoS commitments for the ABT capabilities are set at both cell and block level. The cell level QoS commitment offered by the ABT capabilities is that the cell level QoS offered to the cells in the blocks that are transmitted is equivalent to the QoS commitment offered by either the DBR or the SBR1 capabilities. The block level QoS depends on the specific ABT capability:

- in the ABT/DT capability, blocks may be delayed within the customer equipment; the block level QoS is characterized by an upper bound on the delay experienced by a block before being accepted by the network;

- in the ABT/IT capability, blocks are discarded as a whole in case of congestion within a network element; the block level QoS is then characterized by an objective set for the Block Loss Ratio.

An essential feature of ABT is the use of RM cells to perform rapid adjustment of the rate assigned to a connection as opposed to using classical signalling to renegotiate the parameters of DBR or SBR connections.

2.2.4   Available Bit Rate (ABR)

The above ATM transfer capabilities aim to avoid congestion essentially by preventive control (with the exception, however, of the elastic version of ABT). In contrast, the ABR capability implements reactive control whereby the rate at which users can transmit is dynamically set by the network. The aim is to use all available network capacity and to share it fairly among users who are assumed to have elastic rate requirements, i.e., the rate of a connection is not an intrinsic characteristic but can be freely modulated between declared minimum and maximum values. The ABR mechanism is specified in great detail by the ATM Forum and allows several different options [BF95]. In particular, two possibilities exist for determining the rate allocated to a given connection:

- the rate is determined by the user from a one bit congestion indication instructing the user to increase or decrease its rate according to a defined algorithm;

- the rate is calculated by the network (as the maximum allowed on the different links constituting the connection path) and communicated explicitly to the user.

By adjusting connection rates to be within the available capacity, ABR aims to avoid cell loss. Transfer delay is clearly not guaranteed beyond that corresponding to a minimum cell rate (MCR) which can be specified as part of the traffic contract. ABR is also supposed to be "fair", although the precise meaning of fairness is open to interpretation. Like the ABT capability, ABR relies on the use of RM cells to communicate allocated rates or congestion indications. Indeed the two transfer capabilities have much in common.

The QoS requirements associated with the ABR capability are (loosely) specified as follows: if the user adapts its transmission rate as specified by the feedback indication carried by RM cells, the user data cell stream should experience a small Cell Loss Ratio.

The performance of the ABR transfer capability is discussed in Section 4.2.

2.2.5   Unspecified Bit Rate (UBR)

A further unspecified bit rate (UBR) ATM transfer capability has been defined by the ATM Forum and is currently being considered by the ITU. UBR applies to connections without declared traffic parameters and without QoS guarantees. The understanding is that users themselves implement reactive control by means of higher layer protocols such as TCP to match their input to "best effort" network performance. Use can also be made of the EFCI bit in the cell header to inform the user of network congestion status.

2.3   The traffic contract

In ATM networks, a user must negotiate at connection set-up a traffic contract which includes the ATM Transfer Capability (ATC), the associated traffic parameters and the requested QoS (see [ITU96b]). This traffic contract is intended to enable the network to operate efficiently: ideally, resources are allocated on the basis of the traffic characteristics of the connection in such a way that the negotiated QoS is delivered to each connection, and that network resources are correctly utilized.

Declared traffic characteristics are generally restricted to rule based parameters, as discussed in Section 2.1, although the possibility exists for the declaration of a "service type" from which a more complete statistical characterization can be deduced. Required parameters and their interpretation are determined by the requested ATM transfer capability (see Section 2.2). Cell rates (PCR and SCR) must be declared with an associated CDV tolerance τ. A call request has associated quality of service requirements concerning the Cell Loss Ratio (CLR), Cell Transfer Delay (CTD) and Cell Delay Variation (CDV).

Given the traffic parameters and required QoS, the network must decide if it can accept the new call and still satisfy the requirements of all currently established connections. This is the role of Connection Admission Control (CAC). CAC is discussed at some length in Chapter 5 below. We note here simply that the definition of efficient CAC mechanisms for variable rate connections is still a largely open issue. It proves difficult indeed to establish the necessary practically useful relations linking connection traffic characteristics, network resource allocation and realized performance. Much of the work described in Part III of this report on traffic modelling and performance evaluation is stimulated by the quest for efficient CAC procedures.

Given that a connection is accepted by the CAC mechanism it remains for the network to verify that the user in fact conforms to his side of the traffic contract. Conformance to the negotiated traffic parameters is verified at the ingress User to Network Interface (UNI) and at subsequent Inter Carrier Interfaces (ICIs). In the case of rule based parameters, conformance is easily verified by means of algorithms implemented within policing mechanisms such as the leaky bucket. Conformance rules depend not only on the negotiated traffic characteristics but also on the negotiated ATC. Traffic control actions are of two types: either cells are discarded, or they are downgraded (i.e., identified, using the CLP bit, as low loss-priority cells). Parameter control may be based on either static or dynamic traffic parameters, depending on the transfer capability:

- for DBR and SBR capabilities, traffic parameters are defined at connection set up at each standardized interface and are changed only in the event of contract re-negotiation performed using classical signalling;

- for ABT and ABR capabilities, on the other hand, traffic parameters are dynamically modified during the connection lifetime using resource management cells.

The difficulty of controlling statistical parameters such as the mean bit rate or mean length of a peak rate burst is now well understood [BCT91, Rat91]. The use of rule based parameters solves the parameter control problem but can lead to a very poor description of the traffic streams making it difficult to determine an appropriate resource allocation.

The multiplexing of constant rate connections using DBR transfer capability provides the simplest control options. The guarantee of QoS objectives is nevertheless complicated due to the presence of CDV, as largely discussed in Chapter 3 and Section 5.1.

2.4   Internet integrated services model

The Internet currently offers a single best effort class of service based on the IP protocol. Requirements of different applications are met by the use of higher layer protocols. In particular, TCP implements an end to end error recovery and congestion control protocol suitable for reliable data transmission. Applications without strict semantic integrity requirements can use simpler transport layer protocols such as UDP (User Datagram Protocol). The latter protocol is used for some applications with real time constraints such as interactive voice and videophone communication although perceived quality can be extremely variable due to network congestion. The introduction of new service classes, better adapted to the requirements of the interactive and playback applications identified in Section 1.1, is currently under consideration by the Internet Engineering Task Force (IETF).

These service classes will exploit features of the latest Internet protocol specification IPv6, relying notably on RSVP (ReSerVation Protocol), a feature allowing the reservation of network resources for specific "flows". A flow is loosely defined as a set of packets traversing a network element all of which are covered by the same request for control of quality of service. Service classes offering QoS better than "best effort" are tentatively proposed by the IETF integrated services working group in the form of Internet-Drafts. A number of service classes have been defined based on the identification of flows having specified traffic characteristics and performance requirements:

- guaranteed service class - packets are guaranteed (absolutely) to arrive within a specified delivery time and will not be discarded in the network provided the flow stays within its specified traffic parameters;

- predictive service class - this is a real time service that "provides a fairly reliable delay bound" in that the large majority of packets are delivered within the delay bound;

- controlled delay service class - no guarantees on delay are specified but it is intended that "maximal delays should be significantly better than best effort service when there is a significant load on the network";

- controlled load service class - the network should "provide the client data flow with a quality of service closely approximating the QoS that same flow would receive from an unloaded network element ...".

Note that Internet-Drafts are temporary working documents and are not intended to be used as references. The first and last service classes are currently considered the most serious candidates for standardisation. Discussions on the evolution of the Internet service model may be found in [CSZ92] and [She95].

Implementation of the Internet service classes relies on the notion of traffic contract between user and network: the user specifies his traffic and QoS requirements, the network performs admission control and polices the user traffic to ensure conformance to the traffic specification. This type of preventive traffic control is typical of that required for DBR/CBR and SBR/VBR transfer capabilities, as discussed in Section 4.1. As for the ATC, traffic specifications are made in terms of leaky bucket (or token bucket filter) parameters which can therefore easily be policed by the network.

It is interesting to note the convergence of the Internet and B-ISDN worlds towards a broadly similar set of capabilities, coming from opposite directions: while ATM specifications are being broadened to allow elastic service classes in addition to the initially defined classes with hard QoS guarantees, the

Internet is moving to enhance its current ubiquitous best effort service model
with the introduction of resource reservation.

3 Accounting for cell delay variation

For a given ATM connection cells follow a fixed route and their sequence integrity is respected. However, they may encounter different traffic conditions in shared buffers and therefore experience different cell transfer delays.

- Some cells experience a shorter delay than preceding cells: during a short period of time, the observed peak cell rate may be larger than the source cell rate. This is the *clumping effect*.

- Some cells experience a larger delay than preceding cells. This is the *dispersion effect*.

The term "Cell Delay Variation" (CDV) broadly covers all the phenomena related to the variation of transfer delays for the successive cells of a connection.

In this chapter we first discuss how the clumping and dispersion effects relate to traffic control and the realization of circuit emulation, respectively. We then consider bounds and approximations allowing cell rate and CDV tolerance parameters to be set to account for upstream multiplexer delays. In a final section we successively examine the impact on CDV tolerance of a number of significant factors. Related material on queueing models for CDV is contained in Part III, Section 15.5. Some discussion on CAC methods taking account of CDV is included in Chapter 5.

3.1 Impact of CDV

There are two areas directly affected by CDV, namely the characterization and control of ATM traffic and the design of circuit emulation facilities [GB93].

3.1.1 Cell delay variation and traffic control

As shown in Section 2.1, a cell rate is specified using a conformance algorithm (the GCRA) based on a time period T (inverse of the cell rate) and on a supplementary time parameter τ that quantifies the maximum deviation with respect to a purely periodic cell stream.

The GCRA formalism takes account of cell delay variation in the sense that the amount of clumping affecting a cell stream in which all cells are conforming is limited. Conversely, the GCRA does not limit the amount of dispersion observed on the cell stream: whenever a "late" cell is observed, the GCRA is restarted.

Due to the asynchronous operation of ATM networks, the value taken for τ in the GCRA is only valid at a given interface. This variation has a decisive impact on network dimensioning: traffic characteristics of a cell stream that are enforced at the network ingress may not be valid downstream after the initial cell stream has experienced multiplexing. This also implies that different values of CDV tolerance may be considered at the UNI and at subsequent inter-network interfaces.

CDV tolerances at PCR and SCR levels are needed for defining conformance at standardized interfaces. These tolerances are not necessarily identical: for example, if the traffic is PCR spaced, the CDV tolerance at PCR level is minimal whereas the CDV tolerance at SCR level is hardly modified by a PCR spacer.

3.1.2   Cell delay variation and circuit emulation

The B-ISDN is intended to provide a wide range of broadband services using the ATM network as a backbone. A number of ATM Adaptation Layers (AALs) enhance the basic facilities offered by the backbone network. While some broadband services such as data transfer can tolerate substantial CDV, real time services may rely on circuit emulation facilities. AAL1 has been designed to provide these facilities.

Consider a constant rate application which relies on circuit emulation facilities provided by the B-ISDN using AAL1. The ATM connection set up for such an application would be a DBR connection, i.e., a connection that is solely characterized by its Peak Cell Rate specification. Let T_{CBR} denote the Peak Emission Interval negotiated for the ATM connection. It is assumed that cells are generated by the constant rate source with period T_{CBR} [1].

On the receiving side, AAL SDUs should also be played out with a fixed period corresponding to T_{CBR}. Since cells of the DBR connection experience different delays within the network, a smoothing procedure is implemented in the AAL1 equipment of the receiving side. This procedure consists in buffering a sufficient number of AAL SDUs upon reception before starting to play them out to the receiving application. This buffering is intended to provide an uninterrupted information stream at the receiving side by avoiding any gaps induced by CDV.

More precisely, the AAL SDU received in the first cell of the connection is delayed in the AAL1 play-out buffer before being delivered to the upper layer. Subsequent cells are read out at intervals of T_{CBR} so that, as long as the buffer is not empty, successive AAL SDUs are delivered at the application's rate.

---

[1] The period of the CBR application differs from T_{CBR} in order to take account of the ATM and AAL1 overheads.

Cell losses identified by the AAL1 numbering procedure may be compensated by the insertion of dummy AAL SDUs in the play-out buffer. AAL1 may also insert dummy AAL SDUs if the play-out buffer empties before the end of the connection due to an over-large CDV. If the buffer overflows, AAL SDUs are lost[2].

This section provides some insight into the design of AAL1 buffers under the assumption that cells are neither lost in nor misinserted by the ATM network. Let $\{a_n\}_{n\geq 0}$ be the successive observed arrival times at the buffer and let $\{d_n\}_{n\geq 0}$ be the successive epochs at which information is delivered by AAL1 to the application on the receiving side. Let W_n denote the random component of cell transfer delay between emitting and receiving sides for the nth cell observed at the receiving side. The physical transmission delay can be considered as constant and thus has no impact on the problem addressed here. The variables a_n and W_n are related as follows:

$$a_n = a_0 - W_0 + nT_{CBR} + W_n. \qquad (3.1.1)$$

The designer of circuit emulation facilities only has to set the values of the following two parameters:

- the initial delay R from the reception of the first cell of the connection to the delivery of the first output to the application,

- the size K of the buffer.

The delay R may be either a constant or a random variable depending on the level of the play-out buffer. The delay R can for example be chosen as follows: AAL-SDU 0 is delivered when cell N arrives at the buffer. The corresponding value for the random delay R is:

$$R = a_N - a_0 = NT_{CBR} + W_N - W_0. \qquad (3.1.2)$$

The objective is to dimension the play-out buffer in such a way that the probability it empties or overflows is small. Assume that the first AAL-SDU is delayed by R in the play-out buffer of size K. The inter-arrival time between cell 0 and cell n is given by equation (3.1.1). If the period of the application at the receiving side is always equal to the source period T_{CBR}, the time d_n at which the payload of cell n should be delivered to the user is $d_n = d_0 + nT_{CBR}$ where $d_0 = a_0 + R$.

The buffer is not empty at time d_n if and only if $a_n \leq d_n$ which implies:

$$W_0 + R \geq W_n. \qquad (3.1.3)$$

The buffer is not full at time a_n if and only if $a_n \geq d_{n-K}$ (for $n \geq K$) which implies:

$$W_0 + R \leq W_n + KT_{CBR}. \qquad (3.1.4)$$

---

[2]See ITU-T Recommendation I.363 for a specification of AAL1 procedures.

If R is constant, inequalities (3.1.3) and (3.1.4) define the conditions on R and K. If R is chosen as in (3.1.2), inequality (3.1.3) becomes:

$$NT_{CBR} + W_N \geq W_n \qquad (3.1.5)$$

and inequality (3.1.4) becomes:

$$W_N + NT_{CBR} \leq W_n + KT_{CBR}. \qquad (3.1.6)$$

The above inequalities show that the relevant quantity for dimensioning the buffering procedure (initial delay and buffer size) is the range of end-to-end cell transfer delays, i.e. the difference between the largest and the smallest cell transfer delays under the assumptions that:

- no cell losses or misinsertions occur between the source and the receiving end,

- the buffer neither underflows nor overflows.

Both clumping and dispersion components of CDV then have an influence on the dimensioning of circuit emulation procedures carried out by the AAL1. A rule of thumb for dimensioning the reception buffer in the AAL1 of the receiving side is the following: the buffer should be able to queue at least $2N$ AAL-SDUs where N is such that NT_{CBR} is larger than the $(1 - 10^{-9})$ quantile of the distribution of W. Furthermore, the initial delay taken to deliver the first AAL-SDU of a connection is chosen as the time taken to reach level N in the reception buffer.

3.2   Methods for setting traffic parameter values

In this section we consider the problem of setting the peak emission interval T and the CDV tolerance τ for a stream of given characteristics, the assumed objective being to achieve a negligible non-conformance rate in GCRA(T, τ). For the sake of simplicity, the discussion is mostly presented below in the framework of the choice PCR traffic characteristics. However, the same approach can be directly applied for choosing the SCR characteristics.

3.2.1   Simple approximation

Consider first a CBR application carried by a connection whose corresponding ATM time period is T. If T is chosen as Peak Emission Interval in the GCRA formalism, the value for τ should be chosen in relation to the distribution of the difference $(W_j - W_i)$ between cell transfer times (from the source to the observation point) experienced by two cells belonging to the same connection.

Indeed, as long as all cells up to cell i are conforming to $GCRA(T, \tau)$, we have:

$$
\begin{aligned}
y_i &= c_i - a_i, \\
&= (t_0 + W_0 + iT) - (t_0 + iT + W_i), \qquad (3.2.1) \\
&= W_0 - W_i.
\end{aligned}
$$

where t_0 is the source emission time for cell 0. It is thus seen that, if τ bounds $(W_0 - W_i)$ for all i, all cells are conforming to $GCRA(T, \tau)$. When negotiating a peak rate, it has naturally been proposed to choose as a value for τ, an upper bound for $(W_0 - W_i)$ valid for all i.

The above expression also applies to the case of SCR. In this case, T_{SCR} corresponds to the inverse of SCR and the sum of the Intrinsic Burst Tolerance and the CDV tolerance at SCR level should upper bound $(W_0 - W_i)$. T_{SCR} in that case corresponds to the emission interval of an "equivalent terminal" and is thus arbitrary, as long as it is possible to find a corresponding CDV tolerance parameter.

The exact computation of the distribution of $(W_0 - W_i)$ involves taking care of correlations between successive cell transfer times and the above method is usually not tractable. Moreover, the range for $(W_0 - W_i)$ may be very large or even infinite if W is modelled as a random variable yielding inapplicable values for τ. It has therefore been proposed (see e.g. [Nie90]) to use Equation (3.2.1) to upper bound $(W_0 - W_i)$ in a probabilistic sense by W_{max}, a quantile of the distribution of the observed transfer time for a single cell (e.g., $W_{max} = \sup\{w \mid P\{W \leq w\} \leq 1 - \epsilon\}$, with ϵ being the target value for the probability of transmitting a non-conforming cell). It has also been proposed to consider, as an estimate for τ, the quantity $(W_{max} - W_{min})$ where W_{min} is defined as a lower quantile of W.

Choosing τ as W_{max} provides a rule of thumb method for dimensioning the cell conformance algorithm. Note that the above is not a formal proof that the proportion of non-conforming cells is upper-bounded by ϵ. Indeed, Equation (3.2.1) is valid only when all cells up to cell $(i - 1)$ are conforming.

This method theoretically applies to the case where the negotiated Peak Emission Interval T is equal to the (supposedly) original time period of the source. When T is less than the inter-cell time, Equation (3.2.1) constitutes a conservative worst case basis for estimating τ. This is the only known method that applies to the particular case where it is not assumed that resources are over-allocated to the connection. It turns out, in all known cases, that the simple approximation does provide rather pessimistic estimates for τ. Promising results are obtained by heuristically modifying the values obtained for τ (e.g., by a multiplying factor smaller than 1).

The simple approximation described above and all derived methods rely on the (at least partial) knowledge of the cell transfer time distribution. This distribution is not necessarily known, however, which leads us to look for methods that can be applied with less complete knowledge of the cell stream whose conformance is to be checked. The next section describes such methods.

3.2.2   Renewal approximation

The present section presents some methods for setting a pair (T, τ) using the statistical characteristics of the cell stream to be characterized. All the methods presented here assume that the cell conformance testing queueing model of Section 2.1 is fed by renewal input (although the original stream is obviously not renewal in many cases, i.e., inter-cell times are not independent). The renewal input traffic to the cell conformance queue represents the cell stream modified by the upstream network. Let T_{CBR} denote the mean value of the inter-cell time.

When using the queueing model described in Section 2.1, the analysis of the GI/D/1 queue with limited waiting time (or finite buffer) is needed. This is in general difficult or even intractable. A standard simplification is to approximate the loss probability in the queue with finite waiting time by the probability of exceeding a given level in the corresponding infinite queue.

In order to preserve the stability of the GI/D/1 queue with infinite buffer, the negotiated PCR must be strictly larger than the original peak rate of the CBR connection (i.e. $T < T_{CBR}$). The value for the CDV tolerance τ (which corresponds to the limit fixed for the waiting time in the equivalent G/D/1 model as described in Section 2.1) is then taken to be the level in the infinite queue for which the probability of overflow is less than the target non-conformance probability P_{nc}.

Heavy traffic approximation

If T is only slightly smaller than T_{CBR}, the cell conformance GI/D/1 queue is heavily loaded and it is possible to apply a classical heavy traffic approximation for the virtual waiting time in a GI/D/1 queue (see [Kle75a]). This approach is described in [Bla92]. Note that the present approximation is not related to the load of the upstream network but to the fact that the over-allocation factor $(1 - T/T_{CBR})$ is small, regardless of upstream traffic conditions.

Let P_{nc} denote the non-conformance probability assuming τ is the equivalent buffer size of the cell conformance queueing model. P_{nc} is given by the following formula:

$$P_{nc} = \exp\left(-\frac{\tau}{T_{CBR}} \times \frac{2(1 - T/T_{CBR})}{c_a^2}\right) \qquad (3.2.2)$$

where c_a^2 is the squared coefficient of variation of the inter-cell time. This formula yields a closed form expression for the τ value needed to ensure that the proportion of non-conforming cells is less than 10^{-r} (i.e., $P_{nc} \leq 10^{-r}$):

$$\tau = T_{CBR} \times \frac{c_a^2 r \ln 10}{2(1 - T/T_{CBR})} \qquad (3.2.3)$$

where $(1 - T/T_{CBR})$ is the over-allocation factor. The equivalent buffer size is then given by:

$$\frac{\tau}{T} = \frac{T_{CBR}}{T} \times \frac{c_a^2 r \ln 10}{2(1 - T/T_{CBR})}. \tag{3.2.4}$$

Kingman's bounds

The method presented now is based on estimates for the tail behaviour of the virtual waiting time in the GI/G/1 queue obtained by Kingman [Kin70] and Ross [Ros74] (for more details, the reader can refer to [Kle75a, Section 2.4]). For its application, it is necessary to know not only the squared coefficient of variation for the inter-cell distribution but also a transform of this distribution (generating function or Laplace transform). However, this method applies even if the cell conformance GI/D/1 queue is not in heavy load. This approach has been described in [Kel94a] and [KK94].

Let Y be the inter-cell time for the renewal process. Define v as the strictly positive solution to the equation:

$$E\left(e^{-vY}\right) = e^{-vT}. \tag{3.2.5}$$

From a result given by Kingman, the following upper bound applies for the tail of the virtual waiting time distribution:

$$P_{nc} \le e^{-v\tau}. \tag{3.2.6}$$

Therefore, in order to ensure a proportion of non-conforming cells smaller than 10^{-r} (i.e., $P_{nc} \le 10^{-r}$), it is sufficient to choose τ as follows:

$$\tau = \frac{r \ln 10}{v}. \tag{3.2.7}$$

Comparison of methods based on the renewal approximation

The heavy traffic approximation is now compared with the approximation based on Kingman's bound for an originally renewal cell stream for which an exact solution is known. Consider a Bernoulli process. The exact value for τ is computed using the method given in [DH95] and both approximate renwal process methods are applied for different values of T. Figure 3.2.1 displays the obtained results for a Bernoulli process with mean inter-cell time $(T_{CBR} = 1/p)$ equal to 10.

The need to choose a T value strictly smaller than the original mean inter-cell time T_{CBR} is obvious, when considering Figure 3.2.1: when T tends to T_{CBR}, the value of τ tends to infinity. This phenomenon is observed for the exact result and for both approximations since it is directly related to the GI characteristic of the input stream.

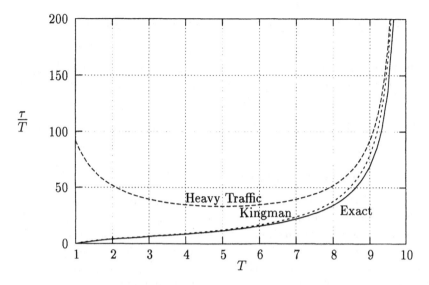

Figure 3.2.1: Comparison of renewal process approximate methods for a Bernoulli process ($p = .1$).

It is also seen in Figure 3.2.1 that both approximations provide upper bounds for τ and that the approximation based on Kingman's bound is always closer to the exact value than that provided by the heavy traffic approximation. Moreover, the approximation based on Kingman's bound is very effective, even if the negotiated value of T is small compared to the original inter-cell time. Conversely, the heavy traffic approximation does not behave correctly when the negotiated value of T is significantly smaller than the original mean inter-cell time.

3.2.3   The special case of tandem queues

If a connection crosses several multiplexing stages before its conformance is checked, it is necessary to evaluate the cumulative amount of CDV caused by these multiplexing stages. This scenario is to be considered, for example, when a user accesses the public network via a private network: the user may have very little control on the procedures that are used within the private network that modify its original cell stream. A similar scenario can also be considered when characterizing the CDV tolerance values at inter-network interfaces when a connection crosses several operator domains.

The upstream network is modeled here as a tandem of FIFO queues. More complex systems involving priorities or fair queueing strategies may be considered in the future but FIFO queueing is a likely option, at least in the near future.

Two methods can be used to estimate end-to-end waiting time:

- Convolution approach: the delays in successive queues are assumed to be independent;

- Recursive approach: the independence assumption is relaxed and a recursive method (as described in Section 15.5 is used in order to statistically characterize the output process of the tandem queues.

CDV tolerance values are then obtained using one of the approaches described above.

Convolution modelling approach

The distribution of the end-to-end delay of a single cell is obtained here as the convolution of successive delay distributions in individual queues. Note that the independence assumption between successive waiting times can only hold if interfering traffic can be considered as independent from one queue to another (see Section 3.3.3). The distribution of queueing time in each queue is needed. These distributions are usually not known unless further assumptions are made, i.e., approximating each queue by a tractable model.

The simplest model is an $M/D/1$ queue which can be considered as valid if the rate of the considered connection is small compared to the total input rate to a queue and if the traffic feeding the queue is "smooth". Under a further assumption of heavy traffic (i.e., each queue is heavily loaded) and assuming worst case delays corresponding to an $M/D/1$ queue at each stage, it is possible to use a classical heavy traffic approximation for the delay distribution $W(x)$ in the $M/D/1$ queue (see Part III, Section 15.1 :

$$W(x) = 1 - e^{-\alpha x}, \text{ where } \alpha = 2(1-\rho)/\rho.$$

In this formula, the time unit for x is the cell transmission time. Assuming a series of M identical queues, the above approximation yields a closed-form formula for the end-to-end delay density function (the Erlang-M distribution):

$$w^{total}(x) = \frac{\alpha^M}{(M-1)!} x^{M-1} e^{-\alpha x}.$$

The above approach is rather pleasing since it yields a closed form expression for the end-to-end queueing delay which in turn can be used to obtain, via the simple approximation, a value for τ. However, it can provide very inaccurate and pessimistic results if either the rate of the considered connection is not negligible compared to the total input rate to a queue or the queues are not heavily loaded. Conversely, the estimate is likely to provide optimistic results if queues are fed with "bursty" traffic.

An alternative approach consists in taking explicit account of the considered connection rate when approximating queueing time in a single queue, computing statistical characteristics (e.g., mean and variance) of the end-to-end queueing time under the independence assumption and, lastly, matching the end-to-end queueing time distribution to a known distribution with the same statistical characteristics, e.g., a Gaussian or a gamma distribution.

A major drawback of the convolution approach is that it naturally leads to unbounded values of CDV tolerance when the number of multiplexing stages grows. In order to illustrate this divergence, consider Figure 3.2.2. This figure displays τ/T versus the number M of multiplexing stages, where τ/T is obtained by different methods. A succession of 30 queues has been simulated, each loaded at 0.85, and the CDV tolerance τ has been obtained by simulation at each stage for a CBR cell stream with $T = T_{CBR} = 5$, the target non conformance ratio being equal to 10^{-3}. The value τ/T obtained by simulation is first compared to that derived from the simple approximation $(\tau \approx (W_{max} - W_{min}))$ where W_{max} and W_{min} are end-to-end transfer time quantiles also obtained by simulation. The last curve corresponds to results obtained using the convolution approach: the queueing delay at each stage is characterized by its mean and variance, and the end-to-end queueing delay characteristics are obtained under the independence assumption. The end-to-end queueing delay distribution is then modeled by a gamma distribution.

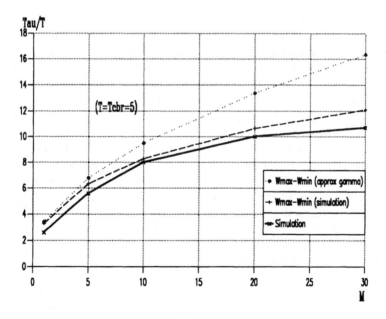

Figure 3.2.2: τ/T as a function of M. Convolution method compared with exact result.

It is first seen in Figure 3.2.2 that, as expected, τ is an increasing function of the number of multiplexing stages. The values computed either directly by simulation, or by using the simple approximation with the end-to-end transfer time obtained by simulation increase moderately as the number of multiplexing stages increases. However, the curve obtained under the convolution approach is significantly steeper than than the other two curves. It turns out that approximating the end-to-end queueing delay with a gamma distribution seems to be accurate only if the number of multiplexing stages is limited (in this example, $M \leq 10$). For a larger number of multiplexing stages, the curves showing the exact τ/T value and the exact $(W_{max} - W_{min})/T$ flatten when the number of multiplexing stages increases, which is not true for $(W_{max} - W_{min})/T$ obtained with the gamma approximation.

Using the recursive approach described below, dependencies between successive queueing delays experienced by a given cell in tandem queues are taken into account.

Recursive approach

As seen above, assuming that delays in successive queues are independent leads to inaccurate results if the number of multiplexing stages is large. A very neat modelling approach has been provided by J. van den Berg and J. Resing in [vdBR92] consisting in recursively deriving the successive inter-cell time distributions along the tandem queues. This approach is described in Part III, Section 15.5 for a similar result, obtained numerically).

The recursive method [vdBR92] strictly applies only if the load of each queue is 1 and if the interfering traffic is batch Bernoulli in each queue. The last assumption means that the case where interfering traffic is heavily correlated is not covered by the proposed analysis.

The recursive method is interesting in the sense that it provides a method to estimate at least the first moments of the inter-cell process at successive stages and even in some cases the generating function for this process. It is then possible to use one of the renewal process approximation methods described previously to derive values for the traffic characteristics of the connection at the egress of the tandem queues under the assumption that the tagged traffic is renewal. For example, this approach has been used in [KK94] in order to study how to choose (T, τ) when the number of queues increases.

In order to evaluate the accuracy of the recursive method, it is necessary to address two subproblems :

(i) does the inter-cell time distribution tend to a limit when the number of stages increases, even when the queues are not overloaded ?

(ii) if there is a limit, how accurate is it to assume that the limiting output process is renewal ?

In order to address the first question, the following simulation study has been carried out: a CBR cell stream crosses 100 queues, and the variance of the output process is estimated at the output of each queue, whose total load is 0.8. Background traffic is Poisson, and each interfering cell stream leaves the system after having crossed one queue. The variance of the output process is plotted in solid lines on Figure 3.2.3 versus the number of stages for different values of T_{CBR}. The value of the variance obtained using the recursive approach is plotted in dotted lines for comparison.

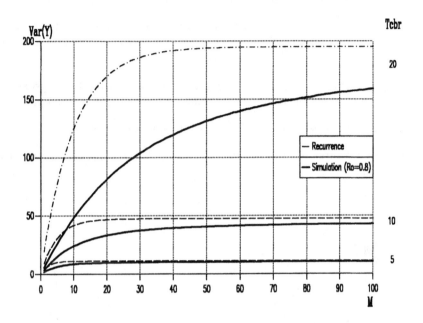

Figure 3.2.3: Variance of the output process along a tandem of queues. Influence of the rate of the tagged CBR connection.

As seen in Figure 3.2.3, the variance of the output process tends to a limit, although the convergence speed critically depends on the rate of the original CBR connection: convergence is rapid for a high bit rate CBR connection and rather slow for a low bit rate connection. In other simulations, we have also observed that the convergence speed depends on the total load offered to each queue: the higher the total load, the more rapid is the convergence. Moreover, it is seen in Figure 3.2.3 that the limit variance found by the recursive approach is an upper bound to the exact variance; this upper bound is tight for high PCR values but less tight for low PCR values.

Validity of the renewal process approximation for tandem queues

Concerning the second question raised above, a simulation study has been carried out in order to test the validity of modelling the output process of tandem queues by a renewal process when looking for CDV tolerance estimates. The framework of this study is as follows: an originally periodic traffic with $T_{CBR} = 5$ crosses a tandem of 30 queues, each loaded at 0.85. Interfering Poisson traffic is fed into each queue, and leaves the system after having crossed one single queue. Under various conditions, the tagged traffic non-conformance ratio is computed at the output of the queues.

The validity of the following methods has been investigated:

- the first method consists in mapping the output process of the tagged cell stream into a gamma process with the measured variance and computing the τ value by simulation,

- the second consists in estimating the τ value by using the approximate method based on Kingman's bound for the previous gamma process,

- the third is based on the heavy traffic approximation.

Note that all three approaches use the variance of the process obtained by simulation.

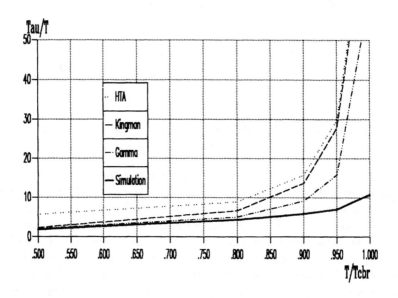

Figure 3.2.4: τ/T as a function of T/T_{CBR}. Comparison with simulation results.

Figure 3.2.4 shows first that the degree of inaccuracy of the various estimates is heavily dependent on the over-allocation factor $(1 - T/T_{CBR})$.

This figure displays τ/T versus T/T_{CBR}, as obtained by simulation and as computed by different approximate methods based on the renewal assumption. The figure represents the area where T/T_{CBR} is larger than 0.5; when T/T_{CBR} is smaller than 0.5, the first and second estimates are very good while the third estimate (heavy traffic) is significantly above the true value for τ. The heavy traffic approximation is clearly the worse. Both the approximation based on Kingman's bound and that obtained by mapping the inter-cell time into a gamma distribution provide rather good and similar estimates if T/T_{CBR} is below 0.8. However, if T/T_{CBR} is larger than 0.9, the three approximations completely diverge. Indeed, although it is possible to compute a finite value for τ if $T = T_{CBR}$ for the real, non-renewal process, this is not the case for any of the approximations which all assume that the process is renewal.

If the over-allocation factor is large enough ($T/T_{CBR} > 0.9$), the approximations based on a renewal process assumption are still pessimistic but they do produce the qualitatively correct behaviour when the number of multiplexing stage increases. This is shown in Figure 3.2.5. This Figure displays τ/T versus the number M of multiplexing stages. The CDV tolerance τ is obtained by simulation at each stage, for a CBR cell stream with $T_{CBR} = 5$, $T = 4$ and for a target ratio of non-conforming cells equal to 10^{-3}.

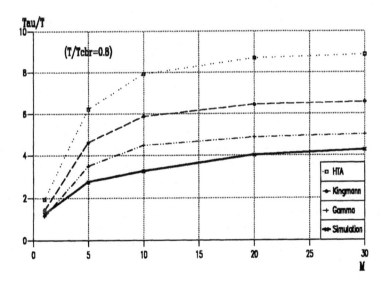

Figure 3.2.5: τ/T as a function of M. Approximations based on renewal assumption compared with exact result.

It is first seen in Figure 3.2.5 that the heavy traffic approximation grossly overestimates the value for τ whereas the approximation based on mapping a gamma distribution is rather accurate. Note that the accuracy is generally better if the over-allocation factor is larger (see also Figure 3.2.4). Furthermore, all four curves flatten when the number of multiplexing stages increases. This shows that all approximate methods, although rather pessimistic, behave qualitatively as they should. Similar results are obtained for the case when the characteristics of inter-cell time distributions in successive queues are computed using the recursive approach described above.

These results show that under the considered assumptions it is possible to effectively provide upper bounds for τ if resources are significantly over-allocated (at least 10% in the examples we have considered).

3.3 Factors influencing the CDV tolerance

The traffic characteristics of a connection depend on several parameters including the statistical characteristics of the source, although these may be significantly modified by the access network. Due to the algorithmic definition of traffic characteristics in ATM networks, we argue here that there is no obvious relation between the statistical characteristics of a source and the traffic characteristics that are negotiated in traffic contracts.

This section attempts to briefly list the major factors that need to be considered when setting traffic characteristic values. The intention is to qualitatively show certain phenomena, and not to provide a quantitative, exact analysis for each case. Therefore, approximate models described previously are used, when applicable. For the sake of simplicity, the PCR framework is chosen, although most results apply directly to SCR.

3.3.1 Load of the access network

The load of the access network is obviously a relevant factor. This is shown on a simple example: consider a low bit rate connection whose waiting time in the access network is taken to be the waiting time in an $M/D/1$ queue (e.g., a 1 Mbits/s connection on a 155 Mbit/s link). The CDV tolerance value is chosen to be the 10^{-9} quantile of this waiting time as described in Section 3.2.1 (simple approximation). Table 3.3.1 provides the values for τ in cell unit times (i.e., in order to obtain a CDV tolerance expressed in seconds, it is necessary to multiply the CDV tolerance value given in Table 3.3.1 by the time it takes to transmit one cell at link rate).

Table 3.3.1: CDV tolerance as a function of the access network load ($T_{CBR} \approx \infty$).

ρ	.2	.5	.7	.8	.85
τ	8	17	31	48	65

Table 3.3.1 clearly shows that the CDV tolerance is an increasing function of the access network load. Similar results have been obtained for different traffic conditions.

3.3.2   Original statistical characteristics

The original rate (mean or peak) of the considered connection generally has a major influence on the CDV tolerance value. The first model used to investigate this influence is the following: the access network is represented by a single queue with Poisson background traffic. The CDV tolerance is obtained, for $T = T_{CBR}$, by using the method described in Section 3.2.1, i.e., it is chosen to be the 10^{-9} quantile of the waiting time of the CBR cells in the M+D/D/1 queue. The total load of the M+D/D/1 queue is constant, but the respective loads of the CBR and background Poisson traffics vary.

Table 3.3.2: CDV tolerance as a function of the connection Peak Cell Rate.

$T = T_{CBR}$	2	5	10	50
τ	22	37	43	47
B_c	23	11	5	2

Table 3.3.2 provides the values of τ and for the maximum burst of conforming back-to-back cells $B_c = 1 + \lceil \tau/(T-1) \rceil$ under the assumption that the total load of the access network is 0.8. The value for τ is obtained by solving the exact M+D/D/1 model (see Part III, Section 15.5. This approach has been used to characterize the "minimum" CDV tolerance that should be allocated to a user [ITU96b] (see Section 2.1). The table shows that for a single multiplexing stage with Poisson background traffic, the CDV tolerance is an increasing function of the Peak Emission Interval but that the burst size of conforming cells on the other hand is largest for a large Peak Cell Rate. Note however that for an extremely high rate (i.e., higher than 50% of the link rate) the maximum burst size of conforming cells eventually decreases with the connection rate since, in the limit, only the CBR connection uses the link. This fact cannot be shown with the M+D/D/1 model which applies only for CBR rates that are integer fractions of the link rate.

Another example is displayed in Figure 3.3.1. This figure shows τ/T versus T_{CBR} for a fixed ratio $T/T_{CBR} = 0.9$ and a non-conformance ratio $P_{nc} = 10^{-3}$. An originally constant rate cell stream with period T_{CBR} crosses tandem queues loaded with (independent) Poisson interfering traffic. τ/T is obtained for different values of M (M is the number of queues along the route).

It appears in Figure 3.3.1 that the rate of the connection for which the maximum of τ/T is attained decreases with the number of multiplexing stages: if a single queue is considered, the maximum is observed for

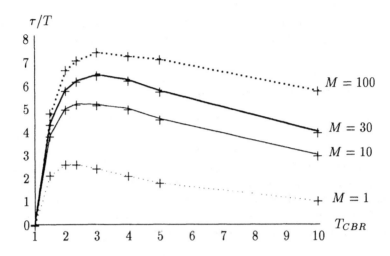

Figure 3.3.1: τ/T versus T_{CBR} for different numbers of multiplexing stages (simulation).

$T_{CBR} = 2$, whereas the maximum is observed for $T_{CBR} = 3$ if a larger number of queues is considered (see the curve obtained for $M = 100$ in Figure 3.3.1).

This result is confirmed by an analytical study of the same problem using both the recursive approach to compute the statistical characteristics of the cell stream along the tandem queues and the heavy traffic renewal process approximation described above. In this study, only heavily loaded queues are considered. Interfering traffic is Poisson, the CDV tolerance τ is computed for a fixed ratio $T/T_{CBR} = 0.9$ for $P_{nc} = 10^{-9}$. Results for several values of T and M are displayed in Figure 3.3.2.

It appears in Figure 3.3.2 that, for a small enough number of tandem queues (here, $M \leq 100$), τ/T has one maximum. The value of T_{CBR} for which this maximum is attained is an increasing function of M. Moreover, when the number of multiplexing stages increases, the curve flattens when T_{CBR} increases impling that the upper envelope of the curve (corresponding here to $M = \infty$) is an increasing function. Indeed, the limit variance obtained using the recursive approach described in [vdBR92] is a decreasing function of the original cell rate. This implies that the limiting curve ($M = \infty$) is monotonic increasing and has no maximum, as appears for the curve corresponding to $M = 1000$.

The practical consequence of this fact is that it is possible to upper-bound the value for CDV tolerance, when the over-allocation factor is known, independently of the number of multiplexing stages. Of course, as seen previously, this upper-bound may be very pessimistic and this method should only be used if no significant knowledge can be obtained on the structure of the access network.

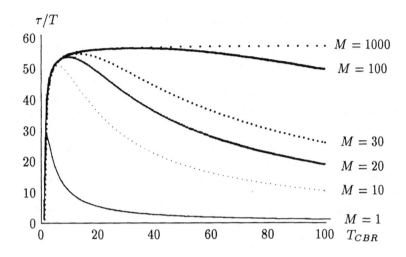

Figure 3.3.2: τ/T versus T_{CBR} under heavy traffic assumption.

3.3.3   Aggregation of cell streams

Using an approach derived from that presented in [KK94], it can be shown
that the relevant quantity is the rate of the traffic that follows the same route
along the tandem queue and not the rate of the particular connection that
follows this route.

This is shown by considering a cell stream that crosses a number of queues
in tandem and a cell sub-stream, within this original cell stream, for which
the traffic characteristics are computed. As before, it is assumed that independent interfering Poisson traffic crosses each queue and leaves the system
once served by a single queue. For a given over-allocation factor, the CDV
tolerance is obtained by first applying a modified recursive approach and then
the approximation based on the use of Kingman's bound. It turns out that
the CDV tolerance depends only on the rate of the original cell stream but
not on the rate of the sub-stream within the original cell stream.

This is a nice result with practical applications: for example, when the
CDV tolerance is modified at an inter-network interface, the Peak Cell Rate
and the over-allocation being fixed, all the connections that follow the same
route (e.g., the VCs of a given VP) can be allocated the same CDV tolerance, independently of their respective rates if the connections were originally
shaped according to their PCR value.

3.3.4   Over-allocation factor

In order to decrease the CDV tolerance, a user may negotiate a higher Peak Cell Rate than the original CBR rate. In this section, the period T_{CBR} of the CBR connection is fixed but the user negotiates a different (higher) Peak Cell Rate (i.e., a value for T smaller than T_{CBR}). The relationship between the over-allocation factor and the CDV tolerance is already apparent in Figures 3.2.1 and 3.2.4.

Figure 3.3.3 also illustrates this question. This figure displays τ versus an over-allocation factor $(1 - T/T_{CBR})$. These results are obtained by simulation for a CBR connection with $T_{CBR} = 5$ that crosses 30 queues with independent interfering Poisson traffic at each queue. The load of each queue is 0.85, and the exact value for τ is obtained (by simulation) for $P_{nc} = 10^{-3}$. This computation is carried out for $M = 1, 10, 30$.

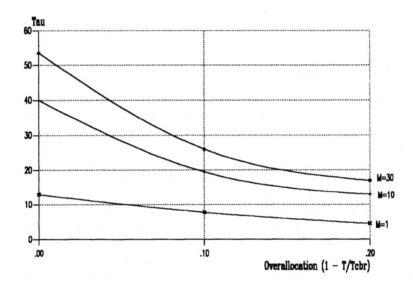

Figure 3.3.3: τ versus over-allocation factor $(1 - T/T_{CBR})$.

It is seen in Figure 3.3.3 that τ is a decreasing function of the over-allocation factor (equivalently, it is an increasing function of T). Furthermore, this property is more apparent if more queues are crossed. This result is also clearly apparent in Formula (3.2.3) which computes a heavy traffic approximation for τ. If ϵ is the over-allocation factor $(1 - T/T_{CBR})$, Formula (3.2.3) yields a value for τ that is proportional to ϵ^{-1} (this also shows by the way that Formula (3.2.3) cannot be used for $T = T_{CBR}$).

The practical consequence of this fact is that the user and the network have a certain degree of freedom when negotiating the Peak Cell Rate and

CDV tolerance: for a given cell stream, there are many possible ways of choosing a pair (T, τ). A value of Peak Cell Rate close to the original period implies a large value for CDV tolerance (but allows a tight resource allocation policy, at least if spacing is performed at the network ingress), whereas a value of Peak Cell Rate significantly greater than the original period (corresponding to resource over-allocation) leads to a smaller CDV tolerance.

The same remark applies to the choice of Sustainable Cell Rate and Intrinsic Burst Tolerance.

3.3.5   A practical example

In this section, we illustrate the following situation: a user, when setting its traffic characteristics, might choose the "best" peak rate and tolerance to minimize some incurred costs. For a given offered traffic, there is potentially a range of peak rates and τ values which would yield a proportion of non-conforming cells compatible with the required quality of service. Which value should the user choose? From the network's point of view, the larger the peak-rate, the greater the amount of resources consumed; if this is reflected in tariffs then the user would want to choose the smallest peak rate possible, subject to the CDV tolerance allowed by the network being large enough. If a PCR spacer is available, the user can set its PCR to any value that is compatible with the QoS requested by the application carried by the connection. Conversely, if no PCR spacer is available, the network may also be able to set an upper bound to the CDV tolerance that can be negotiated by the user.

For a single FIFO multiplexing stage of known buffer size, there is a simple way of setting an upper bound to τ that ensures that no cells are lost in the buffer. Let τ be such that:

$$\tau \leq \frac{N}{\Lambda}$$

where N is the buffer size and Λ is the link rate. The above equation means that the maximum CDV allowed on any cell stream is smaller than the time taken to empty the buffer.

Then, let i connections characterized by their PCR Λ_i and the pair (T_i, τ_i), be multiplexed in the single multiplexing stage. If $\sum_i \Lambda_i < \rho\Lambda$ (where ρ is the link utilization, strictly less than one), the above equation yields

$$\sum_i \Lambda_i \tau_i < \sum_i \Lambda_i \frac{N}{\Lambda} < \rho N < N.$$

This inequality in turn implies that no cell loss is experienced in the buffer. Indeed, during any interval of duration t, the maximum number of cells sent to the buffer by connection i is $\Lambda_i(t + \tau_i)$. Therefore, the maximum number of cells sent by all connections is less than $(\rho\Lambda t + \sum_i \Lambda_i \tau_i)$; since the multiplexer can transmit $\rho\Lambda t$ cells during this time interval, buffering $(\sum_i \Lambda_i \tau_i)$ ensures that no cells are lost.

The above upper bound set to τ may be considered as a limit set by a network operator for any CDV tolerance. Obviously, if the traffic subsequently crosses other multiplexers, their characteristics should also be taken into account when setting traffic parameters that would be valid for the complete route; taking into account the complete route is not an easy task since the characteristics of each source are modified by multiplexing.

The user then needs to choose a T value such that τ can cater for the CDV tolerance corresponding to the application. As an example, let us consider the output of a Fore Systems AVA 200 video camera that uses JPEG compression on "tiles" of 24-bit color pixels, and converts the image into ATM cells using AAL5. In this example the camera generated 25 frames per second. Figure 3.3.4 was generated from a particular 25 frames per second trace of a TV transmission [CLK95] with the camera's spacer disabled, allowing back-to-back cells to be transmitted, giving a potential peak rate of 100Mb/s and a mean of 2.74 Mb/s. The inter-arrival times from the trace were used to calculate the pair (T, τ) using a renewal process approximation method.

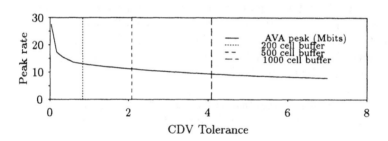

Figure 3.3.4: Peak rate and CDV tolerance for an AVA camera.

From Figure 3.3.4, it is clear that the user could choose to offer any peak rate between 7.9Mb/s and 27.5Mb/s with a corresponding CDV tolerance value. The vertical lines in the figure represent the CDV tolerance obtained with the method described above for different size switch buffers, namely 200, 500 and 1000 cells, with a line rate of 100Mb/s. For a 200 cell buffer, the natural limit of 0.84 ms CDV tolerance implies a peak rate of 13 Mb/s for the AVA camera; if the switch has 500 or 1000 cell buffers then the source could use a peak of 11 or 9 Mb/s respectively. This illustrates well the link between buffer sizes in the switch and allowable CDV.

4   Statistical resource sharing

An essential requirement of a broadband network is that it efficiently share its bandwidth and memory resources between variable rate traffic streams while maintaining some degree of control over the quality of service offered to the user. The nature of the performance guarantees which can be made depends on the applied traffic controls. A first distinction is between preventive and reactive control. The former relies on the network performing admission control based on an *a priori* description of the traffic stream and then policing an admitted connection to ensure that the user abides by the terms of a "traffic contract. Reactive control employs a closed control loop allowing the users' input rate to be adjusted according to the present state of network traffic. The present chapter discusses aspects of preventive and reactive traffic controls and their impact on service quality and the dimensioning of network elements.

Section 4.1 considers open loop preventive control options. This type of control relies particularly on the notion of traffic contract discussed in Section 2.3 whereby the network guarantees certain levels of service quality on condition that the user abides by the description of his traffic stream used for connection admission control. We distinguish two kinds of statistical multiplexing appropriate for this type of control: *rate envelope multiplexing* aims to preserve the bit rate of a connection throughout the network by limiting admission to ensure that the sum of rates at any time only exceeds link capacity with a negligibly small probability; *rate sharing* allows an input rate excess, with surplus traffic being temporarily stored in large buffers, thus achieving higher link utilisation at the cost of longer network delays. The last part of the section considers the possible trade-off between the cell loss rate and the call blocking probability and its impact on multiplexing efficiency.

In Section 4.2 we study the performance of the available bit rate (ABR) reactive control recently defined by the ATM Forum and the ITU. The performance evaluation proceeds by a deterministic analysis in both steady state and transient phases leading to simple formulae for required buffer capacities. Both binary congestion indication and explicit rate options of the ABR capability are considered.

4.1   Preventive traffic control

While reactive control is the general rule in data networks, it was initially thought to be inappropriate for broadband wide area networks due to their high delay-bandwidth product. The delay-bandwidth product determines the amount of data which can be in transit in the network. A high value brings considerable inertia in the realization of rate changes and leads to large buffer requirements. This led to the early definition of preventive control strategies based on a "traffic contract". This is the underlying principle of the DBR and SBR transfer capabilities. The ABT capability is also a preventive control although the possibility of rate negotiation included in the "elastic" version constitutes an element of reactivity.

The notion of traffic contract involves fixing agreed QoS levels for connections of given characteristics. The nature of the QoS guarantees which can be respected with preventive traffic control depends strongly on how statistical multiplexing is performed in network elements. In particular, if the memory for cell buffering provided by ATM multiplexers is intended to absorb only "cell scale congestion" we must perform Rate Envelope Multiplexing (see Section 4.1.1). If a greater amount of memory is provided and used to achieve higher link utilization by absorbing "burst scale congestion", the performance of statistical multiplexing becomes qualitatively different. We refer to multiplexing with large buffers under the general term of rate sharing (see Section 4.1.2). Lastly, in Section 4.1.3, we consider the possible trade-off between cell loss and call blocking probabilities.

4.1.1   Rate Envelope Multiplexing

First consider an ATM network designed to meet the strict delay requirements of real time services assuming, as is presently the case for most switches, that multiplexer queues are served in FIFO order. The need for low maximum delays for services such as interactive voice and video communication means that multiplexer buffers must be small. To avoid cell loss, it is sufficient to maintain the combined arrival rate of multiplexed sources below the multiplex output rate. If an arrival rate overload does occur, the duration of the overload is typically so long that any "small" buffer will remain saturated throughout, whatever its capacity. The above condition on the combined arrival rate is thus, for all practical purposes, also necessary.

This approach has led to the design of traffic controls based on what we term Rate Envelope Multiplexing (REM). The multiplex service rate, which can change in time as a result of network management actions, constitutes the rate envelope and the objective of traffic control is to ensure that the combined input rate of multiplexed connections is within the envelope, at least with a very high probability (e.g., $1 - 10^{-9}$). Figure 4.1.1 illustrates the principle of REM.

The cell loss ratio (CLR) is determined essentially by the probability of an arrival rate overload. The required buffer capacity is such that the probability of saturation is negligible assuming the arrival rate is no greater than the service rate. Typically, 100 cell places suffice to absorb the very short term ("cell scale") congestion due to a concentration of cell arrivals

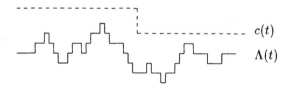

Figure 4.1.1: Principle of Rate Envelope Multiplexing.

from different sources.

Rate envelope multiplexing can be realized either by performing Connection Admission Control (CAC) to limit the arrival rate or by adjusting the allocated multiplexer service rate (e.g., by changing the capacity of a Virtual Path Connection) or both. Let $c(t)$ be the service rate at time t and let the combined arrival rate be $\Lambda(t)$. The negligible overload objective may be expressed:

$$\Pr\{\Lambda(t) > c(t)\} < \epsilon. \tag{4.1.1}$$

In the fluid approximation (i.e., we assume work arrives continuously at rate Λ_t not in discrete units of one cell), no multiplexer buffer (reservoir) is required. REM can then be referred to as "bufferless multiplexing". The cell loss ratio is estimated by the fluid approximation:

$$\text{CLR} \approx E\{(\Lambda(t) - c(t))^+\}/E\{\Lambda(t)\} \tag{4.1.2}$$

REM has the considerable advantage that CLR depends only on connection traffic parameters affecting the stationary probability of the arrival rate. For an on/off source for example, the only significant parameters are the peak and mean rate. (Parameters describing burst and silence lengths and correlations between successive bursts do however have an impact on the duration of overload periods). Delays are also very short allowing tight guarantees on the CTD and CDV QoS parameters.

In the fluid model there is no ambiguity in the definition of rates and the bufferless multiplexing interpretation of REM is clear. Some confusion arises in practice, however, since arrival rates of cell streams are not well defined due to the discrete nature of the flows and, notably, to the presence of jitter. Even CAC for peak rate allocation suffers from the need to account for CDV and this problem is considered in Section 5.1 where, in particular, the notion of "negligible CDV" is introduced. REM as described here should be understood to be applied with the "instantaneous rate" of the multiplexed streams specified with negligible CDV in the sense of Section 5.1.1. This implies, in particular, that considered link rates would be nominal rates, somewhat less than the actual physical rate and determined by the available, cell scale, buffer capacity.

A limitation on the efficiency of REM arises as rate variability increases. A constraint on CLR defines an implicit relation between the characteristics of the offered traffic and the achievable multiplexer occupancy. In particular, consider the following example which illustrates the role of the connection peak rate of on/off sources. N identical on/off sources of peak rate p and mean rate m are multiplexed on a link of nominal capacity c. The CLR is then estimated by:

$$\mathrm{CLR} = \sum_{ip>c} (ip - c) \binom{N}{i} \left(\frac{m}{p}\right)^i \left(1 - \frac{m}{p}\right)^{(N-i)} \times \frac{1}{Nm} \qquad (4.1.3)$$

A constraint on CLR imposes a limit on achievable multiplex utilisation Nm/c. Assume a target CLR of 10^{-9}. The achievable load compatible with the limiting overload probability can be calculated as a function of the source peak rate. This function is plotted in Figure 4.1.2 for several values of N (including the limiting case where the number of sources tends to infinity). This figure illustrates that a high link utilisation is only possible here when the peak rate is a small fraction of the multiplex link rate. For illustration purposes, consider a link of capacity 100 Mbit/s; to accomodate bursts of peak rate 20 Mbit/s ($c/p = 5$) with 10^{-9} CLR would require mean utilization to be limited to around 2%; to achieve 50% utilization with the same QoS objective requires either the sources to be very slightly bursty (i.e., $m/p \approx 1$) when N is small or of very low peak rate ($p << c$) when N is large.

In general, while the achievable link load depends on the precise traffic mix, it may be stated that REM can be efficient for bursty sources with relatively low peak rates but can require a rather low network link utilization if bursty streams with peak rates comparable to the link rate are to be carried.

With the SBR capability, cell loss must be avoided by a judiciously chosen CAC procedure aiming to ensure that the probability of rate overload is sufficiently small. The definition of CAC for REM is discussed in Section 5.2. ABT may be seen as an ATM transfer capability relying on REM where the possibility of a rate overload is eliminated by performing admission control at burst level. It remains necessary, however, to implement CAC procedures ensuring that the probability of burst blocking or delay is then sufficiently small. The effectiveness of REM with connection or burst admission control is discussed in [BT92a] and [DDH94].

4.1.2   Rate sharing

In the fluid approximation there is a clear distinction between bufferless and buffered multiplexing. In practice, since a buffer is always required if only to absorb cell scale congestion, this distinction is somewhat blurred. What we have called Rate Envelope Multiplexing is meant to designate the practical implementation of bufferless multiplexing. The distinguishing criterion is less the buffer size than the connection admission criterion. For REM the

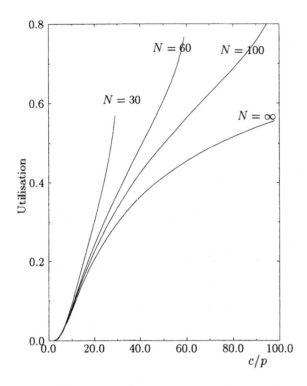

Figure 4.1.2: Achievable load for CLR $= 10^{-9}$ as a function of link rate to peak rate ratio c/p.

criterion consists in ensuring that the probability of an input rate excess over multiplexer capacity should be negligible. In this case, each multiplexed connection may be assumed to dispose of a bit rate equal to its current requirement. If the input rate momentarily exceeds the multiplexer rate (with non-negligible probability), connections may then be said to *share* the available rate. We therefore employ the term "rate sharing" to designate a preventive control statistical multiplexing scheme which is not REM. Rate sharing relies on buffers to absorb momentary input rate excesses. These buffers are typically much larger than those required in REM just to absorb cell scale congestion.

Rate sharing is the kind of multiplexing used in most data networks. In such networks source peak rates are typically of the same order of magnitude as the link rates and, as discussed in the previous section, REM would be extremely inefficient. Rate sharing may be seen to be necessary in broadband networks for connections with a large peak rate, greater than one tenth of the link rate, say. Assuming it is an objective to satisfy CLR, CTD or CDV QoS requirements, the connection admission criterion for rate sharing must take account of the performance of the multiplexer queue.

Given that the traffic offered to ATM multiplexers by connections governed by the SBR/VBR capability is constrained by a leaky bucket (or equivalent), there is scope for deterministic performance guarantees (e.g., delay never exceeds a given threshold) or statistical performance guarantees.

Deterministic performance guarantees

Consider an ATM source controlled by a leaky bucket of rate parameter r_i and token pool size b_i. The number of cells $\nu_i(s,t)$ passing the leaky bucket in time interval (s,t) is then constrained as follows:

$$\nu_i(s,t) \le b_i + r_i(t-s). \tag{4.1.4}$$

If a multiplexer receives cells directly from m such sources, the number of offered cells $\nu(s,t)$ in interval (s,t) is similarly constrained:

$$\nu(s,t) \le \sum_i b_i + \sum_i r_i \cdot (t-s). \tag{4.1.5}$$

Assume that the multiplexer link rate is c and admission control ensures that $c \ge \sum r_i$.

Now, by Reich's theorem, the work in system at time t, W_t, satisfies (see [Cru91]):

$$W_t = \sup_{s \le t}\{\nu(s,t) - c(t-s)\}$$

$$\le \sup_{s \le t}\{\sum b_i + (\sum r_i - c) \cdot (t-s)\}$$

$$\le \sum b_i.$$

Thus, to allocate a rate $c \ge \sum r_i$ and provide a buffer of capacity $B \ge \sum b_i$ avoids any cell loss and delay in a FIFO queue is bounded by $\sum b_i/c$.

The above performance guarantees apply for a multiplexer situated just downstream of the leaky bucket controllers imposing the burstiness bounds (4.1.4). Beyond this first stage, the burstiness of a traffic stream generally changes due to queueing and buffer requirements within the network to ensure zero loss are greater. The evolution of stream characteristics in a variety of network elements is considered in [Cru91]. It is shown, in particular, that a stream satisfying (4.1.4) at the input to a network element introducing a delay never greater than D satisfies on output the inequality:

$$\nu_i(s,t) \le b_i + r_i D + r_i(t-s). \tag{4.1.6}$$

In the particular case of a network element guaranteeing service rate r_i to stream i (i.e., as long as work from stream i is present in the queue it is served at a rate at least equal to r_i in any interval, however small), then the burstiness bounds (4.1.4) are preserved on output [PG93a] (see also Section 6.2).

Statistical performance guarantees

To realize the full potential of statistical multiplexing, it is certainly preferable to be able accept more than the strict minimum of connections determined by the deterministic bounds described above. The difficulty which arises with rate sharing is that the relation between performance, traffic characteristics and network capacity proves very complex. In the following, we do not account for any filtering performed on connections by the access controllers.

Figure 4.1.3 illustrates how the probability of buffer saturation depends on certain traffic characteristics. The graphs are sketches derived from the results of the queueing models presented in Part III of this report. They represent the result of superposing a number of homogeneous on/off sources in an ATM multiplexer although the following discussion clearly applies to more general traffic streams. Note that the "buffer size" and "log Pr{saturation}" axes could be re-labelled "x" and "log Pr{delay $> x$}", respectively.

Figure 4.1.3 (a) shows the effect of increasing the load by adding more sources. First note the characteristic shape of the curves: a first component corresponding to cell scale congestion, where the probability of saturation decreases very rapidly with increasing buffer size, is followed by a second component with a much smaller slope corresponding to burst scale congestion. We refer to these components as cell scale and burst scale components, respectively. The figure shows that as the load increases, required buffer size first increases slowly (as long as this size is determined by the cell scale component) but then increases much more rapidly (when the burst scale component is determinant). A buffer size determined by the cell scale component corresponds roughly to the buffer requirement for REM.

Figure 4.1.3 (b) shows the impact of changing peak rates while maintaining the same load by reducing the burst length in inverse proportion. The cell scale component is unaffected while the burst scale component is translated vertically. This behaviour illustrates the difficulty of using REM for high peak rate sources: the cell scale component progressively disappears as the peak rate increases; to perform REM would imply reducing the load to counter this effect.

Figure 4.1.3 (c) illustrates the impact of the mean burst length assuming silence lengths increase in proportion to maintain the same load. The cell scale component does not change while the slope of the burst scale component is inversely proportional to the mean burst length: a buffer sized by the burst component has the capacity to store entire bursts; as these increase in size the buffer must increase accordingly.

Finally, Figure 4.1.3 (d) shows the effect of changing the probability distribution of the burst length while keeping all other characteristics constant. The slope of the burst scale component decreases as the variance of the burst length increases, showing that performance also depends significantly on this parameter.

The above results apply to on/off connections where the lengths of successive on and off periods are independent and have a distribution with tail probabilities which decrease at least exponentially fast. Correlation between successive intervals (bursts of bursts) will also have a significant impact on the burst component. If the burst or silence length distribution has a heavy tail implying long range dependent traffic the burst component decreases more slowly than exponentially.

The basic lesson to be drawn from the above is that the performance of rate sharing depends significantly on traffic characteristics which are difficult both to measure and to control. It is consequently very difficult to satisfy QoS guarantees, notably for the delay based parameters CTD and CDV. To avoid cell loss, very large buffers must be provided in network elements to avoid the possibility of overflow.

4.1.3   Cell loss and call blocking

There is an obvious trade-off between call blocking and cell loss: when a call arrives we can either reject the call, or accept it and possibly lose cells. Typically we have a call blocking target which is several orders of magnitude lower than the cell loss ratio, such as a call loss of 1 in 10^3 and a cell loss of 1 in 10^6 or better.

This trade-off is illustrated in Figure 4.1.4, which shows expected utilisation and expected cell loss for a link of size 50 where the peak rate is 1, $p = 0.2$, and calls (connections) arrive as a Poisson process with rate ν, chosen so that the offered load ($\lambda = p\nu = 50$) is high, with unit call holding-time. The probability π_i that there are i calls in progress, $i = 0, \ldots, N$ is proportional to $\nu^i/i!$. The call blocking is the probability that N calls are in progress (given from Erlang's formula), the utilisation is the expected number of calls in progress multiplied by their activity, $E_n[np]$, and the cell loss ratio is found by taking expectations of the cell loss, $E[P_{loss}]$. The graphs show what happens when we accept up to N calls, rejecting further call attempts until a call clears down. The load here is very high, so we either have to preserve cell loss, which implies a very high blocking, and low utilisation, or sacrifice cell loss to achieve better utilisation.

Figure 4.1.5 shows what happens when the load is much lower ($\lambda = 25$). Now the maximum utilisation we can possibly get is bounded by 50%, and the cell loss is also bounded, so for certain services we could accept all the calls offered. When the capacity is lower (e.g., 10, or even 1), then the maximum

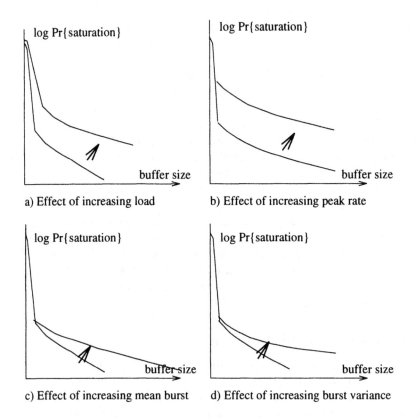

a) Effect of increasing load

b) Effect of increasing peak rate

c) Effect of increasing mean burst

d) Effect of increasing burst variance

Figure 4.1.3: Impact of traffic characteristrics on saturation probability.

utilisation drops dramatically. To increase the utilisation we either have to delay cells, smooth the sources, or use an ABR type service.

There is an obvious way [Key95], to deal with the trade-off between cell and call loss, or between cell loss and utilisation: assume we earn one unit every time we carry a cell, and lose amount y when we lose a cell, for example we might put $y = 10^9$ to represent a cell loss target of 10^{-9}, then in the steady state we want to chose the number N to maximise

$$E\left[np\right] - yE_n\left[M\left(n,p\right)\right] \tag{4.1.7}$$

where the first term is just $\lambda\left(1 - \pi_n\right)$, and M is defined as

$$M\left(n,p\right) = \mathrm{E}\left\{\left(S_n - C\right)^+\right\} = \sum_{i=1}^{n-C} i\mathrm{Pr}\left\{S_n = C + i\right\}. \tag{4.1.8}$$

This approach is used in Section 5.2.4 to model uncertainty about p.

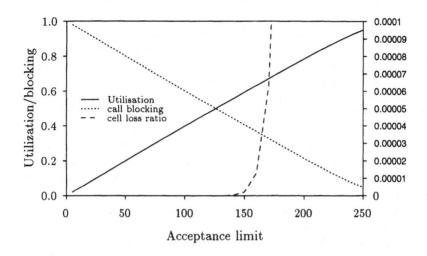

Figure 4.1.4: Utilisation vs cell loss for a high load.

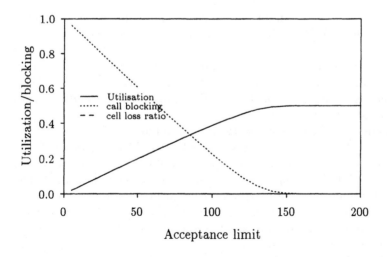

Figure 4.1.5: Utilisation vs cell loss for a moderate load.

4.2   Reactive traffic control

The Available Bit Rate (ABR) service [For96] is one of the best-effort classes of services aimed at satisfying the needs of applications with vague requirements for throughput and delay. The majority of applications found on today's computer networks are of this type, since they have been designed to work over networks providing no guarantee of throughput, loss rate or delay.

Unlike the Unspecified Bit Rate (UBR) service [For96], which provides no service guarantees at all, ABR aims to provide a guaranteed loss rate and employs a link layer flow control mechanism to this end. In 1995 the ATM Forum decided to use a so-called 'rate-based' approach for this flow control mechanism, whereby the network tells the source the rate at which it may transmit cells [1]. The scheme adopted is a reactive closed-loop mechanism [New94] which relies on a regular flow of Resource Management (RM) cells to communicate to the source the availability of bandwidth and the occurrence congestion in the network. The exact mechanism is described in more detail in the next section. Work on recommendations for the ABR service is currently underway in the ATM Forum and the ITU (current versions may be found in [For96] and [ITU96b].

The mechanism being completed by the ATM Forum relies on an end-to-end closed control loop that combines features from the *Forward Explicit Congestion Notification* (FECN) and *Backward Explicit Congestion Notification* (BECN) schemes well-known from conventional packet switching networks with an explicit rate mechanism.

An important performance issue of closed-loop feedback mechanisms is the buffer sizes required in intermediate nodes to achieve zero cell loss. In this section, we derive simple formulae to approximate the maximum buffer lengths for the closed-loop congestion control mechanism proposed in the draft version of the ATM Forum Traffic Management Specification 4.0 [For96]. Here, we look at network configurations with and without priority for the resource management cells. Besides the steady-state, we consider transient phases occurring when the available transmission capacity changes due to the establishment or release of connections.

The remainder of this section is organized as follows. In Section 4.2.1, we briefly describe the rate-based congestion control mechanism. An analytical approach to derive the evolution of the allowed cell transmission rates of the ABR sources and the resulting buffer content is outlined in Section 4.2.2. The derivation of formulae for the maximum required buffer size in a bottleneck link node is presented in Section 4.2.3. Section 4.2.4 discusses numerical results obtained using our analysis and the section concludes with an outlook in Section 4.2.5.

---

[1] The main competitor to this scheme, the credit-based approach [KM95] was rejected by the ATM Forum; however, several companies still plan to implement such a scheme.

4.2.1   The closed-loop rate control mechanism

In the following we explain the operation mode of the rate-based congestion control mechanism as described in [For96]. The ATM Forum only defines the behavior of the *Source End System* (SES) and the *Destination End System* (DES), leaving switch implementation to the vendors. To explain the fundamental behavior of the end systems, the network configuration depicted in Figure 4.2.1 is used.

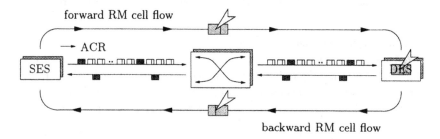

Figure 4.2.1: Basic configuration for rate-based congestion control.

The SES, which is connected with the DES via a number of switches (in Figure 4.2.1 only one switch is shown), is permitted to emit cells according to an *Allowed Cell Rate* (ACR) changing with the congestion status of the network. At connection setup, an *Initial Cell Rate* (ICR), a *Minimum Cell Rate* (MCR) and a *Peak Cell Rate* (PCR) are negotiated. The ACR may vary between the MCR and the PCR, and the SES is allowed to transmit cells with a rate up to the ACR.

To receive feedback information from the network, the SES sends a forward *Resource Management* (RM) cell every N_{rm} data cells. Upon arrival of a forward RM cell at the DES, the cell is returned as a backward RM cell to the SES. Information about the current congestion status of the network resources is provided by the switches and the DES, which may alter the content of the RM cells. At the arrival of a backward RM cell, the SES adjusts its ACR according to the *Congestion Indication* (CI) bit and the *Explicit Rate* (ER) field of the RM cell.

When a backward RM cell is received at time t with CI=1, the ACR must be reduced by at least $N_{rm} \cdot ACR(t)/RDF$ down to the MCR, where RDF is called *Rate Decrease Factor*. If the backward RM cell carries CI=0, the ACR may be increased by no more than $N_{rm} \cdot AIR$, but not beyond the PCR. The factor AIR is called *Additive Increase Rate*. If the ER signaled in an RM cell is lower than the current ACR but higher than the MCR, the ACR is set to this value. To check for RM cell failures and problems with connections having idle phases, a number of time-out mechanisms were developed. Since we do not take into account such situations in this subsection, a description of these mechanisms is omitted (see [For96]).

The CI bit and the ER field in the RM cells are set as follows. Congestion is detected by the network according to the queue length in the switch buffers. Depending on the switch architecture, different actions are performed:

- *Explicit Forward Congestion Indication (EFCI)-based switch:*
  If the queue length exceeds an upper threshold, the switch sets the EFCI bit in the header of data cells to indicate congestion, until the queue length drops below a lower threshold. The DES sets the CI bit in each backward RM cell to the EFCI value of the data cell received last.

- *Explicit Rate (ER)-based switch:*
  ER-based switches are equipped with an intelligent marking and explicit rate setting capability. This allows to selectively reduce the rates of ABR sources by marking the CI bit or setting the ER field in forward and/or backward RM cells according to the degree of congestion.

Both types of switch, which may co-exist in a single network environment, can optionally set the CI bit in backward RM cells to ensure that the source does not increase its rate. Furthermore, backward RM cells may be generated by a switch. These options are not considered in this subsection. A detailed description of switch architectures can be found in [OMS⁺95].

4.2.2 Steady state analysis

Due to the amount of time until the SES receives information about the current utilization of network resources, i.e., the feedback delay, the ACR of each ABR source and the buffer length in the switches show a dynamical behavior. In this subsection, we briefly outline an analytical tool, which has its origin in control theory, to compute the steady-state behavior of the rate control mechanism. Using interim steps of our analysis we derive simple formulae for the maximum buffer content in the next subsection.

The additive increase and the multiplicative decrease of the source rates can be modeled using differential equations. Such an approach has been used e.g., in [BS90, COMM95, OMS⁺95, Rit95b, YH94]. The control mechanisms considered in these papers differ, however, from the one developed by the ATM Forum. Furthermore, no simple formulae for the maximum buffer content are shown in the papers mentioned above. The convergence of feedback mechanisms, which is not necessarily guaranteed, is studied in [FRW92, MS91]. Other papers on feedback control, which are only loosely related to the mechanism investigated here, are e.g., [GA94, KOMM95, YH91].

For further description we make use of the network model depicted in Figure 4.2.2. A number N_{vc} of ABR traffic sources are connected to their destination via a single bottleneck link. The buffer in the ATM switch is assumed to be of infinite capacity and the traffic sources operate in a saturated state, i.e., they have always cells to transmit.

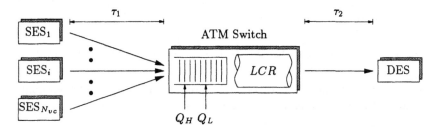

Figure 4.2.2: Network model with a single bottleneck link.

We denote the *Round-Trip Time* (RTT) by τ, the propagation delay between the SES and the ATM switch by τ_1, and that between the ATM switch and the DES, i.e., the propagation delay on the bottleneck link, by τ_2. Thus

$$\tau = 2\left(\tau_1 + \tau_2\right). \qquad (4.2.1)$$

If the buffer length at the ATM switch exceeds Q_H, the switch detects congestion. Congestion is regarded as terminated if the buffer length drops below Q_L. The bottleneck link is assumed to operate with a fixed transmission speed of LCR.

The derivation of the dynamic behavior of the ACR, denoted by $ACR(t)$, and the buffer length $Q(t)$ is described in the following three subsections. We take into account a traffic scenario with N_{vc} identical ABR sources, i.e., all of the N_{vc} connections, which share a common transmission path, have the same values of MCR, PCR, ICR, AIR and RDF. They are therefore in phase during their whole lifetime.

Basic equations

We start with some basic equations for the evolution of the ACR, which can be described in four different phases. The phases are differentiated by the current congestion status of the switch and by whether the buffer is empty or not.

* Phase 1: no congestion is signaled and the buffer is not empty

If no congestion is detected by the switch, the CI bit in the RM cells remains unset and thus, the ACR of each SES is increased by $N_{rm} \cdot AIR$ at each arrival of a backward RM cell until the PCR is attained. In case of a non-empty buffer, these arrivals occur at a constant rate of $LCR/(N_{vc} \cdot N_{rm})$. We can express this increase by the following differential equation

$$\frac{dACR_1(t)}{dt} = \frac{LCR}{N_{vc}} \cdot AIR, \qquad (4.2.2)$$

which is solved by

$$ACR_1(\Delta t) = \min\left\{ PCR,\, ACR(t_0) + \frac{LCR}{N_{vc}} \cdot AIR \cdot \Delta t \right\}, \qquad (4.2.3)$$

with respect to the initial condition $ACR_1(0) = ACR(t_0)$. The interval length Δt is the time elapsed since the starting instant of the considered phase.

* Phase 2: no congestion is signaled and the buffer is empty

In contrast, when the buffer is empty, the arrival rate of the backward RM cells at time t depends on the ACR of the SES τ time units before, i.e., $ACR(t - \tau)$. Now, the differential equation is given by

$$\frac{dACR_2(t)}{dt} = ACR_2(t - \tau) \cdot AIR. \qquad (4.2.4)$$

As solution we obtain

$$ACR_2(\Delta t) = \min\left\{ PCR,\, ACR(t_1)\, e^{\beta \Delta t} \right\}, \qquad (4.2.5)$$

where β is the root of $\beta = AIR\, e^{-\beta \tau}$ and $ACR_2(0) = ACR(t_1)$.

* Phase 3: congestion is signaled and the buffer is not empty

If the switch detects congestion, the ACR is decreased at the arrival of backward RM cells in a multiplicative manner until the MCR is attained. The decrease factor is given by N_{rm}/RDF. Since the switch is assumed to be fully utilized in this phase, the arrival rate of backward RM cells at the SES is the same as in phase 1, resulting in

$$\frac{dACR_3(t)}{dt} = -ACR_3(t)\, \frac{LCR}{N_{vc} \cdot RDF}. \qquad (4.2.6)$$

Equation (4.2.6) is solved by

$$ACR_3(\Delta t) = \max\left\{ MCR,\, ACR(t_2)\, e^{-\frac{LCR}{N_{vc} \cdot RDF} \Delta t} \right\}, \qquad (4.2.7)$$

with the initial condition $ACR_3(0) = ACR(t_2)$.

* Phase 4: congestion is signaled and the buffer is empty

The last configuration may seem unrealistic but it occurs if we give priority to the RM cells. In this case, the length of the buffer has no influence on the forward, and thus on the backward, RM cell flow. Analogous to the previous phases, the differential equation for the ACR is given by

$$\frac{dACR_4(t)}{dt} = -ACR_4(t)\, \frac{ACR_4(t - \tau)}{RDF}, \qquad (4.2.8)$$

which has no explicit solution. However, for small values of τ and reasonable choices of RDF, we obtain the following approximate solution

$$ACR_4(\Delta t) \approx \max\left\{ MCR, \frac{RDF}{\Delta t + \frac{RDF}{ACR(t_3)}} \right\} , \qquad (4.2.9)$$

when phase 4 starts with $ACR(t_3)$.

The evolution of the buffer length during each of the four phases can be computed by the following equation $(i = 1, \ldots, 4)$

$$Q_i(\Delta t) = Q(t_{i-1} + \tau_1) + \int\limits_{x=\tau_1}^{\Delta t} \left(N_{vc} \cdot ACR_i(x - \tau_1) - LCR \right) dx , \qquad (4.2.10)$$

where the delay τ_1 between the SESs and the ATM switch is taken into account. Using the buffer length function, the duration of each phase can be determined depending on the given buffer thresholds Q_L and Q_H.

In the remainder of this subsection we describe the sequence of phases which will occur cyclically for various switch types. A detailed description, which can be found in [Rit95b] for EFCI-based switches without priority for RM cells, is omitted due to lack of space.

The EFCI-based switch

The sequence of phases in steady state, looking at the EFCI-based switch, depends on whether priority is given to the RM cells or not. If we give no priority to the RM cells, i.e., RM cells are queued up together with the data cells, we observe the following sequence:

(1) We start with a non-empty buffer at the ATM switch which is not yet congested. As long as the switch does not detect congestion, the evolution of the ACRs corresponds to *Phase 1*.

(2) If congestion is detected, i.e., the buffer length exceeds Q_H, the decrease of the ACRs is described by *Phase 3*, until the lower threshold Q_L is attained. During this period, the buffer remains non-empty.

(3) When Q_L is attained, the ACRs are increased again, starting with a non-empty buffer at the ATM switch, i.e., *Phase 1*.

(4) Since the sum of all ACRs at the end of the decrease phase is lower than the bottleneck link rate LCR, the buffer might empty during the increase phase. In this case, the increase according to Phase 1 is interrupted by a period where the ACRs increase as described by *Phase 2*.

Hence, for the non-priority case, the cyclic sequence consists of two or four phases, respectively, depending on whether the buffer becomes empty or not. An example evolution for a cycle with four phases is depicted in Figure 4.2.3.

This sequence changes when priority is given to the RM cells. If we assume that the aggregate cell rate of all N_{vc} RM cell flows is lower than the bottleneck link rate LCR, then the arrival rate of the RM cells at the SESs depends only on the ACR of the sources. Since it does not depend on the buffer content we have only two alternating phases:

(1) The increase, which lasts until Q_H is exceeded, is modeled by *Phase 2*.

(2) The decrease phase, which is terminated if the buffer length goes below Q_L, is modeled by *Phase 4*.

In Figure 4.2.4 an example evolution for the EFCI-based switch with priority for RM cells is shown. Here, a phase displacement between the curves for the ACR and the buffer length can be observed, which arises due to the feedback delay.

The ER-based switch

For ER-based switches a number of different ways to signal ERs during a congestion period have been discussed within the ATM Forum. In this subsection, we focus on a method called *Fair Share* (FS), which works as follows. During uncongested periods, no ER is signaled and thus, the ACRs are increased in the manner described for EFCI-based switches. If congestion is detected, the switch computes a FS depending on the bottleneck link rate and the MCRs of the active connections as follows:

$$FS = \frac{1}{N_{vc}} \left(LCR - \sum_{i=1}^{N_{vc}} MCR_i \right) . \qquad (4.2.11)$$

The FS is multiplied by an *Explicit Reduction Factor* (ERF) smaller than 1 and is added to the MCR of each connection. The resulting rates, which guarantee an aggregate cell rate of less than LCR, are signaled to the SESs, i.e.,

$$ER_i = MCR_i + FS \cdot ERF . \qquad (4.2.12)$$

Assuming such a mechanism, the evolution of the ACRs for an ER-based switch without priority for RM cells is given by:

(1) We start again with a non-empty buffer at the ATM switch which is not yet congested. As long as the switch does not detect congestion, the evolution of the ACRs corresponds to *Phase 1*.

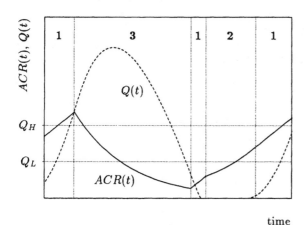

Figure 4.2.3: EFCI-based without priority.

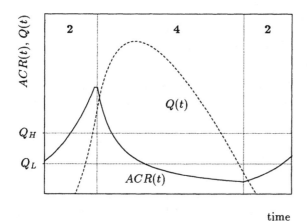

Figure 4.2.4: EFCI-based with priority.

(2) If congestion is detected, the ACRs are set to the ERs computed using equation (4.2.12) and remain unchanged until the buffer length goes below Q_L (*Phase FS*).

(3) Now, the ACRs are increased again according to *Phase 1*. As described for the EFCI-based switch without priority, phase 1 may be interrupted by a period where the evolution of the ACRs follows that of *Phase 2*.

A sample path is depicted in Figure 4.2.5.

If we implement a priority mechanism for the RM cells and make the same assumptions as for the EFCI-based switch, we get the following cyclic sequence, which is illustrated in Figure 4.2.6.

(1) The increase, which lasts until Q_H is exceeded, is modeled by *Phase 2*.

(2) After the detection of congestion, an ER computed as described above is signaled (*Phase FS*). After the buffer length goes below Q_L, the next increase period, which corresponds to *Phase 2*, starts.

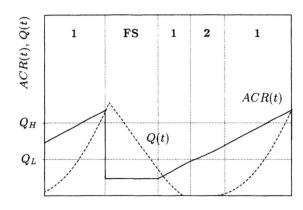

time

Figure 4.2.5: ER-based without priority.

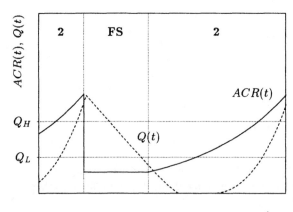

time

Figure 4.2.6: ER-based with priority.

4.2.3   Maximum buffer length

One of the most interesting performance measures of a feedback mechanism is the size of the switch buffers which is required to avoid cell loss. Using the differential equation approach presented above, the buffer requirements during steady state can be approximated. The computation, however, is awkward and may take a considerable amount of time.

To simplify buffer dimensioning and control parameter tuning we now derive upper bounds for the results obtained using the analysis, as presented in [Rit96]. Since the analysis is of an approximate nature, we do not necessarily get upper bounds for the real values. However, a large number of numerical examples show that the approximate and real bounds are in good agreement. Results for the steady state are given in Section 4.2.3 while Section 4.2.3 deals with the transient phase.

The steady state

The derivation of closed-form expressions for the upper bound of the maximum buffer length during steady state is based on the equations presented in Section 4.2.2. In the following subsections, the four different switch architectures presented above are discussed.

The EFCI-based switch without priority   Before we focus on the maximum buffer length, we first look at the maximum value of the ACR which is attained at the end of an increase phase. Assume that the buffer becomes empty during the first part of the increase phase. It starts to fill up again, if the aggregate cell rate of all ABR connections becomes larger than LCR. The evolution of the ACR from this time instant on is given by

$$ACR^*(\Delta t) = \min\left\{ PCR, \frac{LCR}{N_{vc}} \cdot \left(1 + AIR \cdot \Delta t\right)\right\}, \qquad (4.2.13)$$

cf. equation (4.2.3). If we look at the time relative to the switch, we obtain the buffer length

$$Q^*(\Delta t) = \int_0^{\Delta t} \left(N_{vc} \cdot ACR^*(t) - LCR\right) dt = \frac{1}{2} LCR \cdot AIR \cdot \Delta t, \qquad (4.2.14)$$

and thus, the buffer content reaches Q_H after

$$\Delta t = \sqrt{\frac{2 Q_H}{LCR \cdot AIR}}. \qquad (4.2.15)$$

Since the propagation delay between SES and ATM switch is τ_1 and it takes a feedback time of $\tau_f = \tau_1 + 2\tau_2$ until the SES recognizes that congestion is detected, the total amount of time $\Delta t_{ACR_{max}}$ until the ACR stops increasing is given by

$$\Delta t_{ACR_{max}} = \tau + \sqrt{\frac{2 Q_H}{LCR \cdot AIR}}. \qquad (4.2.16)$$

This results in a maximum ACR equal to

$$ACR_{\max} = \min\left\{ PCR, \frac{LCR}{N_{vc}} \cdot \left(1 + AIR \cdot \left(\tau + \sqrt{\frac{2Q_H}{LCR \cdot AIR}} \right) \right) \right\} .$$

(4.2.17)

The maximum length of the buffer $Q_{\max}$ is determined by the length of the buffer after the increase phase plus the cells which have to be queued until the sum of the ACRs drops below LCR, i.e.,

$$Q_{\max} = Q(t_1 + \tau_s) + \int_0^{\Delta t_{Q_{\max}}} (N_{vc} \cdot ACR_3(t) - LCR) \, dt .$$

(4.2.18)

Substituting $ACR_3(t)$ by equation (4.2.7) and performing the integration, we get

$$Q_{\max} = Q(t_1 + \tau_s) - LCR \cdot \Delta t_{Q_{\max}} +$$
$$N_{vc} \cdot ACR(t_2) \frac{N_{vc} \cdot RDF}{LCR} \left(1 - e^{-\frac{LCR}{N_{vc} \cdot RDF} \Delta t_{Q_{\max}}} \right) .$$

(4.2.19)

Note that, since we only consider the part where $ACR_3(t)$ is always larger than or equal to LCR, the maximum function for the MCR is omitted. The buffer length after the increase phase can be upper bounded by

$$Q(t_1 + \tau_s) \le Q_H + \left(\tau + \frac{N_{vc} \cdot N_{rm} - 1}{LCR} \right) \left(N_{vc} \cdot ACR_{\max} - LCR \right) ,$$

(4.2.20)

where the RTT τ and the maximum time until the next backward RM cell is received by the SESs are considered. An upper bound for $ACR(t_2)$ is given by

$$ACR(t_2) \le ACR_{\max} .$$

(4.2.21)

The time interval $\Delta t_{Q_{\max}}$ until the buffer length decreases again can be computed by inverting equation (4.2.7), i.e.,

$$\Delta t_{Q_{\max}} = ACR^{-1}\left(\frac{LCR}{N_{vc}} \right) = -\frac{N_{vc} \cdot RDF}{LCR} \ln\left(\frac{LCR}{N_{vc} \cdot ACR_{\max}} \right) .$$

(4.2.22)

Finally, the maximum expected buffer length is upper bounded by

$$Q_{\max} \le Q_H + \left(\tau + \frac{N_{vc} \cdot N_{rm} - 1}{LCR} \right) \left(N_{vc} \cdot ACR_{\max} - LCR \right) +$$
$$N_{vc} \cdot RDF \left(\ln\left(\frac{LCR}{N_{vc} \cdot ACR_{\max}} \right) + \frac{N_{vc} \cdot ACR_{\max}}{LCR} - 1 \right) .$$

(4.2.23)

The EFCI-based switch with priority The derivation of the upper bound for the EFCI-based switch with priority for the RM cells can be treated similarly. Therefore, we give only the results. The time interval Δt until the buffer content reaches its upper threshold Q_H is determined by the solution of

$$Q_H = LCR\left(\frac{1}{\beta}(e^{\beta \Delta t} - 1) - \Delta t\right), \tag{4.2.24}$$

where β is the constant defined in equation (4.2.5). Thus, it takes

$$\Delta t_{ACR_{\max}} = \tau + \Delta t \tag{4.2.25}$$

to attain $ACR_{\max}$, which is given by

$$ACR_{\max} = \frac{LCR}{N_{vc}} e^{\beta \Delta t_{ACR_{\max}}}. \tag{4.2.26}$$

With the same approach as for the EFCI-based switch without priority, we obtain the following upper bound for the maximum buffer length:

$$Q_{\max} \leq Q_H + \left(\tau + \frac{N_{rm} - 1}{ACR_{\max}}\right)\left(N_{vc} \cdot ACR_{\max} - LCR\right) +$$

$$N_{vc} \cdot RDF \left(\frac{LCR}{N_{vc} \cdot ACR_{\max}} + \ln\left(\frac{N_{vc} \cdot ACR_{\max}}{LCR}\right) - 1\right).$$

$$\tag{4.2.27}$$

The ER-based switch For the ER-based switches working with the FS policy, the computation of upper bounds for the maximum buffer length is simpler. The reason is the immediate reduction of the ACRs to the FS if congestion is detected, which results in an aggregate cell rate lower than LCR. Thus, the buffer content does not increase further in the first part of the rate decrease phase as observed with the EFCI-based switches.

Consequently, taking into account the propagation delay between SESs and the ATM switch as well as the feedback delay, we obtain for the non-priority switch

$$Q_{\max} \leq Q_H + \left(\tau + \frac{N_{vc} \cdot N_{rm} - 1}{LCR}\right)\left(N_{vc} \cdot ACR_{\max} - LCR\right), \tag{4.2.28}$$

and for the priority switch

$$Q_{\max} \leq Q_H + \left(\tau + \frac{N_{rm} - 1}{ACR_{\max}}\right)\left(N_{vc} \cdot ACR_{\max} - LCR\right), \tag{4.2.29}$$

if congestion is signaled in forward RM cells. If backward RM cells are used for signaling, τ has to be substituted by $2\tau_1$ in both equations. Since the increase phase for EFCI-based and ER-based switches is the same, $ACR_{\max}$ is computed by equation (4.2.26) for the ER-based switch without priority, and by equation (4.2.17) for the switch with priority mechanisms.

Transient phases

Besides the steady state, we are also interested in the evolution of the buffer length during transient phases. Therefore, we look at scenarios where the cell rate of the bottleneck link LCR changes at a given time instant. It may decrease if a new real-time connection (CBR or VBR) is established on that link, or increase, if such a connection is released. In this subsection, we consider transient phases occurring if LCR is changed to a new, fixed rate.

The maximum buffer length achieved during a transient phase clearly depends on the time instant when LCR is changed. Thus, we are first looking for the worst-case, i.e., the time instant when the buffer has to absorb the largest peak if LCR is decreased.

We assume that the system is in steady-state. The buffer length attains a maximum value during a transient phase when the bottleneck link rate is decreased precisely at the time instant when the upper threshold Q_H is exceeded.

If we decrease it before Q_H is attained, the switch would detect congestion earlier and thus, the maximum achievable ACR would be smaller. This leads, however, to less cells which have to be queued during the feedback delay and in case of EFCI-based switches, the rate of the aggregate ACRs drops below LCR earlier.

If the bottleneck link rate is changed after the switch detects congestion, the buffer length increases more slowly during the time interval between congestion detection and the change of the bottleneck link rate. This explains the worst-case behavior at the time instant when Q_H is exceeded. For the same reasons, the first peak after the rate decrease will be the largest one. After a transient phase, a new steady state is reached within a number of cycles depending on the control parameter set. Figure 4.2.7 shows an example evolution of the buffer length if LCR is decreased.

If we increase LCR at the moment when Q_H is exceeded, the next peak will also have maximum size. This can be explained by reasoning similar to the above. The evolution of the queue length for this case is depicted in Figure 4.2.8.

These findings simplify the derivation of equations for the transient phase considerably. Since, for the worst-case scenario, the change of LCR to a new value LCR^* has no influence on the maximum values for the ACR corresponding to the first peak, we obtain the same results for ACR_{max} as described in Section 4.2.2. The change of the bottleneck link rate is only reflected in the equations for Q_{max} where the former value LCR has to be substituted by LCR^*.

4.2.4 Performance comparison

In this subsection we compare the performance of the switch architectures investigated aiming at the maximum required buffer size. The steady state

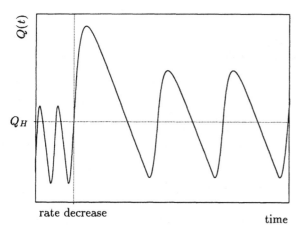

Figure 4.2.7: Transient phase after decrease.

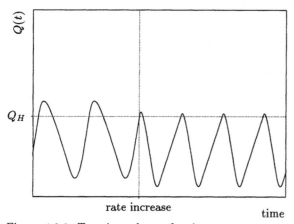

Figure 4.2.8: Transient phase after increase.

and transient phases are taken into account. First, we present some numerical examples to show the appropriateness of our analytical approach.

Figures 4.2.9 and 4.2.10 show bounds for the maximum buffer length as a function of the RTT τ. The control parameters of the feedback mechanism are set to $AIR = 0.1$ *Mbps*, $RDF = 512$, $ERF = 4/5$, $Q_L = 1000$, $Q_H = 5000$ and $N_{rm} = 32$. We used a bottleneck link rate of $LCR = 50$ *Mbps*, $MCR = 0$ *Mbps* and $PCR = 150$ *Mbps*. The range of the RTT corresponds to distances between the SESs and the DES of approximately 1 to 1000 *km*.

Looking at Figure 4.2.9, which takes into account one connection ($N_{vc} = 1$), a good agreement between theoretical and simulation results shown as dots can be observed. Since the differential equation approach is of an approximate nature, the upper bounds for the theoretical results sometimes

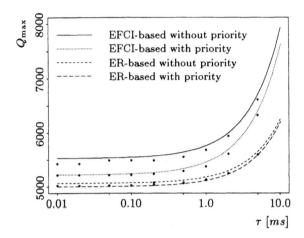

Figure 4.2.9: Influence of RTT ($N_{vc} = 1$).

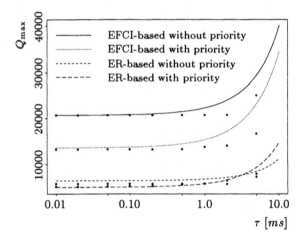

Figure 4.2.10: Influence of RTT ($N_{vc} = 10$).

slightly underestimate the real values. This is given especially for the ER-based switch with priority for RM cells.

Comparing the four switch architectures, the EFCI-based switches clearly require larger buffers than the ER-based switches. The use of priority for RM cells reduces the required buffer space somewhat for the EFCI-based switches but hardly at all for the ER-based switches.

The same holds for a scenario with $N_{vc} = 10$ ABR connections, cf. Figure 4.2.10. Now, however, the performance of the EFCI-based switches becomes worse compared to the ER-based switches. For large RTTs, the real values are slightly overestimated.

The appropriateness of our approach to approximate the size of the first peak in transient phases as described in Section 4.2.3 is verified in Figures 4.2.11 and 4.2.12. Using the same control parameter set as above, the bottleneck link rate is changed from 50 *Mbps* to values ranging from 1 to 100 *Mbps* after the system was in steady state. The RTT is set to 0.1 *ms*. If we compare the performance of the different switch architectures for $N_{vc} = 1$ (Figure 4.2.11) and $N_{vc} = 10$ (Figure 4.2.12), the same relative performance of the switches as noticed before can be observed.

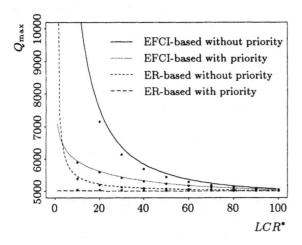

Figure 4.2.11: Influence of LCR^* ($N_{vc} = 1$).

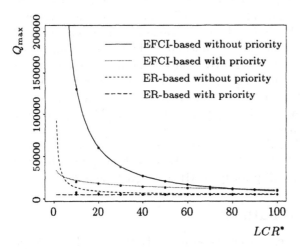

Figure 4.2.12: Influence of LCR^* ($N_{vc} = 10$).

The performance differences are most apparent when the bottleneck link rate is decreased, especially to small values. In case of EFCI-based switches which operate without priority, the required buffer size can be quite large if several ABR connections are active (cf. Figure 4.2.12).

To investigate the influence of the number of active ABR connections N_{vc} on the maximum buffer length, results for a RTT equal to 1 *ms* are shown in Figures 4.2.13 and 4.2.14 for the steady state and transient phases, respectively. The transient phase was generated by halving the bottleneck link rate.

Both figures show a fast, nearly linear increase with the number of connections in case of EFCI-based switches, whereas the maximum buffer length for the ER-based switches remains almost constant. The reduction of the required buffer size due to the use of a priority mechanism for RM cells becomes larger with an increasing number of connections, particularly in transient phases.

Next, the increase of the maximum buffer length with the RTT is shown. In Figures 4.2.15 and 4.2.16 curves for $N_{vc} = 10$ ABR connections are depicted for the steady state and transient phases generated as described above. In principal, we observe the same dependence as in the case of varying the number of connections, i.e., a fast, nearly linear increase for EFCI-based switches and a more distinctive reduction due to the priority mechanism for longer RTTs and transient phases.

The crossing of the curves for the ER-based switches results from the approximation of the maximum ACR, which overestimates the actual values especially for large RTTs if priority for RM cells is used.

The last two figures show the influence on the maximum buffer size of the AIR which belongs to the control parameter set. Again, we study both the steady state and transient phases for a RTT equal to 1 *ms* and $N_{vc} = 10$ ABR connections. Figures 4.2.17 and 4.2.18 show a fast increase of the maximum buffer length with the AIR for EFCI-based switches. The insensitivity of the ER-based switches comes from the immediate reduction to an aggregate ACR below LCR if congestion is detected. Thus, there is only the feedback time required before the queue length starts to decrease. For EFCI-based switches the queue is further increased until the aggregate ACR drops below LCR. This may take a considerable amount of time if the AIR is large and hence high ACRs are attained.

The maximum buffer length stops increasing with the AIR when the AIR is large enough so that the ACR attains the PCR. This effect can be observed for the EFCI-based switch with priority, if AIR is larger than 0.25.

4.2.5   Summary and outlook

To allow for an efficient and economic operation of the rate-based congestion control mechanism for ABR services, the control parameters of this mechanism must be set appropriately taking into account the ATM network environment. A major point of interest is the required size of switch buffers to avoid cell loss.

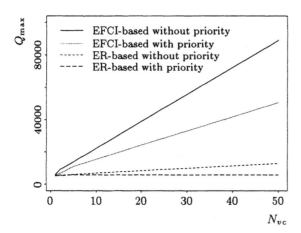

Figure 4.2.13: Inluence of N_{vc} in steady state.

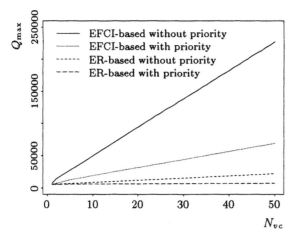

Figure 4.2.14: Inluence of N_{vc} in transient phases.

In this subsection, we presented simple formulae to approximate the maximum required buffer size for various switch types, i.e., EFCI- and ER-based switches with and without priority for RM cells. We derived results for the maximum buffer length during steady state as well as for transient phases. Comparing simulation and theoretical results, a good agreement between both was observed.

In general, the ER-based switches perform better than the EFCI-based switches. This holds particularly for larger numbers of connections and longer round-trip times. The most important feature of the ER-based switches is the minor influence of the choice of the control parameters and the network

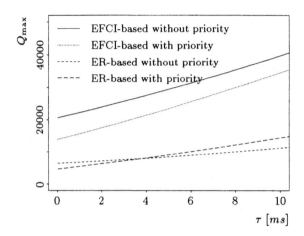

Figure 4.2.15: Inluence of RTT in steady state.

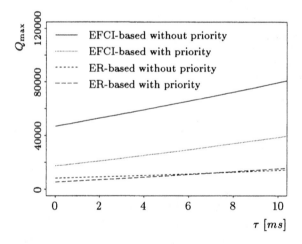

Figure 4.2.16: Inluence of RTT in transient phases.

environment on the required maximum buffer sizes. This enables a stable operation of the network. However, ER-based switches are much harder to implement and are therefore not suitable for low-cost network solutions.

Since the required buffer size increases almost linearly with the feedback delay, a segmentation of end-to-end control loops into smaller control segments should be taken into account for long distance connections.

Regarding the implementation of priority mechanisms for RM cells, a much larger gain can be obtained for EFCI-based switches than for the others. Furthermore, the numerical results show that this kind of switch should be equipped with such a mechanism to reduce the buffer requirements. Oth-

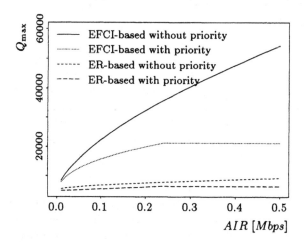

Figure 4.2.17: Inluence of AIR in steady state.

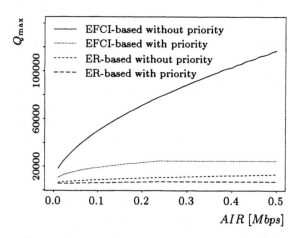

Figure 4.2.18: Inluence of AIR in transient phases.

erwise, extremely large buffers are necessary.

As a consequence, an appropriate tuning of the control parameters for the given network environment is important to achieve a reasonable performance of the EFCI rate-based mechanism. The results derived in this subsection provide an easy to use and powerful tool for this design process.

5 Connection admission control

Connection admission control is a preventive control, which safeguards the QoS of existing connections by refusing additional connections when necessary. CAC is an open loop control, which doesn't modify the characteristics of connections once they are accepted, and so the future consequences of a connection have to be anticipated at connection set up. This contrasts with closed loop reactive controls such as ABR.

It is sometimes helpful to distinguish between two types of multiplexing, *rate envelope multiplexing*, which keeps the rate of accepted connections below the service rate or capacity envelope, and *rate sharing* where buffering or queueing is used to share out the available rate amongst connections, where the sum of the offered rates can exceed the service rate for periods of time. These are explained more fully in Section 4.1.

Typically rate envelope multiplexing is associated with small or *cell scale buffers*, perhaps of the order of 100 cells, whereas rate sharing is associated with larger buffers and burst scale queueing. In the fluid approximation there is indeed a discontinuity in multiplexer queueing behaviour between the bufferless and buffered case. However, in practice, the distinction between rate envelope multiplexing and rate sharing is not always clear cut, and much depends on the time scales induced by the traffic and the degree of buffering. In fact we show later, mirroring the work of Elwalid et al [EMW95] that worst case assumptions for rate sharing can be analysed using bufferless models appropriate to rate envelope multiplexing.

Different interpretations are involved in the application of these ideas to CAC, and some of these are reflected in this chapter. This chapter collects together some of the ideas which have developed in the course of COST242. The chapter proceeds in a step by step fashion: the simplest situation is to peak rate allocate, which is intimately connected with multiplexing CBR streams (see Section 5.1). Even here, the degree of buffering and CDV tolerance can affect how the allocation is performed. In practice sources are unlikely to continuously burst at the peak rate, a fact which can be exploited to extract statistical multiplexing gain. Section 5.2 considers rate envelope multiplexing in some depth, introduces the important notion of effective bandwidth, and explores various ways this could be calculated from knowing the rate distribution, worst case assumptions or measurements. Section 5.3 explores rate sharing, which is in general a harder problem to solve, since much more knowledge is required of the arrival process. However, as for rate envelope multiplexing, we can develop notions of effective bandwidth, or use worst case assumptions or adaptive control to decide whether to accept calls or not. Lastly, in Section 5.4, we explore the relationship between resource allocation and pricing and its impact on CAC.

5.1 CAC for peak rate allocation

A naïve definition of the CAC function for a network link operated with peak rate allocation is to simply admit new connections until the sum of peak rates attains a nominal link capacity. Unfortunately, the presence of CDV complicates things. In general, it is necessary to ensure that the cell loss ratio of admitted connections, given their CDV, is within a certain limit. One approach which would facilitate network operation is based on a notion of "negligible CDV": each multiplexer of link rate c and buffer capacity B is characterized by an allowable load ρ such that the CLR requirement is satisfied as long as the sum of the peak rates is less than ρc and *connections have negligible CDV*.

The definition of negligible CDV is discussed in Section 5.1.1 below. An alternative approach explored in Section 5.1.2, is to perform CAC taking explicit account of declared PCR and CDV tolerance parameters, notably by using the deterministic bounds on required buffer size for zero cell loss.

5.1.1 Negligible CDV

Consider an incoming traffic stream with peak emission interval $T = 1/\text{PCR}$ and CDV tolerance τ. Assume the stream enters the network on an access line with cell transmission time Δ. The definition of negligible CDV can be approached from two angles:

(i) Given a dimensioning rule fixing the admissible load $\rho(c, B)$ of a multiplexer, what (T, τ, Δ) combinations may be said to constitute streams with negligible CDV ?

(ii) Given a rule defining what (T, τ, Δ) combinations should be acceptable, what dimensioning rule must be applied to determine the admissible load $\rho(c, B)$?

To be more explicit, the dimensioning rule in the first approach might be to set ρ such that an M/D/1 queue would have a probability smaller than ϵ of exceeding B ($\epsilon = 10^{-10}$, say). Sources with negligible CDV would then be such that they would lead to a smaller buffer overflow probability if they were substituted for the Poisson process assumed in fixing ρ.

The second approach applies, in particular, when it is necessary to account for the rule in Rec. I.371 [ITU96b] defining the minimum amount of CDV that network elements should be able to tolerate (see (2.1.5)), viz:

$$\tau = \max\{T, 80\Delta(1 - \Delta/T)\}. \tag{5.1.1}$$

Maximum multiplexer load $\rho(c, B)$ should then be limited so that CLR is less than ϵ if all sources have CDV defined by (5.1.1).

The parameters T and τ are defined with respect to the GCRA. This algorithm limits the maximum burst size at the line rate $1/\Delta$ to :

$$\text{MBS}_L = 1 + \lfloor \tau/(T - \Delta) \rfloor. \tag{5.1.2}$$

Assuming all bursts have this size, the interval between the start of successive bursts is equal to $\text{MBS}_L \times T$.

In the following we discuss the feasibility of each of the above two approaches in turn before considering an alternative definition of negligible CDV relying on knowing more about a traffic stream than just the CDV tolerance.

"Negligible CDV" for a given multiplexer dimensioning rule

First note that CAC at each multiplexer is based on knowledge of T and τ alone since only these parameters are conveyed by signalling. It thus appears necessary to assume source traffic is the worst possible for any value of Δ. We assume this worst case traffic is a batch arrival process with batch size b corresponding to MBS in the limit $\Delta \to 0$:

$$b = 1 + \tau/T. \tag{5.1.3}$$

Note that, to simplify, we take the batch size to be a real number.

We consider the specific multiplexer dimensioning rule based on an assumed batch Poisson process with batch size β, i.e., multiplex load ρ and buffer capacity B are such that the buffer overflow probability would then be less than ϵ. For illustration purposes, assume $\epsilon = 10^{-10}$. The buffer size for a load $\rho = 0.8$ would then be $B = 54\beta$.

For a batch periodic traffic the allowed batch size compatible with ϵ will be greater than or equal to β depending on the value of cT, the ratio between the link rate and the source peak rate. Indeed, the product ρcT defines the maximum number of sources which can be multiplexed. For the specific case $\epsilon = 10^{-10}$, $\rho = 0.8$ and, therefore, $B = 54\beta$, we derive the limiting burst sizes $b(cT)$ given in Table 5.1.1 as a multiple of β. These results are derived using the $N * D/D/1$ queue results derived in Part III, Section 15.2. Table 5.1.1 defines acceptable (T, b) combinations for any given value of c. However, if it is necessary to define "negligible CDV" independently of c, it is necessary to assume the limiting value corresponding to large cT, i.e., $b = \beta$.

The idea that each multiplexer has its own notion of what constitutes negligible CDV is certainly not very satisfactory from a traffic engineering point of view. For CAC, it would be necessary to perform a specific test at each stage and we have the unsatisfactory situation where a given source can have negligible CDV on one link of a path but not on another, e.g., PCR = 2 Mbit/s, $b = 7.5\beta$ is acceptable on a 20 Mbit/s link but not on a 40 Mbit/s link (we would require $b \leq 4.7\beta$).

Table 5.1.1: Maximum burst size b given cT for the $M^\beta/D/1$ dimensioning rule.

cT	5	10	20	50	100	400	∞
N	4	8	16	40	80	320	∞
b/β	13.5	7.5	4.7	3.1	2.3	1.5	1

Multiplexer dimensioning for a given rule defining "negligible CDV"

We consider explicitly the rule (5.1.1) given in I.371 [ITU96b] relating τ to T and Δ. First note that in this case it would be unduly pessimistic to assume a batch arrival process as worst case traffic. For fixed T, as the access rate $1/\Delta$ decreases and approaches the value of T, MBS_L given by (5.1.2) with τ given by (5.1.1) increases unboundedly; however, since the access line rate decreases the traffic becomes progressively less bursty. For any fixed multiplexer rate c, there is a value of Δ $(0 < \Delta < T)$ such that the buffer requirement (for given load ρ) would be maximal. In general, performance depends on both T/Δ defining the CDV tolerance and maximum burst size at rate $1/\Delta$ and $c \times \Delta$ defining the number of peak rate bursts which can be superposed in the multiplexer without burst scale delay. Table 5.1.2 gives the value of MBS_L and b derived from (5.1.2) and (5.1.3), respectively as a function of T/Δ.

Table 5.1.2: Maximum burst size MBS and maximum batch size b.

T/Δ	1.33	2	5	10	100
MBS_L	60	41	17	9	2
b	15	21	14	9	2

We consider a superposition of homogeneous periodic burst processes. To analyse multiplexer performance when the source peak rate is smaller than the multiplexer rate, we have used the method described in Part III, Section 17.2.2. The complementary queue length distributions for a number of parameter combinations determined by applying formulae (5.1.1) and (5.1.2) are presented in Figure 5.1.1. The parameters are normalized with respect to the access line cell time Δ. For example, the parameter combination $T/\Delta = 10$ ($\text{MBS}_L = 9$), $c \times \Delta = 20$ might correspond to a 1 Mbit/s stream entering the network on a 10 Mbit/s access line and being multiplexed on a 200 Mbit/s link.

For a small link rate (e.g., $c \times \Delta = 1$, line E in Figure 5.1.1) the probability of saturation decreases rapidly with buffer size. For a greater link rate (e.g., $c \times \Delta = 320$, line A in Figure 5.1.1), the decrease rate is slower (the curve is roughly parallel to that of an $M^9/D/1$ queue, line F) but the probability of zero burst scale queue (the intercept on the y-axis) is small. The maximum

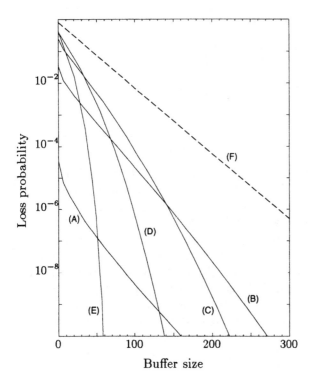

Figure 5.1.1: Loss probability as a function of the buffer size, $T/\Delta = 10$, $\mathrm{MBS}_L = 9$, $\rho = 0.8$.
- Line (A): $c \times \Delta = 320$, $N = 2560$;   - Line (B): $c \times \Delta = 80$, $N = 640$;
- Line (C): $c \times \Delta = 20$, $N = 160$;   - Line (D): $c \times \Delta = 5$, $N = 40$;
- Line (E): $c \times \Delta = 1$, $N = 80$;   - Line (F): $M^9/D/1$.

buffer requirement for traffic characterized by $T/\Delta = 10$ and, for a loss probability equal to 10^{-10}, appears from Figure 5.1.1 to occur for a link rate around 80 times the access rate (line B).

To determine the worst case buffer requirement for given multiplexer load, we have performed a non-exhaustive search over possible T/Δ, $c \times \Delta$ combinations. For a load of 0.8, we find the case $T/\Delta \approx 4$ and $c \times \Delta \approx 80$ to be the most demanding with a buffer requirement of around 400 cells. This gives us one point of the function $B(\rho)$ giving the buffer capacity required to handle any traffic stream satisfying the CDV requirement (5.1.1) assuming the link load to be limited to ρ. For various loads, Table 5.1.3 gives approximate values of $B(\rho)$ and the values of T/Δ and $c \times \Delta$ for which this maximum buffer size is required. The inverse function $\rho(B)$ gives the limiting load ρ compatible with a given buffer size B.

Table 5.1.3: Maximum required buffer size.

ρ	0.5	0.7	0.8	0.9
$B(\rho)$	108	242	400	863
T/Δ	5	4	4	4
$c \times \Delta$	10	35	80	320

The above analysis illustrates the complexity of accounting for the definition of minimal CDV given in Rec. I.371. It is also clear that, even though formula (5.1.1) is derived from a reasonable assumption about the creation of CDV in a FIFO queue, it nevertheless leads to a requirement for potentially very large buffers to account for worst case traffic. In fact, if the CDV were known to have been produced in a simple M+D/D/1 queue (or better), it could be considered "negligible" with respect to a very simple multiplexer dimensioning rule based on the M/D/1 queue, as discussed below.

Periodic streams in a network of queues

The above discussion on the two possible approaches for defining negligible CDV highlights the difficulty of relying only on the CDV tolerance parameter τ. Negligible CDV can be defined more satisfactorily if more is known about the way jitter has been introduced. In particular, mixing exclusively streams which were initially periodic in a network of multiplexers with stable queues does not appear to induce in any stream more jitter than that of a Poisson process of the same rate. This can correspond to a high value of τ, as given, for example, by formula (5.1.1), but the CDV remains negligible in the sense that the M/D/1 queue can still be used as a reference model for multiplexer dimensioning. We are thus led to consider the notion of defining negligible CDV with respect to a standard reference process: a stream has negligible CDV if the performance of a multiplexer would be worse if the stream were replaced by the reference process of the same intensity; the reference process is used to determine the maximum load compatible with a given buffer capacity. The reference process considered in this section is the Poisson process but other processes (e.g., a batch Poisson process) might also be used.

Analytical results on CDV in tandem queues (see Part III, Section 15.5 and the discussion of the "Recursive approach" in Section 3.2.3) allow us to conclude that if a renewal stream on input has negligible CDV when compared to a Poisson stream (e.g., a periodic stream), it continues to have negligible CDV as it passes through a tandem of saturated queues of the same capacity c receiving Poisson cross traffic. The result remains true if the value of c changes from queue to queue: we have verified that the coefficient of variation of the interarrival time remains less than 1 at every stage. The case of non-saturated queues can only be evaluated by means of simulation. Results are typified by those presented in Figure 3.2.3 in Section 3.2.3. This figure demonstrates that the variance of the interarrival time of an initially periodic stream increases from queue to queue but less rapidly than in the case of a saturated queue. The variance tends to a limit which is smaller than that of the saturated tandem.

If we can accept that the variance of the interarrival distribution is a sufficient measure of CDV in the considered through stream, these results demonstrate that an initially periodic stream does not acquire sufficient CDV as it passes through the network to make it worse than a Poisson stream. It can in fact be shown that there is negative correlation between successive inter-arrivals of the considered stream. This might be expected to result in better performance than a renewal assumption reinforcing the likelihood that the stream still has negligible CDV with respect to a Poisson reference stream.

We have not made corresponding simulations but it seems reasonable to assume a change of load from stage to stage would not destroy the property that streams remain "better than Poisson". This would imply that each multiplexer can be dimensioned independently; if it is equipped with a larger buffer, it can accept a correspondingly higher load factor.

The above analytical and empirical evidence lead us to formulate the following conjecture:

> If the CDV is negligible with respect to a Poisson reference process at the network ingress and multiplexing is performed with FIFO queueing, subject to the condition that the sum of PCR values is less than the multiplexer rate, then CDV remains negligible throughout the network.

If true, this conjecture greatly facilitates CAC: the nominal load of a multiplexer of rate c and buffer size B can be set at the maximal load of a rate c $M/D/1$ queue compatible with a sufficiently small probability of queue length exceeding B; connections with negligible CDV are accepted as long as this nominal load is not exceeded on any network link. It remains to specify how the network can ensure that a connection has negligible CDV at the ingress.

Certainly, the simple specification of a CDV tolerance as discussed in the earlier sections of the chapter does not characterize the connection sufficiently. The GCRA is therefore not a satisfactory policing mechanism in this context. One sure way (perhaps the only one) of ensuring negligible CDV is to actively space the cells of the connection. Any spacing function acting simultaneously on a set of connections carried by a single physical link cannot produce perfectly periodic cell streams on output since there will always remain contention between the cells of different connections. However, by virtue of the conjecture, any spacer realizing the virtual queueing system depicted in Figure 5.1.2, will produce streams with negligible CDV: the streams have negligible CDV at the output of the first stage queues (they have an interval at least equal to the peak emission time) and therefore retain this property on output from the final FIFO queue.

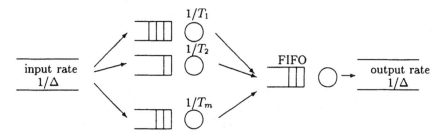

Figure 5.1.2: A spacer creates streams with negligible CDV.

5.1.2   Non-negligible CDV and a "natural" limit

The previous section attempted to define the concept of *negligible* CDV, such that if we denote the rate of the leaky bucket by r_i cells per second (the reciprocal of the interarrival time $1/T_i$) and each stream has *negligible* CDV with respect to a buffer of size B, then QoS constraints will be met provided

$$\sum_i r_i \le C, \tag{5.1.4}$$

where we have used C to denote the capacity we have available, and in this case

$$C = \rho c. \tag{5.1.5}$$

Suppose that the CDV associated with these sources is not negligible, then we have an additional constraint to consider, which relates the maximum burst size of the policers b_i to the buffer capacity B of the multiplexer. The worst-case analysis of Section 4.1.2 shows that no cells will be lost in the shared buffer provided that

$$\sum b_i \le B. \tag{5.1.6}$$

Equation (5.1.4) corresponds to peak rate allocation, where we use the contracted peak rate or bucket leak rate to decide whether to admit calls, subject to the CDV tolerances being within the limits of the buffer, as given by equation (5.1.6).

In summary, we have two conditions to be met: a utilisation constraint and a buffer constraint. In fact, we can replace the two constraints by a single constraint: suppose we choose the bucket depths so that

$$\frac{b_i}{r_i} \le \frac{B}{C}, \tag{5.1.7}$$

which says that the time to empty the leaky buckets is bounded above by the time taken to empty the buffer. Then the buffer equation (5.1.6) is implied by the utilisation constraint (5.1.4). This provides a "natural" bound for CDV: if a user is policed by a leaky bucket with parameter r_i then a natural limit for CDV is to use a bucket depth

$$b_i = r_i \frac{B}{C} \tag{5.1.8}$$

corresponding to a CDV tolerance bounded above by B/C. This clearly shows the effect of buffer sizes in the switches and the importance of the buffer ratio B/b — the larger the buffer in the switch, the larger the CDV that can be allowed. For example, for a 155Mb/s network, a 100-cell buffer switch allows some 0.27ms CDV, 200-cell buffer 0.54ms and so on.

If a source requires larger CDV than the natural bound (5.1.8) then it is *as if* the source actually had a leak rate (peak rate) of

$$C \frac{b_i}{B} . \tag{5.1.9}$$

The constraint (5.1.6) is a worst case bound and can be unduly pessimistic for large numbers of sources. In this case the $N * D/D/1$ or $\sum_i D_i/D/1$ analysis (see Part III, Sections 15.2 and 15.3) is able to provide better bounds, under the assumption that distinct sources are independent, as discussed in the previous section. For instance if we have 1000 constant bit-rate sources with the same rate and zero CDV ($b = 1$), then the cautious approach would say we needed a 1000 cell buffer. Whereas if we use an $N * D/D/1$ analysis and assume the sources are *randomly* phased, we only need an 80 cell buffer to give negligible loss (less than 10^{-12}) at high load (0.9). Indeed, at high load a 133 cell buffer is sufficient to cope with any number of multiplexed sources of size 1 (CBR) using the M/D/1 limit.

In fact the $N * D/D/1$ analysis or $M/D/1$ appromixation provides a bound for much larger CDV tolerances. Suppose, for example, that sources are each policed by a leaky bucket of depth b, and that each source emits through the leaky bucket a cluster of b cells at infinite rate, every b/r units of time. Then a buffer of size $B = 133b$ cells is large enough to cope with the random phasing of the clusters, even when the load is high ($\rho = 0.9$, say). More generally, suppose that r, c, b, B are given. Then we can calculate the number of sources N that can be simultaneously carried, by determining the traffic intensity $\rho = Nr/c$ at which the proportion of cells lost, determined by the $N * D/D/1$ analysis, is negligible (10^{-12}, say). Figure 5.1.3 shows how N varies with the ratios c/r and B/b.

5.2   CAC for Rate Envelope Multiplexing

Peak rate allocation is inefficient for variable bit-rate sources. Consider first, the classic case of *cell-scale* buffering described above for peak rate allocation,

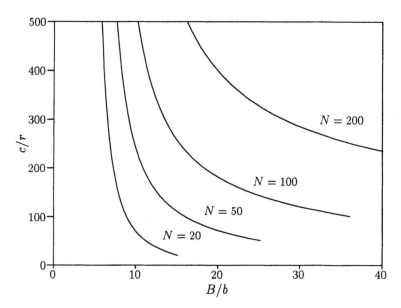

Figure 5.1.3: Relationship between c/r and B/b for fixed N. There is a tradeoff between buffer B and service rate c at the switch, as well as between the bucket depth b and leak rates r at each source.

with only enough buffering to cater for the CDV tolerance. To statistically multiplex we only need to know about fluctuations of bandwidth on a coarse time-scale, ensuring that at any point in time the bandwidth required only exceeds the capacity with small probability (the quality of service constraint). On this "burst-scale" we effectively act as though there is no buffer at all. To emphasise this point: the burst-scale fluctuations are on a longer time-scale than the time to empty a buffer.

As discussed in Section 4.1, rate envelope multiplexing can be realised by limiting the arrival rate or adjusting the service rate to ensure that the combined arrival rate $\Lambda(t)$ is less than the service rate $c(t)$, or exceeds it with defined small probability, ϵ. Using a fluid approximation, the cell loss rate is given by Equation (4.1.2). To proceed further requires some knowledge about the rate distribution. There are broadly three approaches which could be taken:

- the distribution of rate $\Lambda(t)$ might be known or deducible from parameters which are input, or declared;

- a worst-case rate distribution could be deduced from known or policed parameters;

- on-line measurements could be used to estimate some key quantities.

In this section, we consider a number of possibilities derived from these three approaches. First, in Sections 5.2.1 and 5.2.2, we consider alternative definitions of an effective bandwidth for sources of known stastistical characteristics. When only rule-based traffic parameters are known, a conservative CAC strategy is to make worst case traffic assumptions. Possible approaches are discussed in Section 5.2.3. Less conservative resource allocation may be possible if important traffic parameters such as the mean rate can be estimated in real time. Section 5.2.4 describes one possible approach based on decision theory while, in Section 5.2.5, the adaptive CAC method proposed uses the effective bandwidth formulae derived in Section 5.2.2 together with measurements of the instantaneous mean rate.

5.2.1 Effective bandwidth based on Chernoff bound

Effective bandwidths give a number α_i, such that we can meet the grade of service constraint provided that

$$\sum_i \alpha_i \leq C \qquad (5.2.1)$$

instead of equation (5.1.4). There are many models for effective bandwidths – what is important in our context is that the statistical multiplexing element is obtained effectively assuming no buffering, the so-called bufferless model.

Suppose for the moment that we have a single type of traffic, and a number, n, of independent sources of this type. Let $X_i(t)$ be the instantaneous bandwidth of source i. The aggregate requirement of the sources is then the load S defined by,

$$S_n(t) = X_1(t) + \ldots + X_n(t) . \qquad (5.2.2)$$

To meet given quality of service constraints, we require that the load exceed the capacity, C, with small probability. In fact we are usually interested in bounding the cell loss ratio of a stream, P_{loss}, requiring it to be less than $10^{-\gamma}$ for some given number γ, such as 5 or 10, where

$$P_{loss} = \frac{E\{(S_n - C)^+\}}{E\{(S_n)\}}. \qquad (5.2.3)$$

To progress, we need to model the bandwidth requirement. Consider first an on-off model: we assume that the source is either on at the peak rate, r, with probability p, or else off. The mean is rp. The *burstiness*, the peak to mean ratio is $1/p$, and p is the activity of the source — the proportion of time it is active.

The probability that the instantaneous load is s is given by a binomial distribution

$$S_n \sim r \, Bin(n, p) . \qquad (5.2.4)$$

Hence the probability of resource saturation or P_{loss} can be calculated exactly. These quantities are not difficult to calculate, however we can use the Chernoff bound to speed up the calculations, and to obtain some insight.

Consider a heterogeneous collection of sources producing load

$$S = \sum_{j=1}^{J} \sum_{i=1}^{n_j} X_{ji} \tag{5.2.5}$$

where the X_{ji} are independent random variables and should be interpreted as the load placed on an unbuffered resource by a source of class j and n_j is the number of sources of class j. Kelly [Kel91a] discusses this situation and shows using a Chernoff bound that

$$\log P(S \geq C) \approx \inf_{s} \sum_{j=1}^{J} (n_j M_j(s) - sC) \tag{5.2.6}$$

where

$$M_j(s) = \log E\left[e^{sX_{ji}}\right] \tag{5.2.7}$$

is the logarithmic moment generating function of sources of class j. In the case of on/off sources with peak rate r_j and burstiness $1/p_j$ the mean rate of a class j source is $m_j = r_j p_j$ and we have that

$$M_j(s) = \log\left(1 + \frac{m_j}{r_j}(e^{sr_j} - 1)\right). \tag{5.2.8}$$

Corresponding functions for sources with other rate distributions (e.g., Gaussian) may be deduced from the results in Part III, Section 14.3 where the notation $\mu(\cdot)$ is used instead of $M(\cdot)$.

The acceptance region (see [Kel94c])

$$A(n^*) = \left\{ n : \sum_{j=1}^{J} \alpha_j^* n_j + \frac{\gamma}{s^*} \leq C \right\} \tag{5.2.9}$$

with $\alpha_j^* = M_j(s^*)/s^*$ will assure that $\log P(S \geq C) \leq -\gamma$ and where s^* is chosen to attain the infimum in (5.2.6) for some given traffic mix n^*. Figure 5.2.1 illustrates the acceptance regions for several choices of s^* in the situation where the peak rates of classes 1, 2 and 3 are 0.1, 2.0 and 10.0 Mb/s respectively and the mean rates are 0.04, 0.02 and 0.01 Mb/s respectively. The capacity of the resource is $C = 100$ Mb/s and the cell loss probability constraint is $\gamma = 16$. For example, the plane passing through the n_3 axis at 625 has $s^* = 0.333$.

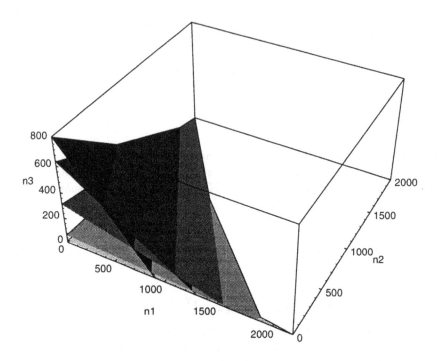

Figure 5.2.1: Acceptance region for heterogeneous sources.

5.2.2   Simple approximate formulae for effective bandwidths

An empirical expression for the effective bandwidth d of a burst scale loss system, i.e., the required bandwidth allocation for a traffic source in essentially all mixtures for a given overall cell loss ratio $P_{loss} = 10^{-9}$, is suggested in [Rob92a] and [Lin91]

$$d = 1.2m + 60\sigma^2/c \qquad (P_{loss} = 10^{-9}),$$

where c is the link rate and m and σ^2 are the mean and variance of the source rate, respectively. If instead of the variance σ^2 only the peak rate h of the source is known, or if the source is of an on/off type, the expression $\sigma^2 = m(h - m)$ is used.

According to [Rob92a], this expression for d holds with an acceptable accuracy in an interesting part of the parameter space, i.e.,

$$h/d \geq 2, \quad m/h \geq 0.05 \quad \text{and} \quad 15 \leq c/h \leq 1000.$$

The values of the coefficients 1.2 and 60 are dependent on the overall cell loss rate $P_{loss} = 10^{-9}$. In the following we shall present a generalization of this effective bandwidth formula. The extension was derived in [RT94a]. General expressions for the coefficients mentioned above are given in terms of P_{loss}. Furthermore, the corresponding effective bandwidth formulae are extended to the whole parameter space so that we have complete effective bandwidth formulae for various cell loss rates.

We start by deriving an expression for the overall cell loss rate P_{loss} if homogeneous traffic sources are considered. In doing so we use the method of equivalent bursts, which means that the total traffic stream is substituted with a superposition of an infinite number of sources generating bursts. The bursts are of fixed length and height, arriving according to a Poisson process. Thus, the overall cell loss rate is

$$P_{loss} = \mathrm{E}\,[\,X - n \mid X \geq n\,] \cdot \mathrm{Pr}\,\{X \geq n\}\,/\mathrm{E}\,\{X\}\,,$$

where X is the number of simultaneous bursts in progress, $n = c/h^*$ and $h^* = \sigma^2/m$ is the bandwidth of a standard burst. As before, m and σ^2 denote the mean and variance of the rate of a single real source.

Using the fact that $X \in \mathrm{Poisson}(a)$, where $a = mr/h^*$ with r equal to the number of real sources integrated to achieve the cell loss rate P_{loss}, the expression for P_{loss} can, after some effort (see [Lin91]), be written as

$$P_{loss} = \frac{n}{(n-a)^2} \cdot \frac{a^n e^{-a}}{n!}. \tag{5.2.10}$$

With equation (5.2.10) as a starting point we shall look for the effective bandwidth of the real source. Introducing the effective bandwidth $d = c/r$ and the variable $q = \sigma^2/mc$ the quantities n and a can be expressed as

$$n = q^{-1}, \quad a = (qy)^{-1}, \quad \text{where} \quad y = d/m.$$

By applying Stirling's formula for $n!$ in (5.2.10) and taking natural logarithms we now get

$$\ln y - (1 - y^{-1}) + 2\,q\ln(1 - y^{-1}) = 1.5\,q\ln q - q\ln(P_{loss}\sqrt{2\pi}).$$

From this equation it is easy to see that y is a function of q and P_{loss}, i.e., $y = y(q, P_{loss})$. By keeping P_{loss} fixed we can draw the graph $y = y(q \mid P_{loss})$. This is done for several values of interest for P_{loss}. These graphs are found to be almost linear over a wide range of q values including the region of interest, say from $q = 10^{-3}$ to 10^{-2}. Thus, the almost linear parts of these graphs can be approximated by

$$y = a(P_{loss}) + b(P_{loss}) \cdot q$$

for different loss rates e.g. $P_{loss} = 10^{-k}$, $k = 5, \ldots, 9$. Using the definitions of y and q above, we get the effective bandwidth

$$d = a \cdot m + b \cdot \sigma^2/c. \tag{5.2.11}$$

To find the coefficients a and b, a straight line has to be adjusted to the almost linear part of each graph $y(q \mid P_{loss})$. There is, of course, some uncertainty involved in drawing these lines. However, we have empirically found that the coefficients are well matched to the formulae

$$a = 1 - \frac{\log_{10} P_{loss}}{50}, \quad b/a = -6 \log_{10} P_{loss}. \tag{5.2.12}$$

So, through equation (5.2.12) we have general expressions for the coefficients for the effective bandwidth formula of equation (5.2.11).

As was pointed out earlier, the effective bandwidth formula holds with good accuracy in a relevant part of the parameter space. When leaving that region, the formula either underestimates or overestimates the bandwidth requirement of the source. To deal with these circumstances we start by rewriting equation (5.2.11) using $\sigma^2 = m(h - m)$ and introducing the quantities

$$x = m/h, \quad z = \frac{-2 \log_{10} P_{loss}}{c/h}, \tag{5.2.13}$$

which together with (5.2.12) yield

$$\frac{d}{ah} = x(1 + 3z(1 - x)). \tag{5.2.14}$$

In the case of homogeneous on/off sources the binomial distribution can be used to find exactly how many sources can be integrated on a link if the parameters m, h, c and P_{loss} are given. Numerical comparisons between equation (5.2.14) and the left graph of Figure 7.1.1 in [Rob92a, page 151] reveal that equation (5.2.14) underestimates the effective bandwidth when c/h is small, i.e., when z is large. To compensate for this underestimation, it can be found empirically that the expression

$$\frac{d}{ah} = x(1 + 3z^2(1 - x)) \tag{5.2.15}$$

should be used instead of equation (5.2.14) when $z > 1$.

From equation (5.2.14) (resp. (5.2.15)) it follows that in case of CBR sources, i.e., when $x = 1$, we get $d = ah$. This amount of bandwidth is, of course, the maximum a source should be assigned. However, it might happen that the right hand side of (5.2.14) (resp. (5.2.15)) becomes greater than 1, which yields $d > ah$. To protect against such an undesired overestimation of the effective bandwidth, the right hand side of (5.2.14) (resp. (5.2.15)) is required to be less than or equal to 1. For equation (5.2.14) this leads to the requirement $x \leq (3z)^{-1}$ and for equation (5.2.15) to the requirement $x \leq (3z^2)^{-1}$. Thus, if in the case where $z \leq 1$ the requirement $x \leq (3z)^{-1}$ is violated or, in the case where $z > 1$ the requirement $x \leq (3z^2)^{-1}$ is violated, the effective bandwidth $d = ah$ is simply assigned to the source. Thus, this complete assignment of effective bandwidth can, as in [Lin94], be summarized with a and z as in (5.2.12) and (5.2.13)

$$
d = \begin{cases}
am(1 + 3z(1 - m/h)), & 3z \leq \min(3, h/m); \\
am(1 + 3z^2(1 - m/h)), & 3 < 3z^2 \leq h/m; \\
ah, & \text{otherwise.}
\end{cases}
$$

5.2.3 Assuming worst case traffic with two leaky buckets

In this section we assume a source is policed using two leaky buckets operating on different time-scales: one leaky bucket polices the peak rate (PCR) with a tolerance CDV, and the other polices the sustained rate (SCR) with a tolerance BT. Rather than using the GCRA formalism, we use the equivalent leaky bucket parameters to describe the source traffic constraints. Let the peak rate be r_i and its associated token pool size bp_i. Let the sustainable rate be scr_i with a burst tolerance translated as a leaky bucket pool size of bs_i. In what follows we assume that

$$
r_i > scr_i \text{ and } bs_i > bp_i . \tag{5.2.16}
$$

There are several approaches that could be taken:

(i) Use a time scale separation argument, with *cell-scale* buffering large enough to cater for CDV (i.e., $B \geq \sum bp_i$), with scr_i and bs_i used as the basis of a burst-level CAC based on effective bandwidths. For instance, at the burst level we can use an on-off model where the source is either on at the peak rate r_i, with probability $p_i = scr_i/r_i$, or else off. In this case the effective bandwidth can be calculated via Equation (5.2.8) or Equation (5.2.4)) and is independent of the burst tolerance bs_i.

(ii) Allocate enough *burst-level* buffers. Suppose that the CDV is small, i.e., $bp_i \ll bs_i$, then it follows from the analysis of Section 4.1.2 that no cells will be lost, and each connection can be allocated its sustainable rate, provided that

$$
\sum_i scr_i \leq C \text{ and } \sum_i bs_i \leq B \tag{5.2.17}
$$

which makes the peak rate and CDV tolerance irrelevant. For homogeneous sources, we can clearly allocate the sustainable rate if

$$
\frac{C}{scr_i} bs_i \leq B \tag{5.2.18}
$$

which Elwalid et al [EMW95] term "bandwidth limited".

(iii) Allocate cell-scale buffers which are large enough to accommodate some of the bursts, or which allow us to define an effective bandwidth: an in-between situation. In other words we have

$$\frac{C}{scr_i} bs_i > B .$$ (5.2.19)

We could "overallocate", by analogy with equation (5.1.9), allocating the source an effective bandwidth

$$C \frac{bs_i}{B} .$$ (5.2.20)

However this ignores the extra information provided by the peak rate policer, and assumes that bursts occur at the line rate. Now assume that the CDV tolerance is small compared with the burst-tolerance ($bp_i \ll bs_i$), which enables us to treat cells as arriving at the peak rate or less. Then the maximum burst of cells that can pass through the sustainable rate policer is

$$\text{MBS}_i = \frac{bs_i}{1 - scr_i/r_i}$$ (5.2.21)

which happens in time MBS_i/r_i. Suppose we serve each source at rate r_i^* (i.e., allocate an effective bandwidth r_i^*), then the queue length of the superposition of the sources is bounded by

$$\sum_i \text{MBS}_i - \sum_i r_i^* \frac{\text{MBS}_i}{r_i} .$$ (5.2.22)

Hence no cells will be lost provided that

$$\sum_i r_i^* \le C$$ (5.2.23)

and

$$\sum_i \text{MBS}_i - \sum_i r_i^* \frac{\text{MBS}_i}{r_i} \le B .$$ (5.2.24)

In the case of homogeneous sources, the best we can do is to exhaust the bandwidth and the buffering simultaneously which, putting $Nr^* = C$ in the above and solving (5.2.24) with the inequality replaced by an equality, gives

$$r^* = \frac{C\,\text{MBS}}{B + C\,\text{MBS}/r} = \frac{rCbs}{bs\,C + B(r - scr)} = \frac{r}{1 + \frac{B(r-scr)}{bs\,C}}$$ (5.2.25)

a result quoted in [EMW95].

The above is a worst case analysis, assuming that bursts of cells from the buckets coincide. It is possible to do better than this, either looking at modulated $N * D/D/1$ queues, or by using statistical multiplexing ideas — see [EMW95] and below.

The distinction between the three scenarios is not clear cut (think for instance, of where we have two very different service classes).

A link with rate sharing

When a source is not "bandwidth limited" (see (ii) above), a "loss-free" effective bandwidth for a policed source, r^*, is given by equation (5.2.25). However this is a worst case analysis. In the case of many sources this is pessimistic, since the analysis assumed that all leaky buckets were full at the same time. If we think of a "worst case" scenario, where the sources transmit as much as possible before they have to stop due to the policing constraints, then (assuming negligible CDV) the traffic offered to the network will comprise bursts of cells at rate r. If this burst of cells is served at rate r^*, then the server will be utilised for a fraction of time

$$p = \frac{r}{r^*}.$$ (5.2.26)

Hence if we have a number of such sources, and assume that the sources are independent, each will offer deterministic loads with rate r^* when they are on, and zero when they are off, where p is the probability that the source is on (see [EMW95]). Therefore if there are n such sources, the probability of resource saturation is given by $\Pr\{S_n \geq C\}$, where now $S_n \sim r^* Bin(n, p)$. Therefore the "bufferless" analysis has a role to play even here.

The following figure, Figure 5.2.2, shows how the effective bandwidths vary in the case of (a) no loss, (b) 1 in 10^9 as the capacity increases for a fixed multiplexer buffer size (200 cells). The figure shows the dramatic effect of allowing a small amount of cell loss (lower effective bandwidth, or equivalently more efficient use of the network). Note that when we allow for a small CLR the greater the capacity, the greater the statistical multiplexing gain, which reflects the fact that the best sources to statistically multiplex are those whose bandwidth is small compared to the line rate. The different lines reflect different burst tolerances, where the sustainable cell rate scr is 0.2, the peak rate r is 1, and bs steps through 1, 5, and 50.

Note also that at large cell capacities, the burst tolerance has negligible effect, and indeed in these cases the effective bandwidth converges to that given from the analysis of Section 5.2.1.

Observe also that for small capacities, the effective bandwidth can actually *increase* as capacity increases. This is because we are holding the buffer capacity fixed, and the buffer constraint bites. Thus, although increasing the capacity allows us to fit more calls on, the effective bandwidth per call can increase in this region.

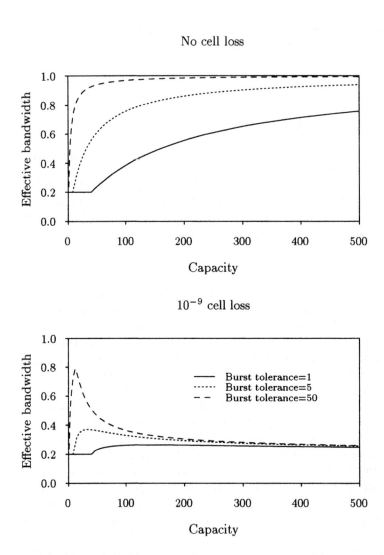

Figure 5.2.2: Effective bandwidths for different burst tolerances and no loss or small cell loss.

5.2.4   A decision theoretic approach to adaptive CAC

It is not possible to police a true mean rate for statistical reasons. The sustainable cell-rate with associated burst tolerance can be interpreted as a short term mean rate. As we have seen in Section 5.2.3, the policed sustainable cell rate can be used as the mean rate in our calculations of effective bandwidths, and so allow statistically multiplexing. But there are problems with this approach: the sustainable rate was originally defined for data sources which send information in fixed length blocks or frames, whereas real-time variable bit-rate sources often use variable length frames. Moreover the user has to be prepared to tolerate losing cells if they exceed this "sustained" contract, whereas in certain instances the multiplexing gain is independent of the burst tolerance. But the real

objection is a fundamental one: how will users know what their mean rate is? For a variable bit-rate video codec, even with the same equipment, the information rate can vary according to the scene being viewed.

But why ask for information if you can do without it? If the network can estimate mean rates, then we avoid a lot of these problems. This is the solution proposed here.

In this approach call acceptance decisions are based on a simple threshold rule: an offered call is accepted if the current load is less than a pre-calculated threshold. The threshold implements an implicit robust estimation procedure, and the decision-theoretic framework facilitates the essential trade-off between the benefits of accepting a call (earning revenue, customer satisfaction) and the disadvantages (threatening quality of service targets).

The schemes we shall consider have the following basic form: when a call is offered it is accepted if the current load, S_n, is less than $s(n)$. Here n is the number of calls currently in progress, and the vector $s = (s(n), n = 0, 1, ...)$ defines the call admission scheme.

Simple Bayesian estimation

We need to estimate the "activity" parameter p, the mean to peak ratio. A simple estimate would just be to estimate this quantity directly by counting cells in an interval of time and dividing by the peak number of cells that could be carried. The appropriate time interval is the time-period of the associated leaky bucket.

To be specific, recall that X_i is the bandwidth requirement of call i, and from now on we normalise capacity and bandwidth so that the peak of each call is 1. The simple estimate of the activity p is just

$$p = \frac{\hat{S}_n}{n}. \tag{5.2.27}$$

However this fails to account for estimation errors. The "large deviation heuristic" says that when rare events happen, they happen in the most likely way. With such an estimate we lose calls through a combination of two (rare) events: an observed low number of active connections, which causes us to underestimate the activity. This causes us to accept more calls than we should, and then we are hit by an unusually large number of active connections in the next time period, which exhausts the capacity.

Indeed, there are good reasons [GK91, GK94] why a better naive estimator is

$$\hat{p} = \tfrac{1}{2} \left(\frac{\hat{S}_n}{n} + \frac{C}{n} \right) , \qquad (5.2.28)$$

half-way between the simple estimator and one which assumes the resource is saturated.

Bayesian Decision Theory provides a suitable framework [DeG86] for taking into account the effects of estimation error. First, we assume that we can represent our "beliefs" about the value of the activity p by a prior probability distribution $f(p)$.

We now make an observation of the load, $S_n = s$ say, which gives some information about the activity, and a simple application of Bayes theorem gives this a posterior distribution, $f(p|s)$, which reflects our changing beliefs. Because S_n has a binomial distribution for our on-off model, and if we further assume that our prior $f(p)$ has a Beta distribution with parameters α and β then the posterior $f(p|s)$ also has a Beta distribution with parameters $\alpha + s, \beta + n - s$. Thus the posterior mean is $(\alpha + s)/(\alpha + \beta + n)$, a weighted average of the prior mean and the simple estimate (5.2.27). Now we have to make a decision: should we accept the next call? Since we want to keep our cell loss below the target value, we require

$$E_{f(p|s)}\left[cell\ loss\right] = \frac{E_{p|s}\left[M\left(n+1,p\right)\right]}{E_{p|s}\left[np\right]} = \frac{\int M\left(n+1,p\right)f\left(p|s\right)\,dp}{(n+1)\int pf\left(p|s\right)\,dp} \le 10^{-\gamma} ,$$

$$(5.2.29)$$

where $M(n,p) = E\{(S_n - C)^+\}$.

This is straightforward to calculate, and gives an acceptance boundary in (n, s) space, the curve $s(n)$. A simple Call Admission Procedure for a given value of n is to observe S_n, and when the next call arrives, accept it if the observed value of S is less that the boundary value $s(n)$. Recall that we have normalised the peak to be 1, hence the cell loss if zero is n is less than C, so we always accept if $n \le C$.

An example is shown in Figure 5.2.3, with a 10^{-9} cell loss target and a capacity of 50. Curves are shown for uniform prior, and three other priors, which both have a mean of 0.2. Note that if we *knew* the true value was 0.2, then the acceptance curve would be a vertical line at $n = 119$. The vague (uniform) prior gives similar results to the curve where we have some idea that the mean is around 0.2. If we have very strong prior information, then the curves can be quite different, as shown by the curve for "strong" prior parameters where the prior distribution (a Beta distribution with parameters $\alpha = 40$, $\beta = 160$) puts almost all the weight on p between 0.1 and 0.3. The more the prior information is concentrated on a small interval around $p = 0.2$, the more the acceptance curve tends towards a straight vertical line at $n = 119$.

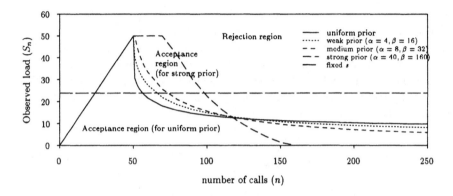

Figure 5.2.3: Acceptance Boundaries.

Remarks

- We can mix calls with varying mean rates safely — we do not have to have calls with exactly the same activities since our procedure will overestimate the cell loss. (If we have calls whose activities are very different, then we can multiplex more if we split the calls into different types, but at the cost of complexity.)

- We could use the method iteratively, using the posterior for stage n as the prior for $n + 1$, as we would in a straightforward Bayesian analysis. We chose not to do this, to avoid the estimation errors mentioned previously, and to allow the value of p to change over time.

- Simulation results on real, heterogeneous traffic have shown that the method does work and guarantees quality of service [GK94].

This approach is simple and conservative, however it can be too conservative. We have defined a call acceptance control, rejecting calls if the observed load is too high, which means that the expected cell loss will be lower than our target. If, on the other hand, we are bombarded by a mass of call arrivals, and run into an extreme overload for a period of time, then the *rate* of admitting calls might be too large. These effects are caused by ignoring the arrival rate.

A defence strategy against high offered loads first described by Bean [Bea93] is the following: *when a call is rejected, accept no further calls for consideration until after one current connection has ended.* This sort of back-off strategy is likely to form part of a switch's defence mechanism anyway, but we give it explicitly here, and assume it is part of the CAC function.

A decision theoretic approach

An adaptive CAC strategy is now an acceptance curve, $s(n)$, where we accept a call if $S_n \leq s(n)$, combined with our back-off strategy. Now assume that calls or connections arrive as a Poisson process of rate ν, so the offered load is $\lambda = \nu p$. Using our assumptions, for any given acceptance curve $s(n)$, we can calculate the probability $\pi(n)$ that there are n calls in progress [GKK95] (note that this will depend upon both λ and p). If we define the acceptance probability to be $a(n)$, the probability that a call is carried

$$a(n) = \sum_{i=0}^{s(n)-1} \binom{n}{i} p^i (1-p)^{n-i}. \qquad (5.2.30)$$

Hence we can now calculate the expected cell loss ratio averaged over n.

We now have the framework to implement a full decision theoretic approach to adaptive call admission. Our overall objective is to maximise the utilisation (carrying as many calls as possible) subject to the grade of service constraint. Suppose we receive unit reward every time we carry a cell and let y be the penalty we incur for losing each cell. Then our objective is to choose the control function $s(n)$ to maximise our expected reward,

$$E_p\left\{E_n\left[np - yM(n,p)\right]\right\} \qquad (5.2.31)$$

the expectation of equation (4.1.7) with respect to the prior taken over p and λ, $f(p, \lambda)$, that is

$$\iint \left[\sum_{n=1}^{\infty} \pi(n; p, \lambda)(np - yM(n; p))\right] f(p, \lambda)\, dp\, d\lambda \qquad (5.2.32)$$

$$= \iint [U(p, \lambda) - (y-1)M(p, \lambda)] f(p, \lambda)\, dp\, d\lambda, \qquad (5.2.33)$$

where U is the utilisation. Expression (5.2.32) is proportional to the expected reward per unit time if each offered cell attracts a reward of one unit while each lost cell incurs a penalty of y units. Thus the constant y measures our trade-off between utilization and cell loss.

Note that the optimisation of (5.2.32) is just a maximization of expected utilization for a given expected cell loss rate, with $y-1$ a Lagrange multiplier attached to the cell loss constraint. Thus y measures the *marginal* cell loss ratio: if a perturbation to the s curve allows additional carried traffic, then each additional offered cell has probability y^{-1} of being lost. In the classical theory of loss networks the use of a *marginal* loss rate in capacity expansion decisions is known as Moe's principle — [BHJ48, pp 216–21].

The dashed line illustrated in Fig. 5.2.4 shows the form of the optimizing s curve for $C = 50$, $y = 10^9$ and $f(p, \lambda)$ uniform on $(0, 1) \times (0, 25)$. The cell loss ratio and utilization achieved by this s curve are illustrated in Fig. 5.2.5 and Fig. 5.2.6 respectively. Note that the cell loss ratio is well controlled over a wide range of values of p and λ: this property is not explicitly sought in the optimization procedure, but is a natural consequence of the form of the objective function (5.2.32) and the use of a uniform prior distribution. Bean [Bea94], in his study of the case $\lambda = \infty$, shows that, when this is the objective, it is possible to choose the s curve so that the cell loss ratio is approximately constant over nearly all values of p, apart from a small interval close to $p = 1$. The solid line in Fig. 5.2.4 illustrates the form of the optimal s curve for $f(p, \lambda)$ uniform on $(0.2, 0.3) \times (0, 25)$ (to the right of the section shown, $s(n) = 0$). As the prior information on the parameter p becomes more precise the optimal s curve approaches a vertical line. If p is known, then the load measurement $s(n)$ conveys no useful information, and the number of calls which may be safely admitted can be pre-computed [Hui88, Kel91a]. Of course such an admission control is highly vulnerable to errors in the specification of p. Later, we consider another extreme, where the s curve is a horizontal line, an extreme which is much more robust against mis-specification of the parameter p.

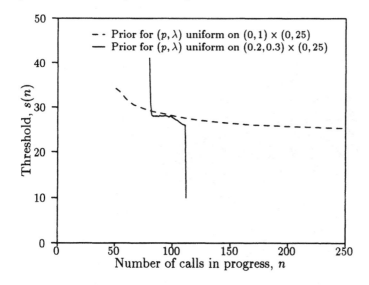

Figure 5.2.4: Optimized s curves. As prior information on burstiness parameter p, becomes more precise, the optimal curve approaches a vertical line.

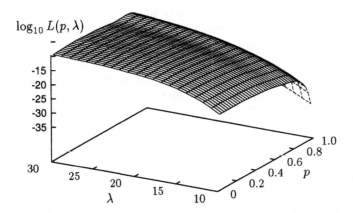

Figure 5.2.5: Cell loss for s the dashed curve from Fig. 5.2.4. Over a wide range of offered load, λ, and burstiness parameter, p, the cell loss ratio is well controlled.

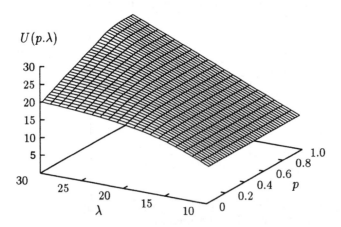

Figure 5.2.6: Utilization for s the dashed curve from Fig. 5.2.4. Utilization necessarily drops below offered load when p is small and λ is large.

Performance

How well does an adaptive scheme do? We know that a simple on-off model is robust against a mix of calls having the same mean, and [GKK95] shows how adaptive methods are able to respond flexibly and robustly to changes in the burstiness of calls. The cell loss ratio is relatively insensitive to the level of the parameter p, but is affected by its rate of change and, in particular, the derivative of $\log p$.

We now compare the performance of an adaptive scheme with the best performance possible when the burstiness parameter p and the offered load λ are known.

If the parameters p and λ are known then the policy maximizing the utilization U for a given cell loss ratio L is of a very simple form. The policy is as follows: accept an offered call if the number of calls in progress, n, is less than a critical value N; reject an offered call if $n > N$; and accept an offered call with probability η if $n = N$. The parameters N and η are chosen so that the cell loss ratio is exactly the desired level L. Thus the optimal scheme has stationary distribution

$$\pi(n) \propto \frac{\nu^n}{n!} \prod_{r=0}^{n-1} a(r) \tag{5.2.34}$$

with

$$a(n) = \begin{cases} 1 & n < N \\ 0 & n > N \\ \eta & n = N, \end{cases} \tag{5.2.35}$$

where N and η depend upon p and λ.

In Figure 5.2.7 we plot the ratio of the utilizations achieved over different values of p and λ, for s the dashed line of Fig. 5.2.4. For each

value of p and λ, the optimal scheme is chosen to achieve the *same* cell loss ratio as is achieved at the point (p, λ) by the scheme s. This plot is necessarily bounded above by 100%, by the definition of the optimal scheme. The important point to note is that, relative to the optimal scheme, the scheme s loses no more than 3% utilization over a very wide range of values of p and λ.

Note that any scheme which uses repeated measurements in order to estimate p and λ could not perform better than the optimal scheme; hence the loss of efficiency through using but a single observation of load is minimal.

Load measurements only

Some of the acceptance curves of Figure 5.2.4 are relatively flat for a range of values, which suggests that it might be possible to ignore the number of call in progress, basing our admission decisions solely on the measured load. For such as scheme, $s(n) = s$ for all n, so that call. In Figure 5.2.8 we

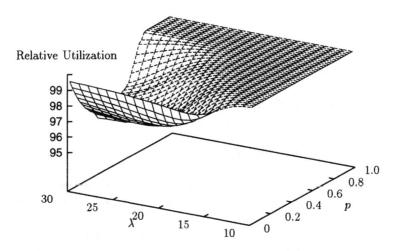

Figure 5.2.7: Utilization relative to optimal scheme where the parameters p and λ are known ($C = 50$).

illustrate the cell loss ratio for a scheme with $s = 26.83$ for $C = 50$, for various values of p and λ. This value of s optimizes the objective function (5.2.32) when $y = 10^9$ and $f(p, \lambda)$ places mass 1 on $(p, \lambda) = (0.2, 25)$: from Figure 5.2.8 we see that the cell loss ratio is fairly well controlled for a range of (p, λ) values about $(0.2, 25)$: compared with Fig. 5.2.5 the cell loss ratio falls away more quickly as p increases.

We now compare the performance of this simple measurement-only scheme with the best performance possible when the burstiness parameter p and the offered load λ are known. In

Figure 5.2.9 we plot the utilisations achieved by the schemes of Section 5.2.4, with $\lambda = C$ and with s values optimiziming the objective function (5.2.32) when $y = 10^9$ and $f(p, \lambda)$ palces unit mass on (p, C). We also plot the utilisations achieved by the optimal scheme, where for each value of p and $\lambda = C$ the optimal scheme is chosen to match the cell loss ratio of the corresponding scheme of Section 5.2.4. The optimal schemes necessarily achieve a higher utilization, but we note that the dominance is not great. Figure 5.2.9 shows that the "load measurement only" schemes are also nearly as efficient as a scheme that knows p and λ precisely. This is striking, and shows that little is lost by basing admission decisions on load-measurements only.

A discussion of the extension of this approach, using aggregated load measurements and a form of differential backoff, to the case of heterogeneous

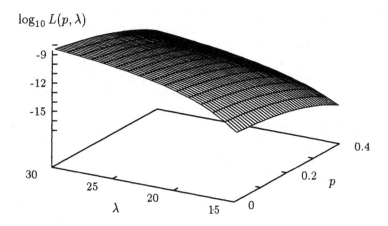

Figure 5.2.8: Cell loss with load measurements only: s horizontal. Note that the region used is smaller than in Fig. 5.2.5.

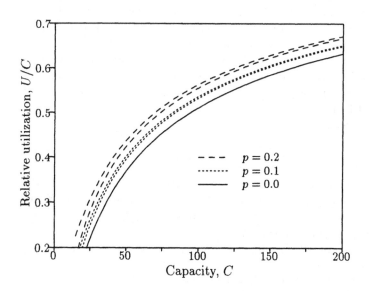

Figure 5.2.9: Utilization relative to optimal scheme where the parameters p and λ are known. For each value of p the upper curve is the optimal scheme and the lower curve is the load only scheme. Note that the relative utilization's natural scale is 0% to 100%, while our scale runs from 20% to 70%.

sources is given in [GKK95].

5.2.5   Adaptive CAC based on measurements

In this section we describe a method for performing CAC based on instantaneous measurements of the total mean rate on the link. We assume here that the effective bandwidth of each type of source is essentially the product of a coefficient and the source mean rate. Each source has a mean rate which is a random variable with a distribution, common for its type and known in advance. The variance of the measured mean for a source of a certain type is also assumed to be known. These background distributions can later be changed within this general method but here we shall suggest some specific variants which might be typical.

We assume we have a limited number of types of source. Each type of source is characterized by its peak rate h_i and the mean rate which, for each individual source, is a random variable Y_{ij} with mean m_i and variance σ_i^2. We also assume that there is a relation between the mean and the variance coming from the Beta distribution with one parameter fixed, i.e., the variance is given by:

$$\sigma^2 = h^2 x(1-x)^2/(3-x), \tag{5.2.36}$$

where $x = m/h$.

The sources with given peak and mean rates are in principal of general VBR type but to identify the measurement accuracy we start with the assumption that sources are of on/off type with exponentially distributed on and off durations. The asymptotic variance for each source with given mean and peak rates is then: $v(t) = 2tmhd$ where d is the mean burst duration in the on/off model. An advantage with this model is that the accuracy of the measurement is expressed just in terms of the mean and peak rates and the measurement length T given in mean burst durations, i.e., $T = t/d$. The variance of the measured mean is then:

$$v = t^{-2}v(t) = 2mh/T \tag{5.2.37}$$

What we measure is then the *mean of the total instantaneous rate* of the link X during the latest period of length T and, given that value of X, we draw conclusions on the effective load of the link to decide whether another call of a certain type can be accepted.

The *effective bandwidth* formula we use here is as follows (see Section 5.2.2):

$$d = 1.2m + 60V/c \tag{5.2.38}$$

where m and V are the mean and variance of the instantaneous rate, given the sources we have. The formula has a more general form, but for the cases we consider for the ideas in this section we must have a limited ratio between link and peak rates (30, say) and a certain size of the burstiness factor to make the statistical multiplexing interesting and for such cases the above expression will do. The coefficients correspond to a CLR of 10^{-9} and we assume a case with small buffers. The only significant parameters are thus the mean, the variance and the link rate c. For an individual on/off source with peak rate h, the equivalent bandwidth d is simply $d = 1.2m + 60m(h-m)/c$. Now, in our approach, the mean rate for an individual source of type i is a random variable Y_{ij} and the effective bandwidth is thus also a random variable

$$d(Y_{ij}) = Y_{ij}(1.2 + 60(h_i - m_i)/c) = \alpha_i Y_{ij}. \qquad (5.2.39)$$

Also, the total effective load of the link is a random variable g, which is a function of all the Y_i's where $Y_i = \sum Y_{ij}$ and thus, with two types of source, this effective load will be the random variable

$$g = g(Y_1, Y_2) = \alpha_1 Y_1 + \alpha_2 Y_2.$$

Now Y_i is the sum of N_i individual sources of type i where N_i is either a fixed number n_i or, in case we don't want to remember how many sources we actually have of each type, a random variable having a certain distribution (e.g., Poissonian).

We now derive the conditional distribution of the equivalent load variable g, given the measured total mean rate X, and we make the reasonable assumption that, due to the large number of sources, they have a multi-dimensional Normal distribution, i.e., from [Fel71],

$$E[g \mid X] = E\{g\} + (X - E\{X\})\mathrm{Cov}\{g, X\}/\mathrm{Var}\{X\},$$

$$\mathrm{Var}[g \mid X] = \mathrm{Var}\{g\} - \mathrm{Cov}\{g, X\}^2/\mathrm{Var}\{X\}$$

It is then necessary to calculate the means and variances of g and X and their covariance. The conditional mean above is the base for our equivalent load given the measurement X, if the conditional variance is small enough. Otherwise, we have to modify the load a bit since, if the conditional variance is too large, the whole method breaks down. Now we start with the cases where the numbers of sources of each type n_i are known. We have,

$$E\{X\} = n_1 m_1 + n_2 m_2,$$

$$\mathrm{Var}\{X\} = n_1 \sigma_1^2 + n_2 \sigma_2^2 + n_1 v_1 + n_2 v_2,$$

$$E\{g\} = \alpha_1 n_1 m_1 + \alpha_2 n_2 m_2,$$

$$\mathrm{Var}\{g\} = n_1 \alpha_1^2 \sigma_1^2 + n_2 \alpha_2^2 \sigma_2^2,$$

$$\mathrm{Cov}\{g, X\} = n_1 \alpha_1 \sigma_1^2 + n_2 \alpha_2 \sigma_2^2,$$

The extension to more than two types of sources is obvious. The conditional variance now has two terms where the first one will vanish in homogeneous cases but otherwise cannot be reduced below a certain level even with very large values of T :

$$\text{Var}\,[\,g\mid X\,] = \left(n_1 n_2 \sigma_1^2 \sigma_2^2 (\alpha_1 - \alpha_2)^2 + (n_1 v_1 + n_2 v_2)\text{Var}\,\{g\}\right)/\text{Var}\,\{X\}\,. \tag{5.2.40}$$

To make the CAC even simpler one could drop the information on the actual numbers of calls of each type and assume that these are random variables of known distribution, e.g., Poissonian with means λ_i. The formulae above can then be used just by replacing n_i by λ_i and σ_i^2 by $\sigma_i^2 + m_i^2$.

Our problem now is that the effective load is a random variable. Suppose that the effective load with respect to $B = 10^{-8}$, called L_8, is exactly the link capacity c (around 120). Then for the corresponding value L_9 (10^{-9} for the same mix) we would have, approximately (see Section 5.2.2): $L_9 - L_8 = (c - 1.2m)/8$ and $L_9 - L_{9-s} = s(c - 1.2m)/8$. Put $b = 10^9 B$ and note that $\ln b = 2.3(L_9 - c)/(L_9 - L_8)$ where $L_9 - c \sim N(\mu, a^2)$, i.e., b has a lognormal distribution : $\ln b \sim N(k\mu, k^2 a^2)$. The choice of μ such that $E\{b\} = 1$ is thus (see, e.g., [Lin68, page 176])

$$-\mu = ka^2/2 = 2.3a^2/2(L_9 - L_8) = 9.2a^2/(c - 1.2m). \tag{5.2.41}$$

In the above expressions a^2 is the previously obtained conditional variance. This is just the quantity we need to modify the conditional mean expression in the effective load.

Thus, in the CAC decision for a source of type i, it is necessary to check if $c - E[\,g\mid X\,] + \mu$ is not less than $\alpha_i m_i$ (or more safely, h_i).

Several examples with, e.g., $c = 120$ (representing cases of that magnitude) have shown that in homogeneous cases with peak rates up to 4 (Mbit/s), a value of T equal to 1000 is sufficient to get the best possible efficiency of the method without the modification, even when the exact number of sources is unknown. For essentially lower peak rates, e.g., 1 or 2, T-values between 100 and 1000 could even be considered. Thus, for homogeneous cases the method works efficiently even without the μ-modification if only the length of the measurement period T is large enough.

Our more complex example here is thus a mix with random numbers of sources of two types: $h_1 = 1$, $m_1 = 0.4$, $\lambda_1 = 100$ and $h_2 = 4$, $m_2 = 0.4$, $\lambda_2 = 50$. We set $T = 1000$, i.e., $\alpha_1 = 1.5$ and $\alpha_2 = 3$. We derive, $E[\,g\mid X\,] = 120 + 2.36(X - 60)$ and $\text{Var}[\,g\mid X\,] = 29.5$.

If the conditional variance is as high as this, the modification is needed, i.e., the effective load is: $120 + 5.65 + 2.36(X - 60) = 120 + 2.36(X - 57.6)$. In other words, we can only accept new calls if $X < 57.6$.

The most criticable points in the above approach are that we trust the proportions in the arrival intensities of the two types of calls and that we assume the distribution of the measurement accuracy is known. In the heterogeneous cases it would thus be better to control the numbers of accepted calls of each type.

5.3   CAC for rate sharing

While many proposals exist in the literature concerning CAC methods, it is notable that the vast majority apply uniquely to rate envelope (or "buffer-less") multiplexing. Since, for rate sharing (i.e., multiplexing using a large burst scale buffer), performance depends on many, often complex, traffic characteristics, it proves difficult to devise admission procedures which are both simple and efficient. In the present section we consider three possible approaches. We first briefly consider in Section 5.3.1 the popular notion of effective bandwidth based, in the present case of large buffers, on the asymptotic behaviour of the queue length distribution. A possible framework for traffic control for high speed data connections is then presented in Section 5.3.2. Finally, noting that the simple leaky buckets used in the definition of current transfer capabilities are often inadequate as traffic descriptors, we consider in Section 5.3.3 the possibility of policing traffic parameters describing the burst structure of a traffic stream.

5.3.1   Effective bandwidth

As indicated in Section 4.1, to predict the performance of multiplexers operated using rate sharing, it is necessary to know a wide range of traffic characteristics. For the particular case of on/off sources, for example, required characteristics would include the distributions of burst and silence lengths (or at least the first moments of these distributions) and some parameters describing the correlation structure of the arrival process, notably in the presence of long range dependence. However, it has been proposed in the literature to use a notion of effective bandwidth similar to that defined for rate envelope multiplexing (see Sections 5.2.1 and 5.2.2) so that the CAC function is simply to compare the sum of effective bandwidths to the nominal capacity of network links. In the case of rate sharing, effective bandwidth is determined from the asymptotic slope of the complementary queue length distribution, as explained in Part III, Section 17.1.1 (see [GAN91, GH91a, EM93, KWC93]).

Two essential characteristics for a definition of an effective bandwidth for CAC purposes are that:

- the effective bandwidth of a traffic stream is independent of the other streams with which it is mixed;

- the sum of the effective bandwidths of two independent traffic streams is be equal to the effective bandwidth of their superposition; this is the additivity property.

In the present case of multiplexing with a large buffer, the above characterics are satisfied for an effective bandwidth determined by the asymptotic slope of the queue length distribution, as described in Section 17.1.1.

Assume for a certain traffic stream i that the complementary distribution of the queue length X in an infinite buffered multiplexer of rate c is approximately:

$$\Pr\{X > x\} \approx e^{-\zeta_i(c)x},$$

where $\zeta_i(\cdot)$ is determined by the statistical characteristics of the traffic stream. To achieve an overflow probability less than ϵ with a buffer of size B, it would be necessary to provide a bandwidth:

$$c_i = \zeta_i^{-1}(-\log \epsilon / B).$$

As shown in Section 17.1.1 , c_i so defined constitutes an effective bandwidth for stream i: if a multiplexer rate is not less than the sum of the c_i of all multiplexed streams, then the probability of buffer saturation will be less than ϵ. There is considerable interest in this approach because the function $\zeta(\cdot)$ is often relatively easy to derive, even for models where the determination of the exact queue length distribution is to complex to be tractable. However, the practical usefulness of this possible CAC procedure depends on the accuracy of (5.3.1) as an approximation.

Note first that, although (5.3.1) is asymptotically correct for Markovian sources, the asymptote slope is not necessarily attained until the buffer is extremely large. Further, the intercept of the asymptote with the probability axis can be quite different to the assumed value of 1, i.e., we can have:

$$\Pr\{X > x\} \sim \eta e^{-\zeta(c)x}$$

with $\eta \ll 1$ or $\eta \gg 1$ (see sketches in Figure 4.1.3). The limitation of this definition of effective bandwidth in this context are discussed in [CLW94].

Secondly, in the case of long range dependent traffic, the asymptotic behaviour of the queue length distribution is not exponential: the tail probabilities decay much more slowly, as a power law or a Weibull distribution (see Part III, Section 17.4.

5.3.2 Control framework for high speed bursty traffic

In this section we envisage a control framework based on SCR and IBT parameters where it is assumed source peak rate is high so that large multiplexer buffers must be provided to absorb burst scale congestion. A typical application for which such a control framework might be appropriate is LAN interconnection. It is difficult in this case to predict network performance since this depends on many traffic characteristics which are very imperfectly captured in the simple SCR parameter set. This in turn makes it difficult to make firm performance guarantees and to devise satisfactory CAC mechanisms.

We consider here a conceptually simple CAC procedure based on providing guaranteed throughput but only "best effort" quality of service with respect to delays. Throughput equal to the sustainable rate SCR is guaranteed by reserving this amount of bandwidth on every link of the connection path. It is further assumed that multiplexer buffers will be large enough to avoid cell loss when connections are shaped to comply with the declared burst tolerance parameters. Exactly how buffers should be dimensioned is an issue addressed below. We first consider how the choice of SCR and IBT parameters affects the network access delays recognizing that these frequently constitute the most significant component of end to end performance [Lel89].

Choosing leaky bucket parameters

The standardized traffic parameters SCR and IBT are defined with respect to the GCRA. We assume a user will implement a leaky bucket on his traffic stream in order to ensure that no cells will be discarded by this GCRA. In this section the cell stream to be controlled is modelled as a variable rate fluid and the leaky bucket is considered as a storage system. The leak rate is denoted by r and "token pool" size by b. We assume that cells arriving when the token pool is depleted will wait in an infinite capacity buffer (see Figure 5.3.1). Note that this buffer may in reality correspond to the action of higher layer protocols which stop forwarding data when a real *finite* buffer approaches saturation.

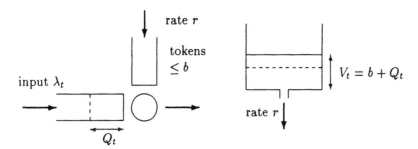

Figure 5.3.1: Buffered leaky bucket and equivalent storage system.

Let the incoming data rate be λ_t. The leaky bucket operates as follows:

- when the token pool level lies between 0 and b it changes at rate $(\lambda_t - r)$;

- when the token pool is empty, incoming data is stored in an infinite capacity buffer which is served (input to the network) at rate r;

- when the token pool level does not increase beyond b.

In the following we neglect discrepancies between the discrete cell based nature of traffic and its fluid representation and assume cell loss in GCRA(T, τ) will be avoided on setting $r = 1/T$ and $b = 1 + \tau/T$. The appropriate choice of leaky bucket parameters (i.e., the SCR parameter set) depends on traffic characteristics and the required quality of service determined here by network access delays.

We assume source traffic is of on/off type where the lengths of successive bursts and silences are statistically independent. This allows us to use the queueing model results derived in Section ref8.4.6. Rather than expressing b in cells we choose the more natural unit in the considered fluid model of the expected volume of one burst. We characterize performance by the function $\theta(u)$ giving the expected rate at which a burst is admitted to the network as a function of burst duration u [RBC93]. Writing $\overline{T_u}$ for the expected delay of the last cell of a burst of length u, expected throughput is:

$$\theta(u) = \frac{u}{(u + \overline{T_u})}.$$

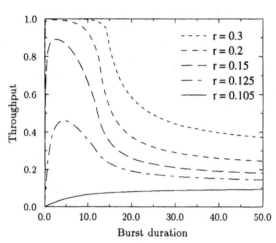

Figure 5.3.2: Throughput against burst duration: b=10, burstiness=10.

Figure 5.3.2 shows the form of $\theta(u)$ for a source with exponentially distributed bursts and silences of means 1 and 9, respectively, as the leak rate r varies; the token pool b remains fixed at 10 (average bursts). The leaky bucket load (mean source rate/r) varies from 0.95 to 0.33 as r increases from 0.105 to 0.3. Source burstiness (peak rate / mean rate) is 10.

The behaviour of $\theta(u)$ is typified by the curve for $r = 0.125$:

- for infinitessimally small u, the throughput is close to zero since the considered burst can arrive when the token pool is empty ($\overline{T_0} > 0$);

- $\theta(u)$ then increases rapidly, however, since $\overline{T_u}$ is small relative to u (short bursts have a high probability of finding a sufficient number of available tokens on arrival);

- as u increases, the probability of finding enough tokens diminishes and the latter part of the burst is constrained to enter the network at rate r causing $\theta(u)$ to decrease;

- the decrease is accentuated as soon as u exceeds $b/(1-r)$ when the rate of the last part of all bursts is systematically reduced;

- throughput tends to r as u increases indefinitely.

The relative significance of the above effects depends on the value of r: for high load ($r = 0.105$) the leaky bucket delays virtually all bursts and effectively acts as a rate reduction mechanism; to achieve near transparency (i.e., significant rate reduction only for exceptionally long bursts), it is necessary to accept a low leaky bucket load (e.g., $r = 0.2$).

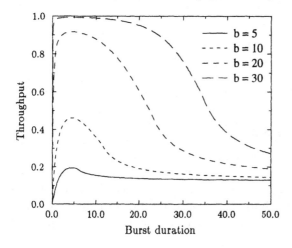

Figure 5.3.3: Throughput against burst duration: $r=0.125$, burstiness=10.

The effect of varying b for fixed r is illustrated in Figure 5.3.3. To attain near transparency in this example with $r=0.125$ it is necessary to equip the leaky bucket with $b=30$ tokens. In realizing the objective of attaining peak rate access for small bursts, there is an obvious trade-off between r and b. We have either high b (30) and low r (0.125), implying large multiplexer memory requirements, or low b (10) and high r (0.2), implying a greater network bandwidth allocation.

The performance of the leaky bucket also depends significantly on the burst and silence length distributions. In Figure 5.3.4 we show how the characteristic $\theta(u)$ varies for different burst length distributions with exponential silences. The chosen distributions are the Erlang-k and a balanced hyperexponential which can produce a range of coefficients of variation below and above 1. The figure illustrates that performance is rapidly downgraded as the variability of burst lengths increases. Almost the same results are obtained for non-exponential silences and exponential bursts and the effects of variability in the two distributions are cumulative.

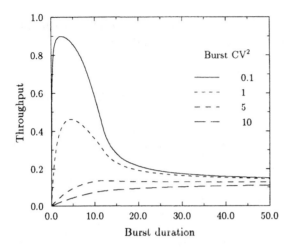

Figure 5.3.4: Throughput against burst duration: $b=10$, $r=0.125$, burstiness$=10$.

Multiplexer buffer dimensioning

As stated at the beginning of this section, the assumed QoS objective is to guarantee throughput (r) with negligible cell loss. It is not possible to guarantee delay criteria such as mean delay or delay quantiles since these depend on traffic characteristics (e.g., burst and silence duration distributions) which cannot be controlled by a simple leaky bucket. To avoid loss we envisage three possibilities:

(i) reserve a quantity of memory equal to the leaky bucket token pool size b for each source;

(ii) determine overall memory requirements assuming worst case traffic compatible with leaky bucket parameters;

(iii) determine overall memory requirements accounting for real traffic characteristics.

The first option is a conservative approach which guarantees zero cell loss. However, the memory will only be used in the extremely unlikely event that all multiplexed sources simultaneously begin to emit a maximum sized burst.

The second possibility assumes the sources are independent but that all emit worst case traffic defined here to mean a succession of maximum sized bursts separated by an interval just long enough to replenish the token pool (see, however, [Dos94] for a discussion on the nature of worst case traffic). In this case, it is possible to evaluate buffer requirements so that the probability of saturation is less than some small number ϵ assuming the phase of the periodic worst case traffic is random. The resulting queueing system is considered in Part III, Section 17.2. In Figure 5.3.5 we show the memory required for a multiplexer handling N homogeneous sources with rate parameter $r = 0.04$

and a multiplexer rate $c = 1.5$, where r and c are expressed in units of the source peak rate. The assumed saturation probability is $\epsilon = 10^{-10}$. The figure expresses the memory requirement per source in multiples of the leaky bucket memory parameter b as a function of the multiplexer load, Nr/c. In this example, accounting for source independence allows a gain of nearly half the memory required for zero loss. In other words, if this second possibility were assumed, the buffer constraint on CAC would allow nearly twice as many connections, or again, the leaky bucket memory parameter for admitted sources could be twice as big.

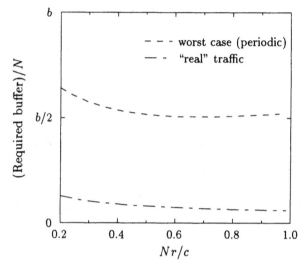

Figure 5.3.5: Comparison between memory allocation schemes: $b=10$, $r=0.04$, $c=1.5$, $\epsilon = 10^{-10}$, mean rate $= 0.01$.

Worst case traffic would not normally be generated by any source. Indeed, in the previous section it was shown that to achieve transparent access, an on/off source with random burst and silence lengths would need to choose a value of b equivalent to around 10 mean bursts and a rate parameter r considerably larger than the actual source mean rate. To appreciate the potential overprovisioning implied by the worst case assumption, consider a hypothetical on/off source with exponentially distrubuted burst and silence durations and a mean rate equal to 0.01 times the peak rate. Fixing $b=10$ (bursts), a rate parameter of $r = 0.04$ is necessary to assure transparent access. Given that the access is transparent, the traffic offered to a network multiplexer is a superposition of exponentially distributed on/off sources and it is possible to predict buffer saturation probabilities using the fluid queue models of Part III, Sections 17.1 and 17.2. The lower line in Figure 5.3.5 corresponds to this evaluation. The memory needed here is between 10% and 20% of the worst case requirement. This difference is to be explained by the shorter burst length (the mean is one tenth of the maximum burst length) and the lower multiplexer load (mean load is one quarter of the assumed worst case load).

The above discussion points to possible inefficiencies of performing CAC using worst case assumptions. It remains impossible, however, to predict real source traffic characteristics if all that is known are the leaky bucket parameters. Furthermore, traffic intensity is known to fluctuate widely over multiple time scales (see Section 1.1.6) so that a given leaky bucket may be transparent in some periods and act as a severe shaper in others. In the following section we have investigated buffer requirements due to given leaky bucket dimensions as a function of traffic intensity varies for a particular type of input traffic.

Impact of leaky bucket shaping

Consider a multiplexer of rate c fed by N homogeneous traffic sources each controlled by a leaky bucket (r, b) where CAC enforces the constraint $Nr < c$. We investigate required buffer size to ensure a saturation probability less than ϵ.

Whatever the nature of the offered traffic, we know that no buffer is required in the two extremes where the mean rate tends to zero or exceeds r. In the latter case the saturated leaky buckets each emit a constant rate stream which simply flows through the multiplexer since we have $c > Nr$. We deduce that the buffer requirement attains a maximum for some load ρ_{max}, $0 < \rho_{max} < 1$. In the following, we seek to characterize this maximum as a function of b and N.

First note that there is a limit leaky bucket load ρ^* such that, for smaller loads, the leaky bucket remains transparent in that the probability of access delay is less than ϵ. In this case, the rate r may be interpreted as an equivalent bandwidth in the sense defined in Section 5.3.1 for a multiplexer with buffer size b and target saturation probability ϵ. In other words, the reservation of rate r for each source is sufficient to guarantee negligible loss in a multiplexer buffer of size b for whatever number of sources. The maximum buffer requirement arises for a load greater than ρ^*.

To accurately characterize the traffic offered to the multiplexer, it is necessary to take account of the smoothing effect of the leaky bucket. Let $A(s, t)$ be the amount of data offered to the leaky bucket in the interval (s, t) and let $D(s, t)$ be the corresponding output. We then have the relation:

$$D(s, t) = (V_s - b)^+ + A(s, t) - (V_t - b)^+ \qquad (5.3.1)$$

The study of $D(s, t)$ for most input processes is very difficult due to the mutual dependence of the three terms in (5.3.1). We have in fact only been able to study the distribution of $D(s, t)$ assuming a simple diffusion input process:

$$A(s, t) = mt + \sigma W(s, t) \qquad (5.3.2)$$

where m is the mean input rate and $W(s,t)$ is the work generated by a standard Brownian motion.

This model may be considered as a heavy traffic approximation for a Poisson input process with exponentially distributed bursts (assumed to arrive instantaneously) of mean size $\sigma^2/(2m)$. This assimilation allows us to express the leaky bucket parameter b in terms of an "average burst size".

The analysis of the multiplexer queue fed by the above leaky bucket controlled sources is performed with large deviation techniques applied to the Beneš method (see [BRS95]).

Figure 5.3.6 shows how the buffer requirement varies with the input load for a saturation probability $\epsilon = 10^{-9}$ and a superposition of $N = 10$ sources. For these results, we assume $c = Nr$ so that the leaky bucket load and multiplexer load are the same. The figure shows that the maximizing load $\rho_m ax$ increases with b. The line labelled "transparent LB" corresponds to the buffer size which would be required if the leaky buckets did not in fact smooth the input (i.e., the limit output as $b \to \infty$). It may be noted that the transparent asumption remains valid for loads considerably higher than the limit ρ^* corresponding to a multiplex buffer size of b. In fact, the leaky bucket proves transparent for the network as long as the access delay probability is not greater than 10^{-3}.

Figure 5.3.7 shows the impact of the number N of multiplexed sources for $b = 2$. The load for which the effect of the leaky bucket cannot be seen (i.e., for which the "transparent LB" line can be used for dimensioning) increases with N to 0.5 corresponding then to an access delay probability of 2×10^{-2}. The notions of leaky bucket transparency from network and customer points of view are thus clearly different.

Numerical studies on the maximum buffer size as a function of b and N lead to the following empirical relation for the buffer size B:

$$B = g_\epsilon b \log N \qquad (5.3.3)$$

where g_ϵ is a factor depending on the required saturation probability. For the considered traffic model we have $g_\epsilon \approx 6$ for $\epsilon = 10^{-9}$ and $g_\epsilon \approx 3$ for $\epsilon = 10^{-3}$. Simulation of a system with on/off input processes suggest that relation (5.3.3) remains valid under more realistic traffic assumptions than those considered here [BRS95].

End to end delays

We consider the end to end performance of a data connection assuming a bandwidth equal to the leaky bucket rate r is allocated on every link of its path. (This means simply that connections are accepted until the sum of their leaky bucket rates exceeds multiplex bandwidth; active connections have access to the entire bandwidth). It is not a question here of using

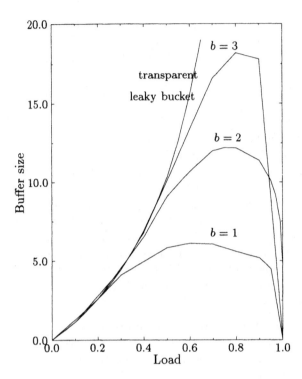

Figure 5.3.6: Multiplex buffer size versus load ($N = 10$ and varying b).

queueing network models to evaluate the performance of a succession of multiplexing stages. We aim simply to evaluate the relative significance of access and network delays and to investigate the scope for performance guarantees concerning response times.

From the leaky bucket performance evaluation described in Section ref8.4.6, it is straightforward to determine the mean access delay of an arbitrary burst (by deconditioning $\overline{T_u}$ with respect to the burst length distribution). Similarly, integration of the complementary queue length distribution of the multiplexer queue with "worst case" and "real" traffic as defined in Section 5.3.2 above readily provides an estimate of the mean delay of an arbitrary *cell* in the multiplex queue. (This should only slightly underestimate *burst* delay when the number of multiplexed sources is large).

Table 5.3.1 presents the mean delays calculated for four source, leaky bucket and multiplex parameter sets. While we have arbitrarily assumed a mean burst length of 1 Mbit, delay for any other value of x Mbits can be deduced simply by multiplying the figures of the table by x.

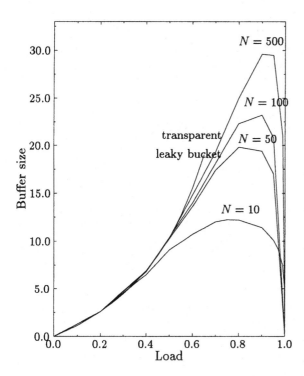

Figure 5.3.7: Multiplex buffer size versus load ($b = 2$ and varying N).

The access delay is very small. This is normal since we have designed the leaky bucket with this objective. The most striking result is the very wide discrepancy between "worst case" and "real" traffic scenarios. The worst case traffic produces a mean delay orders of magnitude greater than the "real" traffic and roughly equivalent to the transmission time at the guaranteed rate r. This illustrates that the worst case scenario is extremely pessimistic. If this were to prevail it would be better to systematically reduce the rate of user traffic at the access ($b=0$). On the other hand, the results for access delays and "real" multiplex delays show that peak rate transmission can be feasible with delays which are small compared to transmission time. It is also relevant to compare these delays with a propagation time of about 5 ms per 1000 km.

There is clearly a dilemma concerning the specification of network response time performance. It is theoretically impossible to guarantee response times except (to a certain extent) by dimensioning to a worst case scenario. The realized response time will then be very much lower than what can be

Table 5.3.1: Mean access and multiplex delays.

source	peak rate (Mbit/s)	100		100	
	mean rate (Mbit/s)	10		1	
	burst length (Mbit)	1		1	
leaky bucket	b (Mbits)	10		10	
	r (Mbit/s)	20		4	
multiplex	rate (Mbit/s)	150	600	150	600
number of sources		7	30	37	150
rate r transmission (ms)		50	50	250	250
peak rate transmission (ms)		10	10	10	10
access delay (ms)		.07	.07	.12	.12
"worst" mux delay (ms)		76.2	43.6	212.8	122.2
"real" mux delay (ms)		3.38	.14	1.42	.01

guaranteed. From a commercial point of view, it seems very unlikely that a network provider will tell its customers it is providing them a service 100 times worse than is actually the case!

5.3.3   Controlling the burst structure

In rate sharing multiplexing the burst structure of the arrival process plays a fundamental role in network performance. In order to prevent congestion situations this burst structure must be controlled. The generic cell rate algorithm is only able to enforce the maximum burst size, however this parameter is not the most representative from the point of view of network performance in rate sharing multiplexing. This is reflected in the fact that the worst case source (when all the bursts emitted by the source are of the maximum size) for a given value of SCR and IBT can be much worse than a typical source with random burst sizes.

In Section 13.1 in Part III, alternative traffic parameters are proposed to characterize the burst structure of the arrival process: the second moment of the number of arrivals in a time interval and the probability of exceeding a certain queue length in a queue fed exclusively by the arrival process. In this section enforcement algorithms for controlling such parameters are presented.

Generalized Leaky Bucket

The proposed enforcement algorithms are based on a generalization of the leaky bucket algorithm [And94]. The leaky bucket ensures that:

$$r(t_1, t_2] \leq G + \frac{L}{t_2 - t_1} \quad \forall\, t_1, t_2 \text{ with } t_2 > t_1$$

where $r(t_1, t_2]$ is the average cell rate in $(t_1, t_2]$, (i.e., $r(t_1, t_2] = \nu(t_1, t_2]/(t_2 - t_1)$ where $\nu(t_1, t_2]$ is the number of arrivals in $(t_1, t_2]$), G is the leak rate which works as a long term bound on the average rate and L is the bucket size which acts as a short-term tolerance for this bound.

The leaky bucket can be generalized to control any other parameter whose value is realised in a time interval. For time averaged parameters, i.e., any function $v(t_1, t_2]$ of the arrival process which satisfies:

$$v(t_1, t_3] = v(t_1, t_2] \frac{t_2 - t_1}{t_3 - t_1} + v(t_2, t_3] \frac{t_3 - t_2}{t_3 - t_1} \quad \text{with } t_1 < t_2 < t_3,$$

the generalized leaky bucket will ensure

$$v(t_1, t_2] \leq G + L/(t_2 - t_1).$$

For parameters measured by averaging over the number of arrivals, i.e., any function $\phi(t_1, t_2]$ which satisfies

$$\phi(t_1, t_3] = \phi(t_1, t_2] \frac{\nu(t_1, t_2]}{\nu(t_1, t_3]} + \phi(t_2, t_3] \frac{\nu(t_2, t_3]}{\nu(t_1, t_3]} \quad \text{with } t_1 < t_2 < t_3,$$

the generalized leaky bucket will ensure

$$\phi(t_1, t_2] \leq G + L/\nu(t_1, t_2].$$

The parameter G is a long term bound for the controlled parameter and L its short term tolerance.

The generalized leaky bucket can be implemented in a similar way to the standard leaky bucket. A generalized leaky bucket for time averages consists of a counter which is incremented during a time interval $(t, t + \Delta t]$ by $v(t, t + \Delta t]\Delta t$ and decremented by $G\Delta t$. The counter for parameters averaged over a number of arrivals is incremented by $\phi(t, t + \Delta t]\nu(t, t + \Delta t]$ and decremented by $G\nu(t, t + \Delta t]$. The counter does not decrease below zero and cells are rejected whenever it reaches the threshold L.

To simplify implementation the counter is only updated at cell arrivals. Let τ_i denote the arrival time of the i-th cell and let b_i denote the bucket content at τ_i just before the arrival. The following operations are performed at τ_i:

- $D_i = (\tau_i - \tau_{i-1})G$ for parameters averaged over time or $D_i = G$ for parameters averaged over the number of arrivals

- $b_i = \max(0, b_{i-1} - D_i)$

- $I_i = (\tau_i - \tau_{i-1})v(\tau_{i-1}, \tau_i]$ for parameters averaged over time or $I_i = \phi(\tau_{i-1}, \tau_i]$ for parameters averaged over the number of arrivals

- $b_{aux} = b_i + I_i$

- If $b_{aux} > L$ the cell is rejected, otherwise $b_i = b_{aux}$ and the cell is accepted.

Controlling the second moment of the number of arrivals in a time interval

In Section 13.1.2 in Part III it is shown that, if selected at the appropriate time scale t, $E\{v(t)^2\}$ has an essential impact on performance. A generalized leaky bucket can be used to enforce $E\{v(t)^2\}$.

Let the time axis be divided into consecutive non overlapping intervals, P_j, of duration t, and let v_j denote the number of cell arrivals in the interval P_j. The function $v(t_1, t_2]$ can be defined as:

$$v(t_1, t_2] = \frac{1}{t_2 - t_1} \sum_{\forall j | P_j \subset (t_1, t_2]} v_j^2.$$

The bucket content is decreased by G units per time unit so that $E\{v(t)^2\}/t$ is long term bounded by G. The decrement for cell i is given by $D_i = (\tau_{i-1} - \tau_i) G$. The total increment corresponding to interval P_j will be $tv(P_j) = v_j^2$ which can be decomposed as a sum of terms for each arrival in P_j as

$$v_j^2 = \sum_{i=1}^{n_j} (2i - 1).$$

Therefore, the increment $I_{j,i}$ for the i-th cell of P_j will be given by:

$$I_{j,i} = (2i - 1) = I_{j,i-1} + 2 \quad \text{with } I_{j,1} = 1.$$

To compute $I_{j,i}$ we thus require an addition and the storage in a register of the increment value of the previous cell.

Let τ_i^0 be the starting time of the interval in which cell i arrives. At each arrival it is necessary to check if the interval in which the last cell arrived has finished and the arriving cell thus belongs to a new interval. If the new cell arrives outside the time interval to which the previous cell belonged, it will be necessary to calculate the new value of τ_i^0. If the successive intervals are consecutive, τ_i^0 is given by:

$$\tau_i^0 = \tau_i - \text{rem}((\tau_i - \tau_{i-1}^0)/t)$$

where $\text{rem}(x/y)$ represents the remainder of the division.

The division $(\tau_i - \tau_{i-1}^0)/t$ is very costly to implement unless t is a power of 2. This means that t has to be approximated to the nearest power of 2. Another equivalent approach for obtaining τ_i^0 without bias is to begin the interval at a random (or pseudo-random) instant in $(\tau_i - t, \tau_i]$. To avoid overlapping of intervals, in the case that $\tau_i - t < \tau_{i-1}^0 + t$, the value of τ_i^0 will be $\tau_i^0 = \tau_{i-1}^0 + t$.

A procedure to choose a random instant in $(\tau_i - t, \tau_i]$ could be to have a counter, s, modulo t/ε (ε is the time unit, e.g., the slot duration) incremented with each new interval. The value $(s+1)\varepsilon$ will be subtracted from τ_i in order to obtain τ_i^0. To reduce dependence between successive samples the values $(s+1)\varepsilon$ and $t - s\varepsilon$ could be used alternately in each interval.

Figure 5.3.8 represents the flowchart of the algorithm where τ_p represents the arrival time of the previous cell of the connection.

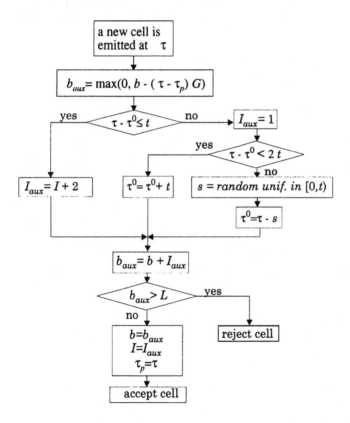

Figure 5.3.8: Algorithm flowchart.

Controlling the burst structure of an arrival process through its queuing behaviour: the burst controller

A second method proposed in Section 13.1.3 in Part III to characterize the traffic generated by an ATM source is based on the queue length distribution of a queue fed exclusively by the source. The burst controller (BC) [And95b] is an algorithm based on the generalized leaky bucket for the enforcement of this distribution.

The burst controller bounds the probability that the work seen by a cell arrival in a fictitious queue with deterministic service rate R and infinite buffer, fed solely by the source, exceeds a certain threshold H. Let $Q_R(H)$ denote this probability.

In this case, the probability we want to bound is averaged over the number of arrivals. Let τ_i denote the arrival time of the i-th cell and let q_i be the amount of work seen at the fictitious queue by the i-th cell. The function $\phi(t_1, t_2]$ will be given by:

$$\phi(t_1, t_2] = \frac{1}{\nu(t_1, t_2]} \sum_{i | \tau_i \in (t_1, t_2]} 1_{q_i < H}$$

The bucket is decreased by G units per arrival, $D_i = G$. For each arrival the increment is computed as $I_i = 1_{q_i < H}$. The amount of work in the fictitious queue is held in a register and is updated at each arrival. Note that no cells actually wait in this queue: it is like a leaky bucket with infinite bucket size. The BC will ensure that $Q_R(H) \leq G$.

In real implementations it will not be necessary to update the bucket counter twice (once for the decrement and once for the increment). Since the decrement is a fixed amount per cell, $I_i + D_i$ will only take two possible values: $1 - G$ or $-G$. These two values can be kept in two registers and the bucket counter update per cell can be performed in one operation instead of two. Note that the queuing system which modulates the increment of the bucket counter has an infinite queue size and, therefore, no cells are rejected by it.

Figure 5.3.9 presents the flowchart of the algorithm. It is slightly more complex than the leaky bucket algorithm but only very simple operations are involved.

Example of multiplexer performance with BC enforcement   As an example of the ability of the BC to enforce the burst structure of an arrival process, we present the effect on queueing performance of the enforcement of two source types with very different characteristics (sources I and II). The objective is to obtain the same queueing performance as that obtained with a smoother reference source (source III). The source characteristics are the following:

- I: on/off source with exponentially distributed on and off periods,

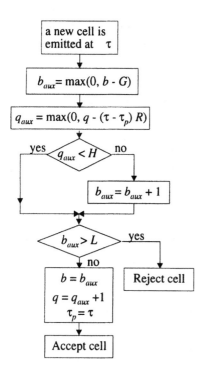

Figure 5.3.9: Flowchart of the BC algorithm.

- II: LAN traffic obtained from Bellcore traces [FL91].

- III: on/off source with exponentially distributed on and off periods. The mean durations of the on and off states are half those of source I. The mean cell rate and peak cell rate are the same for sources I and III.

Intuitively, it is clear that queue lengths obtained when the multiplexer is fed by a homogeneous superposition of type III sources will be half those produced by type I sources. We will consider a multiplexer with $C = 45$ Mb/s and a required QoS given by

$$\Pr\{\text{queue length at cell arrival} \geq M\} < 10^{-3}$$

with $M = 412$ cells. In this scenario, 22 type III sources can be multiplexed. Type I and II sources were enforced to obtain the desired QoS when 22 enforced sources were multiplexed. The configuration is shown in Figure 5.3.10.

According to Section 13.1.3 in Part III the probability of exceeding a queue of M cells in a multiplexer with service rate C fed by N homogeneous sources is related to the probability of exceeding a queue of M/N cells in a queueing system with service rate C/N fed by one source. Thus, we have enforced $Q_{C/22}(M/22)$ of sources I and II to that obtained with source III. Specifically, the BC parameters were set as follows:

- $R = C/22 = 2.038$ Mb/s

- $H = M/22 = 18.73$ cells

- $G = 1.01Q_R(H)$ of source III $= 0.81$.

- $L = 3060$. With this value we obtain a rejection probability in the BC for source III of 10^{-3}

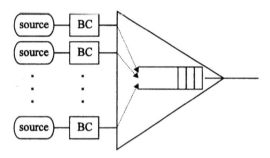

Figure 5.3.10: Multiplexer configuration.

Figure 5.3.11 presents the queue length distribution at cell arrival obtained by feeding the multiplexer with 22 enforced type I sources and Figure 5.3.12 represents the queue length distribution at cell arrival for 22 enforced type II sources. Results for policing and shaping versions of the algorithm are presented. Table 5.3.2 summarizes the probabilities obtained at M. In all cases the traffic control achieved is quite good as is shown by the probability reduction obtained at M.

Table 5.3.2: $\Pr\{$Queue length at cell arrival $\geq M\}$ obtained with the original and enforced sources.

Source type	Original source	Policed source	Shaped source
I	$1.3\ 10^{-2}$	$4.7\ 10^{-4}$	$7.2\ 10^{-4}$
II	$6.1\ 10^{-2}$	$8.9\ 10^{-4}$	$5.3\ 10^{-4}$

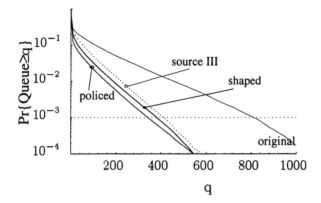

Figure 5.3.11: Queue length distributions with 22 sources of types I, I policed, I shaped and III.

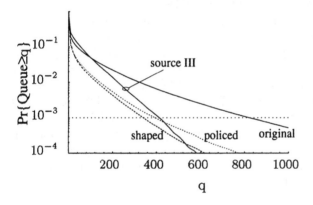

Figure 5.3.12: Queue length distributions with 22 sources of types II, II policed, II shaped and III.

5.4   Relations with effective bandwidths and pricing

In a series of papers [Kel94b, Kel94c, Kel96b, Kel96c], Kelly has argued the relevance of pricing to discussions of traffic characterisation, connection acceptance control and policing. We briefly outline the approach advocated, which is based on the following general definition of effective bandwidth.

5.4.1 Effective bandwidths

Traffic characterisation

Consider a given traffic trace consisting of N packets with arrival times and packet sizes for packet i denoted by t_i and x_i respectively. Write $\mathcal{X} = \{(t_i, x_i);\ i = 1, \ldots, N\}$. A statistical descriptor, known as the *effective bandwidth*, of the trace $\mathcal{X}$ is

$$\alpha(s,t) = \frac{1}{st} \log E\left[e^{sX[\tau, \tau+t]}\right] \qquad (5.4.1)$$

where s is the *space-scale* and t is the *time-scale*. For example, s might be measured in units of bytes^{-1} or cells^{-1} and t might be measured in seconds. Here, $X[\tau, \tau+t]$ is the workload arriving at a resource in the period $[\tau, \tau+t]$ of length t and the expectation is taken over the distribution of such random time periods. Gibbens [Gib96] has computed this statistical desciptor for two well-known traffic traces: namely, the MPEG-1 encoding of the Star Wars movie [Ros95] and for one of the ethernet LAN traces recorded by Bellcore [LTWW94, LWTW94, WTLW95, WTSW95].

Figure 5.4.1 shows the surface, $\alpha(s,t)$ for the MPEG video trace where the space scale is measured in bits^{-1} (using a logarithmic (base 10) axis), the time scale is measured in seconds and $\alpha(s,t)$ is measured in bits per second.

Figure 5.4.2 shows $\log_{10} \alpha(s,t)$ for the Bellcore ethernet LAN trace. At least three regions are discernable in the surface. There is a relatively flat region where s and t are both small. There is a very steep cliff-like region in common with the MPEG-1 surface when t remains very small but s increases. Finally, there is a further comparatively flat region where both s and t are large.

For a fuller discussion on traffic characterisation using effective bandwidths see Gibbens [Gib96].

Link with tariffing

The effective bandwidth of a source is

$$\alpha(s,t) = \frac{1}{st} \log E\left[e^{sX[0,t]}\right] \qquad 0 < s,t < \infty. \qquad (5.4.2)$$

Suppose the arrival process is

$$X[0,t] = \sum_{j=1}^{J} \sum_{i=1}^{n_j} X_{ji}[0,t] \qquad (5.4.3)$$

where $(X_{ji}[0,t])_{ji}$ are independent processes with stationary increments whose distributions may depend upon j but not upon i, and that there is a resource that has to cope with the aggregate arriving stream of work. We interpret n_j

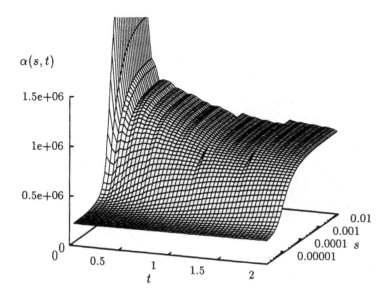

Figure 5.4.1: This figure shows the effective bandwidth surface $\alpha(s,t)$ computed from the Star Wars MPEG-1 video trace. As $s \to 0$, the effective bandwidth approaches the mean rate whereas when s increases the effective bandwidth approaches the peak rate measured over intervals of time t.

as the number of sources of type j, and shall write $\alpha_j(s,t)$ for the effective bandwidth of a source of type j. Thus

$$\alpha(s,t) = \sum_{j=1}^{J} n_j \alpha_j(s,t). \qquad (5.4.4)$$

For a variety of multiplexing models, there is a close relationship between constraints of the form

$$\sum_{j=1}^{J} n_j \alpha_j(s^*,t^*) \leq C^* \qquad (5.4.5)$$

for one or several choices of (s^*, t^*, C^*) and the acceptance region, defined as the set of vectors $(n_1, n_2, \ldots, n_J)$ for which a given performance, described in terms of queueing delay or buffer overflow, can be guaranteed.

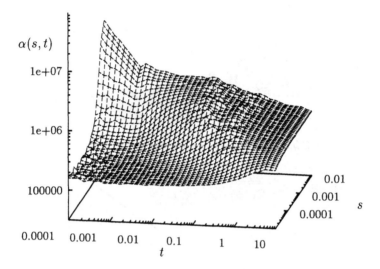

Figure 5.4.2: This figure shows the effective bandwidth surface, $\alpha(s,t)$, for the Bellcore ethernet trace. Each axis has a logarithmic scale (to base 10) with units of s in bytes^{-1}, t in seconds and $\alpha(s,t)$ in bytes per second.

The results of Hui [Hui88, Hui90] establish inequality (5.4.5) as a conservative bound on the non-linear acceptance region for a bufferless model. Kelly [Kel91a] obtained relation (5.4.5) as the linear limiting form of, and as a conservative bound on, the acceptance region for a buffered model with Lévy input. A linear limiting form was established for more general input processes, including the fluid sources studied in detail by Gibbens and Hunt [GH91a] and Elwalid and Mitra [EM93], by Kesidis et al. [KWC93]: this result has a recent generalization by Duffield and O'Connell [DO96]. In several of these models time scales essentially degenerate: the time scale t^* appearing in (5.4.5) approaches zero or infinity. The important recent results of Botvitch and Duffield [BD95], Simonian and Guibert[SG95] and Courcoubetis and Weber [CW96] describe an asymptotic regime where the form (5.4.5) emerges, for finite values of t^*, as a tangent to the limiting acceptance region. Priority models provide further important examples where several constraints of the form (5.4.5) may be needed to approximate the acceptance region. Kelly [Kel96c] provides a recent review, examples and comparison of the above results.

The effective bandwidth of a source depends sensitively upon its statistical characteristics. The source, however, may have difficulty providing such information: for example it may know its peak rate but not its mean rate. One approach might be to measure the effective bandwidth of a connection, perhaps by estimating expression (5.4.2) using an empirical averaging to replace the expectation operator. Is this satisfactory? Suppose a user requests a connection policed by a high peak rate, but then happens to transmit very little traffic over the connection. Then an *a posteriori* estimate of quantity (5.4.2) will be near zero, even though an *a priori* expectation may be much larger, as assessed by either the user or the network. If tariffing and connection acceptance control are primarily concerned with expectations of *future* quality of service, and if sources may be non-ergodic over the relevant time scales, then the distinction matters.

5.4.2   Tariffs and connection acceptance

In this section we describe an approach to tariffing and connection acceptance control mechanisms that can make effective and robust use of both prior declarations and empirical averages. The key idea is the use of prior declarations to choose a linear function that bounds the effective bandwidth (as illustrated in Fig. 5.4.3); tariffs and connection acceptance can then be based upon the relatively simple measurements needed to evaluate this function.

Although an *individual* source may be poorly characterized, certain features of the *aggregate* load on a resource may be known. In this section we assume that the key constraints (5.4.5), and the critical space and time scales appearing in these constraints, have been identified.

Let

$$Z = Ee^{sX[\tau,\tau+t]}, \tag{5.4.6}$$

and rewrite expression (5.4.2) as

$$\alpha(Z) = \frac{1}{st} \log Z, \tag{5.4.7}$$

where the notation now emphasizes the dependence of the effective bandwidth on the summary Z of the statistical characteristics of the source.

Suppose that, before the call's admission, the network requires the user to announce a value z, and then charges for the call an amount $f(z; Z)$ per unit time, where Z is estimated by an empirical averaging. Kelly [Kel94b] discusses tariffs of the form

$$f(z; Z) = a(z) + b(z)Z, \tag{5.4.8}$$

defined as the tangent to the curve $\alpha(Z)$ at the point $Z = z$.

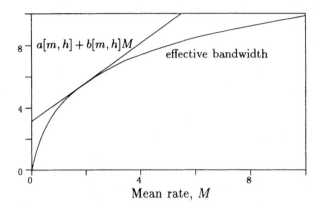

Figure 5.4.3: Implicit pricing of an effective bandwidth. The effective bandwidth is shown as a function of the mean rate, M. The user is free to choose the declaration m, and is then charged an amount $a[m, h]$ per unit time, and an amount $b[m, h]$ per unit volume. The values of $a[m, h]$ and $b[m, h]$ are determined from the tangent at the point $M = m$.

Consider for example the very simple case of an on-off source which produces workload at a constant rate h while in an 'on' state, and produces no workload while in an 'off' state. Suppose the periods spent in 'on' and 'off' states are large: then with M and h the mean and peak of the source, respectively

$$Z = 1 + \frac{M}{h}\left(e^{sth} - 1\right).\tag{5.4.9}$$

(Kelly [Kel94c] provides a numerical illustration of the choice of the parameter st.) If we let z be defined by expression (5.4.9) with M replaced by m then the tariff (5.4.8) may be rewritten as

$$a[m, h] + b[m, h]M,\tag{5.4.10}$$

the tangent to the function

$$\alpha[M, h] = \frac{1}{st}\log\left[1 + \frac{M}{h}\left(e^{sth} - 1\right)\right]$$

at the point $M = m$ (see Figure 5.4.3). Note the very simple interpretation possible for the tariff (5.4.10): the user is free to choose a value m, and then incurs a charge $a[m, h]$ per unit time, and a charge $b[m, h]$ per unit of volume carried.

We now describe how the coefficients defined above can be used as the basis of a simple and effective connection acceptance control.

Suppose that a resource has accepted connections $1, 2, \ldots, I$, and write (a_i, b_i) for the coefficients $(a(z_i), b(z_i))$ describing the choice of a linear bound on the effective bandwidth function (see Figure 5.4.3). Suppose also that the resource measures the load $X_i[\tau, \tau + t]$ produced by connection i over a period of length t (the same length t that appears in the definition (5.4.7)), and let $Y_i = \exp(sX_i[\tau, \tau + t])$. Define the *effective load* on the resource to be

$$\sum_{i=1}^{I}(a_i + b_i Y_i).$$
(5.4.11)

Then a connection acceptance control may be defined as follows. A new request for a connection should be accepted or rejected according as the most recently calculated effective load is below or above a threshold value, with the proviso that if a request is rejected then later requests are also rejected until an existing connection terminates (the backoff mechanism described by [Bea93] and discussed in Section 5.2.4).

Consider the simple case of on-off sources. Let h_i be the fixed and known peak of connection i, write (a_i, b_i) for the coefficients $(a[m_i, h_i], b[m_i, h_i])$ chosen by the user, and let the measured load from connection i be $M_i = X_i[\tau, \tau + t]/t$. Then the effective load on the resource becomes

$$\sum_{i=1}^{I}(a_i + b_i M_i),$$
(5.4.12)

to be compared with a threshold value.

An advantage of the on-off model, both for tariffing and connection acceptance control, is that it bounds other more complex source models. The reader surprised that schemes using only simple load measurements can guarantee strict quality of service requirements should see Gibbens *et al.* [GKK95], where issues of robustness and performance are investigated in some detail.

Suppose that the I sources fall into J classes, with n_j calls of class j, for $j = 1, \ldots, J$, where $\sum_{j=1}^{J} n_j = I$ and suppose also that a_i is negligible. Then the effective load (5.4.12) becomes

$$\sum_{j=1}^{J} b_{(j)} \sum_{i=1}^{n_j} M_i = \sum_{j=1}^{J} b_{(j)} S_j$$
(5.4.13)

where S_j is the total load from calls of class j, and $b_{(j)}$ is the common value of b_i for sources i of class j. The performance of a CAC of this form is discussed in Gibbens *et al* [GKK95, Sec VII].

Both the charging mechanism and the connection acceptance control described above use bounding tangents to the effective bandwidth function. If the same tangents are used for both purposes then the effective load has a natural interpretation as an aggregate charge at the resource over a recent short period. But there is no necessity for identical tangents to be used for charging and for connection acceptance. Thus users choosing a small peak rate might be offered no further choice of tariff, so that for charging purposes the effective bandwidth function is bounded by a single tangent. In contrast, the resource might choose its tangent to the effective bandwidth function at the point where the mean rate is the long-term observed average for traffic with that peak rate.

Distinct effective bandwidth functions might be appropriate for charging and connection acceptance control, since the two areas have quite different time-scales and requirements for precision. Connection acceptance control must use accurately calculated effective bandwidths, based on the buffer sizes, port speeds and other features of current hardware to make decisions on connections as they request connection, otherwise quality of service guarantees on loss rates may be compromised. While charges need to be precisely defined, they influence users' behaviour and software application design over much longer time-scales, where features of hardware may evolve. Thus tariff design might include consideration of the possible effective bandwidth functions appropriate to future hardware and network scale.

Simplicity and predictability of tariffs are other important issues. For example, suppose that users choosing a small peak rate relative to capacity are offered no further choice of tariff, corresponding to the bounding of the effective bandwidth function by a single tangent. Then this tangent might be chosen with predictability in mind: thus the tangent to the effective bandwidth function at the point where the mean rate is equal to the peak rate or a sustainable cell rate has the property that the charge to a user is bounded above in terms of the peak rate or the sustainable cell rate respectively, without overly penalizing users whose mean traffic may be not easily characterized by policing parameters. Other tariffing issues, including possible incentives to users to make several short connections or to split traffic over multiple connections, are discussed in [Kel96a].

6 Weighted fair queueing

Fair queueing originated in the data communications field, initially as a congestion control device preventing ill-behaved users from unduly affecting the service offered to others [Nag87]. In a fluid limit, fair queueing realizes head of line processor sharing, the objective being to share server capacity equally between all customers having packets to transmit. When the fluid service rate is modulated according to weights attributed to the contending traffic streams, we speak of Generalized Processor Sharing (GPS).

Weighted Fair Queueing (WFQ) is a practical implementation of GPS first proposed by Demers, Keshav and Shenker [DKS89] and further discussed and analysed under the name of Packet by packet Generalized Processor Sharing (PGPS) by Parekh and Gallager [PG93a]. In the following we will use PGPS to denote this algorithm reserving WFQ as a generic term. PGPS may be considered as the closest possible approximation to GPS. Its implementation proves somewhat complex, however (see below). A simpler algorithm which is less precise but much easier to realize was proposed independently by Golestani, under the name Self Clocked Fair Queueing (SCFQ) [Gol94], and by Roberts, under the name Virtual Spacing [Rob94]. In fact, this simplified algorithm was first "invented" by Davin in early work on fair queueing performed at MIT [DH90]. A related scheduling discipline is "Virtual Clock" as proposed by Zhang [Zha90] which also provides protection from ill-behaved users and can support diverse throughput guarantees.

6.1 WFQ algorithms

Consider a server of constant rate c handling cells from a number m of traffic streams where each stream has its own queue and cells from a given stream are served in FIFO order. This system is depicted in Figure 6.1.1. Each stream might represent a particular ATM connection or a set of connections grouped together for traffic handling purposes.

6.1.1 Generalized Processor Sharing

Generalized Processor Sharing (GPS) is a fluid approximation to the queueing system of Figure 6.1 where the service rate of any backlogged stream i is proportional to a certain rate parameter ϕ_i [PG93a]. We consider the special case of GPS where, for each stream i, ϕ_i is set to an intrinsic stream rate r_i and admission control is performed to ensure that $\sum r_i \leq c$. The parameter

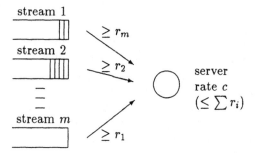

Figure 6.1.1: WFQ system.

r_i can then be interpreted as a minimum bandwidth guarantee. It might be set to an equivalent bandwidth determined to serve a set of VBR connections constituting stream i or to the leak rate of a leaky bucket controlling the burstiness of stream i. Choice of rate parameters is further discussed in Section 6.4.

6.1.2   PGPS

Generalized Processor Sharing is an ideal which cannot be realized in practice because packets are not infinitely divisible. Packet-by-packet GPS [PG93a] is a generalization of the Fair Queueing algorithm proposed by Demers et al. [DKS89]. In this algorithm, variable length packets are served in an order determined so that, as far as possible, they complete service in the order which would prevail if they were in fact served using GPS. In practice, this order can only be determined for the packets present at the end of each service: the server chooses to serve the packet which would finish first in GPS if there were no further arrivals. In the case of ATM cells, it can be shown that for a given arrival process, the difference between the times transmission of a given cell would finish in PGPS and GPS is bounded by one cell transmission time $(1/c)$ [PG93a].

To calculate the hypothetical finishing times necessary for PGPS scheduling, Parekh and Gallager introduce the notion of *Virtual time* [PG93a]. This is a function of real time t which evolves at a rate inversely proportional to the sum of the rate factors r_i of backlogged connections in the corresponding GPS system.

Let $PGPS_i$ be a variable associated with stream i. Applied to ATM cells, the PGPS service algorithm can be stated as follows:

- on a stream i cell arrival

  (i) $PGPS_i \leftarrow \max\{\text{Virtual time}, PGPS_i\} + 1/r_i$
  (ii) time stamp the cell with the value of $PGPS_i$.

- serve cells in increasing order of time stamp.

It is demonstated in [PG93a] that PGPS provides a minimum rate guarantee to the extent that the work accomplished on a given stream queue in any interval differs from that of the equivalent GPS system by at most one cell. The closeness of PGPS to GPS comes at the expense of an algorithm whose implementation would be rather complicated. The complication resides in keeping track of *Virtual time* which requires an evaluation of the number of backlogged streams in the equivalent GPS system.

6.1.3 Virtual clock

A queue scheduling scheme related to PGPS is the Virtual Clock algorithm proposed by Zhang [Zha90]. In this algorithm, the time stamp, denoted VC_i, is calculated with respect to real time t:

- on a stream i cell arrival

  (i) $VC_i \leftarrow \max\{t, VC_i\} + 1/r_i$
  (ii) time stamp the cell with the value of VC_i.

- serve cells in increasing order of time stamp.

This service discipline is simpler than PGPS, it is work conserving, it provides performance bounds similar to those of PGPS (see [HK96]) and it guarantees average throughput for each connection. It has, however, one disadvantage with respect to PGPS, illustrated in the following example:

> a stream emits a message of 1 Mbit at the multiplexer rate of $c = 100$ Mbit/s; although the stream has a rate parameter $r = 1$ Mbit/s, its transmission proceeds at peak rate since it is the only stream with a backlog; however, just before it can emit the last cell of the burst, another stream becomes active and also begins to emit a message at 100 Mbit/s; the last cell of the first burst will have a time stamp roughly equal to the current time $t+1$ second (1 Mbit at 1 Mbit/s). Assuming the second connection also has a rate parameter equal to 1 Mbit/s, this last cell will be further delayed until nearly 1 Mbit of the new message has been transmitted. The message transfer time of the second stream would have been the same if the last cell of the first stream had been emitted without delay. The PGPS scheme would have emitted the last cell just after the first cell of the second burst thus achieving reduced message transfer time.

In general, the Virtual Clock algorithm offers rate guarantees with much less precision than PGPS. As in the above example, realized throughput can be much smaller than r_i for significant periods of time. The Virtual Spacing algorithm described below is as simple to implement as Virtual Clock and achieves nearly the same throughput guarantees as PGPS.

6.1.4   Virtual Spacing/SCFQ

In the following we use the name Virtual Spacing for the special case of SCFQ in the ATM context of constant length packets. The Virtual Spacing algorithm is the same as PGPS except that we replace *Virtual time* by a simpler variable denoted *Spacing time*. *Spacing time* is just equal to the value of the time stamp of the last cell to have been taken from the head of the queue, i.e., in a busy period, the cell currently being transmitted.

Let VS_i be a variable associated with stream i. The algorithm is then:

- on a stream i cell arrival

  (i) $VS_i \leftarrow \max\{\textit{Spacing time}, VS_i\} + 1/r_i$

  (ii) time stamp the cell with the value of VS_i.

- serve cells in increasing order of time stamp.

Since *Spacing time* cannot be greater than the time stamp of any cell already waiting when its value was updated, step 1 of the algorithm implies that the time stamps of the cells of a backlogged stream are arithmetically spaced by the interval $1/r_i$. *Spacing time* only intervenes in the calculation of the time stamp of a cell arriving on a non-backlogged stream and allows the stream to be included at an appropriate place in the transmission schedule.

6.2   Performance guarantees

This section discusses performance of WFQ algorithms in terms of delay bounds and fairness characteristics. We also discuss the adequation of these bounds, notably for dimensioining playback buffers for connections with real time delay constraints.

6.2.1   Leaky bucket controlled streams

Performance bounds can be derived most easily for the idealized GPS service discipline. Suppose stream i is controlled by a leaky bucket of leak rate r_i and token pool size b_i so that the amount of work $\nu_i(s,t)$ arriving in an interval (s,t) satisfies the inequality:

$$\nu_i(s,t) \le b_i + r_i(t-s).$$

Let V_{it} be the amount of work belonging to connection i in the multiplexer queue at time t (i.e., the number of cells in the queue plus the remainder of the cell currently being transmitted). Since the service rate is not less than r_i, we have by Reich's theorem:

$$V_{it} \leq \sup_{s<t} \{\nu_i(s,t) - r_i(s-t)\} \tag{6.2.1}$$

$$\leq b_i. \tag{6.2.2}$$

Now consider the output from the multiplexer in the interval (t,u). Let the amount of connection i work leaving the system in this interval be $\xi_i(t,u)$. This work is composed of the remainder of the cell being transmitted at t, plus the cells completely transmitted in (t,u), plus the transmitted part of the cell currently being transmitted at u. We necessarily have:

$$\xi_i(t,u) \leq V_{it} + \nu_i(t,u)$$

and thus, using (6.2.1):

$$\xi_i(t,u) \leq \sup_{s<t} \{\nu_i(s,t) - r_i(t-s)\} + \nu_i(t,u)$$

$$= \sup_{s<t} \{\nu_i(s,u) - r_i(t-s)\}$$

$$\leq \sup_{s<t} \{r_i(u-s) + b_i - r_i(t-s)\}$$

$$= r_i(u-t) + b_i. \tag{6.2.3}$$

The last inequality shows that the output from the multiplexer conserves the burstiness bound guaranteed at the network input by the leaky bucket. This is an extremely desirable property since it ensures that if all multiplexers on a connection path guarantee the minimum service rate r_i then cell loss can be completely avoided by reserving buffer space equal to b_i. Furthermore, the delay of any cell is bounded by b_i/r_i at each multiplexer, independently of the activity of other connections. In fact, Parekh and Gallager [PG94] have proved the stronger result that, neglecting fixed processing and propagation times, the overall delay through a network of such GPS servers is bounded by b_i/r_i.

It should be noted that (6.2.2) and (6.2.3) derive from the properties of individual connection i alone and do not rely on all connections being controlled by a leaky bucket at the access. If some connection were completely unconstrained at the access, its potential impact on other connections could still be limited by attributing to it a minimal service rate r (≥ 0) and a maximal buffer occupancy. Note finally that to derive the above bounds, we have only used the minimum service rate property of GPS service. The bounds thus apply to any other service discipline offering a minimum rate guarantee including time division multiplexing.

6.2.2   End to end delay

The overall delay bound of b_i/r_i for GPS applies in the fluid regime assuming "cut through" switching at each node, i.e., a cell can begin transmission at a downstream node before it has completed transmission at one or more upstream stages. With the more realistic assumption of store and forward cell switching (i.e., each cell must be received entirely at a given stage before it can be re-transmitted) it is necessary to add the transmission time at each stage to the overall delay. The delay $D_i^{GPS}(K)$ in a network of K stages then satisfies:

$$D_i^{GPS}(K) \le \frac{b_i}{r_i} + \frac{K-1}{r_i}. \qquad (6.2.4)$$

The need to account in the service discipline for the discrete nature of ATM cells introduces supplementary slackness in the above bounds. As previously noted, PGPS achieves the closest approximation to the GPS ideal. It may be shown that the delay $D_i^{PGPS}(K)$ in a network of K stages satisfies:

$$D_i^{PGPS}(K) \le \frac{b_i}{r_i} + \frac{K-1}{r_i} + \sum_{k=1}^{K} \frac{1}{c_k}, \qquad (6.2.5)$$

where c_k is the link rate at stage k [PG94]. Since the c_k are typically much greater than the stream rate r_i, the difference between (6.2.5) and (6.2.4) is small.

A corresponding bound is proved for SCFQ in [Gol95] which translates in the ATM context (i.e., for Virtual Spacing) to:

$$D_i^{VS}(K) \le \frac{b_i}{r_i} + \frac{K}{r_i} + \sum_{k=1}^{K} \frac{m_k - 1}{c_k}, \qquad (6.2.6)$$

where m_k is the number of streams multiplexed at stage k. Note that if all streams have the same rate parameter, the last term in (6.2.6) is approximately equal to the second. The bound then corresponds to two additional cell transmission times at the stream rate per multiplexing stage instead of one for GPS and PGPS.

The usefulness of the above end to end delay bounds depends on the QoS requirements of the traffic stream in question. We distinguish two broad classes of streams depending on whether their connections have strict real time delay constraints or not. For the former, it is essential that the end to end delay be very small and that its variability (jitter) be known, notably for dimensioning the receiver playback buffer.

6.2.3   Jitter and playback buffer dimensioning

Assume the playback buffer operates as a spacer emitting cells with a minimum interval of $1/r_i$. No cells will be lost if the buffer is greater than $r_i D_{max}$

where $D_{\max}$ is the end to end WFQ delay bound. If no cells are lost, the overall delay actually experienced by the cells of a connection in the network and the buffer will be a non-decreasing function of the rank of the cell and will attain a maximum value less than or equal to $D_{\max}$.

Consider the case of a telephone connection handled as an individual stream with rate parameter r_i equivalent (in cells per second) to 64 Kbit/s and a low CDV tolerance parameter (corresponding to $b_i = 2$, say) accounting for initial jitter. Even for GPS service, to account for the maximum possible delay in the playback buffer, it would be necessary to allow 6 ms (i.e., one cell time at 64 Kbit/s) for each multiplexing stage in addition to the initial CDV tolerance. In a large network, this can represent an unacceptably large delay for interactive communications. Typically, the playback buffer would not be dimensioned for the worst possible delay but for a maximum delay equal to a suitably small quantile of the delay distribution. In this respect, PGPS and Virtual Spacing perform better than GPS even though their bounds are greater.

Suppose, for illustration purposes, that all streams are individual 64 Kbit/s CBR connections and all the link bandwidth is allocated ($\sum r_i = c$). In GPS, each cell takes exactly 6 ms to complete transmission. PGPS and Virtual Spacing, on the other hand, behave more like a FIFO queue: cell transmission time is equal to one service time of $1/c$ plus a random delay equal to the waiting time in an $N*D/D/1$ queue (see Part III, Section 15.2). The bounds (6.2.5) and (6.2.6) are determined from a worst case scenario and correspond to the delay of the last cell to be served when all connections emit cells at precisely the same instant. The actual delay of an arbitrary cell is typically very much smaller than this and a playback buffer dimensioned as discussed in Section 3.1.2 would be sufficient. This statement is generally true when multiplexing only streams with low burstiness (i.e., a small value of b_i). Consider now the impact of bursty traffic on the jitter of real time connections.

Since their delay is small, the cells of CBR streams generally arrive to a non-backlogged queue. They consequently have their time stamp derived using the value of virtual time (*Virtual time* or *Spacing time*). Cells of bursty connections, on the other hand, have their time stamps spaced by the reciprocal of their rate parameter. Assume for illustration purposes that all streams have the same rate parameter r but that some are bursty while others are CBR. We consider the operation of Virtual Spacing. It is clear that *Spacing time* is a non-decreasing function of real time since the time stamps of waiting cells, including those added to the queue since the last service instant, are greater than or equal to the current value of *Spacing time*. At the arrival of a cell of a CBR connection, its timestamp is set to *Spacing time* $+1/r$. Now, the time stamp of the first cell of any backlogged stream cannot be greater than this and is generally smaller (being $1/r$ greater than the time stamp of the last cell to have been served which cannot be greater than the current value of *Spacing time*). Consequently, in this example, every CBR cell will have to wait for service behind one cell from every backlogged stream. This is a systematic increase in the delay of CBR streams which hardly profits the bursty traffic.

A solution giving priority to non-backlogged streams consists in modifying the Virtual Spacing algorithm as follows:

- on a stream i packet arrival

  (i) $VS_i \leftarrow \max\{Spacing\ time, VS_i + 1/r_i\}$

  (ii) time stamp the packet with the value of VS_i.

- serve packets in increasing order of time stamp.

A similar modification seems appropriate in the case of PGPS. The worst case delay scenarios used to derive the delay bounds are the same so that these remain unchanged for the revised algorithm.

We have argued that the delay bounds are too loose to be useful for low rate streams with real time constraints (for dimensioning playback buffers, for example)[1]. For bursty connections the delay bounds are even looser in so far as the parameter b_i for such connections is typically much larger than the mean burst size and realized delays are typically very much smaller than these worst case bounds [RBC93]. The more significant feature of WFQ algorithms for such streams is their fairness.

6.2.4   Fairness

The term "weighted fair queueing" implies sharing available bandwidth between active streams in proportion to their rate parameters r_i. This can only be achieved exactly in the theoretical GPS scheduling algorithm. In any packet by packet algorithm, bandwidth can only be shared to within the granularity defined by the packet transmission time. We interpret fairness to mean that, in any given interval $(s, t]$ throughout which stream i remains backlogged, there is a constant $T(s, t)$ such that the number $\eta_i(s, t)$ of stream i cells served in $(s, t]$ satisfies:

$$T(s,t)r_i - 1 < \eta_i(s,t) \leq T(s,t)r_i + 1. \tag{6.2.7}$$

Notice that even with GPS where the *work* performed is exactly proportional to the rate parameters, the *number* of cells transmitted in an interval can only be defined by an inequality equivalent to (6.2.7). The inequality shows that the number of cells taken from the queue tends to a fair share as the

---

[1]Note that the problem is less severe for higher rate video streams than for telephone connections and could be made negligible for the latter if the 'stream' grouped together a sufficient number of connections with an appropriately defined rate parameter.

considered interval increases $(\eta_i(s,t) \gg 1)$. The inequality (6.2.7) can be demonstrated as follows for the Virtual Spacing algorithm.

Let the time stamps of cells served at or immediately before s and t be τ_s and τ_t, respectively. Assume stream i is backlogged in $(s,t]$ and let θ_i be the time stamp of the first cell of this stream to be served in the interval. Since stream i time stamps are spaced by $1/r_i$, θ_i satisfies:

$$\tau_s < \theta_i \leq \tau_s + 1/r_i. \tag{6.2.8}$$

Note that we have $\theta_i = \tau_s + 1/r_i$ when stream i becomes backlogged before s but after the immediately preceding service instant.

Since the time stamps of all stream i cells served in $(s,t]$ are necessarily less than or equal to τ_t (*Spacing time* is non-decreasing), we deduce:

$$\theta_i + (\eta_i(s,t) - 1)/r_i \leq \tau_t \tag{6.2.9}$$

$$\theta_i + \eta_i(s,t)/r_i > \tau_t \tag{6.2.10}$$

Substituting appropriate bounds on θ_i from (6.2.8) we derive (6.2.7) with $T(s,t) = \tau_t - \tau_s$. Note that this result applies to both the original Virtual Spacing algorithm of Section 6.1.4 and the modified algorithm suggested in Section 6.2.3. Further fairness properties are demonstrated in the variable packet length context of SCFQ in [Gol94]. The fairness of PGPS is not directly addressed in [PG93a] but appears to follow immediately from its proven closeness to the GPS ideal.

6.3 Realising WFQ

WFQ is certainly much more complicated than the simple FIFO queues which are used exclusively in most ATM switch architectures. It may however be argued that its substantial traffic control advantages largely justify the increased complexity. A key requirement is to be able to rapidly sort cells in increasing time stamp order.

6.3.1 Sort queueing

We consider an output queueing switch module where service scheduling is performed by a dedicated output controller on each outgoing multiplex. The switch fabric is assumed capable of routing cells to the required output without significantly changing the input traffic characteristics due to cell loss or delay.

The weighted fair queueing algorithms discussed in Section 6.1 rely on service in increasing order of time stamp. The multiplexer output controller must therefore be able to schedule waiting cells so that the cell selected for transmission is always the one with the smallest time stamp. One way of doing this is to sort the cells in a serial memory in order of increasing time stamp; a newly arriving cell is inserted at the appropriate queue place by comparing its time stamp with those of cells already waiting. Devices for rapidly performing this sorting function have been designed by Chao [Cha91]. An alternative design considered in the COST 242 project is based on the following real time sort algorithm [RBS95].

To perform scheduling according to the value of a time stamp, it is not necessary to completely sort messages in increasing time stamp order but only to identify the message with the smallest time stamp at any time when a service can take place. The following algorithm realizes this objective by performing a set of comparison and shift operations in parallel on messages arranged in two tables $A(j)$ and $B(j)$, $0 \leq j \leq m$. Each word of tables A and B is either set to a default maximum value (all bits set to 1) or represents a message and its associated time stamp. The time stamp occupies the k leftmost bits while the remainder identifies the message content (generally a pointer to an address). For the sake of simplicity we neglect the problem of time wrap around (i.e., the fact that the time periodically comes back to zero every $2k$ units) and assume the time stamp unambiguously determines the service order: the message represented by $A(i)$ will be served before the message represented by $B(j)$, say, if $A(i) < B(j)$. All words of tables A and B are initially set to the default value. The sorting algorithm has two phases, one when a new message is inserted, one when the smallest valued message is extracted.

A new message is written to word $A(0)$. Words $A(j)$ and $B(j)$, for $0 \leq j \leq m$, are compared and their contents interchanged if $A(j) < B(j)$. After this operation we therefore have $B(j) \leq A(j)$ for $0 \leq j \leq m$. The words in table A are then shifted one step downwards: $A(j) \leftarrow A(j-1)$ for $1 \leq j \leq m$. $A(0)$ is re-initialized to the default maximum value. Note that the value initially stored in word $A(m)$ is lost.

Messages are extracted on reading from word $B(0)$. By working out particular examples it is easy to convince oneself that this address effectively corresponds to the message with the smallest time stamp. A formal proof is given in [RBS95]. As above, words $A(j)$ and $B(j)$, for $1 \leq j \leq m$, are compared and their content interchanged if $A(j) < B(j)$. The words in table B are then shifted one step upwards: $B(j) \leftarrow B(j+1)$ for $0 \leq j < m$. The value of $B(m)$ is re-initialized to the default value.

An integrated circuit design realizing the above algorithm is proposed in [RBS95]. This circuit turns out to be very similar in conception to so-called systolic sorters used in data processing applications [CM88].

6.3.2   Virtual Spacing algorithm

The Sort Queue associated with each output may have space to contain the entire cell but it would probably be more economical to stock the cell content in a general purpose memory with the Sort Queue entry containing the time stamp and a pointer to the corresponding cell address. This pointer might be the address itself or just the identity of the connection to which the cell belongs: the cell address would then be derived from information contained in a connection context.

If we use a connection context to identify the waiting cells, the Sort Queue need only be long enough to contain one entry for each connection set up on the link. This entry bears the time stamp of the cell which is currently head of the line. When this cell is transmitted, the output controller replaces the time stamp of the Sort Queue entry by that of the cell next in line. This is possible because, with Virtual Spacing, the time stamp of a cell on a backlogged connection can be determined as late as the departure instant of the preceding cell. A cell arriving to a non-backlogged connection, having its time stamp determined by the current value of *Spacing Time*, is entered directly to the Sort Queue.

6.4   Fixing rate parameters

The implementation of WFQ would return to ATM some of its original promise of flexibility necessary for the development of a future safe network. In this section we aim to illustrate this potential by suggesting how different services might be supported. The main distinction is between services with and without real time delay constraints.

6.4.1   Services with real time constraints

Services like voice or video telephony with strict delay constraints and low bandwidth requirements are ideally suited to multiplexing with cell scale congestion only, i.e., using Rate Envelope Multiplexing, as discussed in Section 4.1. This operating mode can be simulated using WFQ by attributing stream parameters for groups of connections with a small value of b and a value of r chosen to ensure the required cell loss ratio (CLR).

Consider a service like the telephone where individual calls appear as on/off VBR connections whose bit rate characteristics are known (i.e., we know the peak rate and the mean rate is known in a statistical sense for the population of telephone calls). It is then possible to accurately predict the stationary probability distribution of the combined instantaneous bit rate of a group of n connections (the number of active connections is binomial). Denote the bit rate at time t by $\lambda_t(n)$. Approximating CLR by the freeze-out fraction, we could fix r such that

$$\frac{\mathrm{E}\{(\lambda_t(n) - r)^+\}}{\mathrm{E}\{\lambda_t(n)\}} < \epsilon \qquad (6.4.1)$$

where ϵ is the target CLR. A stream memory parameter b of a few tens of cell places would be necessary to avoid cell loss when the combined arrival rate is not greater than r. Note that this is not a dedicated memory but rather a device for deciding when to reject cells: the fact that the stream queue exceeds b is evidence that the input rate is currently exceeding the allocated rate r. Any cell loss rate can be fixed, without affecting the service offered to other connections, simply by the choice of the function $r(n)$ denoting the rate threshold above which arriving cells will be lost.

This allocation effectively ensures that the stream CLR is not greater than the target ϵ. However, the rate allocation may be overly generous leading to inefficient link utilisation since no account is taken of statistical resource sharing with other streams; as the rate available to the stream is generally greater than r, the CLR may be *much* smaller than ϵ. For example, if another service with the same QoS requirements were handled by the multiplexer it would be possible to group the connections of both services in a single stream. Clearly, the rate required for the combined stream calculated as above would generally be considerably smaller than the sum of the individual rates.

Consider the effective bandwidth framework for REM discussed in Section 5.2.2 and assume stream i connections are all representative of a certain type. Let $e_i(\epsilon, c)$ be the effective bandwidth of a connection of type i for a link of rate c and target CLR ϵ and let n_i be the number of such connections. If all streams multiplexed in the WFQ system had real time delay constraints and the same CLR requirement ϵ, the allocation of rate parameters

$$r_i = n_i e_i(\epsilon, c)$$

would be sufficient. It is tempting to suppose that, if streams have different CLR requirements ϵ_i, a rate allocation calculated according to

$$r_i = n_i e_i(\epsilon_i, c) \tag{6.4.2}$$

would be sufficient.

In [LT95a], Lindberger and Tidblom demonstrate that this is indeed a conservative rate allocation policy in that realized CLR for each class is less than the respective target value. It is also argued that the allocation (6.4.2) is satisfactory even when some multiplexed streams are bursty with a large buffer parameter b although we note that for periods when the latter are backlogged, the expected CLR of the real time traffic will be greater than the target ϵ.

Multiplexing gains obtained through resource sharing between streams as discussed above can be said to rely on knowledge of traffic characteristics of all streams (e.g., due to leaky bucket access control). However, it is shown in [LT95a] that if one stream (or class of streams) were to set its rate parameter r_i at a value lower than its actual mean rate then, although that stream would experience a loss ratio greater than the target, the other streams would still be protected by the fair queueing mechanism. An incorrect mean rate only affects the stream concerned and any other streams grouped in the same class.

Absolute guarantees on individual stream performance may be difficult to achieve since it would be necessary to make worst case assumptions about competing traffic. The appropriate choice of rate parameters might need to be safer for streams requiring strict guarantees. The appropriate rate parameters for VBR streams having different CLR requirements would then need to be determined using an effective bandwidth derived assuming a link capacity somewhat less than the capacity actually reserved for all these types of VBR streams.

6.4.2   Leaky bucket defined connections

Individual connections defined by leaky bucket parameters r and b are naturally realized in WFQ as streams with corresponding parameters. WFQ allows the network to provide throughput r in a very clear sense and to guarantee negligible cell loss. On the other hand, although delay is bounded as shown in Section 6.2.3, these bounds are too loose to constitute useful performance guarantees. It is argued in [RBC93] that realized delays should typically correspond to peak rate transmission most of the time. However, the need to account for burst scale congestion and the inherent unpredictability of data traffic make it impossible to strictly *guarantee* more than the unduly pessimistic worst case bounds.

In fact, the quality of service on data connections is manifested by end to end delays which depend mainly on the user's choice of leaky bucket parameters since these determine access delays. There appears an obvious trade-off between the respective values of r and b: if r is close to the mean emission rate then b must be very large; conversely, if b is limited (e.g., by the network operator) then the user may need to choose r several times greater than the mean rate (see the examples cited in Section 5.3.2 and [RBC93]). It is noted that, for low access delays, traffic parameter r must be a small multiple of the mean rate while b should be an order of magnitude greater than the mean burst size.

While it is straightforward with WFQ to guarantee the minimal throughput r, it remains to determine the engineering rules for dimensioning multiplex buffers to ensure negligible cell loss given the b value of multiplexed streams. This requirement should be considerably less than the sum of the parameters b. One possibility would be for the network to fix the value of b based on available memory; this value could even evolve dynamically according to current traffic conditions.

6.4.3   Best effort data connections

In a private ATM network, in particular, it may be considered unnecessarily restrictive to impose access control by a leaky bucket: why limit the mean input rate to r_i at times when more capacity is available on network links. The danger of uncontrolled connections saturating network links and affecting the quality of service of other users can be alleviated when WFQ is used by allocating a minimal service rate r and maximum buffer occupancy b, as for leaky bucket controlled connections. A single "connection" of given parameters r and b may be reserved for all best effort traffic, including datagrams. The impact of such connections on other users is thus limited while any spare capacity is immediately available to them. Users would ideally adjust their input rate to prevailing traffic conditions using end to end protocols like TCP. This type of service is standard in current LANs and in data networks like the Internet. In the B-ISDN, the Available Bit Rate (ABR) transfer capability is a means of providing best effort service in which unreserved bandwidth is shared dynamically (ref to ABR section).

ABR connections are generally assumed to share link capacity not currently attributed to connections with QoS guarantees. In the present case of WFQ scheduling, the notion of attributed rate is clear for streams with a small buffer parameter where r is determined for peak rate allocation or REM as discussed above in Section 6.4.1. For leaky bucket controlled streams, on the other hand, the rate reservation is only a long term committment and achieving satisfactory QoS relies on bursts usually arriving to find the link with sufficient capacity available to carry the burst at peak rate. If ABR connections are attributed all unused capacity, such bursts will never have a throughput greater than their reserved rate r.

6.4.4   Virtual Private Networks

The creation of Virtual Private Networks using interconnected VPCs is an interesting possibility afforded by the B-ISDN [WA92]. Users create VPCs of specified capacity between network nodes according to their particular interconnection requirement. Now, at a given node, the sum of the capacities of incoming VPCs can exceed the capacity of any outgoing VPC of the same VPN. With simple FIFO queues, in order to ensure that the traffic offered to such an outgoing VPC does not exceed its designated capacity and consequently interfere with other connections on the same link, it is necessary to implement "output policing" [WA92]. This is a considerable complication to the design of nodes which generally only control the conformity of traffic at their inputs. The generalized use of WFQ constitutes a (partial) solution to this problem:

> in a node terminating the VPCs of a VPN, a WFQ algorithm would simply be applied with the parameters r and b of the outgoing VPCs; it still remains possible for the traffic carried by these VPCs to exceed their allocation but only when the other connections on the link do not fully utilize their own bandwidth allocation.

An output policing mechanism for controlling the VPC peak rate, as considered in [WA92], would probably need to actually space (and not just virtually space) the cells at the designated rate. However, it is not obvious that this strict rate control is preferable to the WFQ solution in the definition of the VPN service. It does not appear any easier to implement.

6.4.5 Link sharing

The considerations in this section are inspired by the discussion on resource sharing in an unpublished Internet Draft on "A service model for an integrated services Internet" by Shenker, Clark and Zhang. A user to user VPC is typically used for a variety of applications materialized by a number of connections identified by a VC identifier. A user may wish to share the bandwidth reserved for the VPC between different groups of applications with minimum rate guarantees calculated to meet specific performance requirements, notably for real time communications. The minimum rates could be guaranteed by applying WFQ with rate parameters defined appropriately for each application group considered as a stream within the VPC. The disadvantage with this approach is that the rate of a momentarily idle stream would be shared between all streams on the multiplexer in proportion to their rate parameters. In order for the rate to be reserved for the VPC, except when all the VPC streams have no backlog, it is necessary to implement a hierarchical form of WFQ.

Let the rate allocated to the VPC stream be r_i and assume this rate is to be distributed to m substreams with individual rate parameters r_{ij} with $\sum r_{ij} = r_i$. A modified Virtual Spacing algorithm realizing the above sharing objectives is as follows. Let VS_i and VS_{ij} be variables associated with stream i and substream ij, respectively. The algorithm is then:

- on a stream i cell arrival

    (i) $VS_i \leftarrow \max\{Spacing\ time, VS_i\} + 1/r_i$

    (ii) $VS_{ij} \leftarrow \max\{Spacing\ time, VS_{ij}\} + 1/r_{ij}$

    (iii) time stamp the cell with the value of $VS_i \| VS_{ij}$ (the concatenation of VS_i and VS_{ij}).

- serve cells in increasing order of time stamp.

It remains to demonstrate that the added complexity is justified by the interest such a link sharing option would have for the user. He would perhaps be better served by setting up an individual VPC for each application.

Part II

Broadband Network Design

Introduction to Part II

It is rather difficult to predict the evolutionary paths that will lead to the implementation of a universal multiservice broadband network. Network economy constraints, evolutionary strategies of the operators and the availability of services in the market can have a significant impact on this evolution.

The following points are particularly worthy of note:

- Local Area Networks (LANs) are now used extensively in the business-corporate sector. At the moment they use different types of technology and mostly support data applications within areas of limited geographical extension. Many new requirements are emerging in connection with LANs:

    (i) to interconnect LANs belonging to multi-site companies and institutions;

    (ii) to provide protocol conversion among LANs using heterogeneous techniques;

    (iii) to extend their range of application in order to also support synchronous video and audio traffic;

    (iv) to connect LANs to the B-ISDN, so that the full range of services is available to the relevant customers.

- A satisfactory return on the huge network investments required to provide broadband access and interconnections to business and residential subscribers can only be achieved with sufficiently high penetration levels and adequate economy of scale and scope. From this perspective, technical solutions able to support both distributive and interactive services are of major interest. In addition to application scalability, it is also of primary importance to achieve satisfactory resource sharing in the access part of the network.

- The core network must be designed to efficiently and reliably carry very large traffic demands with heterogeneous characteristics and requirements. Conflicting objectives have to be balanced in managing such networks. In particular, we can distinguish *economic*, *robustness*, *quality of service* and *fairness* objectives. Specific features of ATM networks can be exploited to cope with such objectives, such as cell priority management and VP and VC switching capabilities. These features can be exploited to design and operate the ATM network as a collection of partially separated logical networks, each one supporting a separate class of traffic. Such a logical decomposition allows us to use a common network infrastructure and transport mechanism even in a multiservice environment. To fully exploit the potential of the network, both provisioning and management processes have to be properly structured, taking account of the inter-relationships between the design and operational variables.

In relation to these topics, Part II of this report deals with significant aspects of multiservice broadband network design, namely:

- networking issues related to LAN's;
- access network sharing techniques;
- VP network design and management in the core network.

It has not been possible to cover all subjects exhaustively in the limited available space. This is particularly true for the first two items, considered in Chapters 7 and 8, respectively. The third item has been more deeply investigated during the COST 242 project and takes all the remaining chapters from 9 to 12.

It should be noted that all the chapters have generally different authors and represent the range of points of view expressed within COST 242. The chapters are self-contained and can be read independently. Performance evaluation methods, the behaviour of shared multirate systems and other dimensioning methods used in Part II, are mostly described in Part III of the report; proper cross-references are given where appropriate.

As to the content, Chapter 7 presents considerations on LANs. Technologies that support multiple services in access networks, such as Ethernet and FDDI, will most probably exist for a long time. Therefore, efficient interworking techniques between legacy LANs/MANs and the B-ISDN are of significant practical importance. A novel Ethernet-compatible medium access protocol, which supports multiple classes of periodic traffic is proposed and the capability of FDDI to support voice-type traffic using its standard timed token ring protocol is presented. DQDB is also discussed in the chapter, as it can play an important role as a B-ISDN access/distribution network due, among other things, to the similarities in segment and cell formats and the support of service integration. ATM-DQDB interworking issues are outlined and the support of isochronous (CBR) traffic by the DQDB protocol is investigated. Also a pure ATM solution, denoted ATM LAN, is examined as an alternative to the legacy shared-medium LANs, providing virtually unlimited scalability in terms of data rates and coverage, as opposed to the very limited scalability of standard LANs. Recent standardization results, in particular at the ATM Forum, and the availability of off-the-self technology makes this approach important for practical networking. An overview of building blocks for ATM LANs is given and some key network design issues are presented.

In Chapter 8, a number of Medium Access Control (MAC) protocols which can be used in access Passive Optical Networks (PON) are described. PON technology can be advantageously used in the access network due mainly to its simplicity and reliability. The reported work was performed primarily in the RACE II project BAF and was reported and discussed in COST 242 meetings. We refer to a number of problems that arise in the PON design and involve traffic performance aspects of different topologies. In particular, for a tree topology two families of protocols are identified, namely the single cell and the multiple cell MAC protocols. The first is characterized by a cell to cell evaluation and resolution of the transmission request. The second relies on a periodic evaluation of the transmission request on the basis of a framed structure. Within the first family we address in particular the Global Fifo, the Circular Permit Programmer, the Belt Permit Programmer and the Priority Access Control Scheme protocols.

In the other chapters of Part II we mostly address design and management problems of Virtual Networks operating on the same ATM platform. In Chapter 9 we consider a general framework for the design and resource management of the core network and suggest an appropriate multi-layer network architecture for ATM networks. In Chapter 10 traffic clustering and multiservice network dimensioning methods are discussed. Chapter 11 considers the topological and dimensioning aspects of logical VP networks and finally, in Chapter 12 we consider network management features related to real time VP bandwidth allocation and associated routing problems.

7 Access network design

Efficient techniques for controlling the access of multiple terminal equipments to a single broadband network interface will play a very important role during the introduction of B-ISDN, for at least two reasons. First, the standardized data rates of the UNI are too high for most existing customer equipments and it would therefore be too costly to connect them individually. Second, a large variety of legacy LANs and MANs, some of them being capable of carrying multiservice traffic, has been deployed worldwide. While some technologies currently used might disappear in the fairly near future, others will most probably exist for a long time, performing a new role as access/distribution network for broadband ISDN. Therefore, efficient interworking techniques between legacy LANs/MANs and the B-ISDN are of significant practical importance.

As for legacy LANs, the IEEE 802.3 CSMA/CD (Ethernet) and the ANSI FDDI are the almost ubiquitously used LAN and second generation LAN technologies, respectively. It is not surprising, therefore, that the first ATM products, aimed at private networks, include Ethernet and FDDI interfaces. This type of networking solution will initially satisfy the immediate needs of faster and more efficient data-only communications. However, solutions that support multiple services in access networks such as Ethernet and FDDI are of interest, too. In Section 7.1, a novel Ethernet-compatible medium access protocol which supports multiple classes of periodic traffic is presented and in Section 7.2 the capability of FDDI to support voice-type traffic using its standard timed token ring protocol is demonstrated. Both approaches are of practical importance since the integration of voice, video and data in a LAN is accomplished without modifying the underlying standardized access protocols.

The Distributed Queue Dual Bus technique (DQDB) was accepted by the IEEE 802 group as a "802.6 Metropolitan Area Subnetwork Standard" in 1990. It uses the Distributed Queue access protocol for data and a TDM-like scheme (pre-allocated slots) for CBR traffic (called isochronous traffic in the context of the standard and in the related literature) to meet the timing requirements for video and voice.

DQDB itself has not been widely introduced as a stand-alone metropolitan area network. However, it is currently in use as a networking basis for providing a new high-speed data-only service called SMDS (Switched Multi-megabit Data Service). DQDB can also play an important new role as access/distribution network to B-ISDN due, among other things, to the similarities in segment and cell formats and the support of service integration. In Section 7.3, ATM-DQDB interworking issues are outlined and the support of isochronous (CBR) traffic by the DQ protocol is investigated. Although accessing B-ISDN's real-time services from a DQDB network requires solving the related protocol and signalling tasks, these issues remain outside of this chapter.

An approach investigated within the framework of ITU-T standardization has been based on using the Generic Flow Control field (GFC) in the ATM cell header. This field was originally reserved for controlling user access at the UNI. Subsequently, it has been realised that the GFC field can be used to control multiple access to a single UNI. Proposals on GFC include that based on DQDB (also known as the Australian proposal) and the ATMR (ATM Ring) proposal by BT and NTT. In Section 7.4 a brief summary and comparison of these methods is given.

Finally, Section 7.5 deals with a pure ATM solution, where ATM is used already in the local access part of the network. This approach is commonly denoted by the term ATM LAN, although there is no restriction to the local area. ATM LAN is an alternative to the legacy shared-medium LANs, providing virtually unlimited scalability in terms of data rates and coverage as opposed to the very limited scalability of standard LANs due to their medium access protocols. Recent standardization results, in particular at the ATM Forum, and the availability of off-the-self technology makes this approach important for practical networking. This section gives an overview of building blocks for ATM LANs, highlights some key network design issues and presents some performance results for relatively simple but already feasible scenarios.

7.1   Support of real time traffic on Ethernet

Standard LANs, in general, have been traditionally viewed as being incapable of carrying multimedia and/or high speed real time traffic due to data rate and protocol limitations. However, due to recent advances in video coding, on the one hand, and the emerging high speed Ethernet versions, on the other hand, data rate demands and resources seem to match quite well for many practical scenarios. Therefore, appropriate integrated media access protocols are needed that (i) support several classes of real time traffic and (ii) are compatible with the original data-only MAC protocol so that stations with integrated functionality can coexist with standard data-only stations.

A novel medium access protocol that supports multiple classes of CBR-type (periodic, real time) traffic is presented briefly in the sequel. This protocol called MCRT (Multi-Class Real Time) provides guaranteed service for different classes of real time users while satisfying the minimum bandwidth requirements of data users. The protocol is also compatible with the Ethernet standard and is transparent to standard "data" stations (i.e., real time user

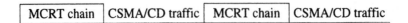

Figure 7.1.1: Channel allocation between MCRT and CSMA/CD users

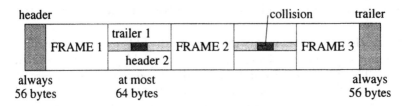

Figure 7.1.2: MCRT chain format

stations can be added to an existing network without requiring any changes to the already operating standard stations).

We specify a synchronous real time application (session) by its total duration and the application cycle length defined as the constant interarrival time between consecutive frame transmissions. Applications having the same application cycle are considered to belong to the same application class.

Users executing the MCRT protocol distinguish between (asynchronous) standard "data frames and synchronous MCRT frames. To guarantee that for the duration of an established virtual circuit of a real time session the MCRT frames do not collide with data frames, MCRT frames of coexisting virtual circuits are concatenated into chains, as shown in Figure 7.1.1. Within a chain, transmission rights are passed from one real time user to another by a procedure termed handshaking, as shown in Figure 7.1.2. The handshaking procedure is based on the use of a special MCRT frame header and trailer transmitted by every MCRT station. Specifically, after locating its predecessor (in the chain) the station initiates a header transmission upon detecting the start of the predecessor's trailer. Similarly the station stops its trailer transmission only when the collision caused by the successor's header transmission is detected. In this way, the "right to use" the channel is passed between successive MCRT stations by means of an enforced trailer/header collision. This procedure maintains a "carrier on" condition during real time chain transmission so that all data stations defer and are thus prevented from disrupting the MCRT chain.

For reasons of simplicity in the design of the station's software/hardware mechanisms for chain organization we choose the cycle length of the application belonging to class i to be $2^i \times$ BTI, with BTI a basic time interval in the network.

Note that the above cycle length ratios match well with the data rates of some important practical voice and video sources since, for instance, 2.048 Mbit/s (which can be a primary rate ISDN or a high quality video), 128 Kbit/s (a normal videoconferencing data rate) and 64 Kbit/s (standard digital voice) correspond to $i = 1, 4$ and 5, respectively.

The BTI can be chosen arbitrarily to suit the set of applications running in the network. A BTI associated timer is implemented at each MCRT station. Timers are maintained loosely synchronized, (up to one round trip propagation delay) only when a real time application from the station is in progress. By relying solely on MCRT frame transmissions for synchronization the protocol remains robust, distributed and devoid of additional specialized hardware.

A number of supplementary protocol rules have been worked out that are omitted here. For instance, the first member of the chain restarts it at the beginning of each cycle by transmitting a sufficiently long preamble to "kill" any data transmission that may occur during this initial time period. Thus, the periodicity of consecutive chains' starting points is ensured. The contiguity within a chain is maintained in cases when a real-time connection is terminated. In such cases the parties send a special last packet called "Terminate" to inform their successors that from now on they should follow the preceding packet. New connections' packets are always added to the end of the chain. Using silence detection, a virtual channel from within the chain is removed for the durations of the silence period and added again to the end of the chain when the subsequent active period starts.

To compute N, the number of supportable parallel sessions, we have to take into consideration the bit generation rate of the real time process, the number of chains in each service cycle, the load of the application cycle length, the real time traffic, the channel bandwidth, the average handshaking duration, the average chain overhead, and the header of the MCRT frame.

The computation of N is quite straightforward for a simple case when only one real-time application is present. As a numerical example, assuming 802.3 standard network for data traffic and an application cycle of 26.2 msec for a 64 Kbit/s rate for the real time application, the maximum number of parallel sessions is as follows:

application class:	0	1	2	3	4
N	124	123	123	122	121

To evaluate the MCRT protocol efficiency we compute the ratio between the total duration of chain and the net real-time traffic throughput. By using the previous network assumptions we obtain:

application class:	0	1	2	3	4
efficiency:	80%	79%	79%	79%	78%

The difference between the total time occupied by the chain and the useful real-time traffic throughput is due to the protocol overhead at the beginning and at the end of a chain as well as within the chain (because of the "handshaking"). Clearly, shorter bus networks and larger cycles improve

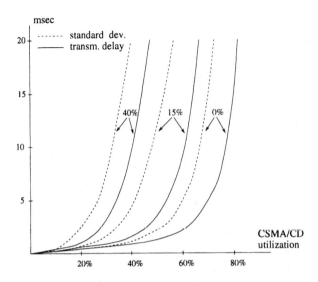

Figure 7.1.3: Average CSMA/CD frame transmission delay and standard deviation as a function of Ethernet users load, with constant MCRT loads

protocol efficiency while the reduction of the basic time interval decreases the number of users in each chain. Thus the maximum header and trailer which appear at the beginning and at the end of every chain act to decrease the protocol efficiency. The introduction of the real time traffic into the network leaves only a portion of the channel available for data transmission. Figure 7.1.3 shows the data packet delay as a function of 0%, 15% and 40% of background real time traffic load. Clearly as the real time traffic increases the average data delay increases too. Furthermore we find that while the real time applications receive virtually constant response time when the MCRT protocol is used, the data applications benefit also. In addition, the maximum combined channel utilization when the CSMA/CD and MCRT protocols are used in conjunction is found to be in excess of that obtained when CSMA/CD is used for both application types as demonstrated by Figure 7.1.4.

In summary, the presented MCRT protocol supports multiple concurrent real time applications providing almost deterministic, collision free service for users with synchronous real time applications while allowing, when no real time chains are present, cycle stealing by standard CSMA/CD stations.

7.2   Support of real time traffic on FDDI

The most widely deployed second generation LAN is FDDI. FDDI has been modified to support service integration in an STM-way, the resulting architecture being FDDI-II currently under standardization. It is very likely that even if FDDI-II is adopted as a standard, LAN manufacturers will not be

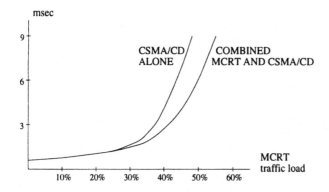

Figure 7.1.4: Average CSMA/CD frame transmission delay in mixed MCRT and CSMA/CD environment compared to CSMA/CD environment alone

enthusiastic about bringing the corresponding products into the market for several reasons.

For the existing standard FDDI (also called FDDI-I to avoid ambiguity in this context) it is possible to prove that it is capable of carrying isochronous traffic using the underlying medium access protocols, i.e., is the timed token ring protocol/synchronous token mode. The data format itself does not match well which results in a certain loss of efficiency because of the relatively large overhead when using ATM-size data units due to the header which contains explicit individual addresses. The advantages of using the same medium access protocol for both data and isochronous traffic with data formats identical or very similar to ATM cells are clearly simplicity and easier realisation of the interworking tasks. It is, however, necessary to ensure that satisfactory quality of service can be granted.

Our approach is characterized by the following key points. We consider a scenario in which an FDDI is connected to B-ISDN via a gateway (see Fig. 7.2.1). ATM-like transmission of isochronous (voice) traffic is used within the FDDI. Asynchronous traffic is transmitted in variable-length frames in FDDI. Using combined "internal" and "external" traffic distribution models corresponding to the LAN-B-ISDN scenario is a key feature of the simulation models.

The main objective is a characterisation of the QoS of the isochronous traffic at the LAN output (i.e., at the input of the B-ISDN network) using standard ATM performance measures, i.e., cell delay, cell delay variation and cell loss.

A realistic network model, motivated by the practical trend of connecting workstations directly to an FDDI ring, contains a relatively large number of stations. Simulation results are presented below for a configuration with 101 nodes, 100 of them being normal MAN stations and the 101th station comprising a normal station and the gateway. These stations are distributed

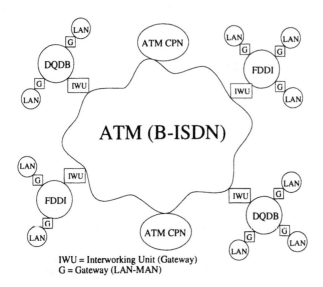

IWU = Interworking Unit (Gateway)
G = Gateway (LAN-MAN)

Figure 7.2.1: FDDI LAN / DQDB MAN - ATM interworking

equidistantly on a 20 km long ring.

All nodes generate both voice and data traffic and have two distinct queues to store generated packets. The external traffic is 20% of the total traffic. In the following analyses, the voice traffic is transmitted using the synchronous token mode and data traffic in the asynchronous token mode. The length of synchronous frames is constant, while the length of asynchronous frames is variable with a maximum of 4500 bytes. Voice sources are modelled at call level.

Performance results for different traffic load conditions show that, with increasing load, the delays are obviously also increasing (see Fig. 7.2.2). Due to different media access protocols the delays of asynchronous packets are higher than those of the synchronous packets. The average synchronous delay does not exceed 2 ms at 80 percent of total traffic load (consisting of 20% asynchronous and 60% synchronous, or 40% asynchronous and 40% synchronous load). The standard deviation of synchronous delay is in the range of 1 ms. This standard deviation is a measure of CDV. The loss of synchronous packets appears first only at 6 ms TTRT value and it does not reach 0.3% even at the highest TTRT (Target Token Rotation Time) value we considered. The simulation results are documented in [SCK93].

We can conclude that the cell delay and cell delay variation at the output of the FDDI (at the gateway) are completely acceptable, provided that the proportion of isochronous traffic is not larger than 60% of the total capacity for which the acceptable range of the target token rotation time has also been determined. With the practical parameters used in simulations, mean and standard deviation of delay (CTD and CDV) are in the range of some

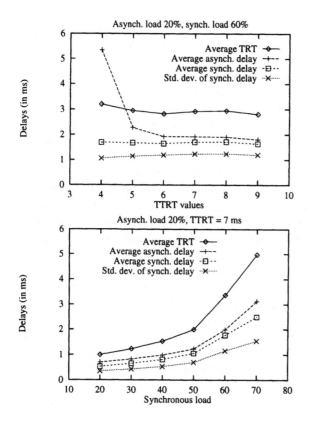

Figure 7.2.2: FDDI simulation results

milliseconds while the cell loss ratio is well below 1%. The target token rotation time must be chosen to be less than or equal to 8 ms. It has also been shown that while satisfying the requirements for isochronous traffic, the data traffic performance is also acceptable.

The results show that the original medium access protocol of FDDI can be used without modifications for collecting voice-type traffic for an ATM interface, using cell-based transmission.

7.3   DQDB as an access network to B-ISDN

DQDB, similarly to other legacy LANs and MANs is expected to serve as access/distribution networks for the ATM WAN and B-ISDN. In this context, a series of interworking problems are yet to be solved if the DQDB technology remains alive in the future. Some of these problems are related to interfacing connectionless services with the connection-oriented ATM/B-

ISDN. Architectures (e.g., the connectionless server approach) and protocols (e.g., LAN Emulation defined by the ATM Forum) have been proposed as general solutions to these tasks.

Interworking of DQDB MANs through the ATM network is an evolutionary step for providing interconnection of a variety of existing LANs. Proper strategies for the interworking units (IWUs), which interconnect the different networks seem to be important. The IWU strategies depend very much on the services the networks might offer and on the type of the interworking networks.

Interworking strongly depends on real time label processing for addressing, switching and routing of information in the interconnected networks and within the IWUs. This dependence is an essential factor for the IWU performance in terms of throughput, delay and traffic control. From a performance analysis point of view an IWU can be regarded as a black box that performs reassembling, segmentation and readdressing functions. Furthermore it is able to buffer data.

In case of connection oriented traffic the IWU has to be transparent, i.e., the only operation performed on an arriving ATM cell before it is transmitted to the DQDB network is a VCI conversion.

In case of connectionless traffic the IWU can also be regarded as transparent, if pipe-lining is used and no buffering of data is performed, so that cells can be transmitted to the local access queue in the DQDB network as soon as they arrive from the ATM network.

An important issue to be dealt with interest of this section in more detail, motivated by the emerging demands for service integration and communication support of multimedia is whether DQDB is capable of collecting isochronous-type traffic for the ATM-based WAN. Originally, the IEEE 802.6 DQDB MAN supports service integration by using the distributed queue protocol for asynchronous (data) traffic and allocating pre-arbitrated slots for isochronous traffic within the same time frame. Pre-arbitrated slots can be used exactly in an STM-way, using one octet (byte) within a slot for a 64 Kbit/s voice channel, which is the originally proposed method. Alternatively, m bytes in every $n \times 125\mu s$ can be allocated, e.g., $m = n = 48$ corresponds to a single standard voice channel.

Let us look at the capabilities of DQDB in carrying isochronous traffic using its principal medium access protocol the distributed queue protocol. We also wish to transmit isochronous information in ATM-like data formats. This is almost the case in DQDB using standard parameters where slots are filled by 52-byte data units called segments containing a 48-byte payload. As noted in the previous section, the advantages of using the same medium access protocol for data, voice and video as well as data formats being identical or very similar to ATM cells are clearly simplicity and easier realisation of the interworking tasks provided that satisfactory quality of service parameters can be guaranteed.

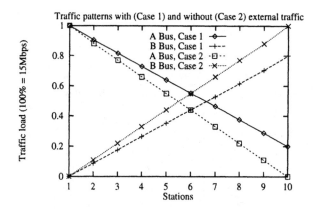

Figure 7.3.1: Traffic patterns on dual bus

Let us consider the same scenario as for FDDI: a DQDB MAN being connected to B-ISDN via a gateway. The main objective is to characterise the QoS of the isochronous traffic at the output of a DQDB (at the input of the B-ISDN network) using standard ATM performance measures, i.e., cell delay, cell delay variation and cell loss.

Each packet arriving from a user to the MAN station has a specified destination that can be either a local MAN address or a remote address. In the case of internal traffic the packets are to be delivered to one of the users of the local MAN. In the case of external traffic, the packet must be delivered to the gateway. The external traffic constitutes a fraction P_{ext} and the internal traffic constitutes a fraction $1 - P_{ext}$ of the total traffic in the DQDB MAN. The internal MAN traffic is assumed to be uniform: each source station selects the destination according to a uniform distribution spanning the remaining $N - 1$ stations. Graphical illustration of the traffic pattern on the two busses is shown in Figure 7.3.1. (Case 1: without external traffic ($P_{ext} = 0$) and Case 2: with external traffic ($P_{ext} = 0.2$)).

The simulation model reflects a practical situation when a DQDB MAN is used as a city-wide backbone with a small number of relatively complex and expensive stations. Thus the number of MAN nodes was taken to be 10, with equal spacing over the bus and with a distance between adjacent stations equal to one slot. The MAN/B-ISDN gateway is always located at the two extremes of the MAN. Data rate and payload values of voice and data segments were selected according to the standard. Asynchronous packets were modelled by a bimodal length distribution, voice sources were modelled at call level.

Examples of simulation results are shown in Table 7.3.1, Figure 7.3.2. It is shown that for a wide range of isochronous traffic that can be as high as 80% of the total bandwidth, CTD, CDV and CLR parameters are fully acceptable, the delay parameters being in the order of tens of microseconds with negligible

Table 7.3.1: DQDB simulation results: delays of segments

| Stations | Asynch. load 20%, isoch. load 80%, ext. traffic 20% | | | | |
| | Average delay | | Average delay in DQ | | Std. dev. of |
	asynch.	isoch.	asynch.	isoch.	isoch. delay
1	4184.737	6.950	64.978	4.960	7.567
2	16560.236	8.216	72.980	6.293	7.272
3	101160.081	11.622	84.577	7.629	9.162
4	306269.054	13.863	101.749	9.820	10.764
5	699034.101	17.930	127.650	12.188	13.939
6	1169084.147	22.974	170.710	15.334	16.372
7	1939764.459	27.619	248.617	19.739	17.340
8	4021215.102	33.515	448.828	24.722	18.823
9	8130124.987	38.695	866.608	31.388	19.104
10	9907768.705	43.789	968.144	36.958	21.006

cell loss. Data traffic performance has also been found acceptable.

The significance of these results is that the DQ protocol can be universally used both for voice and data instead of the originally proposed PA (Pre-Arbitrated) allocation which would complicate the system and require added functionality at the gateway.

7.4 Proposals on the use of the Generic Flow Control field

The success of B-ISDN will depend on the way B-ISDN services are offered to users. An approach investigated within the framework of ITU-T standardization has been based on using the GFC field in the ATM cell header. This field was originally reserved for controlling the access of one single user at the UNI. Subsequently, it has been realised that the GFC field can be used to control multiple access to a single UNI, since it is a cost effective solution for the provision of B-ISDN services in customer premises.

The GFC protocols are shaped by a number of factors, such as the logical configuration, the call bandwidth allocation strategy and of course the concept of fairness. Some early proposals on GFC were developed for a dual bus based on DQDB (also known as the Australian proposal) and others for a logical ring based on the ATMR proposal by NTT and BT. A detailed description of these protocols can be found in [AT91], [AT92], [BT 91] and [BT92b]. Only a brief description will be given here.

The distributed queueing protocol used in the Australian proposal is based on DQDB, with modifications to ensure flexibility and simplicity. The appli-

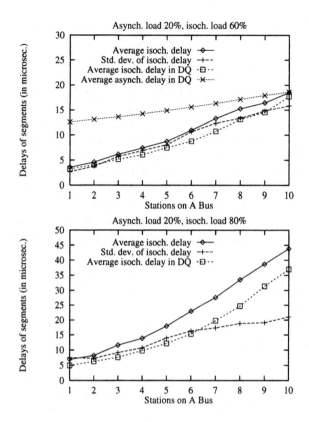

Figure 7.3.2: DQDB simulation results: delays of segments

cable physical topologies are a simpler asymmetric dual bus or the symmetric dual bus which can either be linear or looped. There are three priorities for very jitter sensitive (highest (2)), jitter sensitive (middle (1)) and jitter non-sensitive (lowest (0)) traffic. Each terminal equipment (TE) has a shaper, which filters the cell arrivals at a preassigned rate. Shaping functions for priorities 1 and 2 are based on a fixed credit allocations obtained every fixed time interval (with maximal allowable credit) and a cell-cost to transmit a cell. The proposed protocol uses a bandwidth balancing mechanism similar to that of DQDB for priority 0 traffic to guarantee fairness, and supports destination release as an implementation option to provide extra capacity for local traffic on the bus and to prevent local traffic from reaching the network switch.

The ATMR GFC protocol from NTT and the MSFC protocol from BT are both based on the cyclic reset control mechanism. Performance studies show that ATMR is best for short dual-buses or rings, while MSFC is suitable for long rings or long dual-buses especially where the target for the maximum

CBR jitter is set to a low value. An enhanced protocol combining ATMR and MSFC principles splits traffic in to a high (H) and a low (L) priority class. Accordingly, the transmission of cells is divided into two alternating phases, a H- and a H+L-phase. Delay sensitive services can transmit their cells in both phases, whereas delay insensitive traffic can use the H+L-phase only. There are two terminal types: control TEs (C-TE) and simple TEs (TE). The C-TE sends resetting commands on the ring and thus starts new transmission periods every T units of time. Each TE maintains three kinds of window sizes to indicate the number of cells which the TE is allowed to transmit on the ring during transmission periods. In H-traffic, the number of cells in the short terms restricted within HS and in long term within HL window sizes. In L-traffic, the number of cells is restricted within L window size in an L-reset period. Note that no procedure is proposed allowing values for the transmission period T and the window sizes allocated to the TEs to be chosen.

After the significant initial activity in ITU on accessing B-ISDN using GFC, the issue has been recently removed from the agenda. Many believe that a renewal is expected in the future.

7.5   ATM LAN design

The forces driving wide area networks to an ATM-based architecture already exist in local and, more generally, in enterprise environments. Access to powerful data and image servers and the emerging multimedia applications require high data rate access from desktop associated with both flexibility and scalability. Conventional shared-medium LANs and MANs possess serious limitations: scalability in terms of aggregate speed is limited by the data rate of the medium, conventional LANs lack scalability both in speed and space, MAN scalability in area coverage is also limited. Thus the methods to support integration in legacy LANs and MANs presented in the preceding sections are important alternatives in a limited area and at moderate data rates only.

Because of ATM's inherent scalability, there will be no fundamental difference between an ATM LAN and an ATM enterprise network meaning by the latter a private network that can be a intra-building LAN, a campus network, a MAN or a national or even of global network. An ATM LAN can easily be extended to serve a large area at the cost making the network topology more complex but without changing the underlying switching and transmission techniques. This contrasts with the present practice where it is necessary to move from the existing LAN shared-medium to a switched wide area technology like X.25, frame relay or (narrow-band) ISDN.

Typical ATM LAN scenarios have been considered in several papers in recent years. Investigations have been made on the effect of multiplexing a small number of bursty data streams onto a single ATM link (e.g. [Cam94]) and simple ATM LAN configuration was simulated (e.g. in [FW95]).

An interesting research topic is to compare traditional shared-medium LANs with switch-based architectures, in particular with ATM-based LANs. In [RC94], the performance comparison of a traditional bus-based computer architecture for a multimedia server with an architecture based on a LAN within the computer is presented.

A survey of existing product offerings allows us to identify the following typical building blocks for ATM LANs:

- The ATM Switch, which has only Physical and ATM layer functions. Normally all interfaces have the same data rate. The ATM switch is a key component of an ATM LAN backbone, connecting a number of other ATM devices.

- The ATM Concentrator is a device with different data rates at its input ports as opposed to an ATM switch. Otherwise a concentrator also provides only Physical and ATM layer functions. Another feature is that a concentrator usually cannot provide connections between its input ports and switching (at virtual circuit level) must be accomplished in the ATM switch. Concentrators are used to keep the necessary number of switch ports small.

- The ATM Group Access Device, also commonly called "hub", has at least one IEEE 802 LAN interface. An ATM hub must provide AAL functionality and support different traffic types. In addition to different IEEE 802 LAN interfaces (typically up to four) these devices have several ATM interfaces as well as conventional serial interfaces.

- The ATM Individual Access Device is of a similar functionality as the group access device but only for one user equipment. This device is normally a card plugged into the workstation's bus. It offers signalling for switched virtual connections, multi-protocol encapsulation for interoperability with ATM-connected servers and workstations, hardware support for conversion between data packets and ATM cells, traffic management functions and internetworking for virtual LANs.

In the topological design of ATM LANs the following it is necessary to consider: the distribution of the user population and the expected volume, type and distribution of their traffic. Network flexibility and scalability, reliability and low cost also have to be taken into account.

For a small concentrated user population even a network consisting of only one switch is usually suitable. The introduction of more complicated switching networks is motivated by practical considerations such as a remote user population where the cost of cabling to connect users to the central switch would be high, or the case of a large or growing user population to avoid the use of too complex and thus expensive switches. As an example, consider a network consisting of workstations and file/image servers (see Fig. 7.5.1).

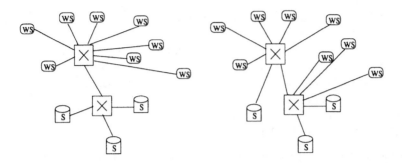

Figure 7.5.1: Two possible network segmentations for the same environment

The size of the population makes the use of two switches necessary. The workstations can establish CBR connections for video-conferencing, and they frequently use the file or image servers generating bursty traffic.

The optimal clustering could be to connect all the workstations to one of the switches, and the servers to the other. In this case delay sensitive CBR traffic will not be affected by the bursty traffic coming from the servers on the link going to the station although the bursty traffic can cause overflow on the inter-switch link. It is possible to reserve bandwidth and police VCC peak rates in an ATM LAN so that the bursty traffic cannot disturb the real time traffic.

If we want to decrease the level of VBR traffic on this link, a symmetric topology can be used, where workstations and servers are equally distributed between the two switches. Thus CBR traffic will appear on the inter-switch link, and will be multiplexed with the VBR traffic. This solution can work only if the quality of the CBR service remains good enough. Thus, to find the proper way of grouping traditional LANs, high-end workstations, data, image and video servers, etc. within an ATM LAN, we have to study the effect on the quality of service parameters of multiplexing a small number of traffic streams.

We simulated a simple ATM LAN with one ATM switch and $N + 1$ stations in a star topology as shown in Fig. 7.5.2. The stations $1, 2, ..., N$ are the normal stations which send (and receive) video traffic to (and from) other stations, while station 0 is an image server, from which the normal stations request image from time to time. This situation represents a video conference process with N participants and one image source containing stored image information to be used during the videoconference. We have assumed that, in addition to point-to-point connections between fully connected videoconferencing workstations there is a point-to-multipoint (multicast) communication between the image server and the workstations. The $N + 1$ stations are spaced equidistantly from the switch and the distance is about 2.5 km. We have examined the network of $N + 1$ stations in 3 cases where $N = 3$, $N = 5$ and $N = 8$, i.e., the number of participants in the conference is 3, 5 and 8

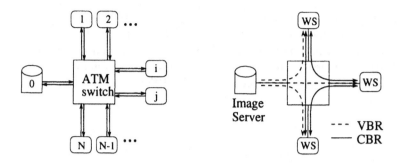

Figure 7.5.2: Network topology and traffic distribution for the simulation

Table 7.5.1: End-to-end delays in case of 5 participants

			1CBR	2CBR	3CBR	4CBR
average delay (msec.)	video	normal st.	0.049	0.049	0.05	0.051
		mixed st.	1.75	1.752	1.691	1.614
		whole sys.	0.383	0.35	0.325	0.293
	voice	normal st.	0.049	0.049	0.05	0.051
		mixed st.	1.742	1.747	1.693	1.61
		whole sys.	0.376	0.351	0.326	0.293
	image (mixed st.)		9.873	9.902	9.849	9.789
maximum delay (msec.)	video	normal st.	0.055	0.058	0.068	0.069
		mixed st.	2.159	2.162	2.158	2.167
	voice	normal st.	0.054	0.057	0.068	0.066
		mixed st.	2.126	2.098	2.094	2.11
	image (mixed st.)		18.665	18.665	18.661	18.662

respectively.

The simulation programs were written in SIMSCRIPT II.5 language and simulations were run on IBM RISC6000 workstations.

For the sake of saving the program runtime, we only examined the case when the image server transmits one image to one of the normal stations called "mixed" during during the time duration beginning when the first packet of the image arrives at the server and ending when the last cell of the image is received by the destination station.

Simulation results have been obtained for end-to-end delays, average queue lengths and cell loss probabilities. En example of end-to-end delay results is shown in Table 7.5.1.

The end-to-end delay of a cell is the time duration between its arrival to the transmitting station and its reception at the receiving one. That is, the end-to-end delay is the sum of the waiting times at the output buffer of the transmitting station, the waiting time at the switch and the propagation time on the links. We have collected data separately for the end-to-end delay of video, stored voice and image traffic at the remaining normal stations and end-to-end delay of video, stored voice traffic on the whole system.

For detailed simulation results we refer to [EGHS96].

8 MAC protocols for access to B-ISDN

A considerable amount of the investment needed for the introduction of the B-ISDN is expected to deal with the access part of the network. Network architectures that allow sharing of the network facilities can reduce investment costs in access networks, and an economical access network clearly attracts residential and small business customers to B-ISDN in an early stage.

In this chapter we describe a number of MAC protocols which can be used for the access network using a Passive Optical Network (PON) technology. This work was performed primarily in the RACE II project BAF and was reported and discussed in COST 242 meetings.

8.1 Access Networks to B-ISDN

Among many transmission technologies applicable to the access network, (ADSL, CATV systems, Active nets, etc) the PON represents a very promising solution. Using this approach, multiple users may share photonic equipment and fibers in the local loop. The advantage of using PON technology in an access network is mainly due to its simplicity and reliability. However, a number of problems arise in the design of these structures, involving both technology and traffic performance aspects.

Different topologies can be used for the access network, as shown in Figure 8.1.1:

- The star topology (1) consists of a set of independent links connected to the line termination node with active elements at each end of a link. The star configuration suits well the existing duct layout.

- The passive optical tree (2) consisting of one active element at the line termination and one active element at each subscriber site (the leaves of the tree) the multiplexing and de-multiplexing functions are performed by passive optical splitters and combiners.

- The single and dual bus architectures (3) enable several customers to share the same transmission medium. The topology can also be folded or double folded, see [AK91]. Depending on the technology involved, each subscriber may access the bus using active or passive components (fast bit flip mechanisms). The bus architecture is less well suited to

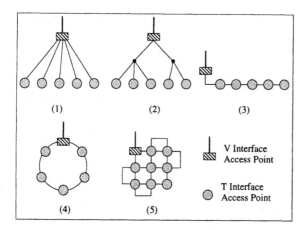

Figure 8.1.1: Different topologies for access networks

the existing duct topology. Moreover, baseband signals will suffer more attenuation and distortion compared to the shorter point-to-point links of a ring, star or tree network. Therefore, in the case of fiber-optic bus networks, less stations can usually be supported. Optical amplifiers may be used to increase the number of stations supported [Wag87], but this increases the overall costs of the network.

- The ring (4) is very similar to the bus configuration; it differs by the extra link between the last node and the line termination. In ring networks, stations are generally connected to the medium using active interfaces. Signals are regenerated and repeated at each node, allowing larger separation between terminals than in bus networks. Advantages of such configuration lie in the possible reconfiguration of the network should a breach occur on some of the links. A disadvantage of the ring is its vulnerability, due to single node failures. Variations to the basic ring topology have been considered to improve network reliability. This topology is not suited to the existing duct layout either.

- Mesh topologies (5) suit high-density areas but are usually unfavourable for use as access networks due to the added complexity in traffic routing at the nodes. Generally it is advisable to keep the complexity and cost of the subscriber terminals very low.

Another important issue in the design of a PON is the question of how bi-directional transmission of information is achieved. Several solutions have been proposed [Moc94]:

- *Space Division Multiplexing* (SDM) : A pair of distinct optical fibers is used. Each fiber conveys either upstream or downstream information.

- *Wavelength Division Multiplexing* (WDM) : With WDM we can handle simultaneous bi-directional signal transmission using different transmission windows (e.g., 1.31 μm and 1.55 μm wavelengths).

- *Frequency Division Multiplexing* (FDM) : We use the same optical window for upstream and downstream transmission, using FDM/TDM modulation techniques for bandwidth sharing.

- *Time Compression Multiplexing* (TCM) : TCM transmits burst-type information in each direction by alternating time periods using the same wavelength over one fiber.

Similar techniques can be used to accommodate service upgrades, allowing the gradual evolution from narrowband services (POTS or N-ISDN) or distributive services (CATV) to ATM provision.

From the above list, the conclusion drawn in the RACE II BAF (Broadband Access Facilities) project was that for an access network a passive optical tree topology suits both the existing duct structures well and offers a reasonable cost/performance trade-off. Furthermore, it does not preclude evolution due to the following reasons (see [BG93]):

- this structure not only shares the exchange equipment but also the fiber infrastructure thanks to the usage of passive optical splitters,

- the optical distribution network being passive, the maintenance cost is reduced and problems of remote powering of street cabinets are avoided.

- passive structures allow an easy implementation of distributive services, on a different wavelength for instance, thus enabling the rapid spread-out of cable TV services,

- easy upgradability by installation of further splitters only when required,

- high degree of resource sharing thus reducing therefore the cost per subscriber significantly.

Bidirectional transmission of information was achieved in the BAF project by using SDM.

All access protocols described in this chapter (except that in Section 8.4.4) are designed for passive tree PONs and are compatible with the Asynchronous Transfer Mode.

8.2   MAC Protocols for PONs with a tree structure

In designing a *Medium Access Control* (MAC) protocol for a tree topology, different problems with respect to upstream transmission (user to network) and downstream transmission (network to user) must be solved. In the upstream direction, the access network performs a multiplexing function. Traffic impairment usually occurs in this direction only, since temporary overload may occur. The aim of the MAC protocol is to prevent collisions between information originating from different network terminations. In the downstream direction, other problems arise including the issues of privacy and security of the communications themselves, as well as privacy on the amount and nature of the communications.

From the traffic performance point of view, the most crucial item involves the definition of suitable mechanisms able to control contention for the same physical medium by several users, avoiding collisions. One difficult point in this problem is the non-negligible round trip propagation delay across the PON. For a 10 km access area this would be of 100 μs.

The BAF project has investigated these problems in connection with the definitions of a number of MAC mechanisms. Two families of protocols were identified, one characterized by a cell by cell evaluation and resolution of the transmission request, and the other characterized by a periodic evaluation of the transmission request on the basis of a framed structure. We will refer to these families as the *single cell MAC protocol* and the *multiple cell MAC protocol*, respectively.

As the MAC protocol determines cell flow at the B-ISDN ingress, it has a major impact on the overall performance of the core network. Hence apart from its main function (avoiding collisions of the cells of different users in the upstream direction) the MAC protocol should also have the following features:

- *efficiency* : the overhead introduced by the MAC protocol should be low:

- *performance* : the delay and delay variation introduced by the MAC protocol should be kept within certain bounds, particularly for services whose delay requirements are guaranteed at connection set up;

- *priority mechanism* : the MAC protocol should be able to distinguish between (preferably different classes of) high and low priority traffic;

- *implementation/cost ratio* : the hardware required to implement the protocol should be neither too complex nor too expensive;

- *fairness* : one subscriber should not be subject to more delay than another;

- *robustness* : when an error occurs (especially when it is introduced by the medium), the MAC protocol should be able to recover from this error.

With respect to performance, the perfect access network operates as a FCFS multiplexer, possibly adding a constant delay due to the roundtrip

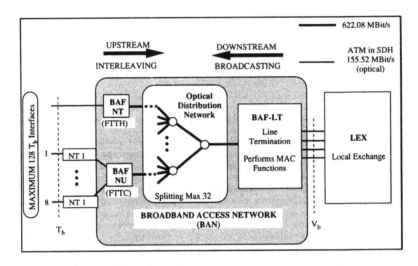

Figure 8.3.1: Architecture of the BAF Access Network

time. Such an access network can then be modelled as a discrete-time queueing system (see e.g. [BK93] or [Slo95b]) in order to answer questions with respect to quantitative aspects.

8.3 Architecture of the Shared Access Network

The architecture considered for the shared access network is depicted in Figure 8.3.1.

The main relevant parts for the MAC protocol definition are:

- the Optical Distribution Network (ODN), realized by a passive optical tree,

- the Network termination consisting either of a BAF Optical Network Unit (BAF-ONU) that can serve up to a fixed number of NT1s (FTTC case) or of a Network Termination that interconnects directly to a terminal (FTTH case). Each NT1 or the BAF-NT provides one T_b interface,

- the Optical Line Termination (OLT) contains the algorithm that distributes the available data rate,

- the splitting ratio of the PON is 1 : 32,

- the access network should provide access to $\pm$ 100 subscribers,

- net bit rate at the network terminations: 149.76 Mbit/s,

- V interface net bit rate 599.04 Mbit/s,

- PON internal gross bit-rate: 622.08 Mbit/s,

- round-trip propagation delay: $100\mu s$.

Access to the shared medium is arbitrated by the MAC protocol, operating on the ATM layer. Since this layer does not support retransmission of lost cells, the MAC protocol should guarantee collision-free access. Together with the fact that the functionality as *master* is assigned to the OLT, the protocols under consideration can be annexed to the class of protocols with *controlled access* and *centralized* control (see [KSY84]).

An access network with a tree topology allows the round trip delay to be equalized for each of the devices which are used to store cells for the upstream direction. This can be achieved by adopting *ranging* procedures which are widely used within satellite networks. The ranging procedure calculates the required electronic delay for each network termination in order to know how long this termination should wait before a cell can be transmitted, starting at the moment the approval to send a cell in upstream direction was received. This gives all terminations the same perception of time, but it is transparent for the MAC protocol for which the distance between OLT and the storing devices appears to be the same.

8.4   Single cell based protocols

In this family of protocols, a device where cells are stored before they are transmitted in the upstream direction has to receive an approval before it is authorized to transmit *one* cell. All single cell based protocols described in this section make use of *Time Division Multiple Access* (TDMA), where time is divided into fixed size slots. A slot is able to transport both an ATM cell and some overhead which is inherited from the physical layer. This overhead is necessary among others things for reasons of fine ranging, bit/byte synchronization and for correction drifts due to temperature changes. The length of the slots differs according to the protocol and varies from 440 bits to 448 bits. For the downstream direction, the overhead is used among others things to exploit the opportunity of transporting *a permit*. This permit represents the address of the network termination that is allowed to transmit an ATM cell in the upstream direction. The permit travels together with a downstream ATM cell from OLT to termination. Because of the broadcast nature of the PON, the destination addresses of ATM cell and permit do not need to be the same. The *engine* producing the permits is referred to as *Permit Distribution Algorithm* (PDA).

Transmission of a permit is the result of a demand for transmission capacity, issued by a network termination. This demand is often referred to as a *request* and each of the described protocols are mainly distinguished by the way they send requests in the upstream direction and by their Permit Distribution Algorithm. This section describes three protocols for tree topologies, the Global FIFO, the BPP and the PACS protocol, and one protocol for a dual bus.

8.4.1   The Global Fifo MAC protocol

In this subsection we describe one of the protocols that was proposed in the BAF project, and that was chosen for the implementation of a prototype (cf. [CGB93]).

Requests

An NT1 advertises its bandwidth requirement by sending requests to the central controller located at the LT. The requests contain information about the state of the buffer in the NT1. We distinguish two types of request:

- *requests coupled with upstream cells.* An upstream cell originating from an NT1 is preceded by a request field which contains the queue length of the NT1 buffer,

- *request Blocks (RB).* Requests coupled with upstream cells have the disadvantage that an NT1 can only reveal its state when it is allowed to send an upstream cell. To overcome this drawback, an RB contains the request of a number of consecutive NT1s. It has the same length as an upstream cell.

Permits

A permit sent to an NT1 indicates that a cell can be transmitted. Two types of permit are distinguished:

- *Permit for an ATM cell.* When the central controller decides that an NT1 can send a cell, it issues a permit containing the address at that NT1. This permit is added to a downstream ATM cell. Remark that since the downstream cell is broadcast, no coupling between the address of permits and ATM cells is needed. The permit class bit CL indicates that this is a permit for an ATM cell.

- *Permits for Request Blocks.* When there are no permits for ATM cells to be sent, a permit for an RB is issued. Such a permit contains the address of the NT1 that is the first to send a request in the request block. The other NT1s that can send requests in the RB can then know the instant of time when they should emit their requests. As the

LT only issues a permit for an RB when it has no permits for ATM cells, no bandwidth is wasted for RBs, which are used to increase the reaction speed of the protocol.

The Permit Distribution Algorithm

This algorithm is executed in the LT and determines how the permits are distributed among the different NT1s. One can think in this algorithm as follows:

- The requests from a certain NT1 contain information about the buffer length of this NT1. From this information one can deduce the number of cells that have arrived to the NT1 since the last request arrived to the LT. These requests for cells are placed in a spacer that enforces a minimum separation (T). As no real cells are spaced, this spacing mechanism can be implemented simply by two counters. The distance T is calculated from the aggregate cell rate of the connections that share the T_b interface. If this aggregate cell rate is R we compute T as:

$$T = min(\frac{\epsilon}{R}, T_{max}); \qquad (8.4.1)$$

  where ϵ is a value around 0.8 and T_{max} is a value around 50. These values were chosen to realize a trade off the cell delay introduced to the cells and the traffic characteristics at the output of the BAF system.

- When the requests for cells leave the spacing devices for each NT they are multiplexed together in a buffer called Global Fifo. This buffer is served each time-slot. If the buffer is not empty, the LT issues requests for cells for the NT1 corresponding to the request that is being served. If this Global Fifo is empty the LT sends requests for RB. In this case the address for the first NT in a RB is chosen by a round robin policy.

8.4.2 The CPP and BPP MAC protocols

The protocols *Circular Permit Programmer* (CPP) and *Belt Permit Programmer* (BPP) were described in detail in [Sta93] and [AV94] respectively. In contrast with the Global Fifo (GF) protocol, these protocols assume that ATM cells for the upstream direction are stored in the ONU instead of in the B-NT1. Both protocols exploit the assumption that the internal gross bit rate of the PON is four times higher than the gross bit rate at the terminations.

Requests

Both protocols use slots with a size of 440 bits which do not contain a request field in the upstream transmission format. In both CPP and BPP, requests are carried in *Request Access Units* (RAUs), composed of the request information of 4 ONUs and are sent in the upstream direction every 17 and 18 slots respectively. The frequency with which an ONU is polled therefore equals 136 slots for CPP and 144 slots for BPP. An RAU conveys 4 fields of 110 bits each, mainly containing the unavoidable bits for the physical layer overhead and a *request field*, dedicated to informing the OLT about the required transmission rights. The functionality of an RAU is very close to that of an RB in the GF protocol.

In the CPP protocol, transmission requirements are expressed in the number of cells that have arrived from each of the connected B-NT1s since the ONU contributed to the previous RAU. The request field of the RAU in the CPP protocol is named RQ and has a total length of 50 bits. This allows a maximum of 10 B-NT1s to be connected per ONU.

The BPP protocol uses the RAU to inform the OLT about the relative arrival instant (with respect to the frame of 144 slots) at which a cell was stored in the ONU buffer. Permits can then be scheduled accordingly. This approximates better the FCFS service policy that governs an ideal ATM multiplexer. This feature reduces the CDV introduced in the access network. However, to be able to perform this feature, the number of B-NT1s that can be connected to a single ONU must be restricted to 3. This still allows $3 \cdot 32 = 96$ customers to be connected.

Permits

The CPP protocol makes use of a *circular* array which consists of 128 entries (i.e. location 0 equals location 128). Each entry is able to contain the address of a B-NT1, to be included in the permit field of each downstream ATM cell. This array is implemented in a RAM and is read out at each downstream time-slot. A counter keeps track of the location to be read next. Upon the reception of an RAU, the RAM reads the number of permits required by the B-NT1s and writes this amount in the array, where it tries to spread the number of required permits over the circular array as fairly as possible.

The PDA of the BPP protocol makes use of a Fifo Permit Multiplexer RAM (FPMR). Every time an RAU arrives at the OLT the FPMR is updated about the arrival history, at 4 ONUs during a timespan of one frame of 144 slots. After receiving the arrival history of all ONUs of a fixed part of a frame, it issues permits accordingly. The protocol is named after the shape of the FPMR that resembles a dynamically growing *belt*.

8.4.3   The Priority Access Control Scheme

The *Priority Access Control Scheme* (PACS) is proposed in [Pan96] and combines the useful properties of both the Global Fifo and the BPP protocols. Just as in the BPP and CPP protocols, transmission requirements are periodically sent in the upstream direction. However, the PDA of this protocol does

not have the aim of rebuilding the arrival characteristics as they occurred at
the entrance of the access network. Just as was proposed in the Global Fifo
MAC protocol, the OLT of this MAC protocol receives the number of cells
stored by the ONUs during the time interval since the last transmission requirements were sent in the upstream direction. In both the Global Fifo and
BPP protocols, however, only requests from a restricted part of the terminations are received by the OLT at the same time. In the MAC protocol
described in this subsection, transmission requirements of all terminations
are advertised simultaneously.

Configuration

The protocol assumes a splitting ratio of 1:16, serving only 16 customers in
the case of fiber to the home (instead of 32 in as in the Global Fifo and BPP
protocols). On the other hand, fewer problems are expected with respect to
limitations of optical power.
The number of terminations connected to a single ONU is restricted to 6, still
allowing a maximum of up to 16*6=96 terminations to be served by the access network in the case of fiber to the kerb. Each ONU contains two buffers:
one for services which are sensitive to delay (traffic classes A and B) and one
for services whose QoS has to be guaranteed, but which are not very sensitive to variations in delay (such as ABR, classes C and D (Connection-less
services)). These buffers are called the *sensitive buffer* and the *non-sensitive
buffer* respectively. An ONU simply forwards an incoming ATM cell to the
appropriate buffer. Each cell waits in the assigned buffer until a permit au-
thorizes the ONU to transmit it in the upstream direction.

Requests

In the upstream direction, 19 slots, possibly containing ATM cells, are alter-
nated with *Request Access Blocks* (RABs). An RAB has a similar functionality as an RB in the Global Fifo protocol and a RAU in the BPP protocol:
advertisement of transmission requirements. A RAB contains information
concerning the number of cells which have arrived at each of the ONUs during the preceding 20 slots. Upon reception of an RAB, the OLT has a clear
overview of the number of cells which have arrived in both the sensitive buffer
and the non-sensitive buffer of all ONUs during a period of 20 slots, and can
now start to grant access to these cells by issuing permits.

Permits

As soon as an ONU receives a permit, it is authorized to send information in
the upstream direction. The nature of this information depends on the kind
of permit received. Three types of permit are distinguished:

(i) *Permits for ATM cells in a non-sensitive buffer.* These permits are only issued if no permits for cells in a sensitive buffer are scheduled. This forces non-sensitive services to make strict use of the spare transmission capacity of the shared access medium.

(ii) *Permits for RABs.* A permit for an RAB is sent downstream every 20 slots.

(iii) *Permits for ATM cells in a sensitive buffer.*

A *Permit Distinguish field* of 2 bits is part of the control field of the downstream slot format and indicates the functionality of the permit.

This method of granting permits guarantees a flexible use of the available data rate and a bounded delay for cells of sensitive services. Moreover, the MAC protocol allows Available Bit Rate and connectionless services to make use of the access network in a way that cannot harm the performance of the delay-sensitive services.

After a permit has been generated, it is stored in a FIFO buffer which is read out every time-slot. Generating permits for the ATM cells, waiting in either a sensitive or a non-sensitive buffer can be achieved very easily, since the sequence of the ONUs within an RAB is fixed. The contribution of each ONU within an RAB can therefore correspond with a pointer to the address of this ONU. However, if permits are generated according to this fixed sequence, an unfairness between the ONUs is introduced. Cells originating from ONUs whose fixed position is located in the last part of an RAB experience more delay than cells from ONUs which occupy the first part of a RAB. To overcome this, an efficient unfairness mechanism has been introduced in [Pan96], retaining the simple manner in which permits can be generated.

Protection mechanism

The idea of expanding the functionality of a MAC protocol by checking whether the transmission rights requested by the terminals correspond with the parameters agreed on during connection set up was adopted from the spacing concept of the Global Fifo protocol. The protection scheme of the PACS protocol, however, is based on the UPC mechanism that operates on a VP basis. This mechanism aims at restricting the influence of malicious users to one single ONU.

Before a permit is generated, the permit distribution algorithm checks whether the number of transmission rights requested by a particular ONU corresponds with the parameters agreed on during connection set up. If the mechanism considers the request to be compliant, the request is ignored (and no permit is issued). The protocol simply relies on its robustness scheme to prevent cells from staying in the buffer for a very long time. For a description of the robustness scheme we refer to [Pan96].

A drawback of forcing a MAC protocol to check whether the terminals behave as they promised they would is that the MAC protocol is not self-contained and, in fact, operates on higher layers than the ATM layer.

8.4.4   A MAC Protocol for a Dual Bus Architecture

In this subsection we describe an access network based on a slotted contra-directional dual bus topology, similar to the earlier DQDB [HCM90]. The principle of this MAC protocol relies on tracking the exact cell arrival times and reproducing a series of permits so that the flows output from the various terminations interleave properly along the bus. We assume that a maximum of N subscribers can be connected to the bus access network.

Two passive slotted contra-directional buses convey frames consisting of ATM cells and the MAC protocol overhead. The overheads encompass as many bits as connected subscribers (i.e., N bits). The subscribers access the upstream bus for write and read data on the downstream bus. A controller node located in the *Central Office* (CO) generates empty frames for the subscribers at a synchronized rate in both directions of transmission and implements the MAC Protocol.

The N MAC protocol upstream overhead bits convey requests while those in the downstream direction convey permits for transmission. The downstream frames also convey ATM cells coming from the core network, while the upstream frames carry outbound ATM cells. Each subscriber is tagged by a number i and each one should use the i-th bit in each header.

The MAC protocol works as follows. Each subscriber may queue for transmission at most one cell per slot. When subscriber i receives a cell for upstream transmission, it marks the i-th bit in the MAC protocol header as "1" in the next upstream frame and withholds the cell for future transmission. The controller receives this request and acknowledges it after an appropriately determined delay by setting the i-th bit in the corresponding downstream frame MAC protocol header. Several requests and permits (one per node) may be written within the same frame using this scheme. When the i-th subscriber observes that a permit has been granted by seeing the i-th bit set to "1" in the downstream MAC protocol header, that particular subscriber may write an ATM cell in the next upstream frame.

The controller adjusts the delays between the reception of a request until the transmission of a permit by appropriate initial ranging procedures in such a way that cells arriving in two contiguous time slots at two different nodes result in a back-to-back upstream transmission, as they would in an ATM multiplexer. This scheme leads to a minimal CDV under given traffic conditions.

Figure 8.4.1 illustrates the structure of the controller node. Requests issued from the upstream bus arrive at the first set of delay lines (1). These delay lines, which equal $(t_N - t_i)$, bring each subscriber virtually to the same distance from the controller node as the farthest one, so that from any subscriber, the propagation delay of the requests up to the multiplexer (2) lasts

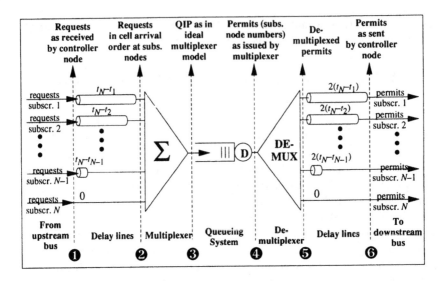

Figure 8.4.1: Model of the controller node functions

t_N slots. The farthest node request does not incur any extra delay. The vision of requests at location (2) is the same as at the access network inlets, but t_N time slots later. The delayed requests are then superimposed, which results in a QIP similar to that in the multiplexer case, at location (3). Permits consist in subscriber node numbers that are stored in a common queueing system. The order of storage when simultaneous requests arrive from more than one subscriber is random. At the multiplexer output, stage (4), the upstream cell transmission sequence is already known, which corresponds to the ATM cell sequence that is going to be received later. The demultiplexed permits are converted to a single series of bits (5), they are then delayed again by $2(t_N - t_i)$ (6), so that the roundtrip delay consisting of the permit propagation from the controller to the target subscriber, added to the return propagation delay of the emitted ATM cell results in a delay of $2t_N$ for every subscriber. These extra delays ensure that no collisions occur in the upstream cell transmissions under correct operation of the MAC protocol. Consequently, a cell sojourns at least $3t_N$ in this access network ($w_{i,min} = 3t_N$).

Clearly, the performance of such an access network matches that of an ATM multiplexer. The controller handles all terminations in a fair and efficient manner. However, the minimum sojourn time for each cell is: $w_{i,min} = 3t_N$.

In conclusion, we have presented in this subsection an access network based on a TDM passive dual contra-directional bus architecture. It constitutes an economical, fair and efficient solution for conveying real-time ATM traffic in the future B-ISDN. The cost of optical components being a key issue for a quick penetration of B-ISDN into the subscriber premises, the cost of the subscriber equipment must be kept low and the write access to the upstream bus is performed using passive components.

Traffic performance of such an access network should correspond to that of an ideal multiplexer model. Consequently, a central controller in the node has been designed that schedules cells as would the ideal multiplexer thus introducing a minimal cell delay variation during the upstream process. The bus architecture enables all subscriber clocks to remain synchronized to the frames and does not require extra preamble as in passive optical tree configurations.

The protocol requires, however, one bit of overhead per subscriber. This could be lowered, at the expense of degraded performance (e.g., share one overhead bit among several subscribers). A fixed minimum delay is also imposed on all upstream transmissions (which appears also in the passive optical tree solutions).

8.5   Multiple Cell Frame Based Protocols

The multiple cell alternative is based on a framed structure of the physical layer. As the name already implies, this family of protocols allows each network termination to send multiple cells at once in the upstream direction. The use of a framed approach allows a sampling of the status of the network termination, independently of the amount of traffic actually generated by either the termination itself or by other terminations. By so doing this, it limits the maximum value of the protocol reaction time (i.e. the time interval before the OLT is aware of fresh arrivals).

8.5.1   The McFred Protocol

In this subsection we describe one of the frame-based protocols, the *MAC FRame baseD* protocol, or abbreviated to *McFred*. A more detailed description of the protocol can be found in [TVC+94]. This protocol assumes the same configuration as the Global Fifo protocol, as explained in Section 8.4.1, i.e., cells are stored in the NT1 buffer before upstream transmission takes place.

Frame Structure

The information flow is organized in the upstream and downstream directions in frames of $125\mu s$. This choice allows $25\mu s$ to receive and process the requests and generate back permits. The frame structure in the upstream direction consists of two parts: The control block containing MAC protocol information and the information block containing the payload cells from the active NT1s.

The control blocks consist of 32 slots, one for each ONU, containing:

- The physical layer preamble for bit and byte alignment, and other physical layer functions.

- The MAC fields, one for each T_b interface.

The MAC field contains, for each NT1, the number of cells arrived during the previous frame. Assuming 149.76 Mbit/s as a net access bit rate for the T_b interface with a frame of $125\mu s$, the maximum value of this number is equal to 44 cells, which can be coded with 6 bits. The remaining two bits can be used, for example, for error protection.

In the downstream direction the MAC protocol information is represented by the transmission permit to the various NT1s. The permits are sent using one byte for each NT1.

The Permit Distribution Algorithm

The permit distribution algorithm works on *a priori* and *per request* base.

- An a priori assignment of the transmission permits is characterized by its simplicity. The drawback is represented by the impossibility of assigning bandwidth with a fine degree of granularity, since the maximum step is represented by one cell per frame. In the algorithm a given amount of pre-allocated permits will be assigned to the NT1s with the granularity derived from the frame length.

- Working on a per request basis allows any granularity to be obtained, because any cell is served just as it comes. The algorithm works by assigning the remaining permits not used a priori on the basis of the actual requests and on the amount of traffic that is expected from each NT1.

The Spacer Unit

The frame structure may produce clusters of cells at full link speed, heavily modifying its traffic profile. This could produce cell discarding in the successive UPC/NPC even if the user is expecting its declaring peak cell rate. Consequently an additional functionality at the LT side is needed to rebuild the original traffic profile on a per VC basis. This functionality is implemented by a Spacer Unit which separates the cells arriving too close together by adding an extra delay. It operates very similarly to the Spacing Policing unit used to perform the UPC function

8.6   Performance Evaluation

Already since the beginning of 1970, when Abrahamson introduced the first
MAC protocol (the ALOHA protocol), performance evaluation has been an
important aspect of the MAC protocol. Until the mid-eighties, information
throughput, channel utilization and (various forms of) delay were the three
most commonly used performance measures. However, after ATM had been
selected as the world standard for broadband ISDN for the public network,
allowing asynchronous and synchronous traffic to share the same resources,
variations in delay became a very important performance measure in all kinds
of telecommunication entities where resources have to be shared.

In this section we present a performance evaluation of the GF, BPP, Pacs
and McFred MAC protocols. All of them use a PON with a tree topology.
Similar studies for the other protocols described in this section can be found
in the bibliography. Performance evaluation is achieved by studying the
complementary distribution function (i.e. $\Pr\{X > x\}$) of two important performance measures: the transfer delay and the *Cell Delay Variation* (CDV),
introduced by each of the MAC protocols.

The *Transfer Delay* of the access network is defined as the time difference
between the sending of a cell at the T_b-Interface and the reception of this cell
at the V-Interface. To characterize CDV, the network performance parameter
known as *1-Point CDV* was chosen. This parameter describes the variation
of the arrival times pattern with respect to the negotiated peak cell rate. It
is measured by observing successive upstream cell arrivals of a tagged source
at the V-Interface and only considers cell clumping, i.e. the effect of cell
interarrival distances which are shorter than T, the reciprocal of the peak
cell rate. The characterization of CDV by means of 1-Point-CDV was given
in [ITU96a] and is recommended for CDV assessment by ITU-T.

All results presented in this section have been obtained from simulation.
In order to facilitate a performance comparison between the discussed protocols, related curves are shown in one figure. Confidence intervals have been
omitted to prevent figures from becoming cluttered. It should be noted that
95 % confidence intervals did not show any deviation larger than 10 % for
values larger than 10^{-3}. The performance of the McFred protocol is measured after the spacing unit and the priority scheme of the PACS protocol
was not used (i.e., all sources were assumed to be delay sensitive).

Studied scenarios

For a performance study of MAC protocols for an access network, it makes
no sense to investigate only one complex traffic mix. For complex systems
such as the PON with a tree structure, a quantitative analysis often hides
problems which might have an important impact on the performance of the
system. Performance studies must therefore be more general in order to be
"service transparent" to a high degree and in order to capture the impact

Table 8.6.1: Used traffic scenarios

Scenario Name	Peak bit rate	Mean bit rate	Mean Burst Length	Total no. of sources	Load	Multi-plexing Gain
1	2 Mbit/s	2 Mbit/s	-	210	0.8	1.0
2	34 Mbit/s	34 Mbit/s	-	12	0.8	1.0
3	50 Mbit/s	5 Mbit/s	100 cells	45	0.42	5.0

on the performance characteristics of all different features of the four MAC protocols for this topology.

Taking this into consideration, the traffic scenarios depicted in Table 8.6.1 have been chosen as input traffic to achieve the goal of this subsection. In all scenarios, the access network is loaded with homogeneous traffic with the given frequencies. The bit rates used in this table are defined at AAL level, ie., overhead introduced by both AAL and ATM layer is **not** included yet. For scenario 1 and 2 an AAL overhead of 1 byte per ATM cell was assumed, while for scenario 3 the AAL overhead was 4 bytes per ATM cell.

Scenarios 1 and 2 correspond to *Constant Bit Rate* (CBR) sources, equally spread over all connected devices. In scenario 3, ATM cells are generated according to an underlying Markov chain, alternating between two states: on and off. When the Markov chain is in the on-state, cells are generated according to a CBR pattern, whereas during the off-state no cells are emitted. Both the number of cells generated during the on-period and the sojourn time in the off-state are chosen to be geometrically distributed. This source model is characterized by three parameters: the peak bit rate, the mean bit rate and the burst length (i.e., the mean number of generated cells during the on-period) and was proposed in [SW86b] as a model for data traffic (e.g., flexible file transfer).

Figures 8.6.1 and 8.6.2 show that the performance of the transfer delay of both the BPP and the PACS protocols is independent of the bit rate of the service. This is in contrast with the GF and McFred protocols, where the transfer delay becomes large if the bit rate of the service becomes smaller. With respect to the 1 Point CDV, Figures 8.6.3 and 8.6.4 show that the BPP, the GF and the PACS protocol introduce higher values of CDV when the bit rate decreases. However, studies have shown that in the case of CBR traffic, upper bounds of the $1 - 10^{-5}$ quantile of the 1-Point CDV for the BPP and PACS protocols appear to be 25 μs and 45 μs respectively, even when the access network is loaded with thousands of 64 kbit/s CBR sources. As one should expect, the spacing unit of the McFred protocol guarantees that the access network introduces very little CDV for sources with all kinds of bit rates.

Figures 8.6.5 and 8.6.6 show the performance in the case of bursty input

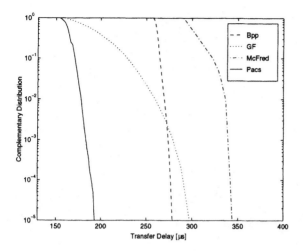

Figure 8.6.1: Transfer Delay in scenario 1

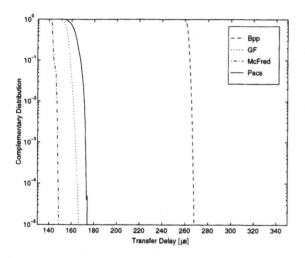

Figure 8.6.2: Transfer Delay in scenario 2

traffic. In this situation temporary overload situations may occur, causing more probability mass in the tails of the distributions. However, even in this situation all protocols show acceptable performance.

To prevent figures from becoming cluttered, results of CPP are not shown. With respect to BPP, the CPP protocol introduces slightly less transfer delay when the access network is loaded with homogeneous CBR sources. In the case of scenario 3, the complementary distribution of the transfer delay would start before the curve of the BPP, but its tail would end approximately 100

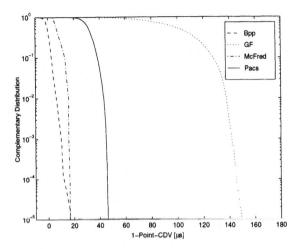

Figure 8.6.3: 1-Point CDV in scenario 1

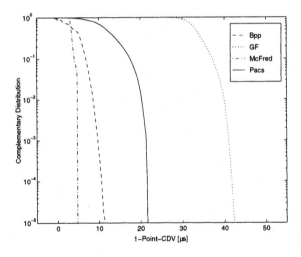

Figure 8.6.4: 1-Point CDV in scenario 2

μs after the tail of the BPP. The 1 Point CDV introduced by CPP varies with the bit rate of the source. For high bit rates (34 Mbit/s), the protocol has the same CDV performance as the PACS protocol whereas for small bit rates there is no clear bound of the $1 - 10^{-5}$ quantile in the case of CBR traffic (for scenario 1 the 1 Point CDV performance approximates the GF curve).

Based on the above results, the BPP and PACS protocols provide the best performance. However, the Global Fifo protocol has been implemented and has proved its capacities in a laboratory environment. All protocols are more or less of the same order of complexity and we do not expect severe problems

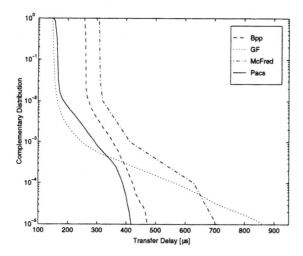

Figure 8.6.5: Transfer delay in scenario 3

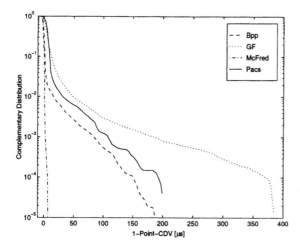

Figure 8.6.6: 1-Point CDV in scenario 3

during implementation of any of the other proposals. The McFred protocol, however, requires a spacing device which clearly will increase implementation costs.

With respect to the overhead each of the protocols requires to operate correctly, we note that McFred operates under least overhead requirements, even less than SDH. The ranking of the rest of the protocols with respect to the required upstream overhead is as follows: GF (5.4%), PACS (8.5%), BPP(9%) and CPP (9.3%). With respect to downstream overhead, the BPP, PACS and CPP all use 3.6 %, while GF uses 5.4% of the bit transport system for overhead requirements. All numbers include overhead inherited from the physical layer. Large upstream overhead values reduce the capacity of the access network at connection level.

9 Generic architecture and core network design

The complexity of multiservice broadband networks requires that the approach to network design and management is based on some systematic framework in which design / management problems and procedures could be systematically interpreted, defined and classified. Such frameworks were proposed for coping with the complexity of networks' *functionality*, e.g., the OSI architecture and the ATM network generic architecture. In this study we consider aspects of a network associated with *resources* and we aim to create a framework for the *design of network resources* (structuring and dimensioning). The framework is associated with a generic, layered architecture and the goal is threefold: (1) to introduce a formal conceptual base for network design problems, (2) to distinguish and classify generic problems and (3) to create a set of network design procedures for the generic problems.

9.1 Z-architecture: generic architecture of network resources

The OSI reference model has been widely accepted to describe the functionality of telecommunications networks. The Z-architecture is a proposal for a complementary model, aimed at the description of physical and logical network resources [LPTB93].

The network as a whole is viewed as a hierarchical layered structure. Each layer is itself a network. The network of layer n is built "on top" of the network of layer $n - 1$ in the sense that resources of layer $n - 1$ are used to build resources of layer n. The bottom layers correspond to networks built of the "simplest" resources of physical nature. The upper layers correspond to networks built of the "complex" resources of logical nature.

Six hierarchically ordered network layers can be distinguished in the Z-architecture (see Figure 9.1.1): Duct Network layer (DN), Physical Network layer (PN), Transmission Network layer (TN), Virtual Network layer (VN), Routing Network layer (RN), Connection Network layer (CN). Each layer corresponds to different types of resources and relations between them. Nevertheless the description of every layer is based on the same, generic objects (models of resources): nodes, links and paths.

The relations between objects of the same layer and between the objects of the adjacent layers (in particular, the way the objects are formed at each

Table 9.0.1: List of abbreviations specific to Chapter 9

BP	Basic Procedure
CN	Connection Network
CON	CONditions
DES	Network DESign
DN	Duct Network
GCL	Guiding Conditions for Link design
GCP	Guiding Conditions for Path design
NC	Necessary Conditions
NET	NETwork resources
NNS	Non-satisfactory Network State
NS	Network State
PN	Physical Network
REQ	REQuirements
RN	Routing Network
SNS	Satisfactory Network State
TN	Transmission Network
TRA	TRAnsition
VN	Virtual Network

layer) are generic. Paths are formed from links by switching consecutive links at intermediate nodes. Links of layer N are formed from paths of layer $N - 1$, at the layers' interface, by means of a multiplexing function grouping or splitting of paths. The interpretation of nodes, links and paths is layer specific; this applies also to the intra-layer switching functions and the inter-layer (interface) grouping/splitting functions.

The framework

Let a *network state* (NS) be defined by the following triple: NS = (NET, REQ, CON), where NET defines network resources and their configuration, REQ defines requirements NET should meet and CON defines conditions concerning the type of available resources and feasible resource configurations in NET, and also the type of services, service quality measures, network optimisation criteria etc. to which REQ refer.

A network state NS is said to be a *satisfactory network state* (SNS) if the network resources and their structure NET fulfill requirements REQ in conditions CON. In short we say that the triple defining an SNS is consistent. If the triple is not consistent we say the network is in a non-satisfactory state

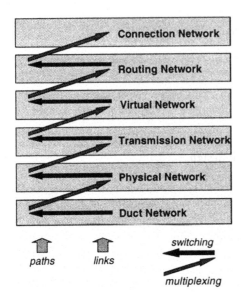

Figure 9.1.1: Z-architecture

(NNS).

A *network design* (DES) is a set of satisfactory network states:

$$\text{DES} = (\text{SNS}_1, \text{SNS}_2, ..., \text{SNS}_s); \qquad \text{SNS}_i = (\text{NET}_i, \text{REQ}_i, \text{CON}_i).$$

A typical example of DES: SNS_1 corresponds to a design regarded as nominal, while the other SNSs are network states corresponding to some predefined failures of resources which change CON_1, require redefinition of REQ_1 and reconfiguration of NET_1. Another example: SNSs of DES correspond to sequential extensions of the network involving the introduction of new resources or services.

The variety of possibilities in constructing DES is very large; the choice depends on the assumptions underlying an adopted network planning policy and the role of design in the planning methodology.

In the most general case, a *DES task*, performed with a DES procedure, consists in finding all three constitutive elements of each SNS from DES, i.e., NETs as well as REQs and CONs. In our research we focus on a special case of the general DES task formulated as follows: for each $i = 1, 2, ..., s$,

Given REQ_i and CON_i,
Find NET_i,
Such that $(\text{NET}_i, \text{REQ}_i, \text{CON}_i)$ is a satisfactory network state.

The above formulation decomposes the DES task into s independent SNS design tasks. As a consequence, the resulting DES does not take into account the feasibility of transitions between SNSs from DES. It is thus necessary, to extend the above formulation of the DES task so that DES satisfies this requirement. This leads to the following formulation of the DES task: for $i, j = 1, 2, ..., s$,

> **Given** REQ_i, CON_i and $TRA_{i,j}$,
> **Find** NET_i,
> **Such that** (NET_i, REQ_i, CON_i) is a satisfactory network state,

where $TRA_{i,j}$ defines the required relationships between the pairs (REQ_i, REQ_j), (CON_i, CON_j) and (NET_i, NET_j) to make transition between SNS_i, and SNS_j feasible. In particular, $TRA_{i,j}$ may define the transition as not feasible or not required.

Layered DES

In order to cope with the complexity of the DES task we shall decompose it into sub-tasks. The idea of the decomposition is closely linked to the Z-architecture.

The network at each network layer may be considered as a separate object for design on the condition that the separate designs are consistent with the designs of the networks of neighbouring layers. If this condition is met, the design of the network as a whole may be decomposed into several designs, each corresponding to a network of a single layer.

Let $DES(n)$ be the network design at layer n:

$$DES(n) = \{SNS_1(n), SNS_2(n), ..., SNS_s(n)\},$$

where

$$SNS_i(n) = (NET_i(n), REQ_i(n), CON_i(n)).$$

Let

$$SNS_i(n+1, n) = (NET_i(n+1, n), REQ_i(n+1, n), CON_i(n+1, n))$$

denote the information concerning the $SNS_i(n+1)$ that has to taken into account in the $DES(n)$ task in order to provide consistency of $DES(n+1)$. An analogous definition applies to $SNS_i(n-1, n)$.

We assume that

$$REQ_i(n+1, n), CON_i(n+1, n), NET_i(n+1, n)$$

and also

$$REQ_i(n-1, n), CON_i(n-1, n), NET_i(n-1, n)$$

are elements of $REQ_i(n) \cup CON_i(n)$.

Let $\text{TRA}_{i,j}(n+1,n)$ and $\text{TRA}_{i,j}(n-1,n)$ denote the information concerning $\text{TRA}_{i,j}(n+1)$ and $\text{TRA}_{i,j}(n-1)$, respectively, that has to be taken into account in the $\text{DES}(n)$ task.

We assume that

$$\text{TRA}_{i,j}(n+1,n), \text{TRA}_{i,j}(n-1,n) \subset \text{TRA}_{i,j}(n).$$

Now the $\text{DES}(n)$ task in the formulation has the following form: for $i, j = 1, 2, ..., s$,

Given $\text{REQ}_i(n)$, $\text{CON}_i(n)$ and $\text{TRA}_{i,j}(n)$,
Find $\text{NET}_i(n)$,
Such that $(\text{NET}_i(n), \text{REQ}_i(n), \text{CON}_i(n))$ is a satisfactory network state.

Generic DES task specification

Each layer consists of nodes (graph vertices) and links (graph edges) and is described by a multigraph $G = (V, E, P)$ where:

- V is the set of nodes
- E is the set of links (sets V and E are disjoint)
- P is a subset of $V \times E \times V$ (i.e. $P \subseteq V \times E \times V$) determining relations between nodes and links:

  - $\forall e \in E \; \exists v, w \in V, (v, e, w) \in P$ (each link links two nodes)
  - $\forall e \in E \; \forall v, w \in V, (v, e, w) \in P \Rightarrow (v, e, w)$ (links are undirected)
  - $\forall e \in E \; \forall v, w \in V, (v, e, w) \in P \Rightarrow v \neq w$ (there are no loops)
  - $\forall e \in E \; \forall v, w, u, t \in V, (v, e, w) \in P \; \& \; (u, e, t) \in P \Rightarrow (v = u \; \& \; w = t) \vee (v = t \; \& \; w = u)$

  (a link cannot link two different pairs of nodes)

A set of layer n paths will be denoted by $\mathcal{P}^n$. A path will be identified with the set of its links: $p = \{e_1, e_2, ..., e_k\}$.

Graph G describing layer n and all its elements is labeled with a superscript n ($G^n = (V^n, E^n, P^n)$, e^n, p^n, etc.). It is assumed that the sets of links of consecutive layers are pairwise disjoint. For $e^n \in E^n$ define:

$$L(e^n) = \{p_1^{n-1}, p_2^{n-1}, ..., p_l^{n-1}\}$$
$$O(e^n) = \{p_1^n, p_2^n, ..., p_l^n\}$$
$$U(e^n) = \{e_1^{n+1}, e_2^{n+1}, ..., e_j^{n+1}\}$$

$L(e^n)$ is the set of paths of the lower layer used to form link e^n; $O(e^n)$ is the set of paths of the same level traversing e^n; $U(e^n)$ is the set of links of the upper layer formed on the paths of $O(e^n)$.

State consistency conditions can be established in terms of Correspondability, Satisfiability and Realisability.

Correspondability (C)

The condition settles layer $n+1$-layer n structural consistency - well defined mapping between links of layer $n+1$ and paths of layer n (so that links of layer $n+1$ correspond to - are built of - layer n paths):

$$\forall e \in E^n \ \forall e' \in E^{n+1}, e \in \cup L(e') \equiv e' \in U(e)$$

Satisfiability (S)

The condition requires that paths of layer n satisfy capacity demands of layer $n+1$ links:

$$\forall e' \in E^{n+1}, S_{p \in L(e')} \operatorname{cap}(e', p) \geq \operatorname{cap}(e')$$

Realisability (R)

The condition provides that capacity of layer n links is sufficient to realise layer n paths:

$$\forall e \in E^n, \operatorname{cap}(e) \geq \tau(S_{\{(e',p):\ e' \in E^{n+1}\ \&\ p \in O(e) \cap L(e')\}} \operatorname{cap}(e', p))$$

where the following notation is used:

– if L is a set family then $\cup L$ denotes the set of all elements belonging to at least one set of L, i.e. $x \in \cup L \equiv \exists P \in L, x \in P$

– $\operatorname{cap}(e)$ is the capacity of edge e

– $\operatorname{cap}(p, e')$ denotes a portion of capacity of e' required on p

– S is a capacity summation operator

– function τ converts the capacity expressed in layer specific units of layer $n+1$ to the equivalent capacity at layer n.

$p(e)$, $\operatorname{cap}(p, e')$, operator S and function τ are layer specific. An exhaustive elaboration for the VN is presented in the next chapter.

For layer n and for a single network state, the generic DES task can be formulated as to find graph G^n and a set of paths $\mathcal{P}^n$ to minimise

$$\sum_{e \in En} \operatorname{cost}(e)$$

such that state consistency conditions (C), (S) and (R), and $CON(n)$, $REQ(n)$ are fulfilled.

All components of the generic specification may be associated with network state description elements NET, CON and REQ. G^n and $\mathcal{P}^n$ belong to $NET(n)$; $\operatorname{cap}(e)$, $\operatorname{cap}(p, e')$, $\operatorname{cost}(e)$, S, τ belong to $CON(n)$. Although $\operatorname{cost}(e)$ belongs to $CON(n)$, a part of its parameters is determined by $SNS(n-1, n)$. The objective is to minimise the total cost and to fulfill the constraints belong to $REQ(n)$. The values of $\operatorname{cap}(e')$ are given by $SNS(n+1, n)$.

9.2   The Virtual Network layer

In this section **VN (Virtual Network)** describes the network of virtual paths in ATM networks. In an ATM network links of VN correspond to ATM VP links. A VN link is formed through asynchronous subdivision of the capacity of a TN (Transmission Network) path (i.e., of a synchronous channel). Each of the obtained asynchronous channels is a VN link; a TN path can support several VN links. Paths in the VN are the familiar ATM VP connections and are formed by setting semi-permanent logical relations between addresses of incoming and outgoing VN links at intermediate ATM switches (ATM cross connects).

The problem consists in determining $\mathrm{DES}(n) = \{\mathrm{SNS}_1(n), \mathrm{SNS}_2(n), ..., \mathrm{SNS}_s(n)\}$ and $\mathrm{SNS}_i(n) = (\mathrm{NET}_i(n), \mathrm{REQ}_i(n), \mathrm{CON}_i(n))$ where n corresponds to VN layer, $\mathrm{SNS}_i(n)$ are satisfactory network states corresponding to the nominal network state ($i = 1$) and some set of network states associated with modified conditions concerning network resources (e.g. due to network element failures) or traffic conditions. These conditions are defined in $\mathrm{REQ}_i(n)$ and $\mathrm{CON}_i(n)$; the required relations between the pair $\mathrm{NET}_i(n)$ and $\mathrm{NET}_j(n)$, to make the transition between $\mathrm{SNS}_i(n)$ and $\mathrm{SNS}_j(n)$ feasible, is defined by $\mathrm{TRA}_{i,j}(n)$.

VN layer's $\mathrm{NET}_i(n)$ is modelled for each $\mathrm{SNS}_i(n)$ with an attributed multigraph $G^n = (V^n, E^n, P^n)$. Each $e \in E^n$ corresponds to a VP link and each $p \in \mathcal{P}^n$ corresponds to a VP connection (virtual path).

Specification of $\mathrm{CON}_i(n)$

In the following we specify formally only the essential elements of CON.

The information "filtered" from the RN layer, $\mathrm{NET}_i(n+1, n)$ specifies the graph G^{n+1} of the RN (Routing Network) layer.

The information intrinsic to the VN layer is represented by K traffic classes, distinguished with a set of attributes. The *VN capacity C* of a VN link e (or VN path p) is expressed as a set of *VN capacity vectors*: $\{\bar{c} = (n_1, n_2, ..., n_K) : n_k \geq 0, k = 1, 2, ..., S\}$. Each VN capacity vector represents a "traffic mix" on e (or p) expressed in terms of the number n_k of calls from traffic class k.

$\mathrm{CON}_i(n)$ defines operators that allow for handling splitting, superposition, etc. of VN capacity. Let C^* express the *boundary* of capacity C (* is a compaction operator):

$$C^* \overset{\triangle}{=} \{\bar{c} \in C : \neg \underset{\bar{x} \in C}{\exists} (\bar{c} \lesssim \bar{x} \wedge \bar{c} \neq \bar{x})\}$$

where the $\lesssim$ operator is defined below:

$$\bar{x} \lesssim \bar{y} \equiv \underset{k}{\forall} \ x_k \leq y_k$$

Capacity superposition is defined as follows:

$$C_1 + C_2 \overset{\Delta}{=} \{\bar{c}_1 + \bar{c}_2 : \bar{c}_1 \in C_1^* \wedge \bar{c}_2 \in C_2^*\}^*$$

Capacity splitting is defined as follows:

$$C_1 - C_2 \overset{\Delta}{=} \{(\bar{c}_1 - \bar{c}_2)^+ : \bar{c}_1 \in C_1^* \wedge \bar{c}_2 \in C_2^*\}^*$$

where the $^+$ operator is defined below:

$$\bar{x} = \bar{y}^+ \equiv \underset{k}{\forall} \; x_k = \begin{cases} y_k & \text{if } y_k \geq 0 \\ 0 & \text{if } y_k < 0 \end{cases}$$

Capacity filtering is defined as follows:

$$C_1 \mid C_2 \overset{\Delta}{=} \{\bar{c} \in N^K : \underset{\bar{x} \in C_1^*}{\exists} \; \bar{c} \lesssim \bar{x} \wedge \underset{\bar{x} \in C_2^*}{\exists} \; \bar{c} \lesssim \bar{x}\}^*$$

The definition of the *capacity comparison* operator is as follows:

$$C_1 \leq C_2 \equiv \underset{\bar{c}_1 \in C_1^*}{\forall} \; \underset{\bar{c}_2 \in C_2^*}{\exists} \; \bar{c}_1 \lesssim \bar{c}_2$$

$\text{CON}_i(n)$ defines principles for capacity translation between TN capacity and VN capacity.

Let function $\eta_{\text{GOS}}(\bar{c})$ determine the TN capacity required for VN capacity vector $(\bar{c})$, for given GOS criteria at VN layer (e.g., cell loss probability).

Let $C_{\text{GOS}}(b)$ determine VN capacity that is accommodated by TN capacity b with required GOS:

$$C_{\text{GOS}}(b) \overset{\Delta}{=} \{\bar{c} \in N^K : \eta_{\text{GOS}}(\bar{c}) \leq b\}^*$$

In the context of VN the state consistency conditions C, S and R (see Figure 9.2.1) are of the following form:

(a) RN - VN Correspondability (C):

$$\underset{e \in E^{VN}}{\forall} \; \underset{e' \in E^{RN}}{\forall} \;, e \in \cup L(e') \equiv e' \in U(e)$$

(b) VN-RN Satisfiability (S):

Note that for a VN link e, $L(e)$, $O(e)$ and $U(e)$ consist of one element each. In particular, each VN path corresponds to a single RN link. Thus if it does not cause any misunderstanding we abbreviate the notation $C(e', p)$ by $C(p)$. The consistency condition reads:

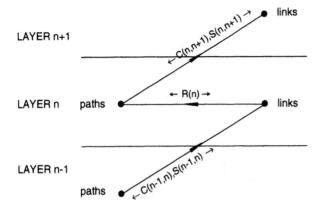

Figure 9.2.1: Layer n consistency conditions

$$\forall_{e' \in E^{RN}} \sum_{p \in L(e')} C(p) \geq C(e')$$

(c) VN Realisability (R):

$$\forall_{e \in E^{VN}} C(e) = C(p), O(e) = \{p\}$$

Apart from the above the *VN cost function* $\mathrm{Cost}_i(\mathrm{VN})$ is defined. It consists of two elements: the cost of utilising TN paths by VN links ("rental price") and the cost of forming/operating VN links and VN paths with switching facilities of the VN layer.

The information "filtered" from the TN layer is given by:

- $\mathrm{NET}_i(n-1, n)$ that specifies graph G^{n-1} and a set $\mathcal{P}^{n-1}$ of paths of the TN layer, and

- $\mathrm{CON}_i(n-1, n)$ that specifies the attributes of TN paths (capacity, cost, etc.), and also consistency conditions between VN layer and TN layer (see Figure 9.2.1):

(a) VN - TN Correspondability (C):

$$\forall_{e \in E^{TN}} \quad \forall_{e' \in E^{VN}} \quad , e \in \cup L(e') \equiv e' \in U(e)$$

(b) VN - TN Satisfiability (S):

$$\underset{e \in E^{VN}}{\forall} \quad C(e, p) = C(e), L(e) = \{p\}$$

(c) VN Realisability (R):

Note that in the TN layer the capacity of a TN path is the same as the capacity of any link of this path, thus the consistency condition reads:

$$\underset{p \in \mathcal{P}^{TN}}{\forall} \quad \underset{e \in E^{VN}: L(e)=\{p\}}{\sum} C(e) \leq C_{GOS}(b(p))$$

where $b(p)$ denotes TN capacity of path p.

The latter condition may be made more specific by taking into account that the superposition of VN paths corresponding to the same RN link should be filtered with the capacity of that RN link (this is because the splitting of C into C_1 and C_2, and next superposing of C_1 and C_2 does not yield C):

$$\underset{p \in \mathcal{P}^{TN}}{\forall} \quad \underset{e' \in E^{RN}}{\sum} \left(\left(\underset{e \in \cup L(e'): L(e)=\{p\}}{\sum} C(e) \right) \mid C(e') \right) \leq C_{GOS}(b(p))$$

Specification of $REQ_i(n)$

The essential elements of REQ:

- $NET_i(n+1, n)$ specifies the capacity of RN links.
- $REQ_i(n+1, n)$ specifies GOS constraints.
- $CON_i(n+1, n)$ specifies structural constraints (e.g., the maximum number of ATM hops on a VN path).

Inherent elements of $REQ_i(n)$ are the $DES(n)$ design objectives. In our formulation the objective is to minimise the total cost of all states of $DES(n)$:

$$Cost(VN) = \sum Cost_i(VN) \cdot Prob(i);$$

where $Prob(i)$ is the probability of SNS_i. Note that this objective is repeated in all $REQ_i(n)$, $i = 1, 2, ..., s$.

Specification of $TRA_{i,j}(n)$

The basic elements of TRA:

- $\text{TRA}_{i,j}(n+1, n)$ defines the equivalence between pairs of links from $\text{NET}_i(n+1)$ and $\text{NET}_j(n+1)$, and also the principles of transforming a link's realisation in $\text{NET}_i(n)$ into the realisation of its equivalent in $\text{NET}_j(n)$. An example of a simple transformation principle in the case of a transition from SNS_i to SNS_j due to traffic changes: keep equivalent set of VN paths, change only their capacity. Two sets of VN paths may be regarded equivalent if, for example, their mapping onto TN paths is the same.

- $\text{TRA}_{i,j}(n-1, n)$ defines the equivalence between pairs of links from $\text{NET}_i(n-1)$ and $\text{NET}_j(n-1)$, and also TN oriented constraints concerning, for example, the minimum time after which the set of TN layer paths from $\text{NET}_j(n-1)$ can be delivered in the transition from SNS_i to SNS_j (this time may depend on the adopted mechanism of paths protection in TN).

TRA also contains internal for layer n principles of transforming VN paths between $\text{NET}_i(n)$ and $\text{NET}_j(n)$ (e.g., only the capacity of VN paths may change but not their mapping on to TN resources). Usually, no special relations between pairs $(\text{CON}_i(n), \text{CON}_j(n))$ and $(\text{REQ}_i(n), \text{REQ}_j(n))$ are required.

9.3   Design procedures

Single layer design procedure

Consider the generic layer n of the Z-architecture. The design task for this layer involves conditions and requirements inherent to layer n and those that are imposed by the neighbouring layers (layer $n+1$ and layer $n-1$). These include the interlayer consistency conditions C, S and R; altogether, a set of five consistency conditions is involved in the layer n design task (see Figure 9.2.1).

The design of layer n has to satisfy, in particular, the five consistency conditions; these constitute a minimal set of necessary constraints. $C(n-1, n)$ and $S(n-1, n)$ are conditions that are to be met while forming layer n links from layer $n-1$ paths; $C(n, n+1)$ and $S(n, n+1)$ are conditions that layer n paths have to satisfy in order to support required links of layer $n+1$. $R(n)$ are capacity related conditions, internal to layer n; structure oriented conditions do not appear explicitly as they are intrinsic to the principle of forming paths from links (concatenation).

A network design for which all consistency conditions are satisfied is referred to as a *feasible* design. In this paper we focus on design procedures for finding feasible designs. Designs that meet extra conditions and requirements (satisfactory designs) can be obtained based on feasible designs; finding a feasible design is the essential part of the task of obtaining a satisfactory design.

A procedure that would be able to solve simultaneously all consistency conditions and find all links and paths is not known in the general case (only in simple cases of little practical interest the problem can be reduced to a solvable linear programming formulation). Therefore the problem has to be decomposed in some way so as to alleviate the need for simultaneous capturing of all consistency conditions, links and paths. In other words, some sort of decomposition (sequential handling) of the conditions, links and paths has to be applied in the design procedure. Decomposition leads however to new kinds of problems associated with the fact that the search of the solution space is not completely under control, i.e., the search is based on incomplete information and thus can lead to ineffective wandering about the solution space.

Decomposition requires distinction of iterated steps in the design procedure. In each step some set of layer n links and some set of layer n paths are designed so that some set of consistency conditions are satisfied. The basic design procedure is as follows:

Basic procedure (BP)

Iterate steps:

Step 1: design some set of links and some set of paths satisfying some subset of conditions $C(n-1, n)$, $S(n-1, n)$, $R(n)$, $C(n, n+1)$ and $S(n, n+1)$

Step 2: remove some subset of designed links, paths and satisfied conditions

In BP all types of consistency conditions (R, S and C) are treated simultaneously, but the amount of links and paths designed in one iteration step can be controlled to provide acceptable complexity.

A special case of BP (BPa) is as follows:

Iterate steps:

Step 1: design some set of links satisfying some subset of conditions

Step 2: design some set of paths satisfying some subset of conditions $R(n)$, $C(n, n+1)$ and $S(n, n+1)$

Step 3: remove some subset of designed links, paths and satisfied conditions

In BPa, the conditions $C(n-1, n)$ and $S(n-1, n)$ are processed (Step 1) separately from conditions $R(n)$, $C(n, n+1)$ and $S(n, n+1)$ (Step 2), and links and paths are designed in separate steps. This separation allows for reduction of the complexity of each procedure step and thus more links or paths can be designed simultaneously in a single iteration step.

The essential difference between BP and BPa is that they provide different means for controlling the search of the solution space for feasible design solutions. In BP the focus is on keeping consistency of designed links and paths in each iteration step. The drawback is that, as limited amounts of links and paths are designed in each iteration step, it is potentially difficult to maintain coherence of solutions of design in subsequent iteration steps.

In BPa, as greater sets of links (paths) can be designed in one iteration step the chances for controlling coherence are potentially greater. The drawback is that it is potentially difficult to design sets of paths consistently with the designed links (links are designed with no relation to consistency conditions concerning paths).

The above mentioned reduction of complexity in BPa is usually sufficiently large to enable designing of all links in Step 1 and all paths in Step 2. If so, in Step 3 *all* designed links, paths and satisfied conditions should be removed. We refer to such a special case of BPa as BPb.

Note that although BPa and BPb are special cases of BP, with all three procedures potentially the same space of solutions, in particular the same set of feasible designs, can be obtained.

To make any of the procedures workable some additional criteria have to be appended to guide the search of the solution space. In algorithmic terms the search method and bounding can be expressed with some additional set of constraints. These are however difficult to define if link and path sets are built stepwise, little by little; in particular, it is difficult to decide which of the available links are most appropriate for building a path. From this point of view, the greater the set of links available at an iteration step the better because subsequent selection of links can be based on a broader range of options, and thus potentially can be performed in a more conscious way in terms of possible consequences for further design steps. This is the main reason for which we advocate BPb.

Note also that if BP does not lead to finding a feasible solution then it might be difficult to decide whether this results from the fact that such a solution does not exist or the fact that additional constraints assumed in the algorithm are too limiting. This problem may, and usually will, occur after a design process has been advanced considerably. In BPb this type of situation may be detected in an early stage of the design process - in the phase of designing links (Step 1).

The final reason for advocating BPb follows from the fact that the choice of a design procedure should support a multilayer network design. As BPb assumes decomposition of consistency conditions within a layer this facilitates creating a procedure encompassing all network layers in a global multilayer, iterative procedure.

As already mentioned, in BPb the set of five consistency conditions is decomposed in Step 1 and in Step 2. This decomposition causes that Step 1 lacks sufficient guidance in the search of the solution space (no hints concerning paths that are to be realised with the designed links). To overcome this shortcoming we can append to the consistency conditions in Step 1 some additional constraints which are implied by the consistency conditions that appear in Step 2. The latter are transformed so as to yield necessary conditions (NC) for designing links in a coherent way with the designing of paths; if NC are not fulfilled no set of paths for a feasible design exists.

Two basic types of $NC(n)$, FNC and CNC, can be obtained from $R(n)$, $S(n, n+1)$ and $C(n, n+1)$. The first is related to a network flow formulation, while the second to a network cuts formulation (the relevant appropriate sets of formulae are reported in [LT95b]).

NCs are necessary, not sufficient conditions. It is thus necessary to provide some criteria which would select among such link-designs which may lead to a feasible network solution. Therefore some additional link-design guiding conditions, referred to in the following as GCL, are required in Step 1. GCL should reflect the "quality" of the design achieved in Step 1 with respect to the requirements of Step 2; the design is of insufficient quality if location and capacities of links do not match the needs of path design. GCL should modify Step 1 in the next iteration in order to achieve satisfactory link-design for Step 2. GCL are of dynamic character - they guide the iterative search of the solution space for feasible designs. In principle, GCL can be modified in each iteration with the account of the current and all preceding, intermediate network designs.

For solution space search efficiency, path-design guiding conditions - GCP - can also be introduced. The idea is to utilise information that can be gathered on the problems that occur during path-design (Step 2) during subsequent iterations. GCP should modify Step 2 in the next iteration in order to avoid bottlenecks (problems) encountered in the current and preceding path-designs. Like GCL, GCP can be modified in each iteration. Possible forms of GCL and GCP are discussed in [LT95b].

To reduce the complexity of link-design, in Step 1 only a subset of all necessary conditions can be considered. However, the subset should be modified in subsequent iterations to take account of problems encountered in previous executions of Step 2. The choice of NC should be correlated with the selection of GCL and GCP.

Taking into account the above settlements, the basic design procedure BPb can be rewritten as follows (see Figure 9.3.1); index i numbers iterations.

Basic procedure (BPb)
Iterate steps:

Step 1: design a set of links satisfying conditions $C(n-1, n)$, $S(n-1, n)$, some subset $NC_i(n)$ of necessary conditions $NC(n)$ and some set of link-design guiding conditions $GCL_i(n)$

Step 2: design a set of paths satisfying conditions $R(n)$, $C(n, n + 1)$, $S(n, n + 1)$ and some set of path-design guiding conditions $GCP_i(n)$

Step 3: remove all designed links, paths and satisfied conditions; select a new subset $NC_{i+1}(n)$ of necessary conditions and new sets

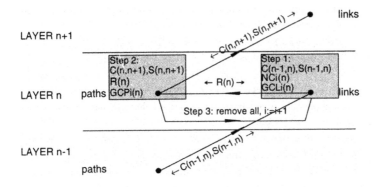

Figure 9.3.1: Layer design procedure

of link-design and path-design guiding conditions $GCL_{i+1}(n)$ and $GCP_{i+1}(n)$.

Multilayer design

If a layer n design is unsuccessful, i.e., no set of layer n links that yields a feasible solution can be found, then some layers below layer n, in particular layer $n-1$, might be redesigned.

The redesign of layer $n-1$ cannot be performed as an ordinary design process of layer $n-1$. This is because consistency conditions $C(n-1,n)$ and $S(n-1,n)$ cannot be formulated since links of layer n are not settled. What can be done however, is to formulate necessary conditions for paths of layer $n-1$ based on transformation (projection) of consistency conditions $C(n,n+1)$, $S(n,n+1)$ taking account of $R(n)$, $C(n-1,n)$ and $S(n-1,n)$. Such projected (necessary) conditions, denoted by $PC(n-1)$, will replace conditions $C(n-1,n)$ and $S(n-1,n)$. With this substitution the $n-1$ layer design proceeds according to the original formulation (BPb).

If this layer $n-1$ design is successful, the design of layer n can be retried. If still no feasible solution for layer n design can be obtained the design process of both layers might be repeated, although some additional (e.g., external) conditions have to be imposed to further direct the design. In particular, the $GCL(n-1)$ and $GCP(n-1)$ could be used for this purpose.

If the layer $n-1$ design is not successful (does not lead to a feasible solution) then $PC(n-1)$ have to be projected further across layer $n-1$. Necessary conditions $PC(n-2)$, have to be produced to replace conditions $C(n-2,n-1)$ and $S(n-2,n-1)$, and layer $n-2$ design has to be performed. Such a process may have to be continued down to the lowest layer. The whole procedure is summarised in Figure 9.3.2.

As to the problem of deriving PCs, note first of all that some kind of projection was already defined in the formulation NCs - consistency conditions

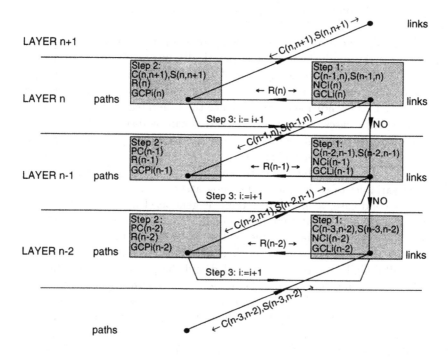

Figure 9.3.2: Multilayer design procedure

$S(n, n+1)$ and $C(n, n+1)$ are projected onto layer n links (bridged over layer n paths). To derive PCs the projection has to be extended further - bridged over layer n links - layer $n-1$ paths (see [LT95b]).

9.4   Implementation issues

In this section we describe the problems involved in implementation of the design procedure BPb introduced in the previous sections. We comment on the core algorithms, discuss in some detail the possible guiding conditions GCL and GCP, and comment on some experience concerning the performance of the procedure.

The task of Step 1 is to find a set of links which fulfill the involved consistency and necessary conditions. Whatever the layer considered, the task consists in solving a set of inequalities of generic type; the class of algorithms appropriate for solving the task depends however on the form of operators S and τ, and also on the form of capacity functions cap() that are used in the inequalities. In the simplest case all operators and functions are linear and continuous; if so, classical linear programming methods can be used. In most complex cases (nonlinear and/or non-continuous operators and functions) the only available algorithms consist in some sort of stochastic solution space search [LTB94], [LT95b].

In stochastic search methods the main problem is their effectiveness. The solution space is usually huge so it is practically impossible to examine all points of the space. In effect some kind of bounding of the space is required; in essence, the guiding conditions GCL fulfill such a role. In algorithmic sense, the GCL are conditions that influence the generated sequence of space points. Two basic cases of sequence generation can be distinguished.

In the first, subsequent points are generated randomly from the set of points that fulfill GCL. Such a generation method is not possible for every form of GCL. It is possible if GCL constrain the range of independently generated variables of each space point separately.

In the second case, a subsequent point is generated by means of a modification of a current point that fulfills GCL. The modification consists in changing the problem variable values within some limits, in a deterministic or random way. Such a new point is checked against GCL. If the new point does not fulfill GCL, the point is discarded and the generation procedure is repeated. This point generation scheme does not require a special form of GCL (as was the case in the point generation scheme described above).

The GCL are useful not only for stochastic search methods but also in the case of linear integer-programming based solution algorithms, for example. In the latter case GCL may be used to reduce the complexity (number of algorithm iterations) [LT95b], [LT94].

In general, there are many possible ways of defining GCL. Considering however that GCL support building links which serve building paths, the GCL should take account of the process of building paths in Step 2. In particular, an observed cause of unsuccessful path-design should be reflected in GCL. A typical cause for unsuccessful design is insufficient link capacity between a pair of nodes. GCL could impose a stronger lower bound on the number and total capacity of links between such a pair of nodes.

The task in Step 2 is to find a set of paths which fulfill the involved consistency conditions and also possibly, in the multilayer design case, the projected conditions PCs. The task is essentially a flow type problem. Note that this was also the case for the Step 1 task with necessary conditions of FNC form. Flows described with FNCs are however aggregated pathwise, whereas in Step 2 each path corresponds to a separate flow. Therefore more individual flows have to be considered and thus the complexity may be greater.

Here again, in the general case (nonlinear and/or non-continuous operators and functions), only stochastic search type solutions are possible. In the special case of linear operators and functions, effective linear integer-programming methods specially adapted to flow type problems can be used. If additionally the set of layer $n + 1$ links has a specific form - all links terminating at one node (startype network) - even more effective solution algorithms exist [AMO93].

Except for the above special cases, stochastic search based solution has to be used (see also Chapter 12), and thus path-design guiding conditions GCP are essential for effectiveness of design. The GCP should provide hints concerning building of paths for every layer $n + 1$ link. Typically, they may bound the set of layer n links from which the set of paths is to be built, and also bound the number of paths and their capacities. Similarly as for GCL, the GCP should build on the experience gained from preceding unsuccessful path-designs.

As in Step 1, two methods of generating subsequent points of the solution space are possible. The choice of the method depends on the form of GCP.

10 Multiservice network dimensioning

Most of this chapter is dedicated to the optimal dimensioning of an integrated ATM switched network subject to both call and cell blocking constraints. In particular, in Section 10.1 the overall problem is formulated in general terms; in Section 10.2 original models are given to solve the basic dimensioning problems, such as how to evaluate equivalent bandwidth parameters for bursty traffics, how to compute QoS fairness among traffics sharing the same link and how to evaluate peakedness of the overflowing traffics. Such models are integrated in a method aimed to extend alternative routing multi-rate models to the dimensioning of ATM networks where traffics are bursty, require different bandwidths and are fully switched at VC level. A further step of the analysis is made in Section 10.3, where the previous models are complemented with others and used to design a still fully integrated ATM network able to perform both VP and VC switching operations. In this case resource dimensioning and control cost are balanced against each other in an optimal way. Finally, in Section 10.4 the case of heterogeneous and variable traffics, implying the consideration of partial integration policies, is outlined. In this last section traffic clustering and resource partitioning are related to QoS requirements and to robustness constraints in different operating conditions.

10.1 Problem formulation

The question "How to integrate/segregate traffics" cannot be answered without considering resource dimensioning, clustering alternatives and management policies. Moreover, any reasonable formulation of the problem should consider traffic variabilities and abnormal network conditions.

To reduce model complexity, traffic variability can be simply expressed by a set A of I possible traffic realisations A_i (reference traffics), i.e., $A = \{A_i\}$, $1 \leq i \leq I$. Each traffic realisation A_i can be characterised by a probability or weight w_i. Since the network carries J different traffic types, every traffic realisation A_i is described by a set $\{A_i^j\}$ of matrices A_i^j, $1 \leq j \leq J$, whose entries represent point-to-point traffics of type j in the realisation i.

From a dimensioning point of view, a possible criterion to derive a final resource dimensioning vector, $R = \{r_k\}$, $1 \leq k \leq K$, from the separate dimensioning, $R_i = \{r_{ki}\}$, where r_{ki} is the optimal size of network element k with reference traffic i, could be based on worst case or average evaluation, i.e.,

$$r_k = \begin{cases} f_1(r_{ki}) = \sum_{i=1}^{I} w_i r_{ki} \\ f_2(r_{ki}) = \max_{1 \leq i \leq I} (r_{ki}) \end{cases}$$

Under such assumptions the following notations can be set:

A = reference traffic set, $A = \{A_i\}$, $1 \leq i \leq I$;

A_i^j = type j traffic matrix belonging to the i-th reference traffic, $1 \leq j \leq J$;

D = vector of j-th traffic call bandwidth, $D = \{d_j\}$;

W = traffic probability vector, $W = \{w_i\}$;

$\mathcal{B}$ = actual call blocking probability matrix, $\mathcal{B} = \{\mathcal{B}_i^j\}$;

$\mathcal{B}_i^j$ = call blocking probability for type j calls of the i-th reference traffic;

B = maximal call blocking probability matrix, $B = \{B_i^j\}$;

B_i^j = maximal call blocking probability for type j calls of the i-th reference traffic;

R_i = resource dimensioning relevant to the i-th reference traffic, $R_i = \{r_{ki}\}$, $1 \leq k \leq K$;

R = final resource dimensioning vector, $R = \{r_k\}$;

C = vector of unitary capacity costs, $C = \{c_k\}$;

$\mathbf{MPS}$ = link resource management policy set;

$\mathbf{TC}$ = traffic clustering alternatives;

Z_d = transmission and switching cost;

Z_c = control cost (a non linear function of R, $\mathbf{MPS}$ and $\mathbf{TC}$).

$\mathbf{MPS}$ can include different call access policies including $\mathbf{CS}$ (Complete Sharing), $\mathbf{TR}$ (Trunk Reservation) and $\mathbf{CP}$ (Complete Partitioning) policies. $\mathbf{TR}$ is characterised by a set of parameters t_j such that a new type j call is accepted if the link free capacity is at least equal to t_j. $\mathbf{PS}$ (Partial Sharing) policy can be seen as an intermediate policy between $\mathbf{CP}$ and $\mathbf{CS}$ policies and can be advantageous when traffic requirements are different.

With these assumptions, and assuming that the network is dimensioned according to a call blocking constraint, network optimisation can be formulated by the following optimisation problem:

Given

Find
$$\boldsymbol{A}, \boldsymbol{B}, \boldsymbol{D}, \boldsymbol{W}, \mathbf{MPS}, \mathbf{TC}$$

Minimising
$$\boldsymbol{R}^*, \mathbf{MPS}^*, \mathbf{TC}^*$$

$$Z_d(\boldsymbol{R}, \mathbf{MPS}, \mathbf{TC}) + Z_c(\boldsymbol{R}, \mathbf{MPS}, \mathbf{TC})$$

with
$$Z_d = \boldsymbol{C} \cdot \boldsymbol{R}^T$$
$$Z_c = g(\boldsymbol{R}, \mathbf{MPS}, \mathbf{TC})$$

where
$$\boldsymbol{R} = \text{either } f_1(\boldsymbol{R_i}) \text{ or } f_2(\boldsymbol{R_i}),$$

Subject to
$$\mathcal{B} \leq \boldsymbol{B}$$

The problem cannot be solved globally due to complexity. It can be solved following a suitable step-by-step solution approach by which traffic and dimensioning methods are merged with appropriate design rules.

10.2 Integrated network dimensioning with bursty traffics

In this section we shall assume that the integrated network we shall dimension is based on link modules of a certain given capacity, e.g., 150 Mbps. This is an advantage for an efficient statistical multiplexing compared to VP based networks discussed in the next sections since we get greater multiplexing zones.

Given these assumptions, the next important question is whether it is worthwhile to have alternative routing in this special type of multi-rate network with given link modules. As we consider a pure loss dimensioned network without delay, the general answer is yes, given that total traffic volume in the network has at least a certain size in relation to the link modules. We shall illustrate this further below.

There are of course a great number of non-hierarchical routing possibilities, but here our main conclusions can still be illustrated sufficiently in a hierarchical network. The examples will in fact be taken from cases with a 2 level and 12 node network where variants with either 1 or 3 nodes at the top level, with or without overflow. These variants can be compared with the same original background traffic matrices.

Each pair of nodes has a traffic mixture for a relevant (busy) period and each type of traffic is essentially characterised by its traffic intensity a and its bandwidth d. This bandwidth can either be a real bandwidth, e.g., for CBR sources, or in case of statistical multiplexing an equivalent (effective) bandwidth of some kind as shown in Part I, Section 5.2. In case we have a traffic control of the Weighted Fair Queueing type the considered rate d

could essentially correspond to the control parameter r. If we accept the busy hour like approach for dimensioning with respect to the GoS, we will have a network with multi-rate traffic and Poissonian arrivals of the original streams. We should not at call level integrate sources of any bandwidth, but do so in reasonable clusters of bandwidths within which the GoS will be in the same magnitude. One such cluster which might dominate the on-demand traffic may, for example, be between 0.4 and 4 Mbps. (These limits are a bit arbitrary but show a typical ratio between the greatest and the smallest bandwidth, and also a reasonable value of the ratio between the link rate and the greatest bandwidth in the cluster to combine with small blocking probabilities in the GoS).

Thus the conditions are quite well related to those for the simple formulae for link blocking in multi-rate models. These formulae are also identical to [Lin94] as it is shown in Part III, Section 18.2.3:

$$B = E((c - d + 1)/d, A/d) \qquad (10.2.1)$$

$$B_0 = E((c - R + 1)/d, A/d) \qquad (10.2.2)$$

where c is the link capacity, d is mean bandwidth in the mix, A is the total traffic volume of the mix, 1 is the smallest and R is the greatest bandwidth in the mix, B is the weighted average blocking for the different types of "call" and finally B_0 is the general blocking probability for all types of call in the case where we use the trunk reservation equalisation strategy.

For our alternative routing network with multi-rate traffic we must extend these results to overflow quantities and we must determine whether it is the case with or without equalisation of the individual blocking probabilities in the mix that we shall extend. Now, there are disadvantages in the case with no equalisation of an overflow stream from a first choice link if , for example, this overflow stream has a different mix to the original stream, and the call types of greater bandwidth in the cluster will have a greater proportion of the lost traffic after each overflow step. Thus we shall here illustrate the overflow extension for the case with the equal blocking constraint within the mix. To obtain this it is important to use the **TR** (trunk reservation) equalisation strategy (all types of calls are blocked at the same states) on all links, i.e., also on the first choice links. By doing so and with original Poissonian arrivals to the network, we have that every type of call in the same mix (the same traffic interest) will be blocked at the same states and thus we get fairness within the mix, though not always between different traffic interests in the network.

Note that though a mix can include an arbitrary number of types of source, the approximate blocking formulae above only depend on the total traffic A and the average bandwidth d of each individual mix and on the greatest bandwidth R, which here is supposed to be general for all mixes. Thus, (A_i, d_i) here denotes these general values for mix i, and not values for an individual type of source.

We now also use the standard mean and variance expressions for overflow traffic from one trunk group with n trunks and traffic intensity a (see [Lin83], for example)

$$M(n, a) = a\, E(n, a);$$

$$V(n, a) = M(n, a)\, [1 - M(n, a) + a/(n - a + 1 + M(n, a))] \qquad (10.2.3)$$

In our integrated network the mean and variance of the overflow stream from the first choice link i are then approximated by (see [Lin94], p. 905)

$$m_i = d_i\, M((c_i - R + 1)/d_i, A_i/d_i);$$

$$v_i = d_i^2\, V((c_i - R + 1)/d_i, A_i/d_i). \qquad (10.2.4)$$

If several such overflow streams are offered to a link of the final route, the corresponding traffic parameters of that superposition process are

$$m = \sum m_i$$

$$v = \sum v_i. \qquad (10.2.5)$$

If we now use Hayward's approximation (see [Lin83], for example) for overflow traffic, the extended peakedness factor is

$$z = v/m$$

and the blocked traffic on the final link will then be

$$m\, E((C - R + 1)/z, m/z), \qquad (10.2.6)$$

where C is the capacity of this link.

We can now start to dimension a star-shaped network with two levels and with either one or a few nodes at the highest level (e.g., 3 in a 12 node network). We then have methods to check which possible high usage links for the direct traffic are worthwhile to have in the network, and then dimension with them in the network for a certain GoS constraint, such as:

No more than 1% of the traffic (either overflow or direct) offered to a final link may be blocked.

This is a practical GoS formulation though it doesn't imply complete fairness between the direct and the overflow traffic. The direct traffic will get lower blocking than 0.01 on the final link, but also the original streams with overflow will here get a lower total blocking since they have more than one choice.

Now these methods can be applied, for instance, to this 12 node network with either 1 or 3 nodes at the high level and logical traffic matrices where the mixes include sources with bandwidths from 1 to 10 (e.g., 1, 2, 5 and 10, which could represent 0.4, 0.8, 2.0 and 4.0 Mbps), with different mean d values for different traffic interests (e.g., 3, 4 and 5). The capacity of the link modules could here be in the magnitude 300, i.e. 30 times the greatest bandwidth. With given proportions between and within the mixes one can let the total traffic volume increase to see when alternative routing becomes worthwhile and which structure to choose. We shall here only draw some general conclusions.

With a given number of nodes (here 12) and an increasing total traffic volume the savings of link modules by alternative routing will be significant above a volume corresponding to a need of a certain number (here round 200) of link modules in the case without overflow. The savings will then increase with even higher traffic volume in the network but with the number of nodes fixed. Of course, the savings should be compared with possible additional costs for a more complex call set up in the overflow case.

Given alternative routing is worthwhile, the corresponding cases with 1 and 3 nodes at the top level will give very similar network costs even if we let the length dependent part of the link module cost vary a bit. An additional reason for choosing alternative routing is thus also that the network is then not so sensitive to the right choice of structure as in the cases without overflow.

A more important reason for choosing the 3 top nodes variant is rather that there might be an upper limit for how many link modules one can connect to the largest node in the network and then this limit will be reached later.

Of course, the conclusions will be different if, as in the next section, one bases the network design on links which are VPs which can be arbitrarily small (and thus one looses important multiplexing efficiency, but on the other hand gets away from the module dimensioning) or if maximal path length is limited due to constraints for delay sensitive traffics.

10.3 Balancing resource dimensioning and control effort

In this section we consider the problem of dimensioning an ATM network, considering both inherent switching modes (VP and VC Switching).

Here, the structure of the network, i.e., the number and the position in the network of the VPs, is given and the main purposes of the analysis are:

(i) to determine how the network cost is shared at the optimum among transmission, switching (VP and VC) and control costs;

(ii) to evaluate the influence of the VP network structure on the cost of the network.

We consider a network with alternative routing and traffics meeting the established end-to-end grade of service constraints at the call level.

The network is managed according to a Complete Sharing policy (**CS**), where all service classes share the same VP network.

Before analysing the results of the numerical evaluations we describe the cost model and the adopted optimisation procedure.

Cost model

The total cost of the network is decomposed into the following three components:

$$Z = Z_d + Z_c = Z(T) + Z(C) + Z(S)$$

where: $Z(T)$ = transmission cost;
$Z(C)$ = switching cost;
$Z(S)$ = set up cost.

The transmission cost depends on the number of ATM links (or transmission paths) necessary to carry the traffic between two nodes. For dimensioning purposes, it is useful to refer this cost to the required equivalent bandwidth amount expressed as a multiple of a certain bandwidth module. The transmission cost can now be expressed by the number of modules on the different transmission sections. Assuming for simplicity that the VP length is proportional to the number of transmission sections and that VP switching is performed in each transmission node, the total transmission cost is given as

$$Z_T = \zeta_T \sum_f N_f t_f$$

where f is the VP index, ζ_T is the transmission cost per section and per module and the quantities t_f and N_f represent the number of links and modules of VP$_f$, respectively.

The switching cost refers to both VP and VC switching operations performed in the network nodes. Assuming a linear relationship between the cost and the average switched bandwidth, the total switching cost is

$$Z_C = \sum_f [\zeta_C + (t_f - 1)\zeta_P]D_f$$

where ζ_C and ζ_P are the switching costs for one bandwidth unit of switched VC and VP, respectively and D_f is the average bandwidth carried on VP$_f$. We assume that one VC switching and $(t_f - 1)$ VP switchings are performed in VP$_f$.

For each end-to-end traffic relation, the set up cost is assumed proportional to the average call rate and to the number of VPs tested during the call set up phase; so

$$Z_S = \zeta_S \sum_r L_r U_r$$

where ζ_S are the set up costs per call and per VP involved in the routing procedure, L_r is the average call rate for traffic relation r and U_r defines the average number of VPs involved in the call set up phase. The last quantity depends on the routing structure.

Optimisation procedure

The design and the dimensioning of the VP configuration of an ATM network addressed in this section is a multidimensional optimisation problem that can be solved resorting to traditional techniques integrating decomposition and a specialised optimisation algorithm [CMST92].

The result of the optimisation depends on the adopted routing structure, in particular on whether it is hierarchical or non-hierarchical.

In hierarchical networks, the routing plan follows some simple rules depending on a predefined node hierarchy. According to the hierarchical principle, the shortest routes (from the hop point of view) overflow onto the longest routes. The last or "final" route is composed of single hop VPs. The number of routes, which can become huge in case of traffic relations having large hop distances, can be limited by an input parameter.

In case of a hierarchical routing scheme the network can be engineered according to the Pratt method [Pra67] modified so that at optimum, with the defined cost functions, the marginal occupancies of the high usage VPs satisfy the following optimisation equation:

$$\zeta_T \frac{t_f}{H_f} = \zeta_T \sum_{i \in S_f} \frac{1}{\beta_i} + (\zeta_C - \zeta_P)(t_f - 1) + \zeta_S t_f$$

where S_f is the set of final VPs receiving the overflow of the high usage VP_f, H_f is the marginal occupancy of high usage VP_f and β_i is the marginal capacity of final VP_i.

The blocking probabilities B_f of the final VPs are adjusted iteratively until the end-to-end GoS constraints are satisfied for each traffic relation and each service.

In case of a non-hierarchical routing structure, any possible path between the origin and the destination nodes can be chosen. The traffic distribution among the routes and the overflow order can be varied, too.

In order to limit the number of routes and to choose the best ones, it can be useful to introduce a cost based metric. It can be seen [GK81] that to optimise the network the following weight X_f can be attached to the VP_f:

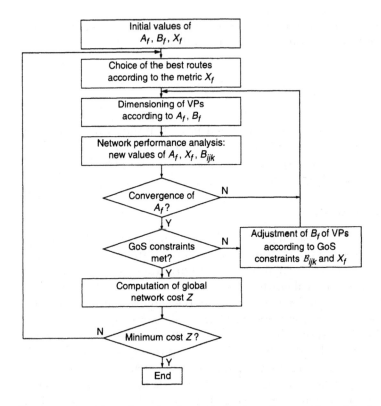

Figure 10.3.1: Flow chart for the optimisation of a non-hierarchical network

$$X_f = \frac{Z_f}{\beta_f}$$

where Z_f and β_f are the cost and the marginal capacity of VP_f, respectively. In this way both the cost and the efficiency of each VP are taken into account. The weight of a route is, of course, the sum of the weights of the involved VPs.

The selection of the routes can be made following different criteria (see also Section 12.2). Hereinafter, the optimisation problem is reduced to that of finding the K shortest paths in the graph weighted according to the metric X_f. The logical overflow order is from the least to the most expensive (following the metric) routes. The number of alternative routes for each traffic relation is limited by an input parameter and can also depend on the weight or on the hop number.

A flow diagram for the switching optimisation in case of a non-hierarchical routing is presented in Figure 10.3.1.

The illustrated procedure has been positively tested comparing the solutions with those achieved applying a simulated annealing method [BT93], i.e.,

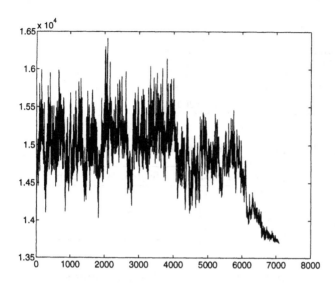

Figure 10.3.2: Cost development in the course of simulated annealing

a probabilistic method usable to find the global minimum of a complicated cost function that may possess several local minima.

To get a feasible approach the last method has been applied in two phases [KV94], respectively called VP optimisation and VP/VC optimisation. In the first phase the variables to be optimised are the capacities of the virtual paths. Virtual channel links are dimensioned with a specific algorithm which ensures that there are sufficient resources in the network to meet the required quality of service. As the first phase approaches the optimum, the second phase starts and the VC capacities are managed to make the final improvement taking the results of the first phase as the initial configuration. Figure 10.3.2 illustrates the convergence of such a simulated annealing algorithm towards the optimum value in the first phase of the optimisation (a seven node, twelve virtual path network is considered). The minimal value of the cost function is obtained in roughly 7000 time steps.

The main asset of simulated annealing is confirmed to be its robustness as it can be applied to almost any kind of problem. On the other hand, simulated annealing is computationally very intensive even for small networks so that, if the cost surface does not contain steep local minima, the above heuristic method can find the solution much faster.

Impact of VP structure and cost balance

The procedure illustrated in Figure 10.3.1 can be used with different VP logical structures. The simplest structure is to offer to each traffic relation a

Table 10.3.1: Service class characteristics

Service	Peak Rate	Mean Rate	Call Duration
T1	64 kbps	64 kbps	100 s
T2	2 Mbps	2 Mbps	2 s
T3	10 Mbps	2 Mbps	200 s

Table 10.3.2: Cost sharing in a 15-node network

VP/VC Strategy	Norm. Tot.	Transm. %	Switch. %	Set up %
Pure VP	1.1	62	36	2
Pure VC	1.2	24	68	8
Opt VP/VC	1.0	34	61	5

dedicated Virtual Path (pure VP strategy). This may correspond to a fully meshed logical network. As in large networks the VP number becomes very huge, this solution is not generally convenient from an economical point of view. The opposite strategy is to introduce only single hop VPs ($t_f = 1$) corresponding to the transmission paths (pure VC strategy). In this way the VP number is very small, but a VC switching is performed at every node; so the switching cost becomes relevant, because the traffic relations which are not linked by physically adjacent nodes are routed through many VC nodes. In general, a better solution corresponds to choosing an optimal number of Virtual Paths and to adopt a routing strategy offering alternative routes to each traffic relation (Opt VP/VC strategy). The determination of an optimal structure of the VP logical network is analysed in Chapter 11. Here a predefined number of alternative structures is analysed by the procedure of Figure 10.3.1 to evaluate the optimal balance of different cost components.

The procedure has been applied to a 15 nodes metropolitan area network having bi-directional VPs described in [MMST94]. The number of VPs in the network is included in the interval [14, 105]. Three classes of service with different characteristics (see Table 10.3.1) and QoS requirements have been offered to the network having a physical link capacity of 150 Mbps. For each service class a different end-to-end traffic matrix has been adopted. A few results [CMST92] are shown in Table 10.3.2 and Table 10.3.3. The strategy Opt VP/VC refers to a hierarchical routing plan with an optimal choice of VPs.

Table 10.3.3: Total cost vs VP number

Number of VPs	10	20	30	40	90
Normalised Total Cost	17821	17679	17395	17381	17610

10.4 Partial integration policies and network design

In the absence of a sophisticated and complex queuing control mechanism at cell level, the complete integration policy of all the traffics considered in the previous sections would imply a highly variable QoS distribution among the services and possibly the provision of the most stringent requirements to every connection, leading to huge overdimensioning of the network (see Part III, Section 18.1). There thus arises the opportunity to properly cluster traffics and to configure a different virtual network for each traffic cluster, so as to provide each service a quality not too much higher than that required.

Another problem that can arise from complete integration is related to control and management of the network. It can be reasonably argued that control operations of ATM networks, such as call admission, routing and policing can be strongly simplified in a network with only quasi-homogeneous traffics, making control functions in some cases feasible and always more robust.

Moreover, it can be also noted that, even when a control method is used for resource sharing purposes at cell level, the problem of traffic integration/separation still remains at call level.

In the following we first consider some possible criteria for clustering traffics and then evaluate the impact of the different partial integration policies on network dimensioning.

10.4.1 Traffic clustering

It is difficult to define exact criteria to cluster traffics in a multiservice network environment as it might be based on the achievement of different goals. In the following we mostly address two of these possible goals. One is relevant to the achievement of a better capacity saving in resource dimensioning when traffics have different call bandwidths and call blocking probability requirements. The other relies on the achievement of a better robustness of the design process. These two items are considered in relation to alternative resource management policies adopted in the network.

Dimensioning with different call bandwidths and GoS requirements

In this section we compare a complete integration policy at call level to certain other types of integration just limited within clusters of bandwidths for on-demand sources with some reasonable GoS constraints in terms of blocking probabilities. Suppose for simplicity that we have calls (connections) with bandwidths in the span from 0.04 to 40 Mbps and in this discussion divide this span into three magnitudes with 0.4 and 4 as the limits for these clusters. Each cluster has some calls with bandwidths equal to the limits, but typical bandwidths are 0.1, 0.2 and 1, 2 and 10, 20 respectively in the three clusters. It is quite obvious that we gain something by (call level) integration at least within each of these clusters. Generally if one dimensions each class separately with the same blocking constraint and then use this total capacity when integrating the three traffic classes then the average blocking will be smaller but the relations between the individual blocking probabilities might not be satisfactory. To obtain fairness one could use trunk reservation, but though this is quite effective when the greatest and the smallest bandwidth ratio is in the magnitude 10, it is not recommended if the ratio is as high as 100 or 1000.

The low bandwidths' cluster would have very efficient integration just by themselves anyway (if it has any considerable traffic volume at all) and one can there probably afford a quite low blocking probability, e.g. 0.001, and then the complete fairness in the class is perhaps not so important. Thus this class can not gain very much by integration with the other classes and it also has very little extra capacity to offer for all the classes to share.

The middle cluster could also afford a reasonably low blocking probability though perhaps a bit higher (0.01, say) and then it is more important to equalise within this cluster. This is quite acceptable here since the greatest to smallest bandwidth ratio is in the magnitude 10.

For the largest bandwidth cluster one can probably not afford a blocking as low as 0.01 and for the very high bandwidths (e.g. 40 Mbps) it may not even be possible to accept connections on demand.

Still, if we were to dimension the last cluster for a higher blocking and then instead use this capacity together with that for the middle cluster, i.e., integrate the two clusters, we would gain a little lower average blocking.

In spite of this fact this integration cannot be recommended for the following reason. The class with the greatest bandwidth can have very low traffic intensities compared to the other classes, and the blocking probabilities are very sensitive to the relative accuracy in the forecasting of these traffic intensities. Since we shall thus probably have to accept a situation with inaccurate traffic intensities for this class, it would be better to have the bad effects of this just in this great bandwidth cluster itself and not also in other clusters where the GoS will be even more badly affected.

Thus, we generally recommend call scale integration just within certain

Table 10.4.1: Dimensioning with robustness constraints

Oper. Cond.		A		B		MPS		
i	W^T	$j = 1$	$j = 2$	$j = 1$	$j = 2$	CP	TR*	PS
1 (Normal)	0.7	10	100	0.001	0.001	338	312	345
2 (Overload 1)	0.1	20	100	0.1	0.001	358	447	356
3 (Overload 2)	0.1	10	200	0.001	0.1	398	412	483
4 (Overload 3)	0.1	20	200	0.1	0.1	418	399	438
$d_1 = 10$, $d_2 = 1$		f_1 (average dim.)				354	**345**	370
		f_2 (worst case dim.)				**418**	447	483

a) equal GoS requirements in normal conditions

Oper. Cond.		A		B		MPS		
i	W^T	$j = 1$	$j = 2$	$j = 1$	$j = 2$	CP	TR*	PS
1 (Normal)	0.7	10	100	0.01	0.001	308	312	299
2 (Overload 1)	0.1	20	100	0.5	0.001	238	447	244
3 (Overload 2)	0.1	10	200	0.01	0.05	401	371	371
4 (Overload 3)	0.1	20	200	0.5	0.05	312	436	323
$d_1 = 10$, $d_2 = 1$		f_1 (average dim.)				311	344	**304**
		f_2 (worst case dim.)				401	447	**371**

b) unequal GoS requirements in normal conditions

* TR-equalisation

clusters of bandwidths of the same or adjacent magnitudes, though the actual limits between these clusters need not be exactly where they happened to be put in this discussion.

Dimensioning with traffic variations and network robustness constraints

To analyse the effect of traffic instability and unpredictable variability of working conditions on traffic clustering, we have considered a simple two-node network with two traffics and four alternative working conditions in which specific loss requirements must be met.

Such overload conditions are identified with uniform and selective 100% overloads whereas a higher loss is allowed for the increased traffic. They are described in Table 10.4.1 where A and B are respectively the traffic and the blocking matrices relevant to all the traffics and operational conditions.

The networks have been dimensioned using the three management policies **CP**, **TR** (set for the equalisation), **PS** (see also Part III, Section 18.2.1). The results are reported in Table 10.4.1. To derive a final dimensioning from the results for each working condition, the functions f_1, f_2 given in Section 10.1, have been used in conjunction with a probability vector W. As it is shown in the table, savings with **TR** can reduce significantly if instability of the offered traffic is considered. Under overload conditions, the **CP** policy is better for worst case dimensioning due to the different GoS requirements of the traffic classes. On the other hand, **TR** has a better performance for average case dimensioning. Finally, **PS** is preferable if traffic classes have different GoS requirements in normal conditions. This holds for both dimensioning criteria.

In the previous analysis only two traffic classes have been considered. The extension to a larger number of traffic classes is not straightforward. Addressing scenarios with different traffic volumes and characteristics (bandwidth, service time, loss requirements) [CCC+91] system behaviour becomes more complex to evaluate. In particular the determination of **TR** parameters becomes a difficult task when different QoS levels are considered, resulting in a less controllable network. Just for example, varying the reservation level of one traffic class can produce unpredictable variations in the losses of the other traffic classes [KW88]. This is another reason not to apply **TR** to distribute QoS differently among the traffic classes.

In conclusion, even if a pure **TR** policy can be beneficial as to resource utilisation, its applicability to heterogeneous traffic scenarios can be constrained by different requirements and by the need of a robust dimensioning. In such situations, a policy like **PS** can be more advisable.

The above conclusions are relevant to the case in which robustness performance is achieved by traffic clustering and network dimensioning. As it shall be shown in Section 12.1, the same objective can also be pursued resorting to some form of dynamic resource management of the network.

10.4.2 Partial traffic integration and network dimensioning

In the previous section we have considered the dimensioning of a single link as a function of the integration policy. Now we extend such considerations at network level in the context of VP/VC switching operation envisaged in Section 10.3.

We consider the same hierarchical 15-node network and traffic considered in Section 10.3. The following policies are compared with each other:

CP: Each service class has its own VP network, completely independent of those used by the other classes. Also the structure of the network is optimised service by service.

CS: All services share the same VP network.

TR: A trunk reservation threshold corresponding to the service with the greatest bandwidth is used in each VP to provide blocking probability equalisation.

PS: The high usage VPs, if they exist, are individually assigned to the service while the final VPs are shared among all the services.

To describe the application of the different integration policies to the hierarchical network, we use the code X_1/X_2, where X_1 refers to the policy used on high usage (HU) VPs and X_2 to the policy used on the final choice (FC) VPs. As we are considering three classes of traffic offered to the network (see Table 10.3.1), we have $X_1 = (x_{11}, x_{12}, x_{13})$ and $X_2 = (x_{21}, x_{22}, x_{23})$. The generic values of the $x_{i,j}$ have the following meaning:

0: traffic j is not offered to the HU or FC VPs

1: traffic j is separated from other traffics

2: traffic j is integrated with some other traffic using the **CS** policy

3: traffic j is integrated with some other traffic using the **TR** policy.

For example the case (1,2,2/1,2,2) means that we are considering the complete integration of broadband traffics T2 and T3 and the separation of narrow band traffic T1 on both high usage and final choice networks. **CS** is represented by (2,2,2/2,2,2), **TR** by (3,3,3/3,3,3) and **CP** by (1,1,1/1,1,1). We also address other policies like (1,1,1/2,2,2), (0,1,1/2,2,2), (1,0,0/2,2,2).

The comparison between the different management policies requires to refer also to the structures of the involved VP networks. The overall VP network has always 14 k ($1 \leq k \leq 3$) VPs depending on the adopted degree of integration. Similarly, the number of HU VPs is equal to $p \cdot k$ ($1 \leq p \leq 90$) depending on the integration policy. If the required end-to-end blocking probability B is set to 1% for all traffic classes, the network costs for different integration policies X_1/X_2 and different VP network structures p are reported in Table 10.4.2. When the maximal blocking probability is chosen equal to 1% for T1 and 5% for T2 and T3 the results are as shown in Table 10.4.3.

As it results from other evaluations, the network is such that with **CS** the total cost is equal to 18499 for the structures with the minimum number of VPs (14) and equal to 22124 for the maximal number of VPs (105). **CS** (2,2,2/2,2,2) is better than the **CP** policy (1,1,1/1,1,1) if we keep the same structure ($p = 30$) for each separated VP network. However, if we remove this constraint and optimise each VP network separately, we obtain a solution (1,1,1/1,1,1)* characterised by a different structure for different traffics leading to a cost lower than in the case of the **CS** policy. In fact, the optimal structure for narrow-band traffic T1 is achieved for $p = 90$ while for T3 traffic the optimal structure is achieved for $p = 20$. It can be noted that, in the last case, the network cost is rather flat with respect to p variations. A further improvement is given by the partial sharing policy (1,2,2/1,2,2), in which only T2 and T3 traffics are integrated, and which constitutes the best considered policy (cost equal to 17206) as the other considered policies imply greater network costs. In the given network the **TR** and **CS** policies seem almost equivalent to each other.

Table 10.4.2: Total costs of a 15-node network ($B_1 = B_2 = B_3 = 0.01$)

Policy Code	VP Network Structure p				
	10	20	30	40	90
(1,0,0/1,0,0)	3174	2951	2843	2764	2579
(0,1,0/0,1,0)	899	898	915	938	1060
(0,0,1/0,0,1)	14093	13920	13956	14046	14554
(1,1,1/1,1,1)	18666	17769	17714	17748	18193
(1,1,1/1,1,1)*		17397			
(1,2,2/1,2,2)		17206			
(2,2,2/2,2,2)	17821	17679	17395	17381	17610
(1,1,1/2,2,2)	18008	17807	17794	17862	18286
(1,1,1/3,3,3)	18243	18319	18525	18802	19868
(0,1,1/2,2,2)	19433	19553	19553	19675	20287
(1,0,0/2,2,2)	20645	20416	20311	20232	20046

When considering Table 10.4.3 with required blocking probabilities $B_1 = 0.01$ and $B_2 = B_3 = 0.05$, all the results are better than the results reported in Table 10.4.2. However in such a case, the best policy is the **PS** policy (1,0,0/2,2,2), whose cost is equal to 16626.

The above numerical results show that when the bandwidth ratios of offered traffics are too high, traffic integration may not be beneficial and other policies like those based on traffic clustering and traffic integration within the clusters can be preferable. Such a consideration holds even more when different GoS constraints or strict prescriptions during network overloads are fixed as design objectives.

Table 10.4.3: Total costs of a 15-node network ($B_1 = 0.01, B_2 = B_3 = 0.05$)

Policy Code	VP Network Structure p				
	10	20	30	40	90
(1,0,0/1,0,0)	3174	2951	2843	2764	2579
(0,1,0/0,1,0)	884	882	897	918	1033
(0,0,1/0,0,1)	13656	13597	13652	13737	14209
(1,1,1/1,1,1)	17714	17430	17392	17419	17821
(1,1,1/1,1,1)*			17058		
(1,2,2/1,2,2)			16916		
(2,2,2/2,2,2)	17546	17646	17114	17081	20941
(1,1,1/2,2,2)	17723	17503	17482	17545	18010
(1,1,1/3,3,3)	18204	18237	18403	18671	19743
(0,1,1/2,2,2)	19156	19384	19218	19353	19981
(1,0,0/2,2,2)	17206	16986	16883	16807	16626

11 Virtual Path network design

The design of Virtual Paths (VPs) plays a very central role in ATM networks.

Such logical networks are located at intermediate level between call admission, traffic enforcement and resource management [HW93]. They can be managed on semi-permanent or on demand basis to optimise switching operations and to improve resource utilisation and operational flexibility [ITU92, Fil89, TKS88, Bur90, Rob92a].

VPs can be used in particular:

- to configure temporary dedicated networks,
- to dynamically assign services to network resources,
- to execute more easily both routing control and management functions (acceptance, routing, monitoring, reconfiguration).

VP network optimisation can involve the consideration of many objectives, e.g. throughput, reward, fairness maximisation or cost minimisation, as well of constraints to be met in different operating conditions.

In the following, we assume that before entering the VP optimisation phase, services have been clustered according to Section 10.4 so as to create clusters carrying traffics with similar characteristics and QoS requirements. Under such assumption, the basic problems of VP network optimisation is the determination of the VP network structure and its dimensioning. These problems can be treated according to the two complementary approaches presented in Section 11.1.

As the same network platform can be used for the creation of different VP logical subnetworks, the way to partition the given physical capacities among the subnetworks also constitutes a relevant design issue. It is addressed in Section 11.2.

11.1 VP network topology optimisation

The first approach [ACS93, CFZ93b] for VP network optimisation considers the VP network as a transport network offering high performance direct connections at low cost per unit of traffic and high performance direct connections between network node pairs, which results in both link economy of scale and simplicity of switching operations. In this simple approach traffic and switching operations at the call level are not directly considered and a multicommodity flow model is employed.

In the second approach, the VP network topology is optimised to balance the VP/VC switching and control operations with network dimensioning [CMST92].

11.1.1 VP network design using a multicommodity flow model

The overall design of the VP network can be decomposed into a three phase process [CFZ93b, CFZ93a]:

(i) Find the optimal structure of the VP network,

(ii) Route VPs on the physical network,

(iii) Map demand on VPs, i.e., load and dimension the VP network using a multicommodity flow (MCF) model.

The first phase means selecting the pairs of nodes that will be connected by VPs, i.e., the VP terminator nodes. This is done by a *clustering algorithm* over the set of all non-neighbour node pairs. The pairs selected as VP terminator pairs are the cluster centres provided by the clustering algorithm. They represent the VP end-nodes of a core transit network. The clustering also defines an initial load assignment for the VPs.

The task of the second phase is to route the VPs on the physical network, i.e., to select a physical VP route to connect the VP terminators found in the first phase. This route selection is done by a randomised algorithm that is based on random choices among the set of all possible shortest paths connecting the given terminators. The objective is to distribute the routes in the network as evenly as possible to avoid overloading links, where the load on a link is interpreted as the total load of the VPs that traverse the link. It can be proven [CFZ94] that in a large network this algorithm produces a VP route-distribution that is close to the best possible.

Having found the VP terminators and routes, a refined loading and dimensioning of VPs is done in the third phase. This can be described as a *multicommodity flow model*, as follows.

For ease of description let us introduce the concept of *logical links*. A logical link between two nodes is either a VP connecting the two nodes or the non-VP fraction of a direct physical link between the two nodes. (The non-VP fraction of a physical link is the part of the link capacity that is not reserved for VPs). We interpret logical links as directed links.

Let us introduce the following notation:

L: set of all logical links in the network;

N: set of network nodes;

$L^{out}(i)$: set of nodes j such that $(i, j) \in L$;

$L^{in}(j)$: set of nodes i such that $(i,j) \in L$;

L_{pq}: set of logical links that use the physical link (p,q);

λ_{ij}: demand from node i to node j;

C_{pq}: capacity of physical link (p,q);

$x_{kl}^{(ij)}$: the part of capacity that is used for (i,j) traffic on logical link (k,l);

$c_{kl}^{(ij)}$: the cost of routing one unit of (i,j) traffic on logical link (k,l).

Now the capacity assignment can be solved by the following linear programming model.

Minimise

$$\sum_{i,j,k,l} c_{kl}^{(ij)} x_{kl}^{(ij)}$$

Subject to

$$\sum_{l \in L^{out}(i)} x_{il}^{(ij)} \geq \lambda_{ij} \qquad \forall\, i,j \in N$$

$$\sum_{k \in L^{in}(l)} x_{kl}^{(ij)} = \sum_{k \in L^{out}(l)} x_{lk}^{(ij)} \qquad \forall\, i,j,l \in N;\ \ l \neq i,j$$

$$\sum_{i,j} \sum_{(k,l) \in L_{pq}} x_{kl}^{(ij)} \leq C_{pq} \qquad \forall\, k,l \in N$$

Here the first constraint guarantees that all demands will be satisfied. The second constraint ensures flow conservation. The third constraint reflects the requirement that the traffic flow cannot exceed the physical link capacities.

Having found the optimising system of the $x_{kl}^{(ij)}$ variables by linear programming, the VP capacities can be computed as

$$V_{kl} = \sum_{i,j} x_{kl}^{(ij)}$$

where V_{kl} is the capacity of a VP connecting node k to node l.

After completing the third phase, the result can be refined by iteratively repeating the three phases, yielding a good sub-optimal solution.

11.1.2   VP network design using traffic and QoS models

As already underlined, in an ATM environment the determination of the structure of a VP logical network assumes a wider scope than in traditional circuit-switched networks, in connection with economy, QoS, fairness, robustness and feasibility of the network.

The complexity of the objectives makes the determination of a VP network structure an off-line process when compared with the other management activities envisaged in Section 12.1 below. However, for different reasons, related to the heterogeneous nature of the traffic offered to the network, or to the lack of knowledge of traffic size, there is still a need to configure or reconfigure the structure of the logical networks by procedures runnable in a short computer time [OY95] and able to manage a large set of operational constraints [YKH95]. The determination of the structure of the logical network depends also on its dimensioning and on traffic and VP routing in the physical network. It appears, therefore, that all these items must be considered all together in a single integrated process.

By using the notations introduced in Section 10.1 the problem can be formulated in the following way:

Given
$$A, B, D, \text{MPS}$$
Find
$$P^*, Q^*$$
Minimising
$$Z_d(P, Q) + Z_c(P, Q)$$
Subject to
$$\mathcal{B} \leq B$$
$$\mathcal{M} \leq M$$
$$\mathcal{L} \leq L$$

where:

P: set of feasible VPs;

Q: set of VP capacities;

$\mathcal{M}$: number of VP or VC switching nodes crossed by a call from origin to destination;

$\mathcal{L}$: number of VPs chained by a call to reach its destination.

M and L are limits related to call set up time and cell delay constraints. To solve the problem we have implemented the procedure shown in Figure 11.1.1 [MST96], which is structured in three main modules pertaining to *Demand Routing*, *Network Dimensioning* and *VP Selection*, respectively. In Demand Routing we first determine a set of paths that are feasible according to the M, L limits. Then the demands are routed according to fixed loading rules independent of traffics. After this point, each VP load is known and the logical network can be dimensioned and costed using the models described in Section 10.3. In the last module VPs are ranked and selected according to traffic-cost sensitive metrics. All the process is iterated until a local minimum is achieved. To guarantee a satisfactory optimisation a search process is

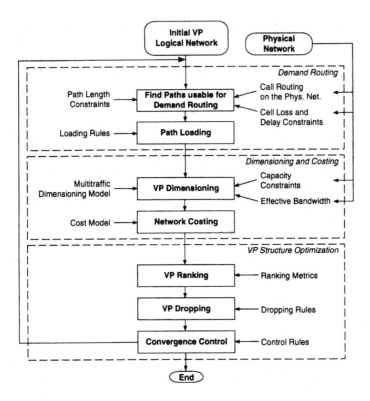

Figure 11.1.1: Determination of the VP network structure

performed in a finite region around the minimum to check that no other better local minimum exists in this region.

In this procedure some feedback from the physical to the logical network appears for different purposes, namely:

- to check that call routing is not too complex and does not overcome the limits established for cell delay;

- to check that call routing on the physical network meets a loopless condition;

- to verify that effective call bandwidths are evaluated correctly, taking into account capacity allocated to each VP.

The procedure has been applied to the same 15-node network and traffic characteristics (Table 10.3.1) already considered in Section 10.3.

Traffic intensities are varied so as to generate two different traffic mixes, mix 1 and mix 2, that have the same total amount of Erlang*Mbit/s, with the low bit rate traffic parcel equal to 65% or 35% of the total traffic. In the

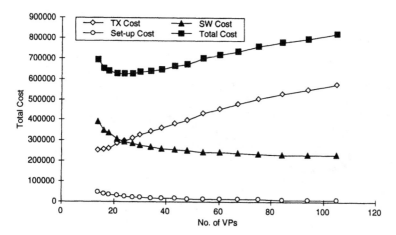

Figure 11.1.2: Cost component and total cost (mix 1)

evaluations we have considered the pure traffic integration policy, **CS**, and the pure traffic separation policy, **CP**.

We have considered also a structure constraint for which networks carrying each traffic can have the same (fixed) or different (adapted) structures.

Figure 11.1.2 shows the network cost shared among its main components for traffic mix 1. In Figure 11.1.3 the total network cost is evaluated for both **CS**, **CP** integration policies and traffic mixes, varying the number of VPs in the network.

The impact of a structure constraint (fixed-adapted structure) is evaluated in Figure 11.1.4. Firstly, the optimal structures relevant to separated traffics T1, T2, T3, are found and their costs summed up to find **CP** adapted structures. These solutions are compared with the **CS** policy sharing the same VP structure for all the traffics. The last Figure 11.1.5 reports the total network cost as a function of the maximal number of tandem VPs in a path, **L**.

Such evaluations confirm in particular that, in comparison with the design of traditional trunk networks, some important differences appear:

a) In circuit switched networks a fixed cost and a modular size are given to each trunk to represent the management effort; they put a limit on the number of trunk groups in the network and influence their location. Such technological limits are relaxed in ATM, even if a fixed cost of VP could be introduced here to represent CAC and policing functions.

b) In traditional networks the set up time does not represent as in ATM virtual networks a very constraining parameter and to achieve reasonable trunk utilisations, traffics can be bundled not necessarily selecting the shortest routing paths.

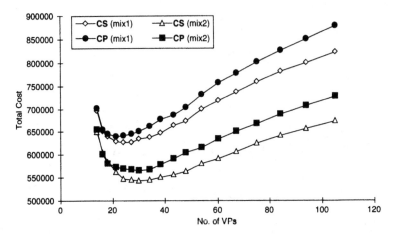

Figure 11.1.3: Total cost vs. different integration policies and traffic mixes

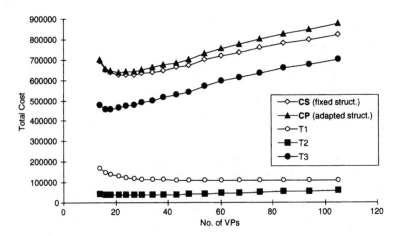

Figure 11.1.4: Total cost vs. VP network structure adaptation (mix 1)

c) In circuit switched networks, real time requirements in delivering information packets are intrinsically met due to the combined peak rate
dedicated resource assignment process. In ATM networks cell delay
and jitter are mostly related to the number of queues crossed by a call,
which depends on both the number of VPs and VC nodes encountered
by a call in its physical routing.

d) Another influencing factor is the effective bandwidth that must be reserved to accept a new call. Such bandwidth turns out to be a nonlinear function of the QoS and depends on the traffic size and on the
shared capacities. This factor can lead to a reduction in the number
of VPs employed in the network in particular when traffics are different and have a limited size. The variations of the effective bandwidth

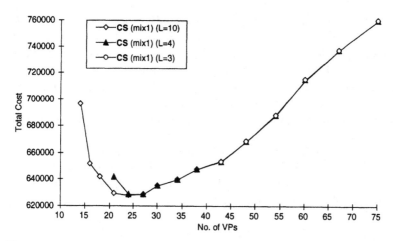

Figure 11.1.5: Total cost vs. different set up time constraints (logical paths lengths)

among the VPs belonging to the path used by a connection can also be significant when it is preferred to reduce the impact on the control effort.

11.2 Capacity partitioning among multiple VP networks

In the previous sections we have recognised that the *complete sharing* of resources in ATM networks can be quite troublesome from the viewpoints of network manageability and QoS guarantees. That is why it can be important to develop schemes for *optimal resource partitioning* for a given network among a set of VP logical networks, each carrying specific traffics. In [FBH⁺94a, FBH⁺94b] the following model is considered.

The network contains altogether J logical links, labeled 1,2,...,J, having capacity C_j. Let $C = (C_1, C_2, \ldots, C_J)$ be the vector of logical link capacities we have to optimise.

The condition that the sum of logical link capacities on the same physical link cannot exceed the physical capacity can be expressed by a linear system of inequalities. Let C_{phys} be the vector of given physical link capacities. Further, let S be a matrix in which the jth entry in the ith row is 1 if logical link j needs capacity on the ith physical link, otherwise 0. Then the condition we require is expressible as $SC \leq C_{phys}$.

We assume a set $\mathcal{R}$ of fixed routes is given in the network. Note that there may be several routes between the same pair of nodes, even on the same sequence of logical links. If several traffic types are to be carried, they are represented by parallel routes, i.e., the physical capacities are shared among a subset of logical links belonging to the same virtual network and the same traffic. The offered traffic (the demand) to a given route $r \in \mathcal{R}$ arrives as a Poisson stream of rate ν_r and the streams belonging to different routes are assumed independent. A call on route r requires A_{jr} units of capacity on link j. If the route does not traverse link j, then $A_{jr} = 0$. The A_{jr} values are collected in a matrix $A = (A_{jr}, j = 1, \ldots, J, r \in \mathcal{R})$. Call holding times are independent and holding periods of calls on the same route are identically distributed with unit mean (this distribution, however, can otherwise be arbitrary).

Our objective is to find the vector C of logical link capacities, subject to the physical constraints $SC \leq C_{phys}$ and $C \geq 0$, such that the *total carried traffic is maximised*. Using the Erlang fixed point model (see, e.g., [Kel91b]) we consider the following optimisation problem:

Maximise

$$\sum_r \nu_r \prod_j (1 - B_j)^{A_{jr}} \tag{11.2.1}$$

Subject to

$$SC \leq C_{phys} \text{ and } C \geq 0,$$

where the dependence of B_j on C_j is defined by the Erlang fixed point equations

$$B_j = E\left((1 - B_j)^{-1} \sum_r A_{jr} \nu_r \prod_i (1 - B_i)^{A_{ir}}, C_j \right).$$

Note that the effect of multirate traffic is approximated here such that the load generated by a call requiring k circuits is approximated by k independent calls needing one circuit each. (This is represented by the thinning factor $(1 - B_j)^{A_{jr}}$). It is known that asymptotically this "homogenisation" leads to the correct blocking probabilities [LH92].

The optimisation problem (11.2.1) is very difficult to solve exactly. On the other hand, it is shown in [FBH+94a, FBH+94b] that it can be well approximated by a convex programming task that can be solved by the following iterative algorithm:

Step 1 Set $B_j := 0$, $a_j := 1$ for each j.

Step 2 Solve the linear programming problem

Maximise

$$\sum_j a_j C_j \tag{11.2.2}$$

Subject to

$$SC \leq C_{phys} \text{ and } C \geq 0.$$

Step 3 Compute new values for the blocking probabilities by

$$B_j := E\left((1 - B_j)^{-1} \sum_r A_{jr}\nu_r \prod_i (1 - B_i)^{A_{ir}}, \tilde{C}_j\right)$$

where $\tilde{C}_1, \ldots, \tilde{C}_J$ come from Step 2 as a solution of the linear programming problem.

Step 4 Set

$$a_j := -\log(1 - B_j), \quad j = 1, \ldots, J.$$

Step 5 If all variables differ from their previous value by less then a given error threshold, then STOP, else repeat from Step 2.

The above iterative procedure has an intuitively appealing interpretation. If a link j at a certain iteration has large blocking probability B_j, then the corresponding a_j coefficient in the objective function will be large. That will inspire the linear programming to increase the value of C_j, since a variable with larger objective function coefficient can contribute more to the maximum. This conforms with the intuition that a logical link with high blocking probability needs a capacity increment. With minor changes the same procedure can be adapted to find the network maximising the expected revenues.

12 Resource management and routing

In the previous chapters various topics of network design have been addressed. Due to the multiservice environment and both variability and uncertainty of the traffics offered to the network, satisfactory resource utilisation and robust performance can be difficult to achieve simultaneously within the design process. To improve performance, some management actions should be introduced into the network with the aim of dynamically adapting the resource assignment to the current traffic levels. To this end, two main aspects of network management are addressed in this chapter. The first concerns the dynamic adaptation of capacities in a VP network and will be considered in Section 12.1. The second considers the traffic and VP routing in ATM networks (Section 12.2). Several routing algorithms are compared and routing schemes maximising network throughput and revenue are considered.

12.1 Adaptive VP capacity management

So far a logical system of virtual paths (VPs) has been considered beneficial to improve network dimensioning and to simplify traffic control and call acceptance. As traffics offered to the network are also uncertain and vary in an unpredictable way even at a short time scale, it is necessary to provide the network with a management system capable of allocating capacities to the VPs in a adaptive way. Such a system should be designed properly, balancing quality of service among the traffics and improving resource utilisation, while maintaining monitoring and control effort at an acceptable level. To this end, hereinafter we address a particular distributed system able to react to both short and long term traffic intensity variations. The system exploits the following operational features:

- VP capacities can be re-assigned at most each t_u time units, where t_u can be constant or variable;

- capacities are assigned to each VP just as a function of the current occupancy of the VPs;

- the relationship between the number of calls active on a VP and the VP capacity is calculated off-line assuring given call blocking probabilities, both in constant and variable traffic conditions;

- conflicts in the assignment process due to limited available link capacities, are solved by minimising the maximal distance between the required and the assigned VP capacity pairs;

  – almost independent controllers operate at the link level, employing a
    decentralised control architecture.

We assume a VP logical network is embedded in a given physical infrastructure and each VP is assigned to a given physical route; each offered traffic
between an O-D pair can be routed on its direct VP according to a model
where both VP network structure and call routing are kept fixed. The maximum capacity C_{max} available on each physical link can be set equal to the
sum of the capacity of VPs using that link, each one calculated by an appropriate dimensioning model according to the offered traffic and required call
blocking probability (the resulting link capacities serve as a reference point
to evaluate the capacity saving S achievable with the reassignment of the VP
capacities).

In managing the system, we have assumed the following simplified control
rules:

  – the VP capacity assignment is performed on each link independently of
    the others so as to reduce the total number of required controllers (the
    link number is lower than the VP number);
  – each VP is managed independently of any other VP sharing the same
    physical link as long as the total capacity required by the VPs does not
    exceed the link physical capacity;
  – the capacity assignment is performed by pre-calculated capacity assign-
    ment tables (CAT) that operate just on the knowledge of the number
    of calls in progress on each VP at the decision instants.

Even though the control action is decentralised at link level, a unique
CAT is used for a given VP, so that the same capacity is assigned to the VP
in each link belonging to the VP route in a decision instant. The possible
resource shortage in a link can make impossible the assignment of the full
capacity given by the CATs to each VP crossing that link, leading to incon-
sistent capacity assignments in the different used links. In this case, as the
maximum number of simultaneous calls in the VP is determined by the minimum capacity assigned on each used link, the decentralised control results
in a resource freezing and wasting. As it has been evaluated in [MPS95], this
drawback can negligible in most cases.

The structure and the operations performed by the link controller are
shown in Figure 12.1.1. To simplify the notation we have considered a peri-
odical control system with period t_u and three traffics carried by three VPs on
the same link. After having determined the capacity assignment table CAT
and t_u for each VP, at every decision instant $k \cdot t_u$, $h \cdot t_u$, $l \cdot t_u$ the controller
observes the number of calls, x, in progress in each VP and reads in CATs
the capacity value to assign. If the sum of capacities assigned to each VP is
not greater than the usable link capacity, then the controller confirms these
capacities, otherwise the link capacity is distributed among the VPs so as to

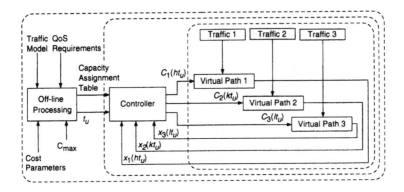

Figure 12.1.1: VP capacity adaptation: link controller structure (three VPs sharing the same link)

minimise the differences between required and assigned capacities [BMPS95]. Due to these features, the core of the control system is mostly related to the capacity management of a single VP operated by the relevant CAT.

12.1.1   VP capacity assignment table (CAT)

Hereinafter we assume that traffics offered to the VPs are characterised by Poissonian arrivals with average interarrival time $1/\lambda$, exponential holding time distribution with mean $1/\mu$ and constant or effective bandwidth b. The capacity assignment table (CAT) is determined by calculating the minimal capacity $C(k)$ to be assigned to the VP so as to satisfy a call blocking probability constraint during the time period t_u. The evolution of the VP occupancies between two successive decision instants $[k, k + 1]$ can be described by a $M/M/C(k)/0$ queue at the transient time [BDMS94]. By this model it is possible to estimate the value of $C(k)$ required to satisfy the call blocking probability in the interval $[k, k + 1]$, once the occupancy of the VP at the instant k is known with no regard to the particular decision instant considered. We can also observe that the evolution of the VP occupancies in the decision instants can be proved to be an ergodic Markov chain. Due to this property, it is theoretically possible to analytically calculate system performance measures such as the call blocking probability and the average assigned capacity. By this model the monitoring frequency $1/t_u$ of a single VP can be chosen so as to minimise an objective function combining both capacity and control costs by applying the Mladineo-Schubert optimisation method [BDMS94] (a linear relationship between the control frequency and the offered traffic size has been proved satisfactory in the numerical computations while considering a whole VP network).

It has been noted that the capacity assignment table is related to a given offered traffic. When it is used for other traffic values, the system perfor-

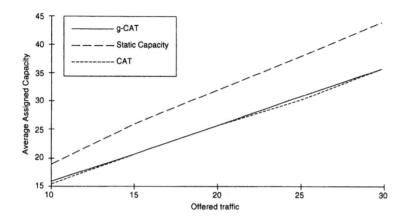

Figure 12.1.2: Average assigned capacity for CAT and g-CAT for a given blocking probability

mance can experience strong degradation; additional features must therefore be provided in the link controller to cope with traffic variations. The unknown traffic intensity could be estimated by simplified methods [VAD96] in which the number of past arrivals are counted and CAT can be adapted accordingly. In order to avoid on line measurements and calculations we can instead exploit the intrinsic adaptivity of the control system to design a generalised CAT (g-CAT) able to satisfy the call blocking probability constraint for a given range of traffic intensities. To this end the traffic is roughly estimated by the current number of active calls and g-CAT is obtained by the interpolation of different CATs, each relevant to a different traffic intensity value. The maximum available capacity C_{max} in this case is given by the sum of the static capacities of all VPs using a link, calculated so as to satisfy the required call blocking probability even in the case of the maximum foreseeable traffic A_{max}.

In the case of unknown and variable traffics the choice of t_u should depend on traffic estimation. Similarly to what we have done for CAT, the length of the time period between two successive decision instants can be varied according to the roughly estimated traffic.

In order to illustrate the performance achieved by g-CAT, we compare the average assigned capacity to the corresponding capacity required by using CAT when the traffic is known. As shown in Figure 12.1.2, the capacity saving degradation due to the use of g-CAT is negligible. In Figures 12.1.3 and 12.1.4, the capacity saving S and call blocking probability are reported for different parameter choices. It can be noted that, if the range of traffic variation is not very wide, even the use of g-CAT with t_u fixed can provide good performance.

We report also the results [MPS95] obtained by applying the proposed

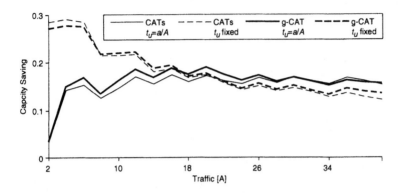

Figure 12.1.3: Capacity saving vs. traffic intensity for CATs relevant to different traffics and g-CAT

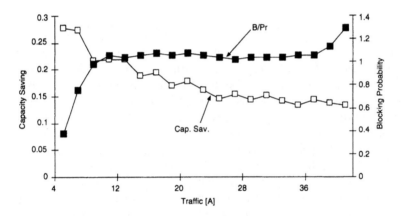

Figure 12.1.4: Capacity saving and normalised blocking probability vs. traffic intensity for g-CAT and $t_u = \alpha/A$ with $\alpha = 4$ and $A = 20$

control method to a 13 VP logical network whose traffic characteristics (maximal traffic intensity A_{max} and bandwidth b), call blocking probability requirements P_r and the required static capacity C_s are reported in Table 12.1.1. We have evaluated the capacity saving S achievable for three offered traffic patterns. With the first pattern the offered traffics coincide with the maximal traffics for which the g-CATs have been calculated. The value of S achievable for each VP is between 15% and 20% and the obtained call blocking probability satisfies the constraint $B \leq 5 \cdot 10^{-3}$ with the exception of VP6 and VP7 for which $B = 7 \cdot 10^{-3}$. With the second pattern traffic intensities are lower than the maximal ones. In this case, the capacity saving increases meaningfully as the available capacity is largely released, still satisfying the constraints. Furthermore, with the third pattern, traffics are even

Table 12.1.1: Characteristics of offered traffics and capacity savings for traffic vectors a), b), c)

VP	A_{max}	b	P_r	C_s	A_1	S_1	A_2	S_2	A_3	S_3
1	22	1	$5 \cdot 10^{-3}$	34	22	18.5%	15	45.9%	20	24.2%
2	16	2	$1 \cdot 10^{-2}$	50	16	16.8%	16	16.6%	12	32.2%
3	14	1	$1 \cdot 10^{-2}$	23	14	17.4%	10	34.8%	20	34.9%
4	20	1	$5 \cdot 10^{-3}$	32	20	19.5%	10	51.5%	11	48.2%
5	12	3	$1 \cdot 10^{-2}$	60	12	16.4%	8	36.7%	8	36.5%
6	24	1	$5 \cdot 10^{-3}$	36	24	17.4%	5	69.4%	12	51.1%
7	40	1	$5 \cdot 10^{-3}$	55	40	16.4%	10	70.8%	20	53.2%
8	16	3	$1 \cdot 10^{-2}$	75	16	16.9%	8	48.3%	8	48.2%
9	30	1	$1 \cdot 10^{-2}$	42	30	16.2%	15	52.7%	15	59.8%
10	16	1	$5 \cdot 10^{-3}$	27	16	20.2%	10	43.1%	10	43.2%
11	12	2	$1 \cdot 10^{-2}$	40	12	16.6%	9	31.2%	6	46.4%
12	14	2	$1 \cdot 10^{-2}$	46	14	7.2%	11	30.1%	16	9.5%
13	26	1	$1 \cdot 10^{-2}$	37	26	15.2%	16	43.6%	13	51.5%
						a)		b)		c)

slightly higher than A_{max}. In this case the required blocking probabilities are achieved at the price of a reduced capacity saving for the relevant VPs, as shown in Table 12.1.1.

In conclusion, the testing and the evaluation of the capacity assignment system show that even with the introduced simplifications and the limited control effort, network capacity consumption can be significantly reduced while meeting the required performance limits.

12.2 Routing in an ATM network

Routing constitutes a very important issue of high speed communication networks and can include several tasks in connection to call admission and flow control. In our context we consider the routing in a restricted sense; in particular we do not deal with priority requirements and multipoint routing.

Routing can be selected according to different objectives, such as to maximising the traffic carried in the network (or the network revenue once a fee is paid by the user for the successfully routed call) or balancing resource utilisation with the aim of protecting the QoS experienced by future incoming calls.

The problem is to find the possible paths in the network connecting each source and destination of traffic patterns that optimises a proper objective function, and it can be formulated as a capacitated non-linear integer valued multicommodity flow problem.

In the following we address three different issues of routing in ATM network:

- allocation of virtual paths (VPs) in the network of ATM physical links (*VP routing*);

- allocation of connections in the network of VPs (*connections routing*);

- allocation of permanent virtual channel connections (VCs) in the network of ATM physical links (*PVC routing*).

In more detail, in Sections 12.2.1 and 12.2.3 we present and compare several routing algorithms usable for VP and PVC routing in an ATM environment, while in Section 12.2.2 we report a solution of a specific connection routing problem aimed to dimension the VP capacities maximising the total traffic carried by the network under given blocking constraints.

12.2.1   Allocation of VPs in the ATM physical network

In this paragraph we consider the allocation of VPs to paths of the ATM network composed of ATM physical links when the traffic (measured in Erlangs) offered between pairs of nodes to be connected by the VPs is known. In our context a set of VPs between a pair of nodes is dedicated to connections of a particular service (or a class of services).

Let D be the set of *traffic demands*. Each traffic demand $d \in D$ is characterised by its:

- end nodes $v(d)$ and $w(d)$

- service class $s(d) \in S$

- required GoS, defined as the maximum allowed blocking probability $\beta(d)$ for the connections of d

- offered traffic $A(d)$ (defined as the connection arrival intensity times the connection mean holding time)

- set of admissible paths $P(d)$ in the considered multigraph G of ATM physical links for realising VPs serving the connections of demand d (note that VPs are dedicated to demands, several VPs can serve one demand, and at most one VP is realised on one path $p \in P(d)$).

Note that we assume service-dedicated VPs and separation between VPs on physical links. In such a case the statistical multiplexing of connections is applied only within a VP, and VPs are established for origin-destination (OD) pairs of nodes and for service classes.

Let E denote the set of (physical) links of multigraph G modelling the ATM network. Each link $e \in E$ has fixed capacity $c(e)$ expressed in capacity units. A capacity unit is defined as a bit-rate [Mbps] (or a cell-rate [cells/sec]) large enough to carry at least one connection of any service class from the set of all service classes S. With each $s \in S$ there is associated a function $f_s(i)(i = 0, 1, 2, ...)$ giving the maximum number of connections of class s that can be supported by i capacity units on a link, i.e., the number of virtual circuits supported by i capacity units on one VP (cf. [GAN91]). A *feasible* (flow allocation) state is defined by finding for each traffic demand $d \in D$ the capacity $y(d, p)$ (expressed in capacity units) of its VP on each path $p \in P(d)$, so that the total capacity of the VPs of d, expressed in virtual circuits, is sufficient to satisfy GoS, i.e.,

$$E\left(\sum_{p \in P(d)} f_{s(d)}(y(d, p)), A(d)\right) \leq \beta(d)$$

and that for no edge e the resulting number of capacity units allocated to it exceeds its capacity $c(e)$ ($E(\cdot, \cdot)$ denotes the Erlang loss formula). The problem, in general NP-complete, is to find a feasible state (flow allocation) which maximises a given objective function, represented by the traffic carried in the network, or achieves the network fairness.

Note that the above formulated problem may be directly applied to dimension and allocate VPs assuming the simplified "direct connections routing", where each traffic demand is provided with a direct "virtual circuit group" composed of a set of VPs used exclusively to carry its traffic. If this is not the case ("transit/alternative connections routing"), a solution to the connections routing problem defines the traffic offered to "virtual circuit groups" and their target blocking probability, specifying the input data for the VP allocation.

Comparison of different VP routing techniques in an ATM network

Two kinds of routing methods have been investigated, based on shortest path allocation [Arv94, GKW91, CBP95] and stochastic allocation [PG95b] respectively.

Shortest path allocation (SP)

A major choice in path allocation is whether this is *network state dependent* or on a *Pre-Computed* basis. When the first method is used, the route computation is done at the time of the request arising from the network management system and is based on current traffic distribution or resource utilisation, leading to a delay due to the calculation time. In contrast, when pre-computation is used, VP set-up is faster, but the lack of knowledge on the precise characteristics of the traffic carried by the VP can be a challenge.

Pre-computation is attractive when the pre-computed routes can be used for a significantly long time period without re-calculating them [CBP95].

In simpler shortest path allocation algorithms a cost is assigned to each link of the network. The VPs are accommodated by the network in decreasing order of their bandwidth requirements. After assigning a path to a OD pair the costs along the chosen path are modified and the procedure iterated.

All these algorithms are computationally efficient and do not require extensive computing capacity. On the other hand they do not guarantee an optimal solution and the result is dependent on the ordering of the VPs. This single cost approach can be generalised assigning a cost vector to each network element [CBP95].

In this case, the individual cost labels can be either *additive*, meaning that the cost of a path will be the sum of those of the links it goes through (it could be the cell loss or delay or the number of hops a VP has to do in the network), or *restrictive*, meaning that the cost of a path going through a list of links is the minimum of their restrictive costs (it may represent the available bandwidth of a link at a certain point in time).

There are different possible implementations of the algorithm depending on the selected allocation rule. In the following we consider two path selection strategies called Least Loaded Path and MAX, respectively, with the following characteristics:

- Least Loaded Path (LLP)
  Paths are selected to distribute the network load in a balanced way, so as to leave more bandwidth for the arriving calls.

- MAX
  Paths are selected considering the gain in terms of traffic carried on a VP due to adding a capacity module to it, minus the sum of (potential) carried traffics on all direct VPs on edges of the considered VP, that would be carried if one capacity unit were added to them [GKW91].

Simulated allocation (SA)

A stochastic VP allocation method was proposed in [PG95a, PG95b]. For each traffic demand $d \in D$ we compute a minimum number $t(d)$ of virtual circuits such that

$$E(t(d), A(d)) \leq \beta(d),$$

i.e., $t(d)$ is the minimum number of virtual circuits required to carry the traffic $A(d)$ with the required GoS. Then, with each traffic demand we associate a Poissonian stream of arrivals, called *VP-demands*, with intensity $\lambda(d, \boldsymbol{y})$ depending on state $\boldsymbol{y}$ of the network (the notion of the state is described below). For each $d \in D$, a VP-demand arrival from its stream requires one capacity unit and is allocated (added) to the VP on an *accessible path* between

$v(d)$ and $w(d)$ (i.e., to a path from $P(d)$ whose edges all have at least one free capacity unit). An arrival is rejected if no path is currently accessible. The essence of SA is that we allow the already allocated capacity units on VPs to be disconnected with rate $\mu(d)$.

The state space Y of the introduced Markov chain is finite with states of the form

$$y = (y(d, p), d \in D, p \in P(d)),$$

The state space Y depends on the arrival intensities $\lambda(d, y)$ and on the rule used to select accessible paths for the allocation of arrivals. Such an *allocation rule* should take into account the impact of a capacity unit allocation on the value of the objective function, i.e., on the total carried traffic. For realistic allocation rules the characterisation of the state space is rather straightforward and the chain is irreducible. Let $x(d, y)$ denote the number of virtual circuits of traffic demand d allocated in state y, i.e., $x(d, y) = \sum_{p \in P(d)} f_{s(d)}(y(d, p))$. A state y is said to be feasible if $x(d, y) \geq t(d)$ for all $d \in D$ and the capacity $c(e)$ of any edge $e \in E$ is not exceeded. Below we assume that a feasible state exists.

To maximise the objective function any kind of the envisaged Shortest Path allocation rules can be considered.

In order to find an allocation pattern of VP-demands we generate a trajectory of the underlying Markov chain. The trajectory is terminated when for each traffic demand $d \in D$, $t(d)$ or more virtual circuits are allocated. If such a flow pattern cannot be found in a reasonable time (such a pattern may not exist at all) we may stop the process after a certain time or by using some other (trajectory dependent) stopping rule.

Let us consider the embedded discrete-time Markov chain (called *allocation chain*) with time epochs corresponding to arrival and disconnection instants of the original (continuous time) Markov chain described above. The number of steps to reach any feasible state (first passage time to the set of feasible states) from a fixed initial state is a random variable called *allocation time* and denoted by T. Notice that T is finite with probability 1, and so is its expected value. This last quantity is called *allocation effort* and is denoted by F.

We note that the allocation effort F of the considered problem depends on arrival intensities $\lambda(d, y)$, disconnection rates $\mu(d)$ and the selection rule.

Two versions of SA have been analysed:

– SA/PP, employing simulated allocation with static search for allocation paths from a predefined set of admissible paths combined with any of the above listed allocation rules

– SA/DP, employing the dynamic (on-line) search of optimal paths at any arrival event (VP-demand) based on the LLP allocation rule.

Table 12.2.1: Comparison of different routing algorithms

	RAN	LLP	MAX	Ref. meth.
TTC	$28625.74 \pm 0.04\%$	$28916.01 \pm 0.03\%$	$28912.89 \pm 0.02\%$	28823.24
TTC/TTO	$0.98 \pm 0.04\%$	$0.99 \pm 0.03\%$	$0.99 \pm 0.02\%$	0.98
$E(B)[\%]$	$1.99 \pm 1.78\%$	$0.99 \pm 2.48\%$	$1.01 \pm 2.28\%$	1.32
$Max(B)[\%]$	$4.93 \pm 1.09\%$	$2.94 \pm 1.20\%$	$2.91 \pm 1.97\%$	7.21
F	$185234 \pm 19\%$	$136693 \pm 17\%$	$140993 \pm 18\%$	-
$T_m[s]$	48.3	35.4	80.6	-

Table 12.2.2: 10-node network, 2 services

	TTC	$E(B)$ [%]	$Max(B)$ [%]	F	time [s]
SA/PP	105155	1.37	1.9	25370	5
SA/DP	105650	0.90	2.3	19060	63
Reference method	106080	0.50	1.8	-	-

Numerical comparisons

We have compared solutions generated by the SA method with a specialised "greedy" method proposed in [Arv94] from the viewpoint of the total traffic carried and connection blocking distribution. The first group of results [PG95b] corresponds to a 20 node network with different traffic patterns by using a predefined set of admissible paths for OD pairs generated with different allocation rules RAN, LLP, MAX (in RAN paths are selected randomly from the set of paths accessible in a given network state). They are reported in Table 12.2.1, where the following notation is used:

TTC: total traffic carried

TTO: total traffic offered

$E(B)$: average blocking probability, i.e., the ratio of the total traffic lost to the total traffic offered : $E(B) = (TTO - TTC)/TTO$

$Max(B)$: maximum of blocking probability (over all demands)

F: allocation effort

T_m: mean simulation time.

The second group of results [PG95a] concerns the comparison between the SA/DP and the SA/PP methods. They are relevant to a 10 node network with 2 services and are reported in Table 12.2.2. Figures 12.2.1 and 12.2.2 show the percentile distribution of traffic carried with a blocking level greater than B for the same traffic pattern in both cases.

The presented results support the following conclusions:

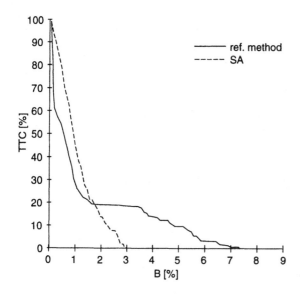

Figure 12.2.1: Percentile distribution of carried traffic (20-node network)

- For the static path search version of SA (SA/PP) the rules LLP and MAX find practically equivalent solutions in terms of carried traffic TTC and maximum individual blocking $Max(B)$.

- With SA/PP, in most cases LLP indicates considerably less allocation effort F to find the solution as compared to a more complicated allocation rule MAX.

- The computation time of SA/PP with LLP is even more considerably less than the one required by MAX, because the LLP rule is much faster than MAX.

- Solutions yielded by SA/PP with the random rule RAN are clearly worse than those of LLP and MAX, although not significantly. Maximum individual blocking for RAN is typically 1.5 to 2 times greater.

- Allocation effort for SA/PP with RAN is in most cases significantly greater than the one experienced with LLP. Thus, even though the allocation rule RAN is faster and the solutions are worse than those of LLP, the latter needs significantly less computation time.

- Out of the three considered rules, SA/PP with LLP is clearly the most effective, both in terms of solution quality and computation time (allocation effort).

- The maximum TTC found by the method of [Arv94] is typically slightly less than the one of SA/PP with LLP or of SA/PP with MAX, and slightly better than the one of SA/PP with RAN.

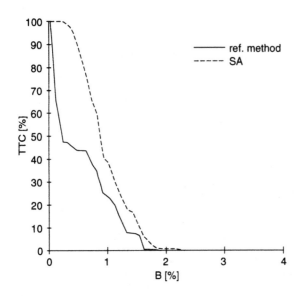

Figure 12.2.2: Percentile distribution of carried traffic (10-node network)

- The dynamic path version SA/DP is capable of producing allocation patterns better than those derived from the static version SA/PP. This is clearly seen from Table 12.2.1 where both indicators $E(B)$ (average blocking over all demands) and $Max(B)$ (maximum blocking over all demands) are substantially lower for SA/DP.

- The allocation effort F for both SA versions is comparable and neither of them outperforms the other.

- The better quality of the results of SA/DP is achieved at the expense of the longer computation time necessary to run the shortest-path procedures for each arriving demand.

- The value of TTC found by the method of [Arv94] is comparable to that of SA/DP (and to that of SA/PP); results for $Max(B)$ are also similar.

- As for the SA/PP version, the greatest difference between the SA/SD approach and the simple allocation method of [Arv94] is that the latter distributes losses much less evenly than SA.

12.2.2 Routing at maximal carried traffic in a logical network

In this section we extend the gradient based hill climbing procedure presented in [FBHA94] and [FBA$^+$95] to set the logical link (VP) capacities such as to maximise the total network revenue.

The starting point of the hill climbing is found by another optimisation which finds the global optimum of a rough, but nicely behaved (concave) approximation of the network revenue. A refinement can be obtained by further hill climbing that uses the partial derivatives of the network revenue function, as they are computed in the following.

The key ingredient is a generalisation of the results of Kelly [Kel89], [Kel90] for the multirate case, with a general smooth blocking function. This allows to compute the partial derivatives of the network revenue without having an explicit formula expressing the dependence of the network revenue on the logical link capacities.

The model is built as follows. For each link k and traffic type i, $1 \leq i \leq I$, we assume there exists a blocking function E_{ik} which, given the link offered traffic demands $\rho_{1k}, .., \rho_{Ik}$ and virtual link capacity C_k, returns the blocking probability of traffic type i on link k. To preserve generality, we do not restrict ourselves to a specific blocking function, any function is allowed that is jointly smooth in all the variables.

To be able to obtain analytical results, we apply the reduced load and link independence assumption that yields the following (fixed point) equations, taking the different traffic types into account:

$$B_{ik} = E_{ik}(\rho_{1k}, .., \rho_{Ik}, C_k) \quad \text{for all } i \text{ and } k \tag{12.2.1}$$

$$\lambda_r = \nu_r(1 - L_r) \quad \text{where} \quad (1 - L_r) = \prod_{i,k}(1 - B_{ik})^{A_{ikr}} \tag{12.2.2}$$

$$\rho_{ik} = \sum_r (1 - B_{ik})^{-1} \lambda_r A_{ikr} a_{ik} \quad \text{for all } i \text{ and } k \tag{12.2.3}$$

where B_{ik} denotes the blocking probability for traffic type i on link k. A_{ijr} indicates the incidence of traffic types, logical links and routes: $A_{ikr} = 1$ if route r uses logical link k and carries traffic type i, otherwise $A_{ikr} = 0$. The parameter a_{ij} defines the bandwidth demand of a type i call on logical link j; ν_r represents the traffic offered to route r and λ_r is the carried traffic on route r.

It can be shown [FBA+95], using the theory of differentiable manifolds, that the first order partial derivatives of the network revenue can be computed as follows, without having an explicit expression for the network revenue function.

Let us define a set of auxiliary parameters c_{ik} by the following system of linear equations:

$$c_{ik} = \sum_h \xi_{hik} \sum_r A_{hkr} \lambda_r \left(w_r - \sum_{j \neq k} \sum_g A_{gjr} a_{gj} c_{gj} \right) \tag{12.2.4}$$

where

$$\xi_{hik} = (1 - B_{hk})^{-1} \frac{\partial E_{hk}}{\partial \rho_{ik}} (1 - B_{ik})^{-1} \qquad (12.2.5)$$

and w_r is the revenue generated by a connection accepted on route r.

These parameters can be interpreted as implied costs. Now the revenue derivative with respect to logical capacity C_k can be formulated as:

$$\frac{\partial W}{\partial C_k} = \sum_h \zeta_{hk} \sum_r A_{hkr} \lambda_r \left(w_r - \sum_{j \neq k} \sum_g A_{gjr} a_{gj} c_{gj} \right) \qquad (12.2.6)$$

where $\zeta_{hk} = -(1 - B_{hk})^{-1} \frac{\partial E_{hk}}{\partial C_k}$.

Similarly, the revenue derivative with respect to ν_r is:

$$\frac{\partial W}{\partial \nu_r} = (1 - L_r)\left(w_r - \sum_i \sum_j A_{ijr} a_{ij} c_{ij} \right) \qquad (12.2.7)$$

Using these results with a specific blocking function, the partial derivatives of the network revenue can be computed and the optimisation can be done by gradient based hill climbing.

The overall method involves optimisation, implied cost computation, fixed point equation solutions and link analysis. Complexity of the method and its applicability are analysed in [MMR96].

12.2.3 Allocation of permanent VCs in the ATM physical network

In this paragraph we consider the allocation of permanent VCs in the ATM physical networks. The task of such a PVC routing is to find the best possible paths for the set of permanent connections of known average bandwidth requirements in order to minimise the network cell loss.

The network is modeled as a graph where the switches are represented by the node set and the links connecting the switches are represented by the set of links E. $c(e), e \in E$ is the capacity, or the available bandwidth of link e. The switches are assumed to use output queueing with finite queues and the FIFO queueing discipline.

Let D be the set of the permanent VCs (connections). $Bw(d)$ means the bandwidth requirement of connection $d \in D$ and $l(d)$ is the tolerable cell loss probability for this connection. The user pays $f(d)$ fee for the acceptance of the connection.

$P(d)$ is the predefined set of candidate paths for connection $d \in D$, $P = \bigcup_{d \in D} P(d)$. For each $d \in D$ let x_{dp} be 1, if the connection is routed to path $p \in P(d)$, and 0 otherwise. For each path $p \in P$ and link $e \in E$ let δ_{ep} be 1 if link e is on path p, and 0 otherwise.

For each link $F(e)$ is the aggregate flow, i.e. the sum of the bandwidth requirements routed through the link, and let L^p be the cell loss probability between the end points on path $p \in P(d)$.

With the above notation the PVC routing task can be formulated as to minimise the network cell loss defined by

$$\sum_{d \in D} Bw(d) \sum_{p \in P(d)} x_{dp} L^p.$$

We have investigated the cases when all the offered connections are accepted. This task is performed through a series of optimisation steps. First, the PVCs are established based on Least Loaded Path (LLP) selection criteria [PJ95, MDJ+95]. Simple shortest path allocation algorithm is used, where the cost assigned to each link of the network is updated after every path assignment. The whole algorithm is stopped if the LLP routing fails to accommodate all the connections. At the end of this initial step the worst case cell loss value of the LLP routing is defined as the tolerable cell loss value for all connections:

$$l(d) = \max_{p,d} \; x_{dp} \, L^p, \quad d \in D,$$

where the L^p values are calculated using the $M/M/1/b$ model or an experimental loss function, discussed later on in this section.

Then, the iterative optimisation is carried out with the objective to maximise the network revenue subject to the constraints that the link capacity limits and the maximum allowable cell loss of the connections are not exceeded. With the above notation the optimisation problem can be stated as follows:

$$Z_{IP} = \max \sum_{d \in D} f(d) \sum_{p \in P(d)} x_{dp} \qquad (12.2.8)$$

subject to:

$$\sum_{p \in P(d)} x_{dp} \le 1, d \in D, \qquad (12.2.9)$$

$$F(e) = \sum_{d \in D} \sum_{p \in P(d)} x_{dp} \delta_{ep} Bw(d) \le c(e), e \in E,$$

$$L^p \le l(d), p \in P(d), d \in D.$$

The IP problem is solved by using the Lagrange-relaxation (LR) algorithm [LY93]. After each iteration step the allowable cell loss probability is updated. The procedure is stopped when not all the connections can be accommodated by the LR algorithm. The result of the last successful optimisation is used to route all the connections.

This algorithm [PJ95, MDJ$^+$95] gives results that do not depend on the order of the connections and produces optimal or near optimal results. Several additional QoS parameters can be taken into account and routing can be adapted to the current network state.

Furthermore, the proposed Lagrange-relaxation method can also be used to select the optimal subset of the offered connections if only a portion of them can be accommodated in accordance with the predefined QoS parameters.

Although the results of the iterative procedure are heuristic, convergence is guaranteed and numerical examples show they are close to the optimum depending on the models used to evaluate QoS and the penalty function used in the relaxation method. Referring to the cell loss probabilities we can distinguish:

- QoS evaluated by the $M/M/1/b$ model
  In this case, to evaluate performances, the ATM nodes are modeled by the $M/M/1/b$ queue. This model is not completely adequate, but closed forms can be used in the computation [GMP89].

- $M/M/1/b$ model augmented by a penalty function
  From the simulation it has been noted that most of the cell loss comes from the nodes where the traffic is routed from a higher capacity link to a lower capacity link. Therefore, a penalty function can be introduced into the $M/M/1/b$ model that penalises the low capacity links outgoing from a node where higher capacity links terminate.

- QoS evaluated by an experimental loss function
  The loss behaviour of single links of different capacities, with different loads and different numbers of users can be simulated and the results achieved by the experiments best fitted by means of different exponential functions replacing the $M/M/1/b$ model in the evaluation.

Numerical comparisons

We have compared the results of the Lagrange-relaxation methods and the simple shortest path (SP) method in two reference networks, I and H, with respectively 10 and 9 nodes [MDJ$^+$95], from the viewpoint of cell loss rate and the load of the most loaded link. These quantities are reported in Figures 12.2.3, 12.2.4, 12.2.5 and 12.2.6. In the case of network H the node to node cell loss rate was practically the same in each case (Figure 12.2.3), however, the maximum load decreases using the Lagrange-relaxation method (Figure 12.2.4), especially by evaluating QoS by the experimental loss functions. The main reason for the marginal improvement is the symmetry of both the network structure and its load pattern.

For the network I a significant improvement is achieved regarding both the node-node cell loss rate and the maximal link load, especially applying experimental loss functions in the Lagrange-relaxation algorithm (Figures 12.2.5

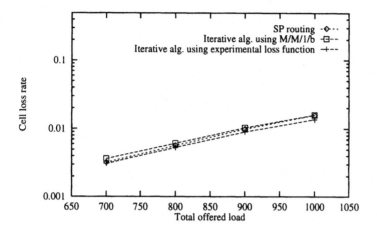

Figure 12.2.3: Cell loss inside the 'H' network

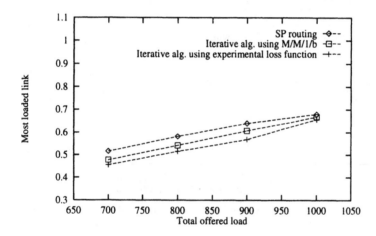

Figure 12.2.4: Maximal link load, 'H' network

and 12.2.6). The simulations justify usefulness of the Lagrange-relaxation discrete optimisation method. However, adequate cell loss and delay models should be developed to be applied in the objective function.

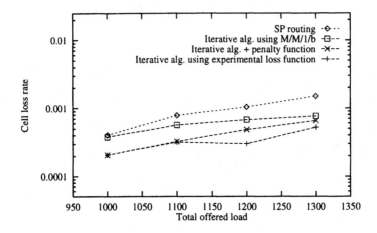

Figure 12.2.5: Cell loss inside the 'I' network

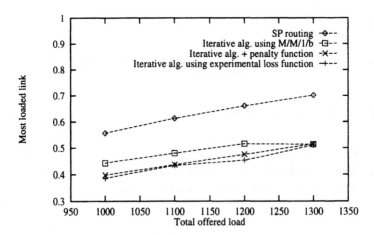

Figure 12.2.6: Maximal link load, 'I' network

Part III

Traffic Models and Queueing Analysis

Introduction to Part III

This part of the book is intended to give an introduction to the theoretical tools for the analysis and design of ATM networks. The final report of the COST 224 project [Rob92a] already contained an extensive exposition of the queueing analysis problems and methods. Part II, Queueing Models for ATM Traffic, of the "old" COST 224 book has been extensively used as a basis for the present review. The material has, however, been substantially changed by updating and enhancing the presentation with newer results and, more importantly, by introducing completely new chapters and/or sections describing the problems addressed in the COST 242 project and their analysis. In order to save space, some of the material of the old book, even when still relevant, has been omitted.

Throughout the presentation use is made of the important problem decomposition by time scales, as first introduced by Hui [Hui88]. This decomposition is based on the qualitatively different nature of the system at three different time scales along with the partial decoupling of the problems at these scales. The hierarchy of time scales – *call scale*, *burst scale* and *cell scale* – is illustrated in the figure below.

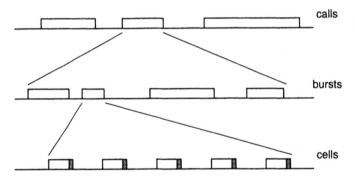

Time scale hierarchy in ATM traffic.

At the cell time scale, the traffic consists of discrete entities, the cells, produced by each source at a rate which is often orders of magnitude lower than the transmission rate of the output trunk. Most ATM source streams, even from variable bit rate sources, are locally periodic when considered at cell scale. Thus the randomness of the total input emerges from the independence of the *phases* of the locally periodic source streams, not from fluctuations of the *rates* of these streams.

At the burst time scale the finer cell scale granularity is ignored and the input process is characterized by its instantaneous *rate*. Consequently, *fluid flow models* appear as a natural modelling tool. At this time scale the problems arise from possible excesses of the total input rate over the output rate. Two different classes of problems can be distinguished in this context depending on the multiplexing option adopted in the system design as discussed in Part I Section 4.1. Using Rate Envelope Multiplexing, buffers are provided only for cell scale queueing and the system appears as a loss system at the burst scale. In the rate sharing option, buffers are provided to absorb at least part of the burst scale traffic fluctuations and we are led to consider burst scale delay systems.

Call scale characterized by the holding time of arriving calls or service demands represents the longest of the time scales. At call scale, an ATM network, even when carrying VBR calls, can be viewed as a multi-rate circuit switched network. Associated with each call is an *effective bandwidth*, a number somewhere between the mean and peak bandwidth requirement which describes the amount of bandwidth which has to be allocated for the call in order to keep the probability of cell losses during the connection below a given upper limit. In many instances, the effective bandwidth can be considered a constant, independent of the traffic mix, and then, indeed, the situation is equivalent to a circuit switched network.

The present part of the book is divided into six chapters after this introduction. The first of them, Chapter 13, deals with the models of ATM traffic on their own merit, separated from the queueing analysis. Already here the paradigm of the time scale hierarchy is utilized and different models are presented to capture the cell scale and burst scale behaviour of the traffic. A large part of the chapter focusses on the models for long range dependent traffic and related estimation and traffic prediction problems, an area which is currently under intensive research. Estimation of traffic at the call scale is also considered.

Chapters 14-17 address the queueing analysis of ATM systems. Typically these problems arise in modelling the behaviour of the queue appearing in an ATM multiplexer situated, for example, at the output links of an ATM switch, see the figure below. Cells coming from a large number of sources through the input links and addressed to the same output link must be buffered at the output because their total arrival rate at the output has considerable random fluctuations.

We usually consider models where a number of sources emit their traffic streams directly into the multiplexer which has one output port. This is an idealisation, because in reality most source streams are already multiplexed into a somewhat smaller number of trunks when they enter a switch. It is obvious that this usually makes no significant difference to the results and, moreover, the input traffic is "more random" and so presents the worse case when the source streams are assumed totally independent, i.e., not mixed together beforehand.

Most ATM switch structures work in a synchronous way so that cells

arrive to the multiplexer at discrete time instants. This suggests the use
of discrete time models. However, in many cases continuous time is the
mathematically more elegant choice. The difference between these models
is usually insignificant and we use them interchangeably choosing the model
which is most convenient for a particular purpose.

In Chapter 14 we give some general tools for the queueing analysis. Notably the Beneš approach, which has proven very efficient for many queueing
problems in the ATM context, is introduced and elaborated for a few generic
problem types. Second, elements of large deviations theory are presented to
the extent needed for later developments.

Chapter 15 addresses queueing problems at the cell scale. Here a few
of the basic queueing models are presented adopting the presentation of the
COST 224 book with some modifications and additions, e.g., the analysis
of the $Geo^N/D/1$ queue, the analysis of the idle and busy periods of the
$N*D/D/1$ queue and the extension of the $\sum D_i/D/1$ queue to batch arrivals.
Two completely new sections have been added. The first deals with the
multiplexing of a periodic and an independent batch stream and the other
considers the models for CDV. The latter includes an analysis of the CDV in
a tandem queue system.

The next two chapters deal with the problems arising from burst level
considerations. Chapter 16 addresses burst scale loss systems and again leans
heavily on the corresponding chapter in the COST 224 book. However, a
novel analysis of the loss processes in unbuffered systems has been added.

In Chapter 17 many aspects of burst scale delay systems are studied.
This is the most voluminous of all the chapters in this part of the book.
It starts with a discussion of Markov Modulated Rate Processes which has
been extended to include a short discussion on effective bandwidths based

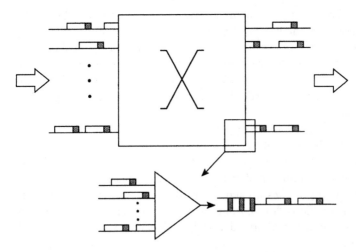

A multiplexer queue at an output port of an ATM switch.

on asymptotic properties, a new example of an MMRP system as well as a general description of their output process. A detailed account is given on the application of the Beneš method for fluid queues with different input traffic models. An alternative modelling approach for queueing systems by discrete time batch Markovian Arrival Processes is briefly reviewed. The queueing problems related to long range dependent input processes are also discussed at length and many new results are presented. The chapter is closed by an analysis of leaky bucket performance.

Call level problems are considered in the final Chapter 18. The multi-rate nature of the traffic mix requires an extension of the methods of the classical traffic theory. One of the basic problems with multi-rate models is how to determine the blocking probabilities of different traffic streams for a single link or, more generally, the end-to-end blocking probabilities of the streams in a whole network. Another addressed issue is how to control these blocking probabilities in order to provide desired grades of service to different traffic classes.

13 Traffic modelling

Various topics connected with traffic modelling and related statistical issues are considered in this chapter. The principles of the characterization of modern teletraffic are discussed in Section 13.1. A number of different model types for cell and burst level studies are presented in the two following sections. Section 13.4 is devoted to long range dependent models. The three remaining sections discuss specific statistical problems: real-time estimation of a varying Poissonian call rate, the estimation of the parameters of long range dependent traffic models, and the prediction of self-similar traffic on the basis of observed past.

13.1 Cell traffic characterization

A key issue in traffic engineering is the identification of parameters for quantifying the statistical characteristics of traffic relevant for network performance. These parameters together with appropriate traffic models are required for performance evaluation and dimensioning methods and also for the design of traffic control functions. Theoretical traffic models are fitted to actual cell flows by matching the values of these parameters in the model and the actual cell stream.

Two approaches can be used to describe a cell arrival process: through the cumulative arrival process $A(t)$ given by the work arriving in $(0, t]$ or through the sequence of arrival times t_n. A description through $A(t)$ is usually preferred since in this case the distribution of $A(t)$ for an aggregation of independent sources is simply obtained from the convolution of the individual source distributions. In the following we will assume that the traffic is stationary and the time origin is arbitrarily chosen, so that $A(t)$ can be considered as the amount of work arriving in an arbitrary interval of length t.

13.1.1 Time scales relevant to queueing performance: a large deviations approach

In order to identify the key traffic parameters, it is important to analyse network performance in order to get some insight into the most significant statistical characteristics of traffic. To simplify the problem, we will look at the queueing performance of a single server system with FIFO discipline, deterministic service times and infinite buffer fed by a superposition of general arrival processes. Our technical approach is based on the theory of large deviations, which provides an elegant way to express (a good approximation of) the unfinished work distribution of a system fed by a general arrival process (aggregation of a large number of independent sources) in terms of the distribution of $A(t)$. This "thermodynamic" way of looking at teletraffic systems has become quite popular — see, e.g., the recent papers [SG95, HK95, DO96]. A summary of the theory is given is Section 14.3.

Let $Q(x)$ denote the probability that the unfinished work in a system with service rate c exceeds x. The starting point is the approximation (and lower bound)

$$Q(x) \approx \sup_t \Pr\{A(t) - ct > x\}.$$

This approximation is more accurate when the number N of sources feeding the queue increases, and it is usually reasonable for moderate values already. For large N, the distribution of $A(t)$ can be approximated by its Chernoff bound:

$$\ln \Pr\{A(t) - ct > x\} \approx -J(t, x + ct),$$

where

$$J(t, w) = \mu^*_{A(t)}(w) = \sup_\beta \left(\beta w - \ln \mathrm{E}\left\{e^{\beta A(t)}\right\}\right)$$

is the entropy function of $A(t)$ at w (see Section 14.3). Therefore, we obtain the result

$$\ln Q(x) \approx -\inf_t J(t, x + ct). \tag{13.1.1}$$

If the multiplexer is fed by N homogeneous and independent sources, the aggregate arriving work will be the sum of the work generated by each source, $A(t) = \sum_{i=1}^{N} A_i(t)$, and $J(t, x + ct) = N J_1(t, x/N + tc/N)$, where $J_1(t, w)$ is the entropy function of one source. Hence, if c and x_1 denote the link capacity per source, $c_1 = c/N$, and unfinished work per source, $x_1 = x/N$, respectively, we have

$$\ln Q(N x_1) = -N \inf_t \left(J_1(t, x_1 + c_1 t)\right) + b(N, c_1, x_1), \tag{13.1.2}$$

where $b(N, c_1, x_1) = o(N)$ for fixed c_1 and x_1.

Equation (13.1.1) implies that $Q(x)$ is basically determined by the distribution of the work arriving in time intervals of the length which minimizes $J(t, x + ct)$, i.e., each queue length is related to a time scale. The statistical properties of the arrival process at that time scale determine the probability of the corresponding queue length. Note that the time scale determining the complementary queue length distribution at x in a system with service rate c fed by N sources is the same as that determining the complementary distribution at kx in a system with service rate kc fed by kN sources. In the following, $t_c(x)$ will denote the interval length which minimizes $J(t, x + ct)$. In most cases $t_c(x)$ will be given by the solution to

$$\frac{dJ(t, x + ct)}{dt} = 0. \tag{13.1.3}$$

Although the exact solution to (13.1.3) is normally very complex, an approximate one can be found by substituting $J(t, w)$ by its second order Taylor series expansion around $E\{A(t)\} = mt$:

$$J(t, w) \approx J(t, mt) + J'(t, mt)(w - mt) + J''(t, mt)(w - mt)^2/2,$$

where the prime denotes $\partial/\partial w$. By the general properties of the entropy function (Subsection 14.3.1), we have

$$J(t, mt) = 0, \quad J'(t, mt) = 0, \quad J''(t, mt) = \frac{1}{v(t)},$$

where $v(t) = \mathrm{Var}\{A(t)\}$. Thus, we obtain

$$J(t, w) = \frac{(w - mt)^2}{2v(t)}.$$

Note that this second order approximation is equivalent to assuming $A(t)$ Gaussian (see Table 14.3.1). Now, $t_c(x)$ is approximately given by the solution to

$$\frac{d}{dt} \frac{(x - (c - m)t)^2}{2v(t)} = 0. \tag{13.1.4}$$

The solution to (13.1.4) seems to be accurate enough in most cases. When $v(t)$ is linear (e.g., a Poisson or batch-Poisson arrival process) the governing time scales are given by (see [AMP94])

$$t_c(x) = \frac{x}{c - m}. \tag{13.1.5}$$

Relation (13.1.5) will also be a good approximation when x is large and $\lim_{t \to \infty} v(t)/t$ is finite or, in other words, when $v(t)/t$ around $t_c(x)$ is almost constant (this is the case for Markovian sources such as on/off sources with exponentially distributed on and off periods).

In Subsection 17.4.2, an expression for $t_c(x)$ is given when the input process corresponds to self-similar traffic. In this case $v(t) = at^{2H}$ and

$$t_c(x) = \frac{Hx}{(1 - H)(c - m)}.$$

Cell and burst scale

In small interval lengths corresponding to the cell scale (i.e., intervals of the order of the cell interarrival time of one source), the aggregate arrival process behaves like a Poisson process since in such small intervals each cell will come from a different source and, therefore, all arrivals will be independent. In longer time intervals several cells from the same source are emitted and, if these arrivals are correlated, the aggregate arrival process stops behaving like Poisson; we are then at the burst scale.

Let t_r be the interval length representing the boundary between cell and burst scale, i.e., the time scale at which the correlation between arrivals becomes significant. Looking at the queue length distribution, queue lengths influenced by time scales smaller than t_r will have a distribution very similar to that of the $M/D/1$ queue. The time scale t_r will be associated with a queue length $q_r(t_r)$ at which the distribution stops behaving like that of an $M/D/1$ queue and greater exceedance probabilities appear due to burst arrivals. The relation between time scales and queue lengths for Poisson processes given by Equation (13.1.5) will be valid for time scales less than or equal to t_r. We have therefore that $t_r = q_r/(c(1 - \rho))$, where ρ is the load of the system.

The time scale t_r can be used as the integration time for the definition of the cell rate, i.e., the rate at time t can be defined as

$$r(t) = \frac{\#\{\text{arrivals in } (t - t_r, t]\}}{t_r}.$$

For practical purposes, a reference value of t_r can be obtained for a given c and an expected load, ρ, using the value of q_r for which

$$\Pr\{\text{queue length in an } M/D/1 \text{ system} > q_r\} = P_{\text{loss}}.$$

This rate definition can be applied then for the measurement of the rate of real cell flows. The resulting rate process could be used as the input for burst scale performance evaluation methods based on fluid flow approximations.

13.1.2   Characterization by moments of the count process

To determine queueing performance we need an approximate description of the function $J(t, w)$ at the values of t related to the queue lengths of interest. Following the same procedure as for the calculation of the time scales, we can expand $J(t, w)$ as a Taylor series with respect to the variable w around mt:

$$J(t, w) = \sum_{n=2}^{\infty} J^{(n)}(t, mt) \frac{(w - mt)^n}{n!}.$$

The n-th derivative $J^{(n)}(t, mt)$ depends on the first n moments of $A(t)$. The more moments we know, the more exact the Taylor series expansion and the better the description of $J(t, w)$ around $w = mt$.

The first two moments seem to be the minimum amount of information needed to describe $A(t)$. A commonly used function related to the first and second moments of $A(t)$ is the Index of Dispersion for Counts (IDC) [HL86, Gus91, AMP94]. It is defined as

$$I(t) = \frac{v(t)}{mt}.$$

The IDC together with the mean cell rate gives all the information about the first and second moment of $A(t)$. Theoretical arrival processes could be fitted to real cell streams by matching their mean and IDC. An example of the computation of the IDC and its limiting behaviour for a general class of arrival processes is given in Subsection 13.2.3.

To check the ability of the first two moments to characterize an arrival process, a study was carried out in [AMP94]. Two points of the IDC curve ($I(t_1)$ and $I(t_2)$ with $t_1 < t_2$) together with the mean arrival rate were used to characterize a traffic flow. The value $I(t_1)$ is intended to capture the variations of the cell rate (stationary distribution of the rate), whereas $I(t_2)$ captures long term correlations, i.e., it is related to the burst structure. The combination of $I(t_1)$ and $I(t_2)$ will capture the shape of the IDC curve in the range of time scales affecting the region of interest of the queue length distribution.

The objective of the study was the characterization of real LAN traffic obtained from measurements made by Bellcore [FL91]. An on/off source model with exponentially distributed on and off periods was constructed by matching the values of m, $I(t_1)$ and $I(t_2)$ to those of the LAN traffic. The queue length distributions of an ATM multiplexer fed by homogeneous mixes of both source types (LAN and on/off) were found by simulation and compared to check the accuracy of the characterization. As an example Figure 13.1.1 shows the IDC and queue length distributions obtained for loads $\rho = 0.6$ and $\rho = 0.8$ (22 and 29 sources respectively) and $t_1 = 8.25$ ms, $t_2 = 100$ ms in a link of capacity $c = 45$ Mbit/s. The values of t_1 and t_2 determine the probability of queue lengths $q_1 = 100$ and $q_2 = 1215$ cells for $\rho = 0.6$, and $q_1 = 50$ and $q_2 = 607$ for $\rho = 0.8$. In the figure dashed lines correspond to on/off traffic and continuous lines correspond to LAN traffic.

It can be seen that the queueing performance obtained with both LAN and exponential on/off traffic is very similar. Close IDCs of LAN and on/off traffic over a certain range of interval lengths means close queue length distributions for a range of queue lengths. Note that queue length distributions do not intercept exactly at q_1 and q_2 because these queue lengths were calculated for LAN traffic; with on/off traffic the exact queue lengths determined by time scales t_1 and t_2 vary slightly.

13.1.3 Characterization by queueing behaviour

With a description based on the first moments of $A(t)$ a good characterization of the distribution is only obtained around the mean. However, the probabilities of interest will be normally at the tail of the distribution and this region is poorly described by the first moments. If the shape of the distribution of the arrival process used to model the real cell stream is not very similar to the real distribution, the results obtained will not be very accurate.

To obtain a good description of the tail of the distribution of $A(t)$ some points of the distribution around the region of interest could be given. The

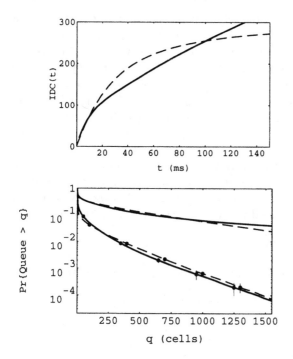

Figure 13.1.1: IDC functions and queue length distributions of measured LAN traffic (continuous lines) and exponential on/off traffic (dashed lines).

probabilities around the given points could be accurately obtained by some kind of interpolation, for example, as in the case of a description by moments, by fitting the distribution of a theoretical model to those points.

An alternative to just using points of the distribution of $A(t)$ could be to use points of the queue length distribution of a system with capacity c_1 and fed by the characterized source. This is justified if we observe that using Equation (13.1.2) an approximate relation can be deduced between the queue length distribution in an infinite buffer with output rate Nc_1 fed by N homogeneous sources (denote it by $Q_{N,Nc_1}(x_1)$) and the queue length distribution seen in a system with output rate c_1 and fed by only one source

$$\frac{\ln Q_{N,Nc_1}(Nx_1)}{N} \approx \ln Q_{1,c_1}(x_1) + b(c_1) \quad \text{for } x_1 \text{ large enough.} \quad (13.1.6)$$

This derives from the fact that for x_1 large enough the function $b(1, c_1, x_1)$, defined by (13.1.2), varies slowly with x_1 with respect to $\inf_t J_1(t, x_1 + c_1 t)$ so that $b(1, c_1, x_1)$ just depends on c_1, $b(1, c_1, x_1) \approx b(c_1)$. This can be intuitively understood if we observe that for x_1 large enough, the interval length at which the infimum of $J_1(t, x_1 + c_1 t)$ is obtained will also be large, and in this long time $A_1(t)$ can be considered as the sum of near i.i.d. random variables given by the work arriving in several time subintervals and, therefore, the Chernoff bound will still be a good approximation for the distribution of $A_1(t)$.

Thus, the relevant traffic characteristics of a cell flow could be measured through the probability of exceeding a queue length x_1 in a system with output service rate c_1, infinite buffer and fed only by the characterized cell flow. The values of c_1 and x_1 selected for the characterization will depend on the expected network conditions like link capacities, buffer sizes, traffic mixes, load, etc.

As an example, to assess the ability of $Q_{1,c_1}(x_1)$ to capture the traffic characteristics relevant for network performance, we have selected three different source types sharing the same mean and peak rate and the same value of $Q_{1,c_1}(x_1)$ at $x_1 = 37.45$ cells and $c_1 = 2.038$ Mbit/s. We compare the queueing performace obtained when homogeneous superpositions of each source type are multiplexed.

The source types are the following:

I: on/off sources with exponentially distributed on and off periods

II: on/off sources with deterministic on and off periods

III: Bellcore's LAN traffic (see [FL91]) shaped to the same peak as sources (I) and (II)

The on and off mean durations of sources I and II have been selected such that the value of $Q_{1,c_1}(x_1)$ coincides with that of the LAN.

The queue length distributions obtained when homogeneous superpositions of $N = 22$ sources of each type were multiplexed in a link of capacity $c = 22c_1$ Mbit/s and infinite buffer were measured by simulation. Figure 13.1.2 represents these distributions together with the ones obtained for one source. It can be seen that, as expected, the distributions for $c = 22c_1$ are scaled curves of the distributions with capacity c_1 and logically coincide in $x = 22x_1 = 824$ cells.

For practical use, a selection of points $(c_1, x_1, Q_{1,c_1}(x_1))$ of the space defined by the family of functions $Q_{1,c_1}(x_1)$ may be enough to obtain an acceptable source characterization for typical network conditions (load, traffic mixes, link speeds, etc.). The ability of a small set of points of the family of functions $Q_{1,c_1}(x_1)$ to characterize the source cell traffic was investigated in [ASV95].

Cell traffic characterization based on queueing performance is consistent with the present traffic descriptor parameters based on the Generic Cell Rate Algorithm (GCRA) presented in ITU-T recommendation I.371. In fact, the GCRA can be seen as a queueing system in which the queue content is measured in units of time. Parameters based on the GCRA, such as PCR (peak cell rate), CDVT (cell delay variation tolerance), SCR (sustainable cell rate) and IBT (intrinsic burst tolerance) are parameters associated to points $(c_1, x_1, Q_{1,c_1}(x_1) = 0)$.

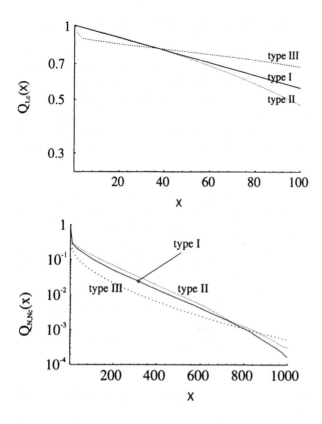

Figure 13.1.2: Queue length distributions for one source and $c_1 = 2.038$ Mbit/s and 22 sources and $c = 22c_1$.

13.2   Cell level models

Cell level traffic modelling is needed for the study of mechanisms which handle with individual cells and are usually supported with special hardware, e.g.,

- switch architectures,

- synchronization of real time services, and

- flow control mechanisms.

There are two basic classes of cell level models: those based on renewal processes and those based on Markov modulated processes. The different models are briefly characterized and discussed in this section. They can be seen in 'live' use at many places in this report.

13.2.1   Renewal streams and their superpositions

ATM traffic analysis tools usually give, among other information, a histogram of the cell interarrival times. As explained at the beginning of Section 13.1, this is usually not sufficient for adequate traffic characterization, in particular when one is interested in the aggregated traffic coming from a large number of sources. For an individual source, however, it may happen that the cell interarrival times can be modelled with a sequence of independent and identically distributed (i.i.d.) random variables U_n, i.e., with a renewal process.

Periodic stream.   The simplest case of a renewal cell stream is a periodic stream. Note that a stationary periodic stream has one random element: the phase, which is uniformly distributed. In measurements of real network traffic, one obtains even for a periodic stream a non-trivial interarrival time distribution, which expresses the cell delay variation caused, for example, by queueing at network nodes. Note that in this case there is some negative correlation between successive cell distances. Considered at cell level, practically all real ATM sources are locally either periodic or idle. If the length of an activity period can be assumed geometrically distributed and successive idle periods are independent, even such a randomly interrupted periodic stream can be modelled as a single renewal process.

Poisson process.   Empirically, there are no grounds for modelling a single ATM source by a Poisson process, i.e., a renewal process with exponentially distributed interarrival times. On the other hand, it is a mathematical fact that the superposition of a large number of independent renewal processes, even periodic ones, converges locally to a Poisson process. Thus, at the cell time scale, a Poisson process, conditioned on the local arrival rate, which is, in fact, a more slowly changing random process, is often a good model for the aggregated traffic coming from a large number of relatively slow sources.

Bernoulli process.   A Bernoulli process is a discrete time process which only takes values 0 and 1, independently for each time slot. Since each cell slot on an ATM link either contains one cell or is empty, a Bernoulli process is often used for traffic modelling at a small time scale. There are no grounds for this except simplicity, since a more natural model for a Poissonian cell stream forced to a pipe of width one cell per one cell time would be the output process of an $M/D/1$ queue, which is an on/off process with geometric off-periods but non-geometric on-periods. Thus, the Bernoulli process has here a bias in the optimistic direction.

13.2.2 Discrete time Markovian arrival processes (D-BMAP)

Another popular class of cell level models are the Discrete time Markov Batch Arrival Processes (D-BMAP), where the distribution of the number of arrivals in a time slot is determined by a Markovian state process. The state process need not have any physical interpretation, and for this reason, Markov modulated models are not very well suited for practical or theoretical traffic characterization at a general level as discussed in Section 13.1 above. On the other hand, these models can be built to satisfy practically any desired traffic characteristics, and there exists a powerful machinery for computing queueing characteristics, for instance, which makes them very useful in research. A summary of their queueing theory is given in Section 17.3. The next section links, for a special case, the second order characterization and Markov modulated arrival processes.

13.2.3 The IDC for D-MAP's

A discrete-time Markovian Arrival Process (D-MAP) is a special case of the D-BMAP introduced in Section 17.3, where the batch size equals one cell. Hence, the transitions are governed by two matrices $\mathbf{H}_0$ and $\mathbf{H}_1$. In this section we derive a formula for the IDC of a D-MAP and a closed formula for its limit. Let $\mathbf{H} = \mathbf{H}_0 + \mathbf{H}_1$ be the underlying Markov chain with stationary probability vector $\overline{\sigma}$, which is assumed to be a row vector in matrix expressions. First we recall the following result for the covariance between the number of arrivals in a slot.

Let $(X_1, \ldots, X_k)$ be random variables, where X_i is the number of arrivals (0 or 1) at time slot i. Then in [BG96], it has been shown that the scalar covariance between X_1 and X_k,

$$\text{Cov}\{X_1, X_k\} = \overline{\sigma}\text{E}\{X_1 X_k\}\overline{e} - \mu_1 \mu_k,$$

where $\overline{e}$ is a column vector consisting of 1's, and μ_1 and μ_k are the scalar means of X_1 and X_k, respectively, is given by

$$\text{Cov}\{X_1, X_1\} = \overline{\sigma}\mathbf{H}_1\overline{e} - (\overline{\sigma}\mathbf{H}_1\overline{e})^2$$

$$\text{Cov}\{X_1, X_{k+1}\} = \overline{\sigma}\mathbf{H}_1(\mathbf{H}_0 + \mathbf{H}_1)^{k-1}\mathbf{H}_1\overline{e} - (\overline{\sigma}\mathbf{H}_1\overline{e})^2, \quad k \geq 1.$$

Note that an extension of this theorem for D-BMAP's was obtained in [Blo93].

The *Index of Dispersion for Counts* (IDC) can be defined for a discrete time process in a similar way as in Section 13.1.2 for a continuous time process. Denote by $[\mathbf{N}^{(k)}(l)]_{ij}$ the conditional probability that in k slots there are l arrivals and at the k-th slot the phase of the process is j, given that at time t=0 the phase was i. Let the corresponding probability generating function be denoted by

$$\mathbf{N}^{(k)}(z) = \sum_{l=0}^{\infty} \mathbf{N}^{(k)}(l) z^l.$$

Clearly $\mathbf{N}^{(k)}(z) = (\mathbf{H}_0 + \mathbf{H}_1 z)^k$. We define the index of dispersion for counts (IDC) as

$$I(k) = \frac{\mathrm{Var}\{N^{(k)}\}}{\mathrm{E}\{N^{(k)}\}},$$

where $\mathrm{Var}\{N^{(k)}\}$ and $\mathrm{E}\{N^{(k)}\}$ denote the scalar variance and scalar mean, respectively, of the variable $N^{(k)}$. Then the IDC is given by

$$I(k) = \frac{\mathrm{Var}\{N^{(k)}\}}{\mathrm{E}\{N^{(k)}\}} = \frac{k\mathrm{Cov}\{X_1, X_1\} + 2\sum_{j=1}^{k-1}(k-j)\mathrm{Cov}\{X_1, X_{1+j}\}}{k\mathrm{E}\{X_1\}}.$$

It is well known that for a renewal process $I(k) = c_1^2$, for all $k \geq 1$, where c_1^2 is the squared coefficient of variation of the number of arrivals in a slot. In particular for a Bernoulli process, $I(k) = 1$ (see, for example, [SW86b]).

Now we derive a closed formula for the limit of the IDC in case of a D-MAP. In view of the previous result, we have that

$$\sum_{j=1}^{k-1} \frac{k-j}{k} \mathrm{Cov}\{X_1, X_{1+j}\} = \sum_{j=1}^{k-1} \frac{k-j}{k} [\sigma \mathbf{H}_1 \mathbf{H}^{j-1} \mathbf{H}_1 \bar{e} - (\sigma \mathbf{H}_1 \bar{e})^2]$$

$$= \sigma \mathbf{H}_1 \{ \sum_{j=1}^{k-1} \frac{k-j}{k} [\mathbf{H}^{j-1} - \sigma \bar{e}] \} \mathbf{H}_1 \bar{e}.$$

In view of [KS67], Theorem 5.1.4 (p.101), the following series is Cesaro-summable:

$$\sum_{j=1}^{k-1} \frac{k-j}{k} [(\mathbf{H}_0 + \mathbf{H}_1)^{j-1} - \sigma \bar{e}],$$

and the Cesaro-limit is given by

$$\mathbf{Z} - \mathbf{I},$$

with $\mathbf{Z} = [\mathbf{I} - (\mathbf{H} - \bar{e}\,\sigma)]^{-1}$ the fundamental matrix of the Markov chain $\mathbf{H} = \mathbf{H}_0 + \mathbf{H}_1$. Hence we obtain

$$\lim_{k \to +\infty} I(k) = \frac{\sigma \mathbf{H}_1 \bar{e} - 3[\sigma \mathbf{H}_1 \bar{e}]^2 + 2\sigma \mathbf{H}_1 \mathbf{Z} \mathbf{H}_1 \bar{e}}{\sigma \mathbf{H}_1 \bar{e}}.$$

Now we give a result concerning the convergence of the covariance towards zero. Assume that the matrix $\mathbf{H}_0 + \mathbf{H}_1$ is diagonizable and that it has one eigenvalue on the unit-circle, namely 1. The convergence of the covariance of the number of arrivals towards zero is geometric, determined by the eigenvalue λ_2 of $\mathbf{H}$ with the largest absolute value, excluding the eigenvalue 1. Thus

$$\frac{\text{Cov}\{X_1, X_{k+1}\}}{\text{Cov}\{X_1, X_k\}} = |\lambda_2|, \quad k \to \infty,$$

or

$$\text{Cov}\{X_1, X_{k+1}\} = c\,|\lambda_2|^k, \quad k \to \infty.$$

Indeed, in view of [Çin75], Theorem 5.1, p.379, there exists a number $\alpha > 0$ such that

$$|\,[\mathbf{H}^n]_{ij} - (\overline{\sigma}_i)_j\,| \le \alpha|\lambda_2|^n.$$

The number α can be determined by means of the spectral decomposition of $\mathbf{H}$. If

$$\mathbf{H}^n = \overline{\sigma}\,\overline{e} + \lambda_2^n \mathbf{B}_2 + \ldots + \lambda_N^n \mathbf{B}_N,$$

then

$$\alpha = \sup_{ij}\{|\,[\mathbf{B}_2]_{ij}\,| + \ldots + |\,[\mathbf{B}_N]_{ij}\,|\}.$$

The case where there are more than one eigenvalue on the unit circle has been studied in [BG96].

13.3   Burst level models

In this section, we discuss several models where traffic is considered as continuous "fluid". As usual, the cumulative traffic in time interval $(0, t]$ is denoted by A_t. When A_t is piecewise differentiable (in some models it is not), the traffic rate at time t is denoted by Λ_t and we have

$$A_t = \int_0^t \Lambda_s \mathrm{d}s.$$

13.3.1   Renewal rate process

This process was studied by Mandelbrot in an economics context and called a "renewal reward process". It is a simple model for a traffic source that changes its rate now and then. It could, for example, be used to describe the behaviour of a variable bit rate video source at large time scales, where the changes in traffic rate are caused by scene changes.

Consider a renewal process $0 < S_0 < S_1 < S_2 < \cdots$ with inter-renewal times $U_n = S_n - S_{n-1}$, $n = 1, 2, \ldots$. When $\text{E}\{U_1\} < \infty$, the process becomes stationary with the choice

$$\Pr\{S_0 \in \mathrm{d}t\} = \frac{\Pr\{U_1 > t\}}{\text{E}\{U_1\}}\mathrm{d}t.$$

Assume this, and let $W_0, W_1, \ldots$ be a sequence of square integrable i.i.d. random variables independent of the renewal process. The *renewal rate process* Λ_t is then defined by

$$\Lambda_t = \sum_{n=0}^{\infty} W_n 1_{\{S_{n-1} \le t < S_n\}}.$$

Obviously, Λ_t is a stationary process with $E\{\Lambda_t\} = E\{W\}$ and $\text{Var}\{\Lambda_t\} = \text{Var}\{W\}$ for all $t \ge 0$. Moreover, the independence assumptions imply that the covariance function of Λ_t is simply

$$\text{Cov}\{\Lambda_0, \Lambda_t\}$$

$$= E\left\{(\Lambda_0 - E\{W\})^2 1_{\{S_0 > t\}}\right\} + E\left\{(\Lambda_0 - E\{W\})(\Lambda_t - E\{W\}) 1_{\{S_0 \le t\}}\right\}$$

$$\tag{13.3.1}$$

$$= \text{Var}\{W\}\Pr\{S_0 > t\} = \frac{\text{Var}\{W\}}{E\{U\}} \int_t^{\infty} \Pr\{U_1 > u\}\, du. \tag{13.3.2}$$

It follows that any function of the form $c(t) = \int_t^{\infty} \phi(s)ds$, where ϕ is non-negative, integrable on $[0, \infty)$ and decreasing, can be realized as the covariance function of a suitably chosen renewal rate process.

The variance of the cumulative traffic process A_t is obtained by integrating the interarrival distribution function three times:

$$\text{Var}\{A_t\} = \frac{2\text{Var}\{W\}}{E\{U_1\}} \int_0^t du \int_0^u dv \int_v^{\infty} ds \, \Pr\{U_1 > s\}. \tag{13.3.3}$$

13.3.2 On/off processes and their superpositions

By an on/off source, we mean a traffic source with alternating activity (on) and silence (off) periods, the transmission rate being constant during each on-period. This is often a good abstraction of a data communication terminal. The usual mathematical model for such a source is the on/off process which can be built up as follows.

Assume that we have a sequence of positive random variables $U_1, U_2, \ldots$ representing the lengths of the on-periods, and another positive sequence $V_1, V_2, \ldots$ representing the off-periods. The basic assumption is that all these random variables are *independent,* the U_i's having a common distribution and the V_i's having another common distribution. (Thus, the model is a simple case of a semi-Markov process.) Denote further

$$I_t = 1_{\{\text{source is on at time } t\}},$$

and let U_0 and V_0 be two more nonnegative random variables, such that I_0, U_0 and V_0 are independent of each other and of the other U_i's and V_i's. Now the process I_t is constructed so that the sequence of subsequent 'on' and 'off' period lengths is

$$\begin{cases} V_0, U_1, V_1, U_2, V_2, \ldots & \text{if } I_0 = 0, \\ U_0, V_1, U_2, V_2, U_3, \ldots & \text{if } I_0 = 1. \end{cases}$$

Assume that the transmission rate in the on-state is h. The traffic rate process is then $\Lambda_t = hI_t$. When U_1 and V_1 have finite expectations, a stationary process Λ_t is obtained by choosing

$$\Pr\{I_0 = 1\} = \frac{E\{U_1\}}{E\{U_1\} + E\{V_1\}} =_{\text{def}} p_{\text{on}},$$

$$\Pr\{U_0 \in dt\} = \frac{\Pr\{U_1 > t\}}{E\{U_1\}} dt,$$

$$\Pr\{V_0 \in dt\} = \frac{\Pr\{V_1 > t\}}{E\{V_1\}} dt.$$

In the special case that U_1 and V_1 are exponentially distributed with parameters β and α, respectively, Λ_t is a two state Markov process, and many quantities can be calculated explicitly. In particular,

$$\text{Cov}\{\Lambda_u, \Lambda_v\} = \frac{\alpha\beta h^2}{(\alpha + \beta)^2} e^{-(\alpha+\beta)|u-v|}, \tag{13.3.4}$$

$$\text{Var}\{A_t\} = \frac{2\alpha\beta h^2}{(\alpha + \beta)^3} \left[t - \frac{1}{\alpha + \beta}(1 - e^{-(\alpha+\beta)t})\right]$$

If the traffic is produced by N similar but independent stationary on/off sources, the number of active sources at time t, say K_t, has the stationary distribution $Bin(N, p_{\text{on}})$ regardless on the distributions of the period lengths.

Densities of the amount of work arriving in an interval of given length are considered in Section 17.2.1.

13.3.3   Poisson burst process

This model describes the traffic produced by "infinitely many sources each of which is active only once", i.e., a limit of the superposition of N on/off sources when N and the silence lengths tend to infinity but the on-period distribution and the total mean rate are kept unchanged.

The traffic thus consists of bursts that begin according to a Poisson process with, say, parameter λ, come with rate h each and have independent lengths U_i with common distribution. Denote the number of bursts going on at time t by K_t. We assume that $E\{U\} < \infty$ and that K_t is stationary. Then, for each t, the distribution of K_t is $Poisson(\lambda E\{U\})$, and

$$\text{Cov}\{K_t, K_{t+s}\} = \frac{\text{Var}\{K\}}{E\{U\}} \int_s^\infty \Pr\{U > u\} du. \tag{13.3.5}$$

Note that this expression is exactly similar to (13.3.2), although the processes and the roles of the periods with lengths 'U_i' are different. This can be intuitively understood on noting that in both cases, dependence between two time points can be "physically caused" only by a period covering both points.

As usual, denote the rate process by $\Lambda_t = hK_t$. The variance of the cumulative arrival process $A_t = \int_0^t \Lambda_s ds$ has, by the remark above, an expression similar to (13.3.3):

$$\text{Var}\{A_t\} = 2\lambda h^2 \int_0^t du \int_0^u dv \int_v^\infty ds \, \Pr\{U > s\}. \tag{13.3.6}$$

(A factor $\text{E}\{U\}$ appears both in numerator and denominator and is cancelled out.)

In the case of $Exp(\beta)$ distributed burst lengths, the process K_t is a birth-death process with birth rate λ and death rate $K_t\beta$. The covariance function of the rate process is

$$\text{Cov}\{\Lambda_u, \Lambda_v\} = \frac{\lambda h^2}{\beta} e^{-\beta|u-v|}, \tag{13.3.7}$$

and the variance of the cumulative arrival process is

$$\text{Var}\{A_t\} = 2h^2 \frac{\lambda}{\beta^2} \left[t - \frac{1}{\beta}(1 - e^{-\beta t}) \right]. \tag{13.3.8}$$

The probability density of A_t is considered in Section 17.2.4.

13.3.4   Gaussian traffic modelling

Consider a superposition of independent and identically distributed cumulative traffic processes $A_i(t)$, $i = 1, \ldots, N$. As usual, we assume that these processes have stationary increments and finite second moments. Let m be their common mean rate, i.e., $\text{E}\{A_i(t)\} = mt$ for all t. For any finite set of time points $t_1, \ldots, t_k$, and for any constants $a_1, \ldots, a_k$, the distribution of

$$\frac{1}{\sqrt{N}} \sum_{i=1}^N \sum_{j=1}^k a_j (A_i(t_j) - mt_j)$$

tends by the central limit theorem to the Gaussian distribution

$$N\left(0, \text{Var}\left\{ \sum_{j=1}^k a_j A_1(t_j) \right\}\right)$$

as N increases. Let B_t be a centered (mean zero) Gaussian process with stationary increments and variance function $v(t) = \text{Var}\{B_t\} = \text{Var}\{A_1(t)\}$.

Then the above shows that

$$\frac{1}{\sqrt{N}} \sum_{i=1}^{N} (A_i(t) - mt) \longrightarrow B_t \quad \text{as } N \to \infty \qquad (13.3.9)$$

in the sense of the convergence of finite dimensional distributions. For an advanced treatment of this topic, we refer to the recent paper by Kurtz [Kur96], where it is also shown that in most cases of interest here, the convergence of finite dimensional distributions can be strengthened to weak convergence in a path space.

The total arrival process $A(t) = \sum_{i=1}^{N} A_i(t)$ can then be approximated by the Gaussian process $\tilde{A}(t) = Nmt + \sqrt{N}B_t$. Note that the Gaussian distribution gives positive probability for negative values, and $\tilde{A}(t)$ is not an increasing process. Usually this does not have serious consequences for the usability of the model.

Thus, whatever the source traffic streams are like, the superposition of a large number of them can be modelled with a Gaussian process which has the same correlation function as the source processes. However, it depends strongly on the statistics of the source streams what number N can be considered sufficiently large.

A Gaussian process, the Brownian motion, also often results in another kind of limit procedure, which is sometimes called "aggregation in time" in contrast to the "aggregation in space" considered above. Most processes $A(t)$ used as traffic models satisfy

$$\frac{A(\alpha t) - m\alpha t}{\sqrt{\alpha}} \longrightarrow aW_t \quad \text{as } \alpha \to \infty, \qquad (13.3.10)$$

where W_t is the standard Brownian motion, a is some constant, and the convergence holds (at least) for finite dimensional distributions. An important class of traffic models for which (13.3.10) does *not* hold are the long range dependent processes introduced in Section 13.4. The related limit results are discussed in Subsection 13.4.5.

13.3.5 The Ornstein-Uhlenbeck model

In this subsection we present a particular Gaussian traffic model, called the Ornstein-Uhlenbeck model, which can be considered as a diffusion approximation for the superposition of a large number of identical, independent rate processes with an exponential correlation function. It was introduced in traffic and queueing theory in [SV91] and [KM91]. For this model, many interesting characteristics can be computed explicitly.

Let the rate process Λ_t be a continuous process satisfying the stochastic differential equation

$$d\Lambda_t = -\beta(\Lambda_t - m)dt + \gamma dW_t,$$

where W_t is the standard Brownian motion. We can also write $\Lambda_t = m + \gamma U_t$, where U_t is the Ornstein-Uhlenbeck diffusion process with parameter β, defined by

$$dU_t = -\beta U_t dt + dW_t.$$

Observing that $d(e^{\beta t} U_t) = e^{\beta t} dW_t$, we obtain the stochastic integral representation

$$U_t = U_0 e^{-\beta t} + \int_0^t e^{-\beta(t-s)} dW_s. \tag{13.3.11}$$

It follows that the conditional distribution of U_t given U_0 is Gaussian with mean $U_0 e^{-\beta t}$ and variance $(1 - e^{-2\beta t})/(2\beta)$. Now it is easy to find the first and second order characteristics of a stationary version of Λ_t:

$$E\{\Lambda_t\} = m,$$

$$\mathrm{Var}\{\Lambda_t\} = \frac{\gamma^2}{2\beta} \overset{\mathrm{def}}{=} \sigma^2, \tag{13.3.12}$$

$$\mathrm{Cov}\{\Lambda(u), \Lambda(v)\} = \frac{\gamma^2}{2\beta} e^{-\beta|u-v|}.$$

Note that this process is obtained as the "aggregation in space" limit of a large number of either renewal rate processes, on/off processes or Poisson burst processes, when the periods appearing in these models are exponentially distributed. Motivated by the last case, we can intuitively think of $1/\beta$ as the "mean burst length" and of σ/β as the "mean burst size" in this abstract model for aggregate traffic.

The integrated process $A_t = \int_0^t \Lambda_s ds$ is Gaussian as well. Substituting (13.3.11) and changing the order of integration we obtain the representation

$$A_t = \int_0^t \Lambda_s ds = mt + \frac{1}{\beta}(1 - e^{-\beta t})(\Lambda_0 - m) + \frac{\gamma}{\beta} \int_0^t (1 - e^{-\beta(t-s)}) dW_s. \tag{13.3.13}$$

Given Λ_0, the distribution of (Λ_t, A_t) is a two-dimensional Gaussian distribution, and it is straightforward to calculate its parameters. Moreover, it follows that the conditional distribution of A_t given both Λ_0 and Λ_t is also Gaussian and has the parameters

$$E[A_t \mid \Lambda_0, \Lambda_t] = mt + \frac{1}{\beta} \frac{1 - e^{-\beta t}}{1 + e^{\beta t}} [(\Lambda_0 - m) + (\Lambda_t - m)],$$

$$\mathrm{Var}[A_t \mid \Lambda_0, \Lambda_t] = \frac{\gamma^2}{\beta^2} \left[t - \frac{2}{\beta} \frac{1 - e^{-\beta t}}{1 + e^{\beta t}} \right]. \tag{13.3.14}$$

In the stationary case, the variance of A_t is easily calculated from (13.3.12):

$$\mathrm{Var}\{A_t\} = \frac{\gamma^2}{\beta^2}\left(t - \frac{1}{\beta}(1 - e^{-\beta t})\right). \tag{13.3.15}$$

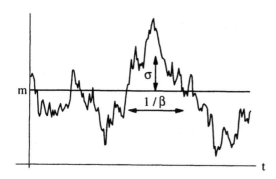

Figure 13.3.1: Effective "burst size" of the Ornstein-Uhlenbeck process.

13.3.6   Markov modulated rate process

The rate process Λ_t is called a Markov modulated rate process (MMRP) if it can be written in the form

$$\Lambda_t = r(Z_t), \qquad (13.3.16)$$

where Z_t is a Markov process and $r(z)$ is a deterministic function. The class of MMRP's is extremely rich. In fact, any reasonable rate process can be represented in this form by choosing an appropriate state space. For practical purposes, however, Z_t must be a structurally simple process: a finite state space process, a birth-death process, Brownian motion etc.

13.4   Long range dependent traffic models

Long range dependence is a rather new topic in teletraffic literature. Therefore, a concise introduction to the notions of long range dependence and self-similarity is presented in Subsection 13.4.1 below. The other subsections are mainly devoted to particular models.

13.4.1   Long range dependence and self-similarity

Assume that a process Y_t, which might represent the traffic rate at time t, is weakly stationary (i.e., stationary as regards its second order characteristics). In particular, the autocovariance function

$$\text{Cov}\,\{Y_s, Y_{s+t}\} = r(t)$$

does not depend on s. Let us assume that $\text{E}\{Y_t\} = 0$, and consider the cumulative process

$$A_t = \int_0^t Y_s \mathrm{d}s.$$

An expression for the variance of A_t is obtained by changing the order of integration:

$$\text{Var}\,\{A_t\} = 2 \int_0^t du \int_0^u dv \, \text{Cov}\,\{Y_u, Y_v\} = 2 \int_0^t du \int_0^u dv \, r(v). \qquad (13.4.1)$$

We see that if

$$a = \int_0^\infty r(v)dv < \infty, \qquad (13.4.2)$$

then $\text{Var}\{A_t\} = L(t) \cdot 2at$, where L is a *slowly varying function*, i.e.

$$\lim_{t \to \infty} \frac{L(\alpha t)}{L(t)} = 1 \quad \text{for all } \alpha > 0.$$

(For Karamata's theory of slow and regular variation, see [Fel71], Sections VIII.8–9.) Further, the variance of the time-scaled and normalized process

$$A_t^{(\alpha)} = \frac{A_{\alpha t}}{\sqrt{2a\alpha}}$$

satisfies

$$\lim_{\alpha \to \infty} \text{Var}\,\left\{A_t^{(\alpha)}\right\} = t,$$

so that its increments are almost uncorrelated for large α, i.e., at large timescales.

On the other hand, assume that the number a defined by (13.4.2) is $+\infty$. A periodic process is one trivial example of such a process. However, the case of interest here is that where the autocovariance function converges to zero at infinity but does this so slowly that the integral diverges. Let us make the somewhat more specific assumption that

$$\text{Cov}\,\{Y_0, Y_t\} = r(t) = L(t)t^p, \quad p \in (-1, 0), \qquad (13.4.3)$$

where L again denotes a slowly varying function. It then follows from (13.4.1) that

$$\text{Var}\,\{A_t\} = \tilde{L}(t)t^{p+2},$$

where $\tilde{L}$ is slowly varying (see [Fel71], the Lemma of Section VIII.9), and

$$\lim_{\alpha \to \infty} \text{Var}\,\left\{ \frac{A_{\alpha t}}{\sqrt{\tilde{L}(\alpha)\alpha^{p/2+1}}} \right\} = t^{p+2}.$$

We see that the correlation structure of the time-scaled and normalized process is again asymptotically time scale independent, but now the variance increases in the power $p+2 \in (1, 2)$. This implies positive correlations between increments, in contrast to the asymptotical absense of correlation obtained in the first case. This introduction motivates the following basic definition (see [Cox84, Ber94]).

Definition 13.4.1 A sequence $X_1, X_2, \ldots$ of weakly stationary random variables is called *long range dependent* if

$$\sum_{n=1}^{\infty} \text{Cov} \{X_1, X_n\} = \infty. \tag{13.4.4}$$

Otherwise it is called *short range dependent*. If $\text{Var}\{X_1 + \cdots + X_n\}$ of a long range dependent sequence grows at speed n^{2H}, where $H \in (1/2, 1]$, then the number H is called the *Hurst parameter* of the sequence. Sometimes we use (13.4.4) for continuous time processes replacing the series by an integral.

Long range dependence is essentially equivalent to some other conditions which could have been chosen as the definition. If $r(k) \sim t^p$, $p \in (-1, 0)$, we have *non-vanishing correlation between large blocks*, e.g.,

$$\text{Corr} \{A_t, A_{2t} - A_t\} = \frac{\frac{1}{2}\text{Var}\{A_{2t}\} - \text{Var}\{A_t\}}{\text{Var}\{A_t\}} \to c > 0 \quad \text{as } t \to \infty,$$

and *singularity of the spectral density at zero*:

$$f(\lambda) = \frac{\text{Var}\{Y_0\}}{2\pi} \sum_{k=-\infty}^{\infty} r(k) e^{-ik\lambda} \approx c\lambda^{-p-1} \quad \text{as } \lambda \to 0. \tag{13.4.5}$$

Markov chains with finite state space and autoregressive processes of finite order have exponentially decaying autocorrelation functions and are thus short range dependent. It follows that practically all processes traditionally used in teletraffic modelling are short range dependent. However, as we shall see below, some usual simple models can be made long range dependent with a slight modification.

We saw above that at large time scales, the correlation structure of the cumulative process is asymptotically independent of the time scale. This observation is related to Lamperti's important theorem [Lam62], which (essentially) says that limit processes obtained by increasing the time scale together with a suitable normalization are always *self-similar* processes, defined as follows.

Definition 13.4.2 A process X_t is called *(strictly) self-similar* with self-similarity parameter (or Hurst parameter) H if, for any $\alpha > 0$, the processes $X_{\alpha t}$ and $\alpha^H X_t$ have the same finite-dimensional distributions. It is called *asymptotically self-similar* if the process $X_{\alpha t}$, suitably normalized, converges weakly to a self-similar process when $\alpha \to \infty$.

When working with second order characteristics only, the following counterpart of the previous definition is used:

Definition 13.4.3 A square integrable process X_t is called *second order self-similar* with self-similarity parameter (or Hurst parameter) H if, for any $\alpha > 0$, the processes $X_{\alpha t}$ and $\alpha^H X_t$ have the same second order characteristics. It is called *asymptotically second order self-similar* if the second order characteristics of $X_{\alpha t}$, suitably normalized, converge to those of a second order self-similar process when $\alpha \to \infty$.

Thus, asymptotic second order self-similarity with $H > 1/2$ is essentially equivalent to long range dependence. Since H is the most commonly used parameter in this context, we summarize the three power criteria of long range dependence in terms of the Hurst parameter H in Table 13.4.1 below.

Table 13.4.1: Three different power criteria for the long range dependence of a weakly stationary sequence (or process) Y_t in terms of the Hurst parameter $H \in (1/2, 1)$. A_t denotes the integrated (cumulative) process and $f(\lambda)$ the spectral density of Y_t. The asymptotics hold as $t \to \infty$ and $\lambda \to 0$.

criterion	$\mathrm{Cov}\{Y_0, Y_t\} \sim t^{\beta_1}$	$\mathrm{Var}\{A_t\} \sim t^{\beta_2}$	$f(\lambda) \sim \lambda^{\beta_3}$
power	$\beta_1 = 2H - 2$	$\beta_2 = 2H$	$\beta_3 = 1 - 2H$

It is easy to see that a non-trivial second order self-similar process with stationary increments must have $H \in (0, 1]$. Strictly self-similar processes with stationary increments and $H > 1$ exist, but they are probably too "pathological" to be useful in teletraffic modelling. In the sequel, our self-similar processes are always assumed to have stationary increments and to be square integrable.

The covariance of the increments in two non-overlapping intervals of a second order self-similar process X_t has the expression

$$\mathrm{Cov}\left\{X_{t_2} - X_{t_1}, X_{t_4} - X_{t_3}\right\} \tag{13.4.6}$$

$$= \frac{1}{2}\left((t_4 - t_1)^{2H} - (t_3 - t_1)^{2H} + (t_3 - t_2)^{2H} - (t_4 - t_2)^{2H}\right)$$

for $t_1 < t_2 \leq t_3 < t_4$.

As a general reference to self-similar processes, see articles in the collection [ET86] and the monograph [ST94], for example.

13.4.2   Rate processes with heavy-tailed period lengths

The models described in this section have silence and/or activity lengths with infinite variance as the element that bring long range dependence into the play. The respective connection between long-range dependence, renewal periods with infinite variance and self-similarity was originally established by Mandelbrot [Man69]. It has indeed often been observed that activity or, still more frequently, silence periods of a traffic source have such "heavy-tailed" distributions (see, e.g., [Vei92, PF95]).

As usual, let Λ_t be a stationary process representing the rate at time t and let $A_t = \int_0^t \Lambda_s \, ds$ be the corresponding cumulative traffic process.

For both the renewal rate process and the Poisson burst process, we have (by (13.3.2) and (13.3.5), with a change of the order of integration)

$$\int_0^\infty \text{Cov}\{\Lambda_0, \Lambda_t\} \, dt = a\text{E}\{U^2\},$$

where U is the generic period length and a is a constant. Thus, Λ_t is long range dependent exactly when U has infinite variance. By (13.3.3), $\text{Var}\{A_t\}$ grows asymptotically at speed t^{2H} with $H \in (1/2, 1)$ if $\text{Pr}\{U > t\}$ decays asymptotically at speed t^{2H-3}.

Consider then a single on/off source process as defined in Subsection 13.3.2, with peak rate taken as unity and mean rate m, and denote $v(t) = \text{Var}\{A_t\}$. Long range dependence can be introduced into this source model by assuming a heavy tail for the distribution of either the silence periods, activity periods or both. We consider distribution functions F such that

$$1 - F(t) \sim ht^{-\beta} \tag{13.4.7}$$

for large t and $1 < \beta < 2$, where h is a positive constant. A typical example is the (translated) Pareto distribution where $1 - F(t) = [\theta/(t + \theta)]^\beta$ for some $\theta > 0$. Consider first the case where the silence period V has at least two finite first moments with $\text{E}\{V\} = 1/\lambda$, while the distribution of activity period U is heavy tailed. Let B denote the distribution function of an activity period and h_b the constant associated with distribution function B in (13.4.7). It can then be shown [BRSV95] that

$$v(t) \sim \frac{2h_b\lambda(1 - m)^3}{(\beta - 1)(2 - \beta)(3 - \beta)} t^{3-\beta} \tag{13.4.8}$$

for large t. Note that $v(t) = O(t^{2H})$ where $H = (3 - \beta)/2$ lies in $(1/2, 1)$, thus confirming that heavy tailed activity periods generate long range dependence and identifying H as the Hurst parameter.

If instead we choose the silence period V to have a heavy tailed distribution with exponent $\alpha \in (1, 2)$ and constant h_a, while the activity period U has at least two finite first moments with $\text{E}\{U\} = 1/\mu$, we obtain

$$v(t) \sim \frac{2h_a\mu m^3}{(\alpha - 1)(2 - \alpha)(3 - \alpha)} t^{3-\alpha} \tag{13.4.9}$$

for large t. This result is structurally similar to (13.4.8). It can in fact be derived by using (13.4.8) for a related problem where the silence and activity periods are exchanged.

Finally, if both on and off periods have heavy tailed distributions, then $v(t)$ is asymptotic to the sum of their corresponding expressions in (13.4.8) and (13.4.9), and the term corresponding to the smaller of α and β will dominate. Defining α (resp. β) equal to 2 if the silent (resp. active) period has its first two moments finite, we can write $H = (3 - \min\{\alpha, \beta\})/2$ in all cases.

13.4.3 Processes generated by deterministic chaotic maps

Erramilli, Singh and Pruthi have investigated the potential of the iteration of deterministic chaotic maps in artificial traffic generation [ES90, Pru95]. The models defined so far produce on/off traffic.

For an example, consider Pruthi's "Double Intermittency Map" on $[0, 1]$. Starting from any $x_0 \in (0, 1)$, a sequence (x_n) is generated by the iteration rule

$$x_{n+1} = \begin{cases} f_1(x_n), & \text{if } x_n \in [0, d) \\ f_2(x_n), & \text{if } x_n \in [d, 1] \end{cases}$$

where

$$f_1(x) = \frac{x}{(1 - c_1 x^{m_1 - 1})^{1/(m_1 - 1)}},$$

$$f_2(x) = 1 - \frac{1 - x}{(1 - c_2(1 - x)^{m_2 - 1})^{1/(m_2 - 1)}},$$

and

$$c_1 = d^{1-m_1} - 1, \quad c_2 = (1 - d)^{1-m_2} - 1,$$

with the model parameters $d \in (0, 1)$, $m_1, m_2 \geq 1$. The source output at time slot n is then defined as

$$y_n = 1_{[d,1]}(x_n).$$

The parameters m_1 and m_2 define the tail behaviour of the distributions of the off and on periods, respectively. The choice $m_1 = m_2 = 1$ gives geometric distributions for both, whereas larger values result in power tails. The values $m_i \in (3/2, 2)$ correspond to finite mean and infinite variance and thus result in a long range dependent process. In the case $m_2 = 1$ and $m_1 \in (3/2, 2)$, Pruthi shows that the Hurst parameter is given by $H = (3m_1 - 4)/(2m_1 - 2)$, and derives a good approximate expression for the invariant density. The theory still has rather few exact results.

Note that the iteration of a deterministic map is a special case of a Markov chain with continuous state space.

13.4.4   Fractional Brownian traffic

A normalized *fractional Brownian motion* (FBM) with Hurst parameter $H \in [\frac{1}{2}, 1)$ is a stochastic process Z_t, $t \in (-\infty, \infty)$, characterized by the following properties:

(i) Z_t has stationary increments;

(ii) $Z_0 = 0$, and $\mathrm{E}\{Z_t\} = 0$ for all t;

(iii) $\mathrm{E}\{Z_t^2\} = |t|^{2H}$ for all t;

(iv) Z_t has continuous sample paths;

(v) Z_t is Gaussian, i.e., all its finite-dimensional marginal distributions are Gaussian.

In the special case $H = 1/2$, Z_t is the standard Brownian motion. We have ruled out the other limiting case $H = 1$ since the corresponding Z_t is a deterministic process with linear paths. This process was found by Kolmogorov [Kol40], but relatively little attention was paid to it before the pioneering paper by Mandelbrot and Van Ness [MN68] (where the FBM also got its present name). They defined the process as the stochastic integral

$$Z_t - Z_s = \tag{13.4.10}$$

$$c_H \left\{ \int_s^t (t-u)^{H-\frac{1}{2}} dW_u + \int_{-\infty}^s \left((t-u)^{H-\frac{1}{2}} - (s-u)^{H-\frac{1}{2}} \right) dW_u \right\},$$

where W_t is the standard Brownian motion. The normalization $\mathrm{E}\{Z_1^2\} = 1$ is achieved with the choice

$$c_H = \sqrt{\frac{2H\Gamma(\frac{3}{2} - H)}{\Gamma(\frac{1}{2} + H)\Gamma(2 - 2H)}}. \tag{13.4.11}$$

Many features of fractional Brownian motions with $H > 1/2$ are different from those of most stochastic processes usually appearing in traffic models. They are not Markov processes and, having non-differentiable sample paths with zero quadratic variation, they do not even belong to the wide class of semimartingales for which there exists an extensive and powerful theory. Figure 13.4.1 presents simulated realizations of Z_t and its increments with different values of H. Note that with high H the path looks considerably smoother than that of an ordinary Brownian motion.

For the use of the FBM as a traffic model element it is pleasant to note that in spite of the strong correlations, it is ergodic in the sense that the stationary sequence of increments $Z_n - Z_{n-1}$, often called *fractional Gaussian noise*, is ergodic (e.g., [CSF82], Theorem 14.2.1).

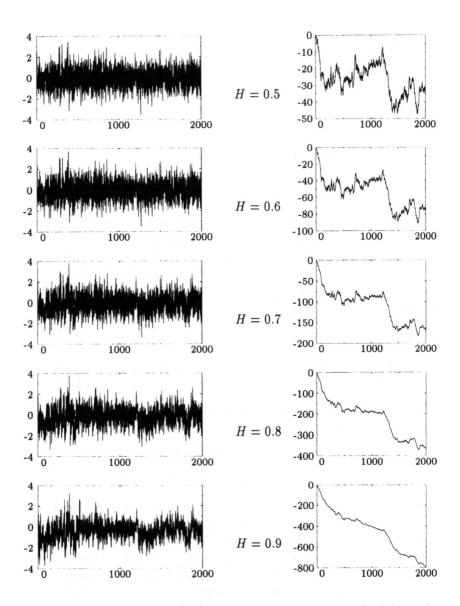

Figure 13.4.1: Simulated realizations of normalized fractional Gaussian noise $Z_{n+1} - Z_n$ and the corresponding cumulative sum Z_n, $n = 1, \ldots, 2000$, with different values of the Hurst parameter H. The underlying pseudo-random sequence is the same in all cases.

As usual, let us denote by $A(s,t)$, $s,t \in (-\infty, \infty)$, the amount of traffic offered to a link in the time interval $[s,t)$, and write $A_t = A(0,t)$. A long-range dependent counterpart to diffusion-based traffic models can now be defined in terms of the FBM as follows:

$$A_t = mt + \sqrt{am}Z_t, \quad t \in (-\infty, \infty), \qquad (13.4.12)$$

where Z_t is a normalized fractional Brownian motion. We call this traffic model *fractional Brownian traffic*. The traffic model has three parameters m, a and H with the following interpretations: $m > 0$ is the mean input rate, $a > 0$ is the index of dispersion at $t = 1$, and $H \in [1/2, 1)$ is the Hurst parameter of Z_t.

Remark 13.4.4 $Z(t)$ is a mathematical object which has no physical dimension and its parameter t is dimensionless as well. Therefore it would be better to write it in the traffic model as $Z(t/t_u)$, where t is physical time and t_u is the physical time unit. This helps, in particular, to avoid confusion by a change of the time unit. However, we shall not do this in order to keep our notation simple. If work is measured in bits and time in seconds, a has the dimension bit·sec. The Hurst parameter H is dimensionless.

A superposition $A_t = \sum_{i=1}^{K} A_t^{(i)}$ of K independent fractional Brownian traffic processes with common parameter H but individual m_i's and a_i's is again fractional Brownian traffic with the same H and the other parameters given as

$$m = \sum_{1}^{K} m_i, \quad a = \sum_{1}^{k} \frac{m_i}{m} a_i.$$

If the a_i's are identical also, then the superposition has the same parameter a as the constituent streams. This motivates the factor $\sqrt{m}$ in (13.4.12): the roles of the three parameters can be separated so that H and a characterize the "quality" of the traffic in contrast to the long run mean rate m which characterizes its "quantity" alone.

Some heuristic insight into the roles of the three parameters can be obtained by fitting fractional Brownian traffic to the asymptotic second order characteristics of an appropriate long range dependent rate process with heavy tailed "on" and/or "off" period distributions (see Subsection 13.4.2). The Hurst parameter is then determined by the tail behaviour of the period length distributions alone, whereas the parameter a depends on the parameters of the "equivalent" rate process in a more complicated way. Such a fitting with the Poisson burst process was discussed in [Nor95].

13.4.5   Fractional Brownian traffic as limit traffic

We now consider the connection between the superposition of a large number N of on/off sources and fractional Brownian traffic. Limit theorems of the type relevant here were first proven by Taqqu in the 1970's. Willinger *et al.* [WTSW95] have found in their source level data traffic analysis evidence for the hypothesis that the self-similarity observed in Ethernet traffic could indeed be explained by the heavy tails of the distributions of the on- and off-period lengths. A detailed presentation of the results in this subsection can be found in [BRSV95].

Denote by $\{\Lambda_t^{(N)}\}$ and $\{A_t^{(N)}\}$ the sum of N i.i.d. copies of the processes $\{\Lambda_t\}$ and $\{A_t\}$ associated with a single source, respectively. Process $\{\Lambda_t^{(N)}\}$ thus represents the bit rate created by the superposition of the N independent on/off sources. In the sequel, we consider the convergence of the workload process $\{A_t^{(N)}\}$ to some limiting work process $\{Y_t\}$ which in turn enables us to bridge the gap to the FBM model.

Limiting work process

Process $\{A_t^{(N)}\}$ has mean Nmt, finite variance $Nv(t)$ and continuous sample paths. The process $\{\Lambda_t^{(N)}\}$ being stationary, $\{A_t^{(N)}\}$ has stationary increments. For each N, the expression $(A_t^{(N)} - Nmt)/\sqrt{N}$ thus defines a centered process with stationary increments, continuous sample paths and variance $v(t)$. In addition, by the Central Limit Theorem, its one-dimensional distributions tend to Gaussian distributions as $N \to +\infty$. Thus it seems plausible that these processes tend to a Gaussian process with these same properties. In fact, recent results from Kurtz [Kur96] prove that such a convergence theorem holds for the general on/off source model. Moreover, Kurtz shows weak convergence properties which are essential for our application to the queueing problem. Such results are technically intricate and for purpose of illustration here, we will call upon existing limit theorems due to Taqqu and Levy [TL86] for certain Renewal Reward Processes (RRP's; in our context, the name "renewal rate process" is preferable, see Subsection 13.3.1) which offer this kind of result for a subclass of cases. Instead of the more general weak convergence results of Kurtz however, the latter is also restricted to L-lim convergence, that is, the convergence of finite dimensional distributions of processes.

We begin by defining an RRP in discrete time. Let the interrenewal time variables $\{U_k\}$, $k \in \mathbf{N}$, define an equilibrium renewal process with renewals at times $t = S_k$ with $S_0 = 0$ and $S_k = U_0 + \cdots + U_k$. The U_k are i.i.d. (except for U_0), with mean $1/d$. Let the i.i.d. variables $\{R_k\}$ be independent of the U_k, and define the process $\{X_t\}$ by assigning for all k the reward R_k to each $t \in (S_k, S_{k+1}]$. $\{X_t\}$ is a stationary RRP. Consider now the sum $\{X_t^{(N)}\}$ of N i.i.d. copies of $\{X_t\}$ and define the cumulative process $\{Y_t^{(N)}\}$ by

$$Y_t^{(N)} = \sum_{s=1}^{t} X_s^{(N)}.$$

The first limit theorem of interest [TL86] states that

$$\text{L--}\lim_{N\to+\infty} \frac{1}{\sqrt{N}} \left(Y_t^{(N)} - Nmt \right) = Y_t, \qquad (13.4.13)$$

where $\{Y_t\}$ is a zero-mean Gaussian process with stationary increments and $Y_0 = 0$.

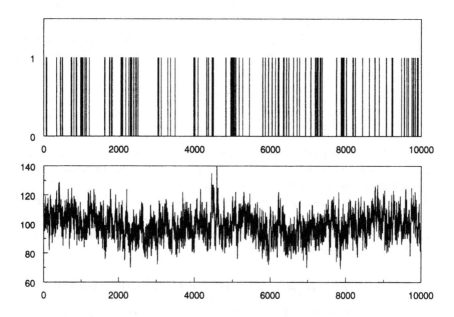

Figure 13.4.2: Simulated 0/1-valued renewal rate processes with on-probability 0.1 and shifted Pareto distributions with mean 10 and $\beta = 1.4$, which corresponds to $H = 0.8$. Above: one realization. Below: superposition of 1000 realizations.

For our purposes here, we restrict the R_k to be binary variables such that $\Pr\{R_k = 1\} = m$, $\Pr\{R_k = 0\} = 1 - m$. Figure 13.4.2 illustrates this case. The similarities with the on/off arrival process are then clear. A reward $R_k = 1$ generates an active period, X_t corresponds to the instantaneous bit rate Λ_t and $Y_t^{(N)}$ to $A_t^{(N)}$. There remains, however, the important difference that the R_k are i.i.d., whereas the active and silence periods must alternate. The sequence $\{R_k\}$, nevertheless, can be thought of as consisting of "random" active and silence periods in the obvious way. Assume now that the discrete distribution of variable U is heavy tailed with exponent $\alpha \in (1,2)$ and associated coefficient h_d in (13.4.7). It is easily shown that active and silence periods then have heavy tailed distributions as well, with identical exponent α. The limit process $\{Y_t\}$ derived in (13.4.13) then also satisfies [TL86]

$$\text{Var}\{Y_t\} \sim \Sigma^2 t^{(3-\alpha)/2} \qquad (13.4.14)$$

for large t, where

$$\Sigma^2 = \frac{2h_d d m(1-m)}{(\alpha-1)(2-\alpha)(3-\alpha)}.$$

Convergence to FBM

The process $\{Y_t\}$ is not a FBM but, in view of (13.4.14), it is long range dependent because its variance grows as t^{2H} for large t and some $H > 1/2$. By accumulating the process over time and rescaling, we give to the dependence structure at all scales this same long-term character and thereby generate self-similarity. This is the content of the following result in the discrete time context [TL86],

$$L-\lim_{T\to+\infty} \frac{1}{\sqrt{h_d}T^H}Y_{Tu} = \sqrt{\frac{2d\,\mathrm{Var}\{R_0\}}{(\alpha-1)(2-\alpha)(3-\alpha)}}Z_u, \qquad (13.4.15)$$

where $\{Z_u\}$ is a FBM with Hurst parameter $H = (3-\alpha)/2$. As mentioned above, such a convergence result can be extended [Kur96] to weak convergence and also generalised to continous time.

On the basis of these limit results, it can be conjectured that fractional Brownian traffic is a rather generally applicable model for data traffic aggregated from a large number of independent work stations and other long range dependent sources.

13.4.6   A two-component FBM-based traffic model

Even for long range dependent traffic aggregated from a large number of individual communication streams, fractional Brownian traffic in the simple form considered in Subsection 13.4.4 need not be a satisfactory model at small time scales. For example, an addition of strongly variable but short range dependent traffic to the mixture could not be described in that model in another way than by reducing the available bandwidth. The simplest way to avoid such inflexibility problems would be to add a Brownian component to the model, i.e., to introduce the Gaussian traffic model

$$A_t = mt + \sqrt{b}W_t + \sqrt{a}Z_t, \qquad (13.4.16)$$

where W_t is a normalized Brownian motion and Z_t is a normalized FBM, independent of W_t.

But what if two such traffic streams with different Hurst parameters for their FBM parts are aggregated? In order to avoid adding several FBM's to (13.4.16), we suggest for further study the idea to use the above model with a suitable *fixed* Hurst parameter H of Z_t, chosen "once and for all". Only the remaining three parameters m, a and b are then used for fitting the model to measurement data. Of course, the model will always have the same ultimate Hurst parameter, but it can be expected to have a satisfactory fitting flexibility in a restricted range of time scales. Note that "busy hour"-like time periods are needed for dimensioning anyway and long range dependence should not really be modelled to cover longer periods than in the magnitude of 10^4 seconds because non-stationary variation dominates in still larger time scales (days, months).

To fit the three parameters, estimates of the mean rate m and of $v(t) = \text{Var}\{A_t\}$ for two different t's are needed. If $v(t)$ is known for, e.g., $t = 1$ and $t = 10^4$, then the parameters a and b are determined by the equations

$$v(1) = a + b, \quad v(t)/t = b + at^{2H-1} \text{ with } t = 10^4. \qquad (13.4.17)$$

Trusting the model, we could use an estimate of $v(10^2)$ instead of $v(10^4)$, but it is important that we don't underestimate $v(10^4)$. (On the other hand, the estimation of $v(10^4)$ is problematic because of the non-stationary variation at large time scales.)

A nice property of the parametric class (13.4.16) with H fixed at a general value is that a superposition of such processes will stay in the class, and the three free parameters are additive when independent traffic streams are aggregated. Being second order characteristics, they can also be estimated for each source separately without assuming that the source streams are Gaussian.

This model offers also an extension of the "equivalent burst model" introduced in [Lin91] (see also [Rob92a], pp. 152–153). Suppose that we have the superposition of all streams represented by the parameters m, b, a (sums of the individual ones) and in addition by the variance of the instantaneous rate σ^2, which is also the sum of the corresponding individual values. One can now introduce an "equivalent burst model", fitting A_t (compare [Rob92a], page 152), where the total rate process is the superposition of bursts with height h and mean duration d, and where the number of such bursts has a Poisson distribution with mean $\lambda = \lambda_1 + \lambda_2$. One can now also split m into $m_1 + m_2$, where $m_1/m_2 = a/b$ and with the same h we also have $\sigma_1^2/\sigma_2^2 = a/b$. Now, with exponentially distributed on-periods for the fast varying part, the fitting relations would be

$$h = \frac{\sigma^2}{m}, \quad \lambda_2 = \frac{2m_2^2}{b}, \quad d_2 = \frac{b}{2hm_2}. \qquad (13.4.18)$$

Thus we have modelled the fast varying part of the total traffic process as an equivalent burst process with the parameters m_2, h and d_2. For the slowly varying part we know the mean m_1, the variance of the instantaneous rate σ_1^2, and its type of long range dependence.

Now, whatever value the general H has, 0.75, 0.8 or even 0.9, it is certainly essentially greater than 0.5, and thus one can conclude that the Z_t-part is varying very slowly in comparison to the W_t-part. Thus we could dimension with the marginal distribution of this slowly varying part as a background process for the more fast varying process, i.e. at each "occasion" we have a certain number of bursts on from the slowly varying part. Buffers can hardly compensate for the bursts with such heavy tail distributions. The marginal distribution, however, can be modelled, for example, with the equivalent burst method, if m_1 and σ_1^2 are known, as a Poisson(λ_1) number of on-going bursts of height h, where $\lambda_1 = m_1/h$ and $h = \sigma_1^2/m_1$.

The dimensioning with respect to a certain CLR in the GOS will now be simpler. For example, a mid-size buffer (say, 10^3 ATM cells) which could improve the efficiency of the link capacity will only do so for the fast varying part whose capacity is the random variable "c minus the marginal r.v. of the slowly varying part". Thus, the impact of the general value of H here is essentially that it affects the proportion between the parts with the fast and slow variations.

13.4.7   Fractional ARIMA processes

In this subsection, we give the definition of the class of Fractional Auto-Regressive Integrated Moving-Average (FARIMA) processes [Hos84]. They were introduced as an extension to the traditional ARMA and ARIMA classes, which are only capable of modelling the short-range portion of the correlations of empirical data sets.

Let us start from the discrete time random walk $X_t = \sum_1^t a_k$, where a_t is a Gaussian white noise process, i.e., a sequence of independent standard normal variables. Denoting by B the backshift operator $BX_t = X_{t-1}$, we have $(1 - B)X_t = a_t$. This equation defines X_t as an ARIMA(0,1,0) process. The integrated (cumulative) process $Y_t = \sum_1^t X_k$ is then an ARIMA(0,2,0) process satisfying $(1 - B)^2 X_t = a_t$, and so on. Now, we can generalize this definition for fractional orders d and say that X_t is a FARIMA(0,d,0) process if it satisfies

$$(1 - B)^d X_t = a_t, \qquad (13.4.19)$$

where the fractional difference operator $(1 - B)^d$ is defined by the binomial series

$$(1 - B)^d = \sum_{k=0}^{\infty} \binom{d}{k} (-B)^k.$$

With $d \in (-\frac{1}{2}, \frac{1}{2})$, the process is stationary, and we have the inverse relation $X_t = (1 - B)^{-d} a_t$. Taking the cumulative sum of FARIMA(0,d,0) we obtain FARIMA(0,d+1,0). The spectral density of FARIMA(0,d,0) can be shown to be

$$f(\lambda) = (2 \sin \frac{\lambda}{2})^{-2d}, \quad \text{for} \quad 0 < \lambda \leq \pi. \qquad (13.4.20)$$

Thus, $f(\lambda) \sim \lambda^{-2d}$ as $\lambda \to 0$, which shows that FARIMA(0,d,0) with $d \in (0, 1/2)$ is long range dependent with $H = d + 1/2$.

More generally, a FARIMA(p,d,q) process is defined by the equation

$$\phi(B)(1 - B)^d X_t = \psi(B)a_t,$$

where ϕ and ψ are polynomials of orders p and q, respectively. (The classical ARMA(p,q) class is the FARIMA($p,0,q$) class.)

If a good approximation of both the short-range and the long-range correlations is mandatory, the FARIMA(p,d,q) processes with non-zero p and/or q may be used. In these cases, one may determine an estimate $\hat{d}$ of d with one of the methods described below for the estimation of H, and generate either a new sequence or a new spectrum from the original data set, where the long-range correlations are removed under the assumption that $\hat{d}$ is an appropriate estimate of d for the underlying process, and compute the remaining $p + q$ parameters with standard ARMA techniques [SS95, BJ76]. For instance,

$$\hat{f}_{ARMA(p,q)}(\lambda) = \hat{f}_{FARIMA(p,d,q)}(\lambda)(2\sin\frac{\lambda}{2})^{2\hat{d}}. \qquad (13.4.21)$$

13.4.8   Pseudo self-similar models

In this subsection we consider traffic models which are based on Markov chains with finite state space but exhibit approximately self-similar behaviour over several timescales. For lack of a better term, we call half of the growth rate of Var$\{A_t\}$ in a log-log plot the *local Hurst parameter* of A_t. The motivation for applying Markovian models even when the real traffic might be long range dependent is that powerful tools have been developed for calculating performance measures for both finite and infinite capacity queues - for recent progress, see, for example, [Luc91, LR93, Blo89, BBM93]. Since these models are, however, necessarily short range dependent in the strict mathematical sense, we call them pseudo self-similar.

We present two approaches which have been developed within COST242. The theory of these models bears certain resemblance to the *theory of decomposability*, developed by Courtois in the mid-70's [Cou77]. Pioneering work was provided by Simon and Ando who studied several cases in economics and physics [SA61, Sim62, Sim69]. The basic observation behind this theory is that many systems can be analyzed as a hierarchy of components and subcomponents with strong interactions within components at the same level and lower interactions with other components. Such systems are called *nearly completely decomposable,* and they have been observed in computing, economics, biology, genetics and social sciences.

The "message" of the theory is that aggregation of variables in a nearly decomposable system must separate the analysis of the short term and long term dynamics. There are two major theorems. The first says that a nearly decomposable system can be analysed by a completely decomposable system if the intergroup dependencies are sufficently weak compared to intragroup ones. The second says that the results obtained in the short term will remain approximately valid in the long term, as far as the relative behavior of the variables of the same group is concerned.

A Markov modulated on/off source

This model is a simple discrete time Markov modulated model for the process of cell arrivals on a slotted link. Call X_t the random variable representing the number of cells (assumed to be 0 or 1) during the tth time slot, i.e., during time interval $[t-1, t)$. Let $Y_t \in 1, 2, 3, ..., n$ be the modulator process, assumed to be a discrete time homogeneous Markov chain with transition matrix $\mathbf{A} = (a_{ij})$. Let $\phi_{ij} = \Pr[X_t = j \mid Y_t = i]$. We consider the stationary case only and denote the stationary distribution of Y_t by $\pi_i = \Pr\{Y_t = i\}$. Since the process X_t is $\{0, 1\}$-valued, its moments are equal and given by

$$\mathrm{E}\{X_t^k\} = \mathrm{E}\{X\} = \vec{\pi}\Lambda\vec{e}, \quad k = 1, 2, \cdots, \quad t = 0, 1, 2, \cdots \quad (13.4.22)$$

with $\vec{\pi} = (\pi_1, \pi_2, ..., \pi_n)$, $\vec{e}$ a column vector consisting of 1's, and Λ defined as

$$\Lambda = \mathbf{diag}(\mathrm{E}[X \mid Y = 0], \mathrm{E}[X \mid Y = 1], ..., \mathrm{E}[X \mid Y = n]). \quad (13.4.23)$$

If N_m represents the number of arrivals in a window of m time slots, the variance of N_m can be written as

$$\mathrm{Var}\{N_m\} = m\mathrm{E}\{X\} - m^2(\mathrm{E}\{X\})^2 + 2\sum_{i=1}^{m-1}(m-i)(\vec{\pi}\Lambda\mathbf{A}^{i+1}\Lambda\vec{e}). \quad (13.4.24)$$

Now, we propose a family of models for which

$$\mathbf{A} = \begin{pmatrix} 1 - \frac{1}{a} - (\frac{1}{a})^2 - \cdots - (\frac{1}{a})^{n-1} & \frac{1}{a} & (\frac{1}{a})^2 & \cdots & (\frac{1}{a})^{n-1} \\ \frac{b}{a} & 1 - \frac{b}{a} & 0 & \cdots & 0 \\ (\frac{b}{a})^2 & 0 & 1 - (\frac{b}{a})^2 & \cdots & 0 \\ \cdots & \cdots & \cdots & \cdots & \cdots \\ (\frac{b}{a})^{n-1} & 0 & 0 & \cdots & 1 - (\frac{b}{a})^{n-1} \end{pmatrix}$$

and

$$\Lambda = \begin{pmatrix} 1 & 0 & 0 & \cdots & 0 \\ 0 & 0 & 0 & \cdots & 0 \\ 0 & 0 & 0 & \cdots & 0 \\ \cdots & \cdots & \cdots & \cdots & \cdots \\ 0 & 0 & 0 & \cdots & 0 \end{pmatrix}.$$

Thus, the Markov chain is characterized by 3 parameters: $a > 1$, $b \in (0, a)$ and n. The expectation $\mathrm{E}\{X\}$ is independent of a and given by

$$\mathrm{E}\{X\} = \frac{1 - (1/b)}{1 - (1/b)^n} \quad (13.4.25)$$

with $\lim_{n \to \infty} \mathrm{E}\{X\} = 1 - 1/b$ when $b > 1$. Note that the limiting infinite state space system is still positively recurrent if $b > 1$. The process X_t is a discrete time on/off process with geometrically distributed on-periods and "almost" heavy tailed off-periods.

For given n, the model can be fitted to a given mean m and local Hurst parameter H_l. The parameter b is determined by $m = (1-(1/b))/(1-(1/b)^n)$, which is easy to solve numerically. The remaining parameter a is found, for example, iteratively with the Newton-Raphson method, estimating the slope of the variance plot with least squares. The domain where the process is self-similar can be enlarged by increasing n. Table 13.4.2 shows some values found with the algorithm described.

Table 13.4.2: a and b as a function of H_l and n, $E\{X\} = 0.05$.

H_l	$n = 4,\ b = 0.4418$	$n = 5,\ b = 0.5764$	$n = 6,\ b = 0.6737$
0.70	2.65	2.60	1.99
0.75	4.97	4.38	3.22
0.80	10.27	6.70	4.80
0.85	28.49	12.97	8.21
0.90	–	85.61	50.02

The system is nearly completely decomposable in the sense of Courtois, when the sets $\{1, \ldots, n-1\}$ and $\{n\}$ are taken as the subsystems. Figure 13.4.3 shows that the variance/time plot of $X(t)$ is similar to that of a positively dependent self-similar process over 4 time scales. The graph corresponding to the decomposable Markov chain is also plotted in order to show that the influence of state n on the variance curve in this case starts at time scale 10^4. In the case $n = \infty$, $b > 1$, $X(t)$ shows second order self-similar behaviour over all timescales.

Figure 13.4.4 shows variance-time plots, that is, the function $\text{Var}\{N_m/m\}$, calculated with (13.4.24). Note that the domain where the local Hurst parameter is larger than 0.5 becomes wider with increasing H; it seems that the Markov chain needs more states for pseudo self-similarity with a low H_l than with a high one! The knee where the pseudo self-similarity region ends is particularly clearly seen at the curves with $H_l = 0.75$ and $H_l = 0.8$. For higher H_l's, the knee is no longer visible — more computing time would have been required to extend the figure to the right. For the highest H_l's, the variance-time plot is somewhat wavy.

Figure 13.4.5 presents the "visual test" of the self-similarity of $X(t)$. At each timescale, we observe bursty periods separated by less bursty subperiods as in measured LAN traffic [RB96, RB95, LTWW94].

The simulation of such a Markov chain requires care, in particular because of the large differences between the transition probabilities. If $a = 10$ and the Markov chain (8 states for example) is in the first state, $a_{12} = 10^{-1}$ but

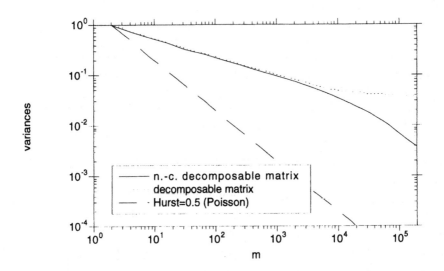

Figure 13.4.3: Behaviour of a nearly decomposable matrix and a decomposable matrix versus window size m.

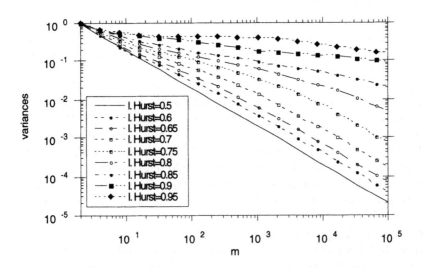

Figure 13.4.4: Variance-time plots for the pseudo self-similar Markov process with 5 states, with different local Hurst parameters.

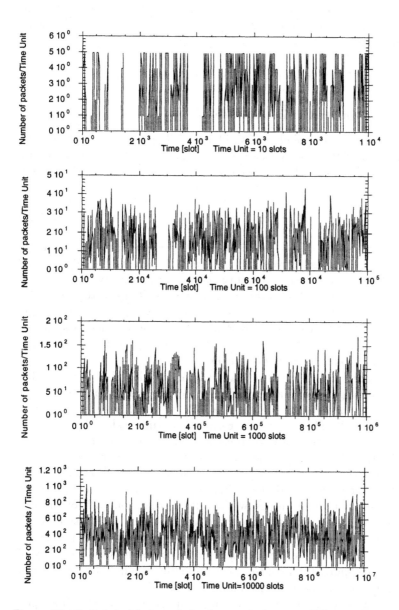

Figure 13.4.5: 5-state Markov chain time evolution at 4 different timescales, $H_l = 0.78$.

$a_{18} = 10^{-7}$. The ratio between the two extreme transition probabilities is of the order 10^6. If this Markov chain is simulated with classical methods, the random number generator must be very good. Rather than relying on this, probabilities for the rare jumps can be generated in several steps, i.e., using the decompositions

$$\Pr[Y_{t+1} > k+1 \mid Y_t = 1] = \frac{1}{a}\Pr[Y_t > k \mid Y_t = 1].$$

A Markov modulated Poisson process

Our second approach to pseudo self-similar modelling uses a continuous time Markovian Arrival Process (MAP) [Luc91, LMN90, Neu89]. The MAP is a Markov renewal process whose transition probability matrix $\mathbf{F}(x)$ is of the form

$$\mathbf{F}(x) = \int_0^x e^{\mathbf{D}_0 u}du \, \mathbf{D}_1, \tag{13.4.26}$$

where the matrices $\mathbf{D}_0 = [D_{0ij}]$ and $\mathbf{D}_1 = [D_{1ij}]$ are a stable matrix and a non-negative matrix, respectively, whose sum is an irreducible infinitesimal generator $\mathbf{D}$ with stationary probability vector $\vec{\pi}$, respectively.

We will explain the idea of the construction of MAPs with apparently self-similar behaviour qualitatively using Figure 13.4.6 — for a more comprehensive mathematical treatment, see [And95a]. We consider here the underlying Markov chain of a MAP $\mathbf{D}$, without considering the special structure of the two component matrices $\mathbf{D}_0$ and $\mathbf{D}_1$.

What we are looking for is a Markov chain model of a physical phenomenon exhibiting self-similar features. At the very macroscopic level what we see is just a constant rate, i.e., at a very large time scale there is hardly any variability at all. This of course corresponds to time scales where the self-similar behaviour has disappeared. Imagine that zooming in on the process, i.e., considering a more coarse time scale, the source now behaves roughly like a two state source, we model this behaviour with a two state Markov chain with sojourn times in each state which are chosen according to the time scale currently under consideration. Now zooming in again on each of the two states we can again imagine that under a closer look each of the states again exhibits a two-state behaviour but with a time constant which is smaller. Doing this again and again we obtain a model consisting of a hierarchy of nested two-state sources. Of course, for a finite dimensional Markov chain it is only possible to do this a limited number of times.

With the above description, it is not immediately clear what will happen whenever the process changes state on any level different from the most coarse. To explain this, consider again the eight state chain in Figure 13.4.6 with states denoted 1-8. The set of states 1-4 are the substates of one of the long term states and 5-8 the substates of the other. Each of these two sets of substates (states 1-4 and 5-8, respectively) can again be divided in two

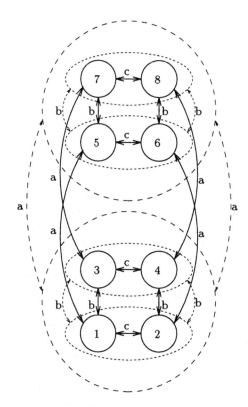

Figure 13.4.6: Joint Markov chain of three independent two-state Markov chains. Each of the transition intensity variables a, b and c can take on two different values depending on the direction of the transition, i.e., $a = c_{13}$ or $a = c_{23}$, $b = c_{12}$ or $b = c_{22}$ and $c = c_{11}$ or $c = c_{21}$.

sets of substates, namely states 1-2 and 3-4, respectively 5-6 and 7-8. Each of these 4 sets of states can be seen as substates of the medium term states. Finally, each of these four sets of substates can be divided in two physical short term states. To exemplify we look at possible transitions from state 1 in Figure 13.4.6. From state 1 there are three possible transitions: a short term transition that will take us to state 2, a middle term transition to state 3 and finally a long term transition that will take us to state 5. To obtain analytical simplifications, we make some further assumptions, namely that the sub-Markov chains of any two corresponding states are identical. In that sense the state of the long term process has no influence on the behaviour of the subprocesses, remembering that we are currently only considering the underlying Markov chain $\mathbf{D}$ of the MAP. Clearly the underlying Markov chain under consideration is a superposition of a number of sub-Markov chains. To avoid confusion it is emphasized that the MAP we are considering is not necessarily itself a superposition of two state MAP's — whether this is true or not depends solely on the way we choose the matrix $\mathbf{D}_1$. The underlying Markov chain of each of the two state sources can be described in the following manner.

$$\mathbf{D}_i^* = \begin{pmatrix} -c_{1i} & c_{1i} \\ c_{2i} & -c_{2i} \end{pmatrix}.$$

Hence the superposition can be described as a Kronecker sum, (see, e.g., [Bel70])

$$\mathbf{D} = \mathbf{D}^* = \bigoplus_{i=1}^{d} \mathbf{D}_i^*,$$

where d is the number of superpositions (or depth of the model).

Here an illustration has been given of a methodology for constructing self-similar behaviour over several time scales using two state sources as building blocks. However, this approach can easily be generalized to more general source models, i.e., more complex sources such as that obtained by super-posing a number of homogeneous two state sources. This generalization, the General Multi-Level Process (GMLP), was originally presented in [Jen95] (see [And95a] for a reference in English). For simplicity, work until now has mainly been focused on the special case of the superposition of two state sources described above and here we will also restrict ourselves to this special case. For this particular case, it turns out (see [AN96]) that the structure of the underlying chain permits substantial analytical simplifications in the general matrix expression for the variance-time curve and the covariance function, for example. Similar simplifications can also be obtained for more general GMLP models applying, for example, results from [AMS82] and [Ide89].

We now need to address the question of how to split the matrix $\mathbf{D}$ into the matrices $\mathbf{D}_0$ and $\mathbf{D}_1$. Since we are interested in modelling processes with inherent variability we restrict the class of MAP's to the class of Markov Modulated Poisson Processes (MMPP) thus reducing the problem of choosing the unknown matrix $\mathbf{D}_1$ to the problem of choosing the vector $\vec{\lambda}$ of arrival intensities. Here we will present two different constructive methods which can be applied to obtain $\mathbf{D}_1$. (Both methods were originally presented in [Jen95]; see [And95a] for a reference in English.) Later we will describe a method for parameterizing one of these approaches in order to construct a process eluding a specified local Hurst parameter over several time scales. The idea of both methods is related to the self-similar interpretation of the matrix $\mathbf{D}$.

Addition method To explain the addition method we will consider the nested hierarchy of models previously introduced. Given the fundamental rate of the process, λ^* say, we have to split this rate in two when considering a two state process. We will denote the fundamental rate in each of these (aggregated) states by $\lambda^* + \delta_{11}$ and $\lambda^* - \delta_{21}$ respectively, assuming $\delta_{i1} \geq 0$. The fraction of time spent in each of the two (aggregated) states is π_{11} and

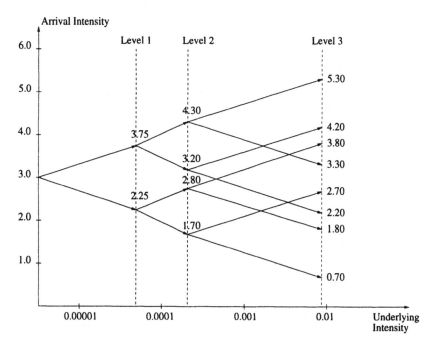

Figure 13.4.7: Illustration of how the arrival intensities are found with the addition method.

π_{21}, respectively. Thus, to preserve $\lambda^\star$, we have the following relationship for some number d_1

$$(\delta_{11}, \delta_{21}) = d_1(\pi_{21}, \pi_{11}).$$

Once this is performed the procedure can be continued until finally the rate associated with each state in the Markov chain is obtained. Since the values λ_i, $i = 1, \ldots, 2^d$, are nonnegative the restriction $\lambda^\star \geq \sum_{i=1}^{d} \delta_{2i}$ must hold. Figure 13.4.7 illustrates the method in the case of symmetrical sources, i.e. $c_{1i} = c_{2i}$.

By a closer examination of the addition method it is seen that the process obtained by applying this method is nothing but a superposition of heterogeneous Interrupted Poisson Processes (IPP) and a Poisson stream, the Poisson source having intensity $\lambda^\star - \sum_{i=1}^{d} \delta_{2i}$ and the IPP's given by $\mathbf{D}_i^\star$'s with arrival intensity $\delta_{1i} + \delta_{2i}$ in the on-states. This property will be crucial to the derivation of the fitting method described later.

Multiplication method The basic idea of the multiplication method is as for the addition method, here, however, the intensities of the nested levels are obtained from the previous levels by multiplication rather than addition. Figure 13.4.8 exemplifies the multiplication method for symmetrical sources and for a multiplication constant which is independent of the levels. In this

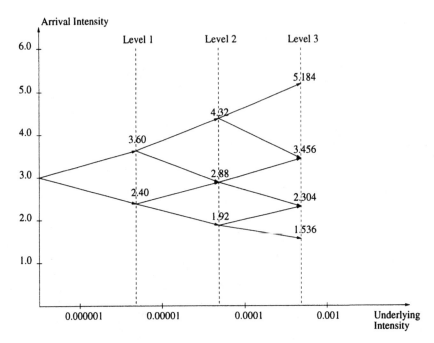

Figure 13.4.8: Illustration of how the arrival intensities are found with the multiplication method.

case the arrival intensities of the state of the Markov chain is given by a distribution which is obtained by a binomial distribution by a translation and simple scaling. For the multiplication method there is no simple way to describe the process given by $\mathbf{D}$ and $\vec{\lambda}$ as a superposition of simple two-state sources, for example.

Using the multiplication method with level independent multiplication factor it could be argued that there is only one parameter to be estimated. The dimension of $\mathbf{D}$ 2^d is given by the number of time scales under consideration, which also implies the values c_{ij}. The fundamental rate of the MMPP is given by the intensity of the process and finally the only parameter left is the multiplication factor p.

Fitting   Clearly, even with the relatively simple two state based models, there are many parameters to be fitted. It is therefore convenient to limit the number of parameters by initially choosing the underlying time constants to satisfy $c_{11} = c_{21} = a^{i-1}c_{1i} = a^{i-1}c_{2i}$ for $i = 1, .., d$, i.e., the underlying time constants are logarithmically spaced with a factor a between each level. This approach seems to be in good agreement with the idea behind the construction of the models. A procedure is then needed to find the arrival intensities $\vec{\lambda}$, a and the absolute values of, e.g., c_{11}.

In [AN96], a fitting procedure for the special case of the addition method was described (superposition of heterogenous IPPs and a Poisson source). The algorithm is based upon fitting the intensities $(\vec{\lambda})$ using the covariance function. This is convenient since the covariance function of a superposition of subprocesses can be found by summing the individual covariance functions. It is shown in [AN96] how the arrival intensities can be obtained in a nice recursive manner when fitting to a covariance structure of the form $\text{cov}(k) = ck^{-\beta}$, where c is a positive constant, $0 < \beta \le 1$ corresponding to second order self-similar behaviour. The parameter c_{11} and a fine tuning of the largest time constant $c_{1d} = c_{2d}$ can then easily be found from a least squares fit (one parameter at a time) to the derivative of the log-log plotted IDC curve.

The fitting procedure seems to work very well over the entire range $[\frac{1}{2}, 1)$ of the local Hurst parameter. In [AN96], three models of depth 6 (64 states) producing a local Hurst parameter of 0.60, 0.75 and 0.90, respectively, were found. The goodness of the fits can be seen in Figure 13.4.9 displaying the derivatives of the log-log plotted IDC curves. Thus, it is possible to produce a local Hurst parameter within a very narrow margin over several time scales (about 4) by applying the fitting algorithm to a structurally simple 64 state model.

Queueing behaviour   The very special structure of the Markov chain can be exploited in queueing analysis. In [AJN95] and [And95a], models of this type display queueing behaviour similar to that of measured LAN data (e.g., [ENW94]). In [AJN95], it was shown that MMPP models with exactly same fundamental rate and IDC could exhibit substantially different queueing behaviour indicating that fitting to first and second order characteristics alone does not imply a good fit of queueing behaviour.

13.5   Real-time estimation of call arrival intensity

Tasks related to advanced management of telecommunications networks, such as optimal dynamic routing or dynamic allocation of bandwidth for virtual paths in an ATM network, call for a method to estimate the current call arrival intensities. Typically one can assume that the arrival process constitutes a doubly stochastic Poisson process, i.e., a Poisson process where the intensity λ_t is also a stochastic process. What is observed is the point process of call arrivals (counting process N_t), and the task is then to estimate λ_t based on these observations. Such a real-time traffic estimation problem is essentially a filtering problem. In this section we present two approaches to this problem. If the nature of the intensity process λ_t is fully known in advance, it may be possible to calculate a Bayesian estimator, which exploits to a maximum degree all the available information (and in that sense is opti-

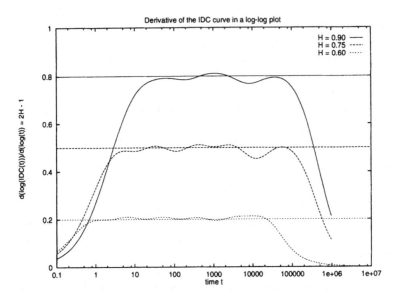

Figure 13.4.9: The derivative of the log-log plotted IDC curves for the three fits to respectively $H = 0.6$, $H = 0.75$ and $H = 0.90$.

mal). On the other hand, the most straightforward estimators are based on a sliding window. This latter approach belongs to the area of linear filtering theory.

13.5.1   Bayesian estimator

In this subsection we assume that $\lambda_t = \Lambda + \Sigma X(t)$, where Λ and Σ are constants, and $X(t)$ is a standard Ornstein-Uhlenbeck process (SOU; see Subsection 13.3.5). The stationary distribution of λ_t is normal $N(\Lambda, \Sigma^2)$, and its autocorrelation function is the same as for the underlying SOU process, $\mathrm{Corr}\{\lambda_s, \lambda_t\} = e^{-|t-s|}$. An intrinsic flaw in the model is that negative values $\lambda_t < 0$ may occur. However, if $\Sigma \ll \Lambda$, the probability of such an event is negligible. Our purpose here is to present an approximate algorithm for the calculation of the Bayesian estimator (the mean of the conditional distribution function of λ_t given the past arrival history N_0^t). Any details omitted here may be found in [AV95].

Bayesian distribution

Let us first consider the evolution of the conditional distribution in an interval Δ void of new arrivals. Let $\hat{\lambda}_t = \mathrm{E}[\lambda_t \mid N_0^t]$, and denote the nth central moment by $m_n(t)$,

$$m_n(t) = \mathrm{E}\left[(\lambda_t - \hat{\lambda}_t)^n \mid N_0^t \right], \qquad n \geq 0.$$

Obviously, $m_0(t) = 1$ and $m_1(t) = 0$. For $n < 0$, define $m_n(t) = 0$. Now it is possible to derive the following moment equation for any $n > 0$ and $t \in \Delta$:

$$m_n'(t) = n(n-1)\Sigma^2 m_{n-2}(t) \tag{13.5.1}$$
$$-n\Sigma(\hat{X}_t + \hat{X}_t')m_{n-1}(t) - nm_n(t) - m_{n+1}(t),$$

where $\hat{X}_t = E[X_t \mid N_0^t] = (\hat{\lambda}_t - \Lambda)/\Sigma$. Thus the time derivative of the nth moment depends on, besides the moment itself, two of the lower moments and one higher moment.

Another question is how the moments of the conditional distribution of λ_t change at the instant t of an arrival. Denote the nth moment around the origin by $\widehat{\lambda_t^n}$,

$$\widehat{\lambda_t^n} = E[\lambda_t^n \mid N_0^t], \qquad n \geq 0.$$

It is possible to derive the following equation for any $n \geq 0$:

$$\widehat{\lambda_t^n} = \widehat{\lambda_{t-}^{n+1}}/\hat{\lambda}_{t-}. \tag{13.5.2}$$

The chain of equations is coupled again, the nth moment depending on the left limits of the first and one higher moments.

Approximate solution

The chain of coupled equations (13.5.1) may be truncated by assuming that also the *a posteriori* distribution of λ_t is Gaussian. Then all the odd moments vanish (in particular, $m_3(t) = 0$). Solving the resulting pair of ordinary differential equations yields the following formulae (valid in an interval Δ void of new arrivals):

$$\begin{aligned}
\hat{\lambda}_t &= \hat{\lambda}_0 e^{-t} + (\Lambda - \hat{\sigma}_0^2 e^{-t} - \Sigma^2(1 - e^{-t}))(1 - e^{-t}), \\
\hat{\sigma}_t^2 &= \hat{\sigma}_0^2 e^{-2t} + \Sigma^2(1 - e^{-2t}),
\end{aligned} \tag{13.5.3}$$

where $\hat{\sigma}_t^2 = E[(\lambda_t - \hat{\lambda}_t)^2 \mid N_0^t] = m_2(t)$. The other chain of coupled equations (13.5.2) may be truncated by the same normal approximation as above, resulting in the following formulae (valid at an arrival instant t):

$$\begin{aligned}
\hat{\lambda}_t &= \hat{\lambda}_{t-}(1 + (\hat{\sigma}_{t-}/\hat{\lambda}_{t-})^2), \\
\hat{\sigma}_t^2 &= \hat{\sigma}_{t-}^2 (1 - (\hat{\sigma}_{t-}/\hat{\lambda}_{t-})^2).
\end{aligned} \tag{13.5.4}$$

Equations (13.5.3) and (13.5.4) constitute the estimation algorithm.

13.5.2   Window based estimation

In this subsection we assume only that the intensity process λ_t is stationary (with mean Λ and variance Σ^2). Our purpose is to study estimators based on a sliding window, i.e., on a count of past arrivals within a window divided by its length, possibly with different weights in different parts of the window. We will see that the accuracy of a sliding window estimator depends only on the second order properties of the underlying process model. For any details omitted here we refer to [VAD96].

Standard window estimator

Let N_t be the counting process of the arrivals. A window based estimator X for the arrival intensity λ_0 at time 0 is

$$X = \int_0^\infty W(t) \, dN_t = \sum_i W(t_i), \qquad (13.5.5)$$

where $W(t)$ is the window function such that $\int_0^\infty W(t) \, dt = 1$ and the t_i are the arrival instants. (For notational simplicity we use here reversed time.) X itself is, of course, a random variable with mean $E\{X\} = \Lambda$.

In choosing the window function one has to balance between two trends: if the function is concentrated around time 0 the estimator is based on a few recent arrivals and can be very noisy; on the other hand broadening the window too much means that the estimator is based on outdated information.

Improved window estimator

The standard window estimator X can be improved by taking it to be of more general linear form $Y = aX + a'$, where a and a' are arbitrary constants. We wish to choose the window function $W(t)$ and parameters a and a' such that the mean squared error $E\{\Delta^2\} = \text{Var}\{\Delta\} + E\{\Delta\}^2$, where $\Delta = Y - \lambda_0$, is minimized.

Consider first the choice of parameter a'. Since the variance of Δ is not affected by the constant a', the optimal choice of a' makes the estimator unbiased. It follows that $a' = (1 - a)\Lambda$. With this value we have

$$Y = \Lambda + a(X - \Lambda). \qquad (13.5.6)$$

Next we consider the optimization of parameter a. It is possible to prove that

$$\frac{E\{\Delta^2\}}{\Sigma^2} = 1 + a^2(\beta I_1 + I_2) - 2aI_3, \qquad (13.5.7)$$

where $\beta = \Lambda/\Sigma^2$. I_1, I_2 and I_3 stand for the integrals

$$I_1 = \int_0^\infty W^2(t) \, dt, \quad I_2 = \int_0^\infty V(t)\rho(t) \, dt, \quad I_3 = \int_0^\infty W(t)\rho(t) \, dt,$$

where $\rho(t) = \text{Corr}\{\lambda_u, \lambda_{u+t}\}$ and $V(t) = 2\int_0^\infty W(u)W(u + t) \, dt$. The standard window estimator X is obtained by setting $a = 1$:

$$\frac{E\{\Delta_X^2\}}{\Sigma^2} = 1 + \beta I_1 + I_2 - 2I_3. \qquad (13.5.8)$$

A better estimator Y is obtained by minimizing (13.5.7) with respect to a, yielding

$$\frac{\mathrm{E}\{\Delta_Y^2\}}{\Sigma^2} = 1 - \frac{I_3^2}{\beta I_1 + I_2}, \qquad (13.5.9)$$

with the optimal value attained at

$$a^* = \frac{I_3}{\beta I_1 + I_2}. \qquad (13.5.10)$$

Note that the results depend on Λ and Σ only through the parameter $\beta = \Lambda/\Sigma^2$.

Optimal window function

So far we have considered the optimization of parameters a' and a. In order to find the truly optimal window estimator the form of the window should also be subject to optimization. It is shown in [Sny75] that the optimal window function $W(t)$ satisfies the following integral equation

$$\beta W(t) + \int_0^\infty W(u)\rho(t - u)\, \mathrm{d}u = \rho(t), \qquad t \geq 0.$$

This integral equation arises naturally in the study of linear filters for the detection of signals in the presence of additive noise, see [Van71], for example. In fact, Snyder [Sny75] describes how determining the optimal window may be recast as such a detection problem. In addition, he discusses how such problems may be treated with various tools from linear filtering theory. The solution may also be obtained directly by transform methods [VAD96].

Optimal window size

Here we consider a simpler problem in which the form of the window function is given but it contains one free parameter, T, characterizing the temporal extent of the window. A further minimization is performed with respect to T. Consider an exponential window,

$$W(t) = \frac{1}{T} e^{-t/T}.$$

The standard window estimator (13.5.5) is

$$X = \frac{1}{T} \sum_i e^{-t_i/T}.$$

Given X, the optimal estimator Y (with fixed T) is calculated from (13.5.6) and (13.5.10).

In particular, for the exponential correlation function $\rho(t) = e^{-|t/\tau|}$ (consider for instance the Ornstein-Uhlenbeck process model presented earlier), we have (with fixed T)

$$\frac{E\{\Delta_X^2\}}{\Sigma^2} = 1 + \frac{b}{x} - \frac{1}{1+x}, \qquad \frac{E\{\Delta_Y^2\}}{\Sigma^2} = 1 - \frac{1}{1+x} \cdot \frac{x}{x + b(1+x)}, \quad (13.5.11)$$

with the optimum obtained at $a^* = x/(x + b(1 + x))$. In these formulae $x = T/\tau$, and b is the dimensionless parameter

$$b = \frac{\beta}{2\tau} = \frac{1}{2} \cdot \frac{1}{(\Sigma/\Lambda)^2} \cdot \frac{1}{\Lambda\tau}.$$

As the expressions in (13.5.11) are rational functions of x, the minimization with respect to this parameter can be performed analytically giving the optimal window size for the standard estimator X and for the optimal estimator Y,

$$x_X^* = \frac{\sqrt{b}}{1 - \sqrt{b}}, \qquad x_Y^* = \sqrt{\frac{b}{1+b}}.$$

Finally, substituting these optimal window sizes into (13.5.11) gives the mean squared errors,

$$\frac{E\{\Delta_X^2\}}{\Sigma^2} = 1 - (1 - \sqrt{b})^2, \qquad \frac{E\{\Delta_Y^2\}}{\Sigma^2} = 1 - \frac{1}{(\sqrt{b} + \sqrt{1+b})^2},$$

with the optimal a being $a^* = 1 - \sqrt{b/(1 + b)}$.

Furthermore, it is proved in [VAD96] that, for the exponential correlation function, the exponential window estimator Y with the optimal parameter values is the optimal estimator also with respect to the form of the window function.

13.5.3 Numerical results

Numerical results of the accuracy $\sqrt{E\{\Delta^2\}}/\Sigma$ of different window based estimators as a function of b are shown in Figure 13.5.1. We assumed the exponential correlation function. In addition to a exponential window function ("exp"), the accuracy of an ordinary straight case window ("box") has been studied. For comparison, the accuracy of a Bayesian estimator is also shown ("Bayes"). The accuracy of the Bayesian estimator is not solely a function of b but depends separately on $\Lambda\tau$ and Σ/Λ [AV95]. Here we have plotted the curve for the case where Σ/Λ is kept fixed to 0.3 and $b = 1/(2(\Sigma/\Lambda)^2(\Lambda\tau))$ is controlled by the value of $\Lambda\tau$.

The Bayesian estimator is the best one, as it should be by construction. The estimators with an exponential window function are more accurate than their straight case window counterparts. For very small b, the accuracies of

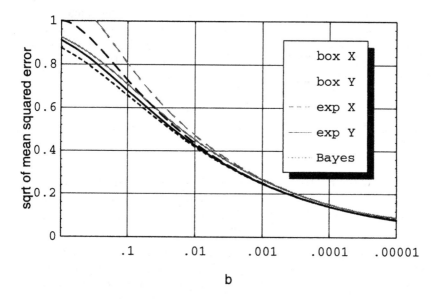

Figure 13.5.1: Accuracy of the estimators.

the estimators differ only depending on the window function. For values of b closer to 1, the more important factor is whether one uses the standard window estimator or the improved one. In general, the differences are relatively small. However, it should be noted that these results were obtained with the *optimal* window size. With a wrong window size the accuracy can deteriorate.

13.6   Parameter estimation from long range dependent traffic data

Long range dependence has important implications for parameter estimation. As will be shown in this section, many estimators are more biased and converge more slowly than in the traditional short-range dependent case. As a general reference to the topic, the reader is referred to Beran's monograph [Ber94]. A review in the teletraffic context is included in [LTWW94].

13.6.1   Estimation of the mean rate

The mean rate of a stationary traffic stream over time interval $[0, T]$ is usually estimated by the sample mean

$$\hat{m}_T = \frac{A_T}{T},$$

where A_t again is the cumulative arrival process. With long range dependence, however, the estimation of the mean rate may become so inaccurate that the whole notion of mean rate must be interpreted in a different way than, say, the mean rate of a Poisson process. To illustrate this, let us assume that the true traffic obeys the fractional Brownian model

$$A_t = mt + \sqrt{ma}\,Z_t.$$

The variance of $\hat{m}_T$ is

$$\mathrm{Var}\,\{\hat{m}_T\} = maT^{2H-2},$$

so that bringing the relative standard deviation below ϵ requires that

$$T > \left(\frac{1}{\epsilon}\sqrt{\frac{a}{m}} \right)^{1/(1-H)}.$$

The left hand side increases rapidly with respect to H. For example, if we have $m = 2$ Mbit/sec and $a = 200$ kbit·sec (typical values in today's LAN interconnection — at least by the example of Subsection 13.6.6), then the observation time required for the very modest accuracy of a standard deviation 10% would be $10^{1/(2-2H)}$ seconds which is about five minutes with $H = 0.8$ but more than 27 hours with $H = 0.9$! In the latter case even the meaning of the mean rate parameter becomes problematic, since the non-stationary day-time variation has to be taken into account.

13.6.2   The inaccuracy of sample variances

Let us again restrict ourselves to the simplest case, fractional Brownian traffic, and assume that the mean rate m is estimated by the sample mean as above. The two remaining parameters, a and H, can be read from the variance function

$$v(t) = \mathrm{Var}\,\{A_t\}.$$

However, the estimation of this curve is difficult for large H, since the sample variances are then both strongly biased and very slowly convergent.

Let us first consider the bias. Subtracting the estimated drift from A_t, we obtain a realization of the process

$$B_t = \sqrt{ma}\left(Z_t - \frac{t}{T}Z_T \right), \quad t \in [0, T],$$

that is, of a "fractional Brownian bridge". The fact that the increments of B have smaller variances than those of A becomes more important with increasing H. To simplify the notation, let us choose $T = 1$ and consider the means of squared increments

$$S_n^2 = \frac{1}{n}\sum_{k=1}^{n}(B_{k/n} - B_{(k-1)/n})^2.$$

With straightforward calculation, we get

$$E\left\{S_n^2\right\} = \frac{1}{n}\sum_{k=1}^{n}\text{Var}\left\{B_{k/n} - B_{(k-1)/n}\right\}$$

$$= \frac{ma}{n}\sum_{k=1}^{n}(n^{-2H} - c(n,k)n^{-2}) \qquad (13.6.1)$$

$$= \text{Var}\left\{A_{1/n}\right\} - man^{-2},$$

where the numbers

$$c(n,k) = 2n\,\text{Cov}\left\{Z_{k/n} - Z_{(k-1)/n}, Z_1\right\} - 1, \quad k = 1,\ldots,n$$

obviously sum to n. For $H \in (1/2, 1)$, the variances of the subsequent intervals differ. The numbers $c(n,k)$ depend essentially only on k/n and have biggest variation when H is about 0.72 — see the left plot of Figure 13.6.1. The value of H has a strong impact on the biasedness of S_n^2 as an estimator of $\text{Var}\{A_{1/n}\}$. The right plot in Figure 13.6.1 shows that the bias is considerable even for very large n when $H = 0.9$.

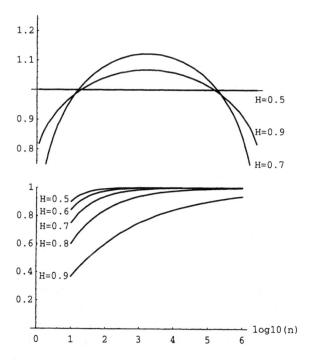

Figure 13.6.1: Top: the numbers $c(100,k)$ of (13.6.1) for $H = 0.5$, 0.7 and 0.9. Bottom: approximate shape of the ratio $E\{S_n^2\}/\text{Var}\{A_{1/n}\}$ for $H = 0.5$, 0.6, 0.7, 0.8 and 0.9.

Let us then consider the convergence of the sample variance to the true variance. We simplify the problem by assuming that the mean rate is known exactly, so that the "bridge effect" is absent, and consider the unbiased variance estimators

$$\hat{S}_n^2 = \frac{1}{n} \sum_{i=1}^{n} (Z(i) - Z(i-1))^2.$$

Using the fact that for a centered bivariate Gaussian vector (X, Y) we have

$$E[Y \mid X] = \frac{\mathrm{Cov}\{X, Y\}}{\mathrm{Var}\{X\}} X,$$

$$\mathrm{Var}[Y \mid X] = \left(1 - \frac{\mathrm{Cov}\{X, Y\}^2}{\mathrm{Var}\{X\}\mathrm{Var}\{Y\}}\right) \mathrm{Var}\{Y\},$$

a straightforward calculation gives

$$\mathrm{Var}\left\{\hat{S}_n^2\right\} = \frac{2}{n^2} \sum_{i=1}^{n} \sum_{j=1}^{n} \mathrm{Cov}\left\{Z(i) - Z(i-1), Z(j) - Z(j-1)\right\}^2$$

$$= \frac{2}{n} + \frac{4}{n^2} \sum_{i=2}^{n} \sum_{j=1}^{i-1} \mathrm{Cov}\left\{Z(i) - Z(i-1), Z(j) - Z(j-1)\right\}^2.$$

$$(13.6.2)$$

For large n, we have (see, e.g., [ST94], p. 335)

$$\mathrm{Cov}\left\{Z(i) - Z(i-1), Z(j) - Z(j-1)\right\} \approx H(2H-1)|i-j|^{2H-2}. \quad (13.6.3)$$

Approximating the double sum in (13.6.2) by the double integral

$$H(2H-1) \int_1^n dt \int_1^t ds |t-s|^{4H-4}$$

yields the approximation

$$\mathrm{Var}\left\{\hat{S}_n^2\right\} \approx \frac{2}{n} + \frac{4}{n^2} \frac{(H(2H-1))^2}{|4H-3|} \left|\frac{n^{4H-2}-1}{4H-2} - n + 1\right|, \quad H \neq 3/4. \quad (13.6.4)$$

(For the boundary value $H = 3/4$, the corresponding formula is slightly more complicated, containing a logarithmic expression.) Thus, the value $H = 3/4$ separates the region where the relative standard deviation goes to zero as $n^{-1/2}$ from a region where the convergence is still slower and depends on H:

$$\sqrt{\mathrm{Var}\left\{\hat{S}_n^2\right\}} \sim \begin{cases} n^{-1/2}, & H < 3/4 \\ n^{-2(1-H)}, & H > 3/4. \end{cases} \quad (13.6.5)$$

Again, there is a big difference between the cases $H = 0.8$ and $H = 0.9$: the sample size required for a standard deviation of 10% is less than 10^4 for the former but greater than 10^6 for the latter! It can be further shown that the estimator is asymptotically normal for $H < 3/4$ but not for $H > 3/4$ (see, e.g., [Ber94], p. 70, and the references therein).

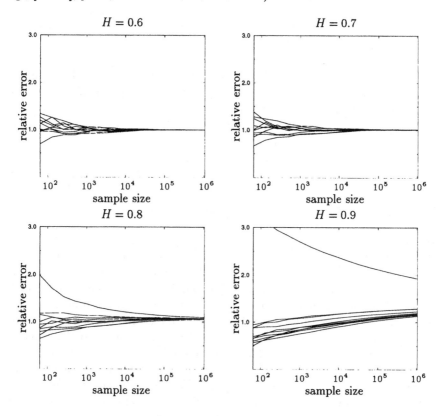

Figure 13.6.2: Relative error of variance estimates with different resolutions from 10 simulated FBM samples.

Figure 13.6.2 illustrates the (in)accuracy of the 'naive' variance estimation by showing, for several values of H, the relative accuracy of the variance estimate with different sample sizes (resolutions) and for 10 randomly chosen realizations. For each H, the same pseudorandom sequences are used. Note that there happens to be one 'bad' realization for which the sample variances are totally erroneous for high H but not for $H = 0.7$ or lower. (The phenomenon that the curves exceed 1 for large sample sizes in the cases $H = 0.8$, 0.9 is probably an artefact caused by the inexact simulation method.)

13.6.3   Variance based estimation of parameters H and a

Assume that fractional Brownian traffic A_t is observed over time interval $[0, T]$ and that the mean rate is estimated by $\hat{m}$ and that the sums of squares

$$S_k^2 = \sum_{i=1}^{2^k} \left(A(i2^{-k}T) - A((i-1)2^{-k}T) - \hat{m}2^{-k}T \right)^2$$

are calculated for $k = k_0, \ldots, k_1$. Note that bigger k corresponds to higher resolution (and a larger data set). Then the remaining parameters H and a can be consistently estimated by, e.g.,

$$\hat{H}_k = \frac{1}{2} \log_2 \frac{2 S_{k-1}^2}{S_k^2} \qquad (13.6.6)$$

and

$$\hat{a}_k = \frac{S_k^2}{\hat{m}2^k (2^{-k}T)^{2\hat{H}_k}}. \qquad (13.6.7)$$

By consistence we mean, as usual, that the estimators converge to the right values when k increases. This convergence is, however, slow for high H since $S_k^2/2^k$ is then a poor estimator for the corresponding variance.

For any stationary traffic process, the numbers $\hat{H}_k$ can be used to get a quick look at the dependence structure of the process. If the process is self-similar, the $\hat{H}_k$'s are approximately equal for all k's for which the variance estimator has reasonable accuracy. Figure 13.6.3 shows the local H estimates calculated from the variances shown in Figure 13.6.2. Besides the variance of the estimator, the systematic underestimation for high H and too small sample size is clearly seen. Note, in particular, the differences between the cases $H = 0.8$ and $H = 0.9$. Another interesting detail in the case $H = 0.9$ is that the lines are separate but almost parallel at high resolutions: it seems that because of their strong correlation, the local H estimates appear to give a more definitive answer than they in fact do.

13.6.4   Estimation of H by R/S analysis

The R/S-statistic, alias *rescaled adjusted range*, defined below is particularly attractive because of its relative robustness against changes in the marginal distribution, even for long-tailed or skew distributions. On the other hand, for marginal distributions which are close to normality a dramatic loss in efficiency is reported and, to the best of our knowledge, no detailed analysis of robustness of R/S statistics has taken place yet. Given an empirical time series of length N ($X_k : k = 1, \ldots, N$), the whole series is subdivided into K non-overlapping blocks. Now, we compute the rescaled adjusted range

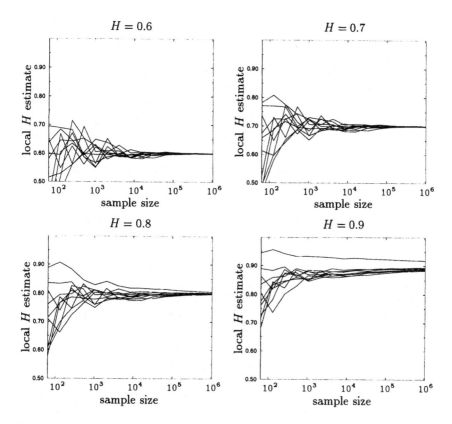

Figure 13.6.3: Local dyadic Hurst parameter estimates with different resolutions from 10 simulated FBM samples.

$R(t_i, d)/S(t_i, d)$ for a number of values d, where $t_i = \lfloor N/K \rfloor (i - 1) + 1$ are the starting points of the blocks which satisfy $(t_i - 1) + d \leq N$. We have

$$R(t_i, d) = \max\{0, W(t_i, 1), \ldots, W(t_i, d)\} - \min\{0, W(t_i, 1), \ldots, W(t_i, d)\},$$

where

$$W(t_i, k) = \sum_{j=1}^{k} X_{t_i+j-1} - k \cdot \left(\frac{1}{d} \sum_{j=1}^{d} X_{t_i+j-1}\right), \quad k = 1, \ldots, d.$$

$S^2(t_i, d)$ denotes the sample variance of $X_{t_i}, \ldots, X_{t_i+d-1}$. For each value of d, we obtain a number of R/S samples which decreases from K for larger values of d. We compute these samples for logarithmically spaced values of d, i.e., $d_{l+1} = m \cdot d_l$ with $m > 1$, starting with a d_0 of about 10. Plotting $\log R(t_i, d)/S(t_i, d)$ versus $\log d$ results in the R/S plot, also known as the *pox diagram*.

Next, a least squares line is fitted to the points of the R/S plot omitting the R/S samples of the smallest values of d (because they are dominated by short-range correlations) and those of large values of d where the number of samples is less than, say, 5 (because they are statistically insignificant). The slope of the regression line for these R/S samples is an estimate for the Hurst parameter H. Both the number of blocks K and the number of values d should not be chosen too small. In addition, some care has to be taken when deciding about the end of the transient, i.e. which of the small values of d should not be taken into consideration for the regression line. In practice, it should be checked whether different parameter settings lead to consistent H estimates for $X^{(m)}$ with different aggregation levels m.

13.6.5 Estimation of H by Whittle's maximum likelihood estimator

If more information on the H-estimate, such as confidence intervals, or on the estimator itself, such as efficiency and robustness, are needed, then periodogram-based estimators are used. The main idea of these methods is to assume a certain parametric process class, for instance FARIMA(p,d,q), and to fit the parameters of this process to the given empirical sample. The fitting should be optimal in the sense that the periodogram of the sample and the spectral density of the process minimize a given goodness-of-fit function.

As mentioned above, the spectral density of self-similar processes obeys a power law near the origin. Thus, the first idea to determine the Hurst parameter H is simply to plot the periodogram (defined in (13.6.9)) on a log-log grid, and to compute the slope of a regression line which is fitted to a number of low frequencies. This should be an estimate of $1 - 2H$. In most of the cases this will lead to a wrong estimate of H since the periodogram is *not appropriate* to estimate the spectral density [SS95]. More sophisticated methods have to be applied to obtain useful estimates of H.

Several periodogram-based estimators can be found in the literature. Here we will focus on a MLE (Maximum Likelihood Estimator) as presented in [Ber92, WTLW95]. This is based on Whittle's approximate MLE for Gaussian processes [Whi53]. For Gaussian sequences, this estimator is asymptotically normal and efficient [FT86, Dah89].

The spectral density of the self-similar process is denoted by $f(\lambda; \theta)$, where the parameter vector of the process $\theta = (\theta_1, \ldots, \theta_M)$ is structured as follows. θ_2 denotes the Hurst parameter H. $\theta_1 = \sigma_\epsilon^2$ is a scale parameter, where σ_ϵ^2 is the variance of the innovation ϵ of the infinite AR-representation of the process, i.e., $X_j = \sum_{i=1}^\infty \alpha_i X_{j-i} + \epsilon_j$. This implies $\int_{-\pi}^\pi \log(f(\lambda; (1, \theta_2, \ldots, \theta_M)))d\lambda = 0$. If necessary, the parameters θ_3 to θ_M describe the short-range behavior of the process. For fractional Gaussian noise and FARIMA(0,d,0), only σ_ϵ^2 and H have to be considered. With $\eta = (\theta_2, \ldots, \theta_M)$, the Whittle estimator $\hat{\eta}$ of η minimizes the goodness-of-fit function

$$Q(\eta) = \int_{-\pi}^{\pi} \frac{I(\lambda)}{f(\lambda; (1, \eta))} d\lambda, \qquad (13.6.8)$$

where $I(\cdot)$ denotes the sample periodogram defined by

$$I(\lambda) = \frac{1}{2\pi N} |\sum_{j=1}^{N} X_j e^{ij\lambda}|^2. \qquad (13.6.9)$$

$\hat{H}$ is given by $\hat{\theta}_2$ and the estimate of σ_ϵ^2 by

$$\hat{\sigma}_\epsilon^2 = \int_{-\pi}^{\pi} \frac{I(\lambda)}{f(\lambda; (1, \hat{\eta}))} d\lambda. \qquad (13.6.10)$$

The approximate 95%-confidence interval of $\hat{H}$ is given by

$$\hat{H} \pm 1.96 \sqrt{\frac{V_{11}}{N}}, \qquad (13.6.11)$$

where $V = 2D^{-1}$ and the matrix D is defined by

$$D_{ij} = \frac{1}{2\pi} \int_{-\pi}^{\pi} \frac{\partial}{\partial \theta_i} \log f(\lambda) \frac{\partial}{\partial \theta_j} \log f(\lambda) d\lambda. \qquad (13.6.12)$$

For implementation details, see the appendix of [Ber94], where an $S+$ listing of the Whittle estimator is provided. Given some knowledge in numerical mathematics, no special library functions are necessary to implement the above formulae. However, Fast Fourier Transforms, vector, and matrix functions would make the programming more convenient.

In practice, there are two problems which may have an effect on the robustness of the estimator:

- Deviations from the assumed model spectrum: i.e., deviations at higher frequencies lead to a bias in the estimate of H. One possible solution is to estimate H only from periodogram ordinates at low frequencies. For large data sets, one can also aggregate the data over non-overlapping blocks of length m and compute several $\hat{H}^{(m)}$ for the $X^{(m)}$.

- Deviations from being Gaussian: better fit to a Gaussian process can often be achieved by transforming the data but then it has to be proven that the estimate of H for the original and the transformed data sets are identical. For instance, this is the case for the log-transformation $Y = \log(X)$.

13.6.6   Examples

We illustrate the statistical methods described above by applying them to two sets of real-world data.

Internet traffic, transmitted over ATM

The first data set, called "kesk", is a time series of the amount of traffic transmitted from the Tampere University of Technology to the Finnish University Network backbone during a period of six working hours (between 9 and 15) and registered second by second [Haa96]. The traffic is usual Internet traffic which is gathered by a campus FDDI and then transmitted over an ATM link. The raw data are shown in Figure 13.6.4. The traffic looks stationary over the measurement period — if the irregular large "waves" of size 15–60 minutes are interpreted as a consequence of long range dependence and not of non-stationarity. Since there are no *a priori* obvious deterministic explanations (like the lunch break) for these changes in traffic intensity, modelling with a long range dependent stationary process is reasonable.

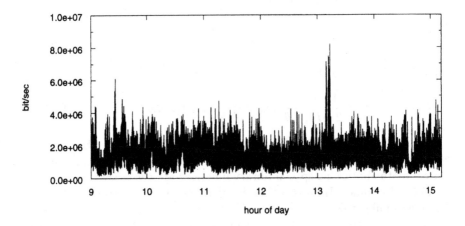

Figure 13.6.4: Data set "kesk": Total traffic from Tampere University of Technology to the Finnish University Network on July 19, 1995, with resolution one second. From [Haa96].

We start by looking at the marginal distribution of the amount of traffic transmitted in one second. Both the empirical distribution function and the QQ-plot in Figure 13.6.5 show a good fit to a lognormal distribution. Since most of the statistical theory assumes the process to be Gaussian, we log-transformed the sequence before applying the H estimators, except for the estimates based on sample variances. (It is known that this transformation leaves the Hurst parameter unchanged; see, for example, [HDLK95, Ber94].)

Figure 13.6.6 shows the R/S plot of "kesk" together with a least squares regression line. The plot hints to considerable self-similarity, and the slope of the regression line gives the H estimate 0.9. The local dyadic variance based estimates (13.6.6) of H are shown in the left graph of Figure 13.6.7. Because of the large value of H, the variance estimates can be suspected to be too

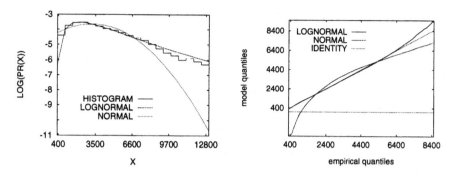

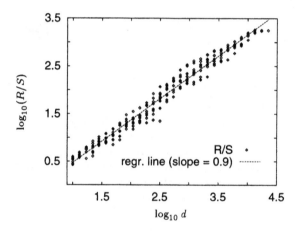

Figure 13.6.5: The closeness of the marginal distribution (i.e., the distribution of the amount of traffic transmitted in one second) of the data set "kesk" to the normal and lognormal distribution.

Figure 13.6.6: The R/S plot of the data set "kesk", with the regression line corresponding to $H = 0.9$.

low and also to suggest a too low value for H — cf. Figure 13.6.3. Therefore, $H = 0.9$ is compatible with the variance based estimates as well. The related estimates for the fractional Brownian traffic parameter a, shown in the right graph of Figure 13.6.7, are still less reliable, and it is not wise to conclude more than that the order of magnitude of a in "kesk" seems to be 10^5 bit·sec.

Assuming the log-transformed "kesk" as fractional Gaussian noise, Whittle's maximum likelihood estimator suggests the value $H = 0.86[0.86,0.87]$ (here and below, the interval in brackets is a 95% confidence interval). Finally, let us fit a FARIMA model to "kesk". The first try is the simplest, FARIMA$(0,d,0)$. Applied to this model, Whittle's estimator gives the value $d = 0.44$, corresponding to $H = 0.94[0.93,0.95]$: model assumptions seem to have a surprisingly strong influence on the estimate. Figure 13.6.8 shows that

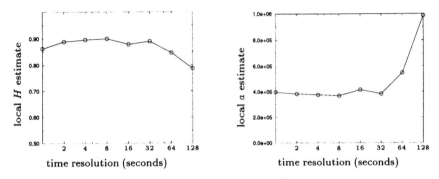

Figure 13.6.7: Local H and a estimates for the data set "kesk".

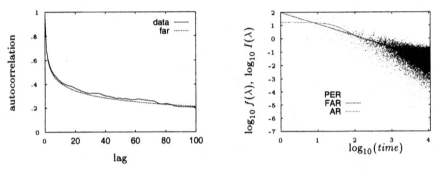

Figure 13.6.8: The sample autocorrelation function and periodogram of the data set "kesk". The ACF and spectrum of the fitted FARIMA(0,0.44,0) process are plotted also. The spectrum of an AR(p) process, where the order $p = 75$ is determined using Akaike's Information Criterion (AIC), is shown for comparison.

FARIMA(0,0.44,0) approximates "kesk" with a satisfactory accuracy with respect to both the periodogram and the autocorrelation function, and it is not necessary to use a more complicated model like FARIMA(p,d,0) with $p > 0$.

As a conclusion of our analysis of "kesk", we can say that the traffic in question can be roughly modelled in time scales larger than second as, for example, fractional Brownian traffic with H closer to 0.9 than to 0.8 and a being a few hundred kbitsecs, but it is not possible to give "the" right value of either parameter on the basis of our data. Further, the marginal distribution at resolution one second is rather far from Gaussian when the mean rate is about 2 Mbit/s, and the Gaussian model would be too optimistic, underestimating the peaks.

Jurassic Park, MPEG-I coded video

Our second example is the "dino" sequence produced with MPEG-I encoding from the movie "Jurassic Park" as described in Section 1.2 of Part I. This time series needs to be modelled with a FARIMA(1,d,0) process. Again, the GOP sequence was log-transformed before the estimators were used.

The R/S plot is shown in Figure 13.6.9. The slope of the regression line gives the H estimate 0.85. Assuming FARIMA(1,d,0), the Whittle estimate for H is 0.90 [0.82,0.97] and the autoregression (AR) parameter is 0.51 [0.43,0.60]. The autocorrelation function and the periodogram are shown, together with the exact model values, in Figure 13.6.10.

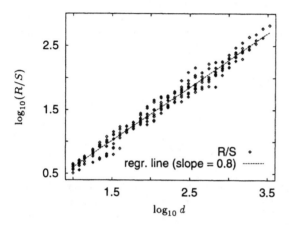

Figure 13.6.9: The R/S plot for the data set "dino", with a regression line corresponding to $H = 0.85$.

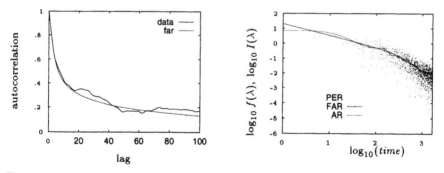

Figure 13.6.10: The sample autocorrelation function and periodogram of the data set "dino". The ACF and spectrum of the fitted FARIMA(1,0.4,0) process are plotted also. The spectrum of a fitted AR(p) process, where the order $p = 21$ is determined using Akaike's Information Criterion (AIC), is shown for comparison.

13.7   Prediction of fractional Brownian traffic

One consequence of the positive correlations of long range dependent traffic is that non-trivial traffic prediction is in principle possible at several timescales. It is interesting to study how such a prediction should be made and what can be expected to be gained from it. Let us consider the case of fractional Brownian traffic and assume that the three parameters m, a and H have been reliably estimated during a long observation period. The problem of relatively short term traffic prediction, say, predicting the value of $A(t, t + h)$ on the basis of observing $A(t - T, s)$ for $s \in [t - T, t]$, then reduces to the problem of finding more or less explicit expressions for the predictors (conditional expectations)

$$\widehat{Z}_{h,T} = \mathrm{E}[\,Z_h \mid Z_s, \ s \in (-T, 0]\,], \quad h > 0, \ T \in (0, \infty].$$

It would be natural to represent the predictors as integrals over the observed part of the process in the form

$$\widehat{Z}_{h,T} = \int_{-T}^{0} g_T(h, t)\mathrm{d}Z_t, \tag{13.7.1}$$

where $g_T(h, t)$ is an appropriate weight function. When the integrand is a smooth deterministic function, an integral w.r.t. Z_t can be defined simply as a limit of Riemann sums in L^2. As a general treatment on integration with respect to Gaussian processes that are not necessarily semimartingales, the reader is referred to [HC78]. We note here only the following formula for the covariance of two such integrals: for $f, g \in L^2(\mathrm{I\!R}; \mathrm{I\!R}) \cap L^1(\mathrm{I\!R}; \mathrm{I\!R})$ we have

$$\mathrm{E}\left\{\int_{\mathrm{I\!R}} f(s)\mathrm{d}Z_s, \int_{\mathrm{I\!R}} g(t)\mathrm{d}Z_t\right\} = H(2H - 1)\int\int_{\mathrm{I\!R}^2} f(s)g(t)|s - t|^{2H-2}\,\mathrm{d}t\,\mathrm{d}s.$$

It was shown in [GN96] that for each $h > 0$ and $T \in (0, \infty]$, the representation (13.7.1) holds with

$$g_T(h, -t) = \tag{13.7.2}$$

$$\frac{\sin(\pi(H - \tfrac{1}{2}))}{\pi} t^{-H+\frac{1}{2}}(T - t)^{-H+\frac{1}{2}} \int_0^h \frac{\sigma^{H-\frac{1}{2}}(\sigma + T)^{H-\frac{1}{2}}}{\sigma + t}\,\mathrm{d}\sigma$$

for $T < \infty$, and

$$g_\infty(h, -t) = \frac{\sin(\pi(H - \tfrac{1}{2}))}{\pi} t^{-H+\frac{1}{2}} \int_0^a \frac{\sigma^{H-\frac{1}{2}}}{\sigma + t}\,\mathrm{d}\sigma. \tag{13.7.3}$$

L.C.G. Rogers has derived, in the case $T = \infty$, another formula where the integral is a usual one and Z_t appears as a factor in the integrand [Rog95]:

$$\mathrm{E}[\,Z_h \mid Z_s, \ s \le 0\,] = -\frac{\sin(\pi(H - \tfrac{1}{2}))}{\pi} \int_{-\infty}^{0} \frac{\left|\dfrac{t}{h}\right|^{-H-1/2}}{h + |t|} Z_t\,\mathrm{d}t. \tag{13.7.4}$$

It is interesting to note that the weight function in (13.7.2) goes to infinity both at the origin and at $-T$ when T is finite — see Figure 13.7.1. Intuitively, the non-monotonicity can be understood so that the "closest witnesses" to the unobserved past have special weight.

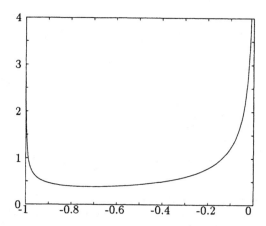

Figure 13.7.1: The prediction weight function $g_1(1,t)$ for $H = 0.9$. The function approaches infinity at both ends of the interval.

The variance of the predictor $E[Z_a \mid Z_s, \ s \in (-T,0)]$ is a concrete measure of the statistical unpredictability of fractional Brownian traffic. It was shown in [GN96] that

$$\text{Var}\{E[Z_h \mid Z_s, \ s \in (-T,0)]\} \tag{13.7.5}$$

$$= \text{Var}\{Z_h\}\, H \int_0^{T/h} g_{T/h}(1,-s)\left((1+s)^{2H-1} - s^{2H-1}\right)\, ds.$$

Moreover, for $T = \infty$ we have a closed form expression in terms of the gamma function:

$$\text{Var}\{E[Z_h \mid Z_s, \ s \leq 0]\} \tag{13.7.6}$$

$$= \text{Var}\{Z_h\}\left(1 - \frac{\sin(\pi(H-\tfrac{1}{2}))}{\pi(H-\tfrac{1}{2})}\frac{\Gamma(\tfrac{3}{2}-H)^2}{\Gamma(2-2H)}\right).$$

The relative variance of error $\text{Var}\{Z_h - \widehat{Z}_{h,\infty}\}/\text{Var}\{Z_h\}$ is plotted in the left graph of Figure 13.7.2 as a function of H. Note that, as a consequence of self-similarity, this quantity is independent of h. It is seen that the predictive force of the past is not very high unless H is rather large. The past before 0 explains half of the variance of Z_h when H is about 0.85, which is a rather typical value for daytime Ethernet traffic according to the Bellcore measurements [LTWW94]. The right graph of the same figure depicts the relative variance of error $\text{Var}\{Z_1 - \widehat{Z}_{1,T}\}/\text{Var}\{Z_1\}$ as a function of T with $H = 0.9$. It is seen that for the prediction of Z_h, it makes relatively little difference whether we know Z on $(-h, 0)$ or $(-\infty, 0)$.

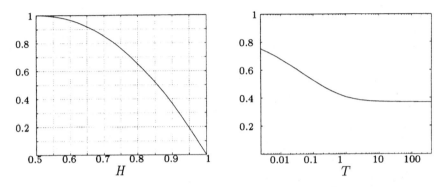

Figure 13.7.2: Left: the relative variance of error $\text{Var}\{Z_h - \widehat{Z}_{h,\infty}\}/\text{Var}\{Z_h\}$ as a function of the self-similarity parameter H. Right: the relative variance of error $\text{Var}\{Z_1 - \widehat{Z}_{1,T}\}/\text{Var}\{Z_1\}$ as a function of T; $H = 0.9$.

Thus we have two rules of thumb for the short term statistical predictability of fractional Brownian traffic:

- the past before t explains about half of the variance of any single future value A_u, $u > t$;

- one should predict (with the appropriate non-uniform weights) the next second with the latest second, the next minute with the latest minute etc.

14 General tools for queueing analysis

This chapter gathers some general results and theoretical concepts which are used repeatedly in subsequent chapters. We first include a brief statement of some useful relations linking performance parameters. The Beneš method, largely developed within the COST group, is then introduced and applied to queueing systems with discrete, fluid and diffusion input. The last section summarizes a number of important results from the theory of large deviations particularly useful for predicting the probability of such rare events as cell loss in ATM systems.

14.1   Some useful relations

In this section we present some elementary relations which prove useful in analyzing the queueing processes or in interpreting the results.

Little's formula

$$\overline{X_t} = \lambda \cdot W, \qquad (14.1.1)$$

where $\overline{X_t}$ is the mean number in the system, λ is the mean arrival rate, and W is the mean time in the system. Little's result applies to any stationary system. It can be readily shown to hold for any realization by expressing the cumulated time the customers spend in the system in a long interval of time T in two different ways: the number of arrivals times the mean time spent in the system and the time integral of the number of customers in the system [Kle75b]. The "system" may be taken to mean the "waiting room" or the "waiting room + server" with two different interpretations for (14.1.1).

Loss formula

$$P_{loss} = 1 - \frac{1 - \phi}{\rho}, \qquad (14.1.2)$$

where P_{loss} is the probability an arbitrary customer is lost, $\rho = \lambda \cdot t_s$ is the offered load (t_s being the mean service time) and ϕ is the probability that the server is idle. To prove this relation for a stationary system consider again a long interval of time T. The mean number of served customers in this interval is clearly $(1 - \phi)T/t_s$. Therefore, the mean number of lost customers is $\lambda T - (1 - \phi)T/t_s$. When this is related to the mean number of arrived customers, Equation (14.1.2) follows. We have implicitly assumed a single server system. The result, however, is valid for multiserver (S-server) systems as well with ρ being interpreted to be the mean load per server $\lambda t_s/S$ and ϕ being the probability an arbitrarily chosen server is idle. Equation (14.1.2) was used in [Hüb90] in a more limited context but, as we have seen, it is generally valid.

Unfinished work vs. number in the system

The quantity we are mainly focusing our attention on is the unfinished work in the system, V_t. Occasionally, we may wish to consider the number in the system, X_t. In a *constant service time single server system*, the following relation is trivially true (the unit of work is the service time of one customer):

$$X_t = \lceil V_t \rceil.$$

Therefore, we have the following identity between the complementary distributions:

$$\Pr\{X_t > n\} = \Pr\{V_t > n\}, \quad \text{for } n \text{ an integer.} \tag{14.1.3}$$

Cell loss probability vs. queue length tail probability

An upper bound for the cell loss probability in a discrete time $G/D/1$ system can be obtained as a simple generalization of Wong's result [Won90] for a discrete time $N*D/D/1/K$ system as follows. Let X_t denote the queue length in slot t and X_t^∞ the corresponding queue length in a hypothetical infinite capacity queue with the same realization of the arrival process. Then

$$\rho \cdot P_{loss} \le \Pr\{X_t^\infty > K\} \tag{14.1.4}$$

with ρ denoting the load factor. This follows by noting that the number of cells arriving when $X_t^\infty \ge K$ sets an upper bound to the number of lost cells. But, as time is slotted, each such arrival prolongs the so called overflow time (i.e., when $X_t^\infty > K$) by one time slot. Thus, in a long period of time, the total number of lost cells is bounded above by the cumulated overflow time. When this is related to the number of arrivals in the same period we arrive at inequality (14.1.4).

Cell delay vs. buffer contents

The results presented in this subsection hold for a discrete-time queueing model with service times (cell transmission times) equal to one slot, a FIFO service discipline, and either finite or infinite storage capacity, and, as such, are especially useful in the performance evaluation of ATM-based communication networks [VB95]. In the infinite capacity case, it is assumed that the equilibrium condition $\rho < 1$, is satisfied. The buffer occupancy at the beginning of the k-th slot is represented by X_k. Then, under the assumption that the evolution of this random variable can be described by the system equation

$$X_{k+1} = (X_k - c)^+ + b_k,$$

where b_k denotes the number of cells entering the buffer during slot k (not counting cells that are rejected, for instance due to buffer overflow) and c is the maximum number of cell transmissions per slot, it can be shown that a simple relation exists between the steady-state probability mass function of the buffer occupancy X and the cell delay W (the number of slots an arbitrary cell resides in the buffer, including its transmission slot). The following one-to-one relationship can be derived in case $c = 1$:

$$\Pr\{W = n\} = \Pr\{X = n\}/\rho, \qquad n \geq 1, \tag{14.1.5}$$

independent of the details of the cell arrival process (uncorrelated or correlated, Markovian or non-Markovian, stationary or non-stationary) except for the utilization

$$\rho = \lim_{k \to \infty} \mathrm{E}\{b_k\}/c.$$

In the general case $c \geq 1$, the previous relation can be extended to

$$\Pr\{W = n\} = \frac{1}{\rho c} \sum_{i=-(c-1)}^{c-1} (c - |i|)\Pr\{X = cn + i\}, \qquad n \geq 1,$$

as was also shown in [SB93] through a generating functions approach.

A result equivalent to (14.1.5) holds for fluid queues. Let $Q(x)$ be the complementary distribution of the virtual waiting time in a fluid queue and let $R(x)$ be the equivalent distribution "seen" by an "arriving molecule" (the notion of conditioning on an arrival in a fluid sytem is clarified below). Kontovasilis and Mitrou [KM94] show that for Markov modulated input, we have:

$$R(x) = Q(x)/\rho, \quad x > 0. \tag{14.1.6}$$

In fact the result holds for any stationary input process Λ_t.

Let V_t denote the virtual waiting time at time t of a stationary fluid queue with input rate process $\{\Lambda_t\}$ and unit service rate. In this fluid context, the expectation of a function $f(V_0)$ of the workload at time 0 "conditioned on there being an arrival at time 0" is naturally defined as

$$-\int_0^\infty f(x)R(dx) = \frac{\mathrm{E}\{f(V_0)\Lambda_0\}}{\mathrm{E}\{\Lambda_0\}} = \rho^{-1}\mathrm{E}\{f(V_0)\Lambda_0\}. \tag{14.1.7}$$

Let u be a fixed real number. Notice that $\dot{V}_s = (\Lambda_s - 1)1_{\{V_s > 0\}}$, which yields

$$e^{iuV_t} - e^{iuV_0} = \int_0^t \frac{d}{ds}\left(e^{iuV_s}\right) ds = \int_0^t iue^{iuV_s}(\Lambda_s - 1)1_{\{V_s > 0\}} ds$$

Taking expectations in this equation, by joint stationarity of the processes $\{\Lambda_t\}_{t\in\mathbb{R}}$ and $\{V_t\}_{t\in\mathbb{R}}$, one sees that necessarily, for all $s \in \mathbb{R}$,

$$E\left\{e^{iuV_s}; V_s > 0\right\} = E\left\{e^{iuV_s}\Lambda_s; V_s > 0\right\}.$$

The left-hand side of this expression is also equal to $-\int_{0+}^{\infty} e^{iux}Q(dx)$, while the right-hand side, in view of (14.1.7), equals $-\rho\int_{0+}^{\infty} e^{iux}R(dx)$. Equality of these two terms holds for all $u \in \mathbb{R}$; since Fourier transform characterizes measures, we see that the two measures Q and ρR coincide on $(0,\infty)$, as stated in (14.1.6).

14.2 The Beneš approach to virtual waiting time

For a queueing system with deterministic service time, the *virtual waiting time*, alias *unfinished work in system*, turns out to be a good choice for a central object of study. For example, if the transmission time of one cell is taken as the time unit, then the queue length in cells is simply the integer part of the virtual waiting time. A theorem due to Beneš (see [Ben63], [Bor76]), expressing the virtual waiting time of a $G/G/1$ queue, is repeatedly used below. We state this result in a generalized form, which also allows continuous ("fluid") input and finally extend it to a non-differentiable input process which requires special care.

14.2.1 A generalized Beneš formula

Consider a service system with an unlimited buffer. Assume that the system is stationary so that 0 represents an arbitrary time instant. The server capacity is assumed to be 1 unit of work per unit of time.

In order to determine the amount of unfinished work in system at time 0, we look backwards in time to see what has arrived and what has been processed. Let $A(t)$, $t \geq 0$, denote the amount of work arriving to the system in the interval $[-t, 0)$. To ensure stability, we require that

$$\lim_{t\to\infty} \frac{A(t)}{t} < 1 \text{ a.s.} \tag{14.2.1}$$

Let V_{-t} be the amount of work still in the system at time $-t$. In queueing terms, V_{-t} is the virtual waiting time at $-t$. Define $\xi(t) = A(t) - t$, $t \geq 0$, to be the excess work arriving in $[-t, 0)$. Then V_{-t} is given by Reich's formula

$$V_{-t} = \sup_{u \geq t}(\xi(u) - \xi(t)). \tag{14.2.2}$$

In particular,

$$V_0 = \sup_{t \geq 0} \xi(t). \tag{14.2.3}$$

The virtual waiting time can thus be obtained as the maximum of a stochastic process. We are mainly interested in the complementary distribution function of V_0:

$$Q(x) = \Pr\{V_0 > x\}.$$

As an immediate corollary from (14.2.3) we have $\Pr\{V_0 > x\} \geq \Pr\{\xi(t) > x\}$ $\forall t \geq 0$, and thus get the lower bound

$$Q(x) \geq \sup_{t \geq 0} \Pr\{\xi(t) > x\}.$$

To derive an exact relation we partition the event $\{V_0 > x\} = \{\exists t \geq 0 : \xi(t) > x\}$ according to the last exit time $T^x = \sup\{t \geq 0 : \xi(t) = x\}$ of the excess work process $\xi(t)$ from the level x, leading to the generic Beneš principle

$$\Pr\{V_0 > x\} = \Pr\{T^x \in [0, \infty)\} = \int_0^\infty \Pr\{T^x \in du\}.$$

The last exit time T^x is illustrated in Figure 14.2.1.

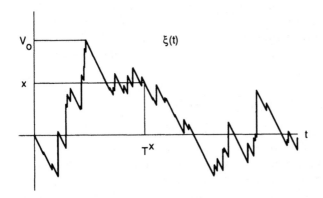

Figure 14.2.1: A realisation of the process $\xi(t)$ with the last exit time T^x.

Except for some very artificial arrival processes, the last exit time T^x is almost surely uniquely determined by the condition $\{\xi(T^x) = x$ and $V_{-T^x} = 0\}$. If further the process $\xi(t)$ is piecewise continuous and differentiable, we can deduce the Beneš formula:

$$Q(x) = \int_{u>0} \Pr\{\xi(u) \geq x > \xi(u + du) \text{ and } V_{-u} = 0\}. \tag{14.2.4}$$

This result covers all realisable queueing systems and has proven extremely useful in the context of ATM traffic theory. Although the integral can seldom be calculated exactly, practical approximations with good accuracy are often available. In particular, simple omission of the condition $\{V_{-u} = 0\}$

yields for $Q(x)$ an upper bound which often is both numerically calculable and rather close to the correct value.

For modelling purposes it is sometimes of interest to consider processes where $\xi(t)$ is not differentiable. Such cases arise, for instance, when $A(t)$ is modelled by a Brownian motion or fractional Brownian motion. In these cases (14.2.4) is not applicable as such and a more careful analysis of the condition $T^x \in du$ in (14.2.1) is required, see Section 14.2.4. First we give more concrete forms of (14.2.4) in the cases of a $G/D/1$ queue and a general fluid buffer.

14.2.2   The constant service time queue

In a constant service time queueing system, work arrives discontinuously in quanta of equal size, in our case ATM cells, which we choose as the unit of work. Thus,

$$A(t) = \nu(t) = \ \# \text{ of cell arrivals in } [-t, 0).$$

We then have $\xi(t) = \nu(t) - t$, and the probability on the right hand side of (14.2.4) is concentrated on the values of u such that $x + u$ is an integer. We can thus replace the integral by a summation:

$$Q(x) = \sum_{n > x} \Pr\left\{\nu(n - x) = n \text{ and } V_{-(n-x)} = 0\right\}. \tag{14.2.5}$$

As we noted above, the joint probability is in general difficult to calculate exactly because of the difficulty of taking the condition $V_{-(n-x)} = 0$ into account. It is, however, possible to derive a good approximation by applying the identity $\Pr\{A \cap B\} = \Pr\{A\} - \Pr\{\overline{B}\}\Pr[A \mid \overline{B}]$ giving

$$Q(x) = \sum_{n > x} \left\{\Pr\left\{\nu(n - x) = n\right\} - \rho \cdot \Pr\left[\nu(n - x) = n \mid V_{-(n-x)} > 0\right]\right\}.$$

Here the conditioning event is the complement of that in (14.2.5) and can be approximately replaced by the requirement that there is one arrival at $-(n - x)$, with the result

$$Q(x) \approx \sum_{n > x} \left\{\Pr\left\{\nu(n - x) = n\right\}\right. \tag{14.2.6}$$

$$\left. - \rho \cdot \Pr\left[\nu(n - x) = n \mid \text{one arrival at } - (n - x)\right]\right\}.$$

The same result can also be derived by a local stationarity argument, see [Rob92a]. Note that only quantities related to the arrival process appear, and in that sense this is a closed form solution.

14.2.3   The general fluid queue

As another application, assume that $A(t)$ is an absolutely continuous function, so that the system behaves like a fluid reservoir with constant leak rate whenever its content is non-zero. Let Λ_{-t} be the arrival rate at time $-t$. We then have

$$\Lambda_{-t} = \frac{dA(t)}{dt} = 1 + \frac{d\xi(t)}{dt}.$$

Writing

$$\xi(u + du) = \xi(u) + (\Lambda_{-u} - 1)du$$

in (14.2.4), we deduce

$$Q(x) = \int_{u>0} \int_{0 \leq \lambda \leq 1} \Pr\{x \leq \xi(u) < x + (1 - \lambda)du,$$

$$\lambda \leq \Lambda_{-u} < \lambda + d\lambda \text{ and } V_{-u} = 0\},$$

which we can rewrite as:

$$Q(x) = \int_{u>0} du \int_{0 \leq \lambda \leq 1} d\lambda\, (1 - \lambda) \tag{14.2.7}$$

$$\cdot \frac{d^2}{dx\,d\lambda} \Pr\{\xi(u) \leq x,\ \Lambda_{-u} \leq \lambda \text{ and } V_{-u} = 0\}.$$

See [SV91] for an alternative and more formal derivation of this result.

14.2.4   The case of non-differentiable input

It is possible to apply the Beneš approach also in the case where the cumulating traffic arrival process $A(t)$ has locally unbounded variation, like a diffusion process for example. Although such behaviour is completely unnatural for a traffic process, it is sometimes encountered as a nuisance property of a model that is logically simple in other respects, like a diffusion or a self-similar process.

The special problems can be illustrated by the simplest case of this type,

$$A(t) = mt + W_t,$$

where W_t is the standard Brownian motion. Assume that the traffic is offered to a buffer with constant leak rate c. The storage occupancy process V_t can be defined by Reich's formula (14.2.2) as usual, and it is non-negative. The queueing system has, however, some paradoxal properties.

The output within an interval $[s, t]$ is defined in the natural way as

$$U(s, t) = A(s, t) + V_s - V_t.$$

We then have from (14.2.2) that

$$U_t =_{\text{def}} U(0,t) = V_0 + ct - \sup_{s \le t}(cs - A(s)),$$

so that U is the difference of two increasing processes and thus has paths that are differentiable almost everywhere. However, its derivative is c at almost every time point, although the long run mean output rate is of course m — the output proceeds at full rate except that it is interrupted by singular negative output in a nowhere dense but non-denumerable (perfect, Cantor-like) set of timepoints. This is the mathematical explanation for the counterintuitive feature that *the storage is almost never empty*, a fact that makes the formula (14.2.4) inapplicable in this case.

To simplify the notation still further, assume that $c - m = 1$, in which case we are looking for the distribution of the maximum of a standard Brownian motion with negative drift -1. It is well known that this distribution is exponential with parameter 2, but our problem is to find out what shape the Beneš approach takes in this case. Denote again by $T^x = \sup\{t : W_t = x + t\}$ the last exit time of the process $W_t - t$ from the level x. In order to find the density function of T^x, consider a short interval $[t, u]$ and decompose the probability of the event $\{T^x \in [t, u]\}$ as follows:

$$\Pr\{T^x \in [t, u]\} = \Pr\{\exists s \in [t, u] : W_s = x + s\} \qquad (14.2.8)$$

$$\cdot \Pr[W_v < x + v, \forall v > u \mid \exists s \in [t, u] : W_s = x + s].$$

The homogeneous Markovian structure of Brownian motion implies that the second factor is of the form $\gamma(u - t)$. If we divide by $u - t$ and let $u \searrow t$, the left hand side converges to the density of T^x at t. On the right hand side, a meaningful limit is obtained by dividing the denominator $u - t$ into two copies of $\sqrt{u - t}$. It is not difficult to show that

$$\lim_{u \searrow t} \frac{1}{\sqrt{u - t}} \Pr\{\exists s \in [t, u] : W_s = x\} = \frac{2}{\pi\sqrt{t}} e^{-\frac{x^2}{2t}},$$

and that the result remains the same if W is replaced by a drifted Brownian motion. It then follows from (14.2.8) that there exists also a finite positive limit

$$L = \lim_{u \searrow t} \frac{\gamma(u - t)}{\sqrt{u - t}},$$

and

$$\Pr\{T^x \in dt\} = \sqrt{\frac{8}{\pi}} \Pr\{W_t \in x + t + dt\} \cdot L.$$

Since T^0 is finite and positive with probability 1, we have

$$1 = \sqrt{\frac{8}{\pi}} L \int_0^\infty \Pr\{W_t \in t + dt\} = \sqrt{\frac{8}{\pi}} L \int_0^\infty \frac{1}{\sqrt{2\pi t}} e^{-\frac{t^2}{2}} dt = \sqrt{\frac{8}{\pi}} L.$$

Thus, we have deduced the classical result

$$\Pr\{T^x \in dt\} = \frac{1}{\sqrt{2\pi t}} e^{-\frac{(x+t)^2}{2t}} dt = \Pr\{W_t \in x + t + dt\}, \qquad (14.2.9)$$

and an integration finally gives that $\Pr\{T^x \in (0,\infty)\} = e^{-2x}$. As a side-product, we obtained the rather unobvious result

$$\lim_{u \searrow t} \frac{1}{\sqrt{u-t}} \Pr[W_v < x + v, \forall v > u \mid \exists s \in [t,u]: W_s = x + s] = \sqrt{\frac{\pi}{8}}.$$

The above reasoning resulted in an explicit formula because the second factor of (14.2.8), divided by $\sqrt{u-t}$, converged to a constant. In the general case, this is replaced by a function of $L(x,t)$ which is difficult to compute or even estimate. However, the first factor is usually the dominating one, and a reasonable approximation can be obtained by replacing $L(x,t)$ by a normalizing constant. When the local behaviour of $A(t)$ is that of a fractional Brownian motion with parameter H (see Subsection 13.4.4), the first factor in (14.2.8) has to be divided by $(u-t)^H$ and the second by $(u-t)^{1-H}$ in order to obtain a finite positive limit when $u \searrow t$. See [NSVV95] for details.

14.3   Elements of the theory of large deviations

The theory of large deviations provides a very powerful methodology for obtaining reasonably accurate estimates for probabilities of rare events. We give here a short summary of these techniques in the one-dimensional case, omitting most proofs and strict formulation of theorems. We refer to [SW95] and [Buc90] as more comprehensive expositions of the theory.

14.3.1   Free energy and entropy functions, and exponential families

Let X be a random variable, defined on some probability space $(\Omega, \mathcal{F}, P)$, with moment generating function $\psi(\beta) = E\{e^{\beta X}\}$. Its logarithm

$$\mu(\beta) = \ln \psi(\beta)$$

has several names. Let us follow the vocabulary of Ellis [Ell85] and call it the *free energy function* of X. (The theory can be seen as some kind of abstract thermodynamics; the number β is then related to temperature). It is easy to see that μ is a convex function. In particular, the set

$$\mathcal{D}_\mu = \{\beta : \mu(\beta) < \infty\}$$

is convex, which in our one-dimensional case means that it is an interval containing at least zero. $\mu(\beta)$ is analytic in the interior of $\mathcal{D}_\mu$.

The convex conjugate $\mu^*(x)$ of $\mu(\beta)$, defined by

$$\mu^*(x) = \sup_{\beta}(\beta x - \mu(\beta)), \tag{14.3.1}$$

is called the *entropy function* of X. In view of the supremum operation in (14.3.1) classically related to the Legendre transformation, function μ^* is also known as the *Cramér-Legendre transform* of μ. It is also referred to as the *large deviation rate function* of variable X. Note that if the Laplace transform $\varphi^*(\beta) = \psi(-\beta)$ is considered, we also have

$$\mu^*(x) = -\inf_{\beta}(\beta x + \ln \varphi^*(\beta)) \tag{14.3.2}$$

as an alternative expression to (14.3.1). Function μ^* is obviously non-negative. Being the supremum of convex functions, i.e., straight lines, it is a convex function, and

$$\mathcal{D}_{\mu^*} = \{\, x \ : \ \mu^*(x) < \infty \,\}$$

is a convex set, i.e., an interval, which can be shown to be non-empty. If $\Pr\{X = a\} = 1$, then $\mathcal{D}_{\mu^*} = \{a\}$. Both functions μ and μ^* are always lower semicontinuous.

The third thing we define is the family of probability measures P_β, called the *exponential family of distributions* associated with X:

$$\mathrm{d}P_\beta = \frac{e^{\beta X}}{\psi(\beta)}\mathrm{d}P = e^{\beta X - \mu(\beta)}\mathrm{d}P, \quad \beta \in \mathcal{D}_\mu.$$

We write $E_\beta\{\cdot\}$ for an expectation taken with respect to P_β. The expectations $m(\beta) = E_\beta\{X\}$ play an important role. When β is in the interior of $\mathcal{D}_\mu$, we have

$$\mu'(\beta) = m(\beta) = E_\beta\{X\}, \quad \mu''(\beta) = m'(\beta) = \sigma^2(\beta) = \mathrm{Var}_\beta\{X\}.$$

The latter equality shows that if X is not constant, then $m(\beta)$ is strictly increasing in the interior of $\mathcal{D}_\mu$, so that the mapping $\beta \mapsto m(\beta)$ has a well-defined inverse $x \mapsto \beta_x$. Moreover, it is then seen by differentiating that for x in the interior of $\mathcal{D}_{\mu^*}$ we have

$$\mu^*(x) = \beta_x x - \mu(\beta_x) \tag{14.3.3}$$

and

$$\frac{\mathrm{d}}{\mathrm{d}x}\mu^*(x) = \beta_x. \tag{14.3.4}$$

If $\mathcal{D}_\mu$ is open and X is not a constant, the mapping $\beta \mapsto m(\beta)$ is one-to-one from $\mathcal{D}_\mu$ to the interior $\mathcal{S}^o$ of the convex hull of the support of the distribution of X.

Many important parametric families of distributions are in fact of this exponential type, for example binomial (and Bernoulli as a special case),

Table 14.3.1: Shifted mean and entropy function for some standard distributions.

Distr.	$m(\beta)$	$\mu^*(x)$
$\mathrm{Bin}(n, p)$	$\dfrac{np}{p + (1-p)e^{-\beta}}$	$x \log \dfrac{x}{p} + (n-x) \log \dfrac{n-x}{1-p} - n \log n$
$\mathrm{Exp}(\lambda)$	$\dfrac{1}{\lambda - \beta}$	$x\lambda - 1 - \log(x\lambda)$
$\mathrm{Poisson}(a)$	ae^{β}	$x \log \dfrac{x}{a} + a - x$
$\mathrm{N}(m, \sigma^2)$	$m + \sigma^2 \beta$	$\dfrac{1}{2} \left(\dfrac{x-m}{\sigma} \right)^2$

Poisson, exponential and Gaussian. The functions $m(\beta)$ and $\mu^*(x)$ have in these cases the simple expressions shown in Table 14.3.1.

For all distributions appearing in the table, $\mathcal{D}_\mu$ is open. As an example where this is not the case, consider the shifted Pareto distribution $\Pr\{X > x\} = (1+x)^{-\gamma}$, $x > 0$, where γ is a positive parameter. Then $\mathcal{D}_\mu = (-\infty, 0]$. The measures P_β can still be defined for $\beta < 0$, and $m(\beta)$ maps $(-\infty, 0)$ onto $(0, \mathrm{E}\{X\})$.

If X is a sum of independent random variables, then $\mu(\beta)$, $m(\beta)$ and $\sigma^2(\beta)$ are obtained by summing the respective quantities for the constituent terms, i.e., if Y and Z are independent, then $\mu_{Y+Z}(\beta) = \mu_Y(\beta) + \mu_Z(\beta)$, etc.

14.3.2 Probabilities of large deviations

The Cramér-Chernoff theorem

Let us now see how the above notions can be used in the study of the probabilities of certain types of rare events. The most basic tool is the elementary inequality

$$\Pr\{X \geq x\} \leq \mathrm{E}\left\{ e^{\beta(X-x)} 1_{\{X \geq x\}} \right\} \leq e^{\mu(\beta) - \beta x}, \quad \text{for } \beta \geq 0. \qquad (14.3.5)$$

Assume that $\mathrm{E}\{X\}$ exists and $x \geq \mathrm{E}\{X\} \geq -\infty$. Then the supremum in (14.3.1) is obtained within values $\beta \geq 0$, and minimizing the right hand side of (14.3.5) with respect to β yields the important Chernoff upper bound

$$\Pr\{X \geq x\} \leq e^{-\mu^*(x)} \quad \text{for } x \geq \mathrm{E}\{X\}. \qquad (14.3.6)$$

The upper bound (14.3.6) is often in the correct order of magnitude. Indeed, if β_x is defined, we can write the *equality*

$$\Pr\{X > x\} = e^{-\mu^*(x)} \cdot E_{\beta_x}\{e^{-\beta_x(X-x)} 1_{\{X \geq x\}}\} \qquad (14.3.7)$$

and note that the expectation at the right is not very small for most usual distributions since x is the mean of X with respect to P_{β_x}.

When we consider a large number of independent and identically distributed (i.i.d.) random variables, the Chernoff bound becomes asymptotically exact in the following sense. Let $X_1, X_2, \ldots$ be i.i.d. copies of X and denote $S_n = X_1 + \cdots + X_n$. For any set $B \subseteq \mathbb{R}$, define

$$\mu^*(B) = \inf \{ \mu^*(x) : x \in B \}. \tag{14.3.8}$$

Then the famous Cramér-Chernoff theorem gives the upper bound

$$\limsup_{n \to \infty} \frac{1}{n} \ln \Pr \left\{ \frac{S_n}{n} \in B \right\} \leq \mu^*(B) \quad \text{if } B \text{ is closed} \tag{14.3.9}$$

and the lower bound

$$\liminf_{n \to \infty} \frac{1}{n} \ln \Pr \left\{ \frac{S_n}{n} \in B \right\} \geq \mu^*(B) \quad \text{if } B \text{ is open.} \tag{14.3.10}$$

When $x > \mathrm{E}\{X\}$ and x is in the interior of $\mathcal{D}_{\mu^*}$, the upper and lower bounds together give

$$\lim_{n \to \infty} \frac{1}{n} \ln \Pr \left\{ \frac{S_n}{n} \geq x \right\} = \mu^*(x), \tag{14.3.11}$$

which suggests the use of the approximation $\Pr(S_n/n \geq x) \approx e^{-n\mu^*(x)}$. Such an approximation, which holds rigorously logarithmically as stated in (14.3.11), will be noted as

$$\Pr \left\{ \frac{S_n}{n} \geq x \right\} \asymp e^{-n\mu^*(x)}. \tag{14.3.12}$$

The symbol $\asymp$ implies restricted logarithmic equivalence. The Cramér-Chernoff theorem holds for any distribution of X_1, although its message sometimes reduces to a triviality.

Improved approximations

The approximation (14.3.12) can be improved by applying the central limit theorem to estimating directly the correcting factor in (14.3.7) (cf. Bahadur and Rao [BR60]). Now assume that we have a large number of independent, not necessarily identically distributed random variables, say $X_1, \ldots, X_n$, and choose $X = \sum_1^n X_i$. We then have $\mu(\beta) = \sum_1^n \mu_i(\beta)$, so that

$$\mu^*(x) = \beta_x x - \sum_1^n \mu_i(\beta_x).$$

(Note that all μ_i's have the same β_x as their argument. This is in fact connected to an abstract counterpart to the law of the equalization of temperatures.) Suppose now that the distributions of the X_i's are such that their sum $X = \sum_1^n X_i$ is approximately normally distributed with mean m and variance σ^2. If x is much larger than m, then the normal approximation may not be good for the tail probability $P(X \geq x)$. However, for the shifted probability P_{β_x} the normal approximation is usually good at x, since now x is the mean of the distribution. Now we can approximate the correcting factor $E_{\beta_x}\{\exp(-\beta_x(X - x))1_{\{X \geq x\}}\}$ in (14.3.7) by assuming X Gaussian:

$$E_{\beta_x}\{e^{-\beta_x(X-x)}1_{\{X \geq x\}}\} \approx \frac{1}{\sqrt{2\pi}\sigma(\beta_x)} \int_0^\infty e^{-\beta_x t - \frac{1}{2}(t/\sigma(\beta_x))^2} dt$$

$$= \frac{1}{\sqrt{2\pi}\sigma(\beta_x)} \int_0^\infty e^{-u} \left(\sqrt{\beta_x^2 + \frac{2u}{\sigma^2(\beta_x)}}\right)^{-1} du$$

$$\approx \frac{1}{\sqrt{2\pi}\beta_x\sigma(\beta_x)}. \qquad (14.3.13)$$

Substituting (14.3.13) into (14.3.7) we obtain the following formula, which is both accurate and often easily calculable:

$$\Pr\{X \geq x\} \approx \frac{e^{-\mu^*(x)}}{\sqrt{2\pi}\beta_x\sigma(\beta_x)}. \qquad (14.3.14)$$

This approximation was derived and used in an admission control algorithm by Hui ([Hui88], [Hui90]). Using (14.3.4) for the derivative of $\mu^*(x)$ and taking into account that the exponential $e^{-\mu^*(x)}$ is asymptotically dominant for large x, an equivalent formula can be derived from (14.3.14) for the probability density $f(x)$ of X, when defined, as

$$f(x) \approx \frac{e^{-\mu^*(x)}}{\sqrt{2\pi}\sigma(\beta_x)}. \qquad (14.3.15)$$

For the estimation of the density of the workload entering a fluid queue, formula (14.3.15) has also been derived in [BGRS94] on the basis of the Laplace inverse formula

$$f(x) = \frac{1}{2\pi i} \oint e^{zx + \ln f^*(z)} dz$$

relating density f and its Laplace transform f^*. The integral is here taken over a suitable contour in the complex plane. Expression (14.3.15) is then obtained using the saddle-point method to estimate the contour integral for large values of x (see also [SG95, App.1] for a further discussion on the asymptotic estimation of densities).

A similar idea can be used to approximate the probabilities of single values in a discrete convolution. (This technique was effectively applied in [VR89]). Suppose that the above X_i's are of lattice type, that is, they take values in the set $\{kd : k = 0, \pm 1, \pm 2, ...\}$ where the number d defines the lattice spacing. This is, in particular, the case when each X_i is a sum $X_i = \sum_{\ell=1}^{R} d_\ell \xi_\ell$ where variables ξ_ℓ are independent and where the d_ℓ are given integers; here d is the Greatest Common Divisor (GCD) of all d_ℓ. We have to estimate the probability $\Pr\{X = kd\}$, where kd is far from the mean m. Now, if the variances of the X_i's are all small compared with the variance of X, then the local central limit theorem applies at kd with respect to $P_{\beta_{kd}}$, so that we have

$$P_{\beta_{kd}}(X = kd) = P_{\beta_{kd}}(X < (k+1)d) - P_{\beta_{kd}}(X < kd) \approx d \cdot \frac{1}{\sqrt{2\pi}\sigma(\beta_{kd})}$$

with good precision. Thus we obtain the estimate

$$\Pr\left\{\sum_{i=1}^{n} X_i = kd\right\} = E_{\beta_{kd}}\left\{\psi(\beta_{kd})e^{-\beta_{kd}X}1_{\{X=kd\}}\right\}$$

$$\approx \frac{d}{\sqrt{2\pi}\sigma(\beta_{kd})} \cdot e^{-\mu^*(kd)}. \tag{14.3.16}$$

Further elements on these improved estimates can be found in [Buc90, pp.117-121].

The Gärtner-Ellis theorem

It is not necessary to consider a sum of independent random variables in order to obtain accurate limit results with the theory of large deviations. The following approach was created by Gärtner and Ellis. Consider a sequence of random variables $X_1, X_2, ...$, which correspond to the sums S_n of the Cramér-Chernoff theorem, having the free energy functions $\mu_1(\beta), \mu_2(\beta), ...$, respectively. The basic assumption is that there is a *scaling sequence* a_n such that $a_n \to \infty$ and the limit

$$\lim_{n \to \infty} \frac{\mu_n(\beta)}{a_n} = \mu(\beta)$$

exists (but need not be finite) for every β. As a limit of convex functions, μ is automatically convex, and the set $\mathcal{D}_\mu$ where it is finite is again an interval. Let us make the following additional assumptions:

(i) 0 is an interior point of $\mathcal{D}_\mu$

(ii) μ is differentiable in the interior of $\mathcal{D}_\mu$

(iii) $|\mu'(\beta_n)| \to \infty$ when β_n approaches the boundary of $\mathcal{D}_\mu$.

The corresponding "entropy function" μ^* is again defined by (14.3.1) and, for sets, by (14.3.8). We then have the upper bound

$$\limsup_{n \to \infty} \frac{1}{a_n} \ln \Pr\left\{ \frac{X_n}{a_n} \in B \right\} \leq -\mu^*(B) \qquad \text{if } B \text{ is closed}$$

and the lower bound

$$\liminf_{n \to \infty} \frac{1}{a_n} \ln \Pr\left\{ \frac{X_n}{a_n} \in B \right\} \geq -\mu^*(B) \qquad \text{if } B \text{ is open.}$$

If x is in the interior of $\mathcal{D}_{\mu^*}$ and μ^* is strictly increasing at x, the upper and lower bounds together yield

$$\lim_{n \to \infty} \frac{1}{a_n} \ln \Pr\left\{ \frac{X_n}{a_n} > x \right\} = -\mu^*(x).$$

14.3.3 Maximum of a sequence with negative drift

A task often encountered in queueing theory is to find the distribution of the maximum of a process with negative drift. The theory of large deviations provides an efficient general approach to this problem. Our presentation is based on the results by Duffield and O'Connell [DO96], but instead of stating theorems, we outline the main ideas as a method.

Consider a sequence X_n of random variables with mean zero. In a typical application, X_n would have stationary increments, but we need not make such an explicit assumption. Denote by $\mu_n(\beta)$ the free energy function of X_n. Our basic assumption is that, for an appropriate increasing function $f(n)$, the limit

$$\lim_{n \to \infty} \frac{\mu_n\left(\frac{f(n)}{n} \beta \right)}{f(n)} = \mu(\beta)$$

exists for every β and that $\mu(\beta)$ is lower semicontinuous and differentiable and that $\mathcal{D}_\mu$ contains a neighborhood of zero. Moreover, we assume that $f(n) = n^\gamma L(n)$, where $L(n)$ is a slowly varying function, i.e., for any $\alpha > 0$, $L(\alpha n)/L(n) \to 1$ as $n \to \infty$.

Define the corresponding entropy functions $\mu_n^*(x)$ and $\mu^*(x)$ as usually. The Gärtner-Ellis theorem then applies to the sequence $(f(n)/n)X_n$. In particular,

$$\liminf_{n \to \infty} \frac{1}{f(n)} \ln \Pr\left\{ \frac{X_n}{n} > y \right\} \geq -\mu^*(y)$$

for each y such that $y > 0$ and y is in the interior of $\mathcal{D}_{\mu^*}$. Now, for any such y and for any constant $a > 0$, we have

$$\liminf_{x \to \infty} \frac{1}{f(x)} \ln \Pr\left\{ \sup_n (X_n - an) > x \right\}$$

$$\geq \liminf_{x\to\infty} \frac{1}{f(x)} \ln \Pr\left\{X_{\lceil x/y\rceil} - a\lceil x/y\rceil > x\right\}$$

$$= \liminf_{x\to\infty} \frac{1}{f(x)} \ln \Pr\left\{\frac{X_{\lceil x/y\rceil}}{\lceil x/y\rceil} > a + \frac{x}{\lceil x/y\rceil}\right\}$$

$$\geq \left(\liminf_{x\to\infty} \frac{f(\lceil x/y\rceil)}{f(x)}\right)\left(\liminf_{n\to\infty} \frac{1}{f(n)} \ln \Pr\left\{\frac{X_n}{n} > a + y\right\}\right)$$

$$= -\frac{\mu^*(y+a)}{y^\gamma}.$$

Since the original expression does not depend on y, we obtain the nice lower bound

$$\liminf_{x\to\infty} \frac{1}{f(x)} \ln \Pr\left\{\sup_n(X_n - an) > x\right\} \geq -\inf_{y>0} \frac{\mu^*(y+a)}{y^\gamma}$$

$$=_{\text{def}} -\theta. \tag{14.3.17}$$

It is more difficult to derive the corresponding upper bound result. The following "brute force" reasoning works in some cases at least. Assume that the number θ defined in (14.3.17) is positive and start with the inequality

$$\Pr\left\{\sup_n(X_n - an) > x\right\} \leq \sum_{n=1}^{\infty} \Pr\left\{X_n - an > x\right\}$$

$$\leq \sum_{n=1}^{\infty} e^{-\mu_n^*(x+an)}$$

$$= e^{-\theta f(x)} \sum_{n=1}^{\infty} e^{-\mu_n^*(x+an)+\theta f(x)},$$

where the second step is an application of Chernoff's upper bound (14.3.6). Thus, a sufficient condition for the upper bound result

$$\limsup_{x\to\infty} \frac{1}{f(x)} \ln \Pr\left\{\sup_n(X_n - an) > x\right\} \leq -\theta$$

is that

$$\limsup_{x\to\infty} \frac{1}{f(x)} \ln \sum_{n=1}^{\infty} e^{-\mu_n^*(x+an)+\theta f(x)} \leq 0. \tag{14.3.18}$$

In the case that $X_n = U_1 + \cdots + U_n$, where the U_i's are i.i.d. with $E\{U_i\} < 0$ and free energy function $\nu(\beta)$, say, the above reasoning becomes simpler. The choice of the scaling function is $f(n) = n$, $\mu_n(\beta) = n\nu(\beta)$, and $\mu(\beta) = \nu(\beta)$. The number θ is determined as the minimum of $\nu^*(y)/y$. Using (14.3.4) yields that θ is in fact the positive root of

$$\nu(\theta) = 0.$$

Further, it is not necessary to check the condition (14.3.18), since simple reasoning shows that the upper bound in fact holds for every x instead of being just asymptotical. Indeed, define for $x > 0$ the stopping time $T_x = \inf\{n : X_n > x\}$. Then

$$\Pr\left\{\sup_n X_n > x\right\} = \Pr\{T_x < \infty\} = \sum_{n=1}^{\infty} E_\theta\left\{e^{-\theta X_n + n\nu(\theta)} 1_{\{T_x = n\}}\right\}$$

$$\leq e^{-\theta x} \sum_{n=1}^{\infty} P_\theta\{T_x = n\} \leq e^{-\theta x}.$$

14.3.4   Conditioning on a large deviation

Let again $X_1, X_2, \ldots$ be i.i.d. random variables and let $S_n = X_1 + \cdots + X_n$. We want to estimate the conditional probability

$$\Pr\left[X_1 \in dx \mid \frac{S_n}{n} \geq z\right],$$

where $z > E\{X_i\}$. By (14.3.14), we have

$$\Pr\left\{\frac{S_n}{n} \geq z\right\} \approx \frac{e^{-n\mu^*(z)}}{\sqrt{2\pi\beta_z}\sigma(\beta_z)}, \quad \Pr\left[\frac{S_n}{n} \geq z \mid X_1 = x\right] \approx \frac{e^{-(n-1)\mu^*(z')}}{\sqrt{2\pi\beta_{z'}}\sigma(\beta_{z'})},$$

where $z' = (nz - x)/(n - 1)$. Using these approximations, we get for large n

$$\Pr\left[X_1 \in dx \mid \frac{S_n}{n} \geq z\right] \approx \Pr\{X_1 \in dx\} \, e^{n\mu^*(z) - (n-1)\mu^*(z')}$$

$$\approx \Pr\{X_1 \in dx\} \, e^{\mu^*(z) - (n-1)\mu^{*\prime}(z) \cdot (z - z')}$$

$$= \Pr\{X_1 \in dx\} \, e^{\beta_z x - \mu(\beta_z)}$$

$$= P_{\beta_z}(X_1 \in dx),$$

where the manipulation of the exponent was based on the mean value theorem and on the relation (14.3.4).

We see that, loosely speaking, conditioning on a large deviation of the totality is approximately the same as making the (smallest) exponential shift to a distribution where the condition is a typical case. This result is often called *Boltzmann's law*, referring to its origin in statistical mechanics.

15 Cell scale queueing

We consider an ATM multiplexer (Fig. 15.0.1) where a number of incoming streams are directed towards the same outgoing multiplex. At the cell scale, the cell rate from each source can be regarded to be fixed. For most of the time, the total cell rate from all the sources is less than the capacity of the outgoing multiplex. Even then, however, the multiplex cannot always transmit all cells immediately as they arrive because several arrivals can occur more or less simultaneously due to the asynchronous nature of the system.

To avoid cell losses a buffer is needed in the multiplexer. The arrangement most commonly conceived is that of output buffering, shown in Fig. 15.0.1. This is more efficient than input buffering where one congested output multiplex would form a bottleneck also for the transmission of cells addressed to other outgoing multiplexes.

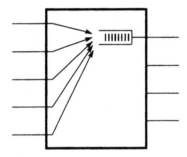

Figure 15.0.1: Output buffering in an ATM multiplexer.

The short term behaviour of the queue of the output buffer forms the topic of the present chapter. In Sections 15.1 to 15.3 we introduce three basic queueing models which are often used to describe the cell scale queueing problem. These are in order of increasing complexity: the $M/D/1$ (and related $Geo^N/D/1$), $N*D/D/1$ and $\sum D_i/D/1$ models. These all are single server constant service time models, as is appropriate for a fixed cell size. In Sections 15.4 and 15.5, we consider models designed to describe the cell delay variation (CDV) and the effects of cell phasing of different sources. These models are based on a superposition of a periodic stream and a batch Bernoulli background stream. The evolution of CDV through a tandem of queues is also considered. Finally, in Section 15.6, we discuss priority queueing models.

15.1   The $M/D/1$ and $Geo^N/D/1$ queues

As usual, we denote by $M/D/1$ the queueing system with Poisson arrivals
and deterministic (constant) service time. It is of course a simple special case
of the $M/G/1$ system and was studied already by Erlang in 1909.

The $M/D/1$ queue is an appropriate model of an ATM multiplexer queue
if we can assume that the cells arrive as a Poisson process. Poisson input can
be considered as "totally random" and therefore it is a natural choice if the
(constant) arrival rate is the only known characteristic of the traffic. Also,
in the case where the arrival process is a superposition of a large number of
periodic processes, it is locally very similar to a Poisson process. Thus, if the
load of the system is low, the Poisson model is rather generally applicable.
On the other hand, as shown in the next Section, the $M/D/1$ model gives
far too large tail probabilities for the queue length distribution in a heavily
loaded system with periodic inputs.

15.1.1   Virtual waiting time distribution

We choose the service time as our time unit so that the arrival intensity is
equal to the server load ρ.

It holds for any $M/G/1$ system that the distributions of the number of
customers in system as seen by arriving or departing customers or random
observers are all equal, and the unfinished work V in system (virtual waiting
time) has the same distribution as the waiting time (assuming FIFO service;
see [Kle75b]).

The Laplace transform of the distribution of unfinished work V in the
$M/D/1$ system is thus obtained by the well known Pollaczek-Khinchin formula for the waiting time:

$$V^*(r) = \mathrm{E}\left\{e^{-rV}\right\} = \frac{1-\rho}{1 - \rho\dfrac{1-e^{-r}}{r}}. \tag{15.1.1}$$

Taking derivatives at the origin we deduce

$$\mathrm{E}\{V\} = \frac{\rho}{2(1-\rho)}, \quad \mathrm{Var}\{V\} = \frac{\rho(1-\left(\frac{\rho}{2}\right)^2)}{3(1-\rho)^2}. \tag{15.1.2}$$

A finite sum expression for the distribution of V can be derived by inverting
the Laplace transform (15.1.1), by solving an integral equation ([Sys86]) or by
using the Beneš method introduced in Section 14.2. Let us consider the latter
and recall the notation of Section 14.2. The complementary distribution
function $\Pr\{V > x\}$ is denoted by $Q(x)$. As observed in [Rob92b], we have
by (14.2.5)

$$Q(x) = \sum_{n>x} \Pr\left\{\nu(n-x) = n \text{ and } V_{-(n-x)} = 0\right\}$$

$$= \sum_{n > x} \frac{(\rho(n - x))^n}{n!} e^{-\rho(n-x)} (1 - \rho) \qquad (15.1.3)$$

$$= 1 - (1 - \rho) \sum_{n=0}^{\lfloor x \rfloor} \frac{(\rho(n - x))^n}{n!} e^{-\rho(n-x)},$$

where the first equation follows from the independence of $\nu(t)$ and V_t, and the second is obtained by using the identity (cf. [Sys86], p. 220)

$$\sum_{n=0}^{\infty} \frac{(a + nb)^n}{n!} e^{-(a+nb)} = \frac{1}{1 - b}.$$

Result (15.1.3) is valid for all values of $x \geq 0$ and for all loads $0 \leq \rho \leq 1$. However, it poses a numerical problem for large values of x when ρ is close to 1 due to the appearance of large alternating, very nearly cancelling terms in the sum. The problem can be avoided by the algorithmic solution derived in [Vir95]. The last expression of (15.1.3) is of the form $Q(x) = 1 - (1 - \rho)e^{\rho x} P'_{\lfloor x \rfloor}(x)$, where $P'_n(x)$ is a polynomial of degree n. The trick is to use "local coordinates" by defining $P'_n(x) = P_n(x - \lfloor x \rfloor)$. Thus

$$Q(x) = 1 - (1 - \rho)e^{\rho x} P_{\lfloor x \rfloor}(x - \lfloor x \rfloor).$$

The polynomials $P_n(x)$ can be calculated recursively as follows. Given the coefficients $P_n = (a_0, a_1, \ldots, a_n)$ of $P_n(x) = \Sigma_i a_i x^i$, the coefficient vector P_{n+1} of the next polynomial is obtained from

$$P_{n+1} = (\Sigma_i a_i, -\beta a_0/1, -\beta a_1/2, \ldots, -\beta a_n/(n + 1)),$$

where $\beta = \rho e^{-\rho}$. The recursion starts from $P_0 = (1)$. This recursion is numerically stable. Note also that each polynomial P_n needs only to be evaluated for small arguments $0 \leq x - \lfloor x \rfloor < 1$.

15.1.2 Asymptotic results

To derive the very small buffer saturation probabilities required in ATM systems, we must calculate the tail of the virtual waiting time distribution. Asymptotic formulae constitute an efficient calculation means.

We start with a simple approximate formula, which is an accurate heavy traffic limit. Assume the basic notation of Section 14.2. It is well known that $(\nu(mt) - m\rho t)/\sqrt{m\rho}$ converges weakly to the standard Brownian motion $W(t)$ when $m \to \infty$ (see, e.g., Theorem VIII.3.11 of [JS87]). Now we can apply the general formula (14.2.2) using the approximation

$$\nu(t) \approx \sqrt{\rho} W(t) + \rho t$$

and then using the formula for the probability that the Brownian motion hits a line ([SW86a], formula (2.2.8)):

$$\Pr\left\{\sup_{t\geq 0}\frac{W(t)}{at+b}\geq 1\right\}=e^{-2ab}.$$

Let us deduce heuristically a heavy traffic limit theorem when $\rho=\rho_n$ approaches one in such a way that there exists a finite positive limit

$$a=\lim_{n\to\infty}\sqrt{n}(1-\rho_n).$$

Now

$$\Pr\left\{\frac{1}{\sqrt{n}}V^{(n)}>x\right\}=\Pr\left\{\sup_{t\geq 0}\left[\frac{1}{\sqrt{n}}(\nu^{(n)}(nt)-\rho_n nt)-(1-\rho_n)\sqrt{nt}\right]>x\right\}$$

$$\xrightarrow{\ \ }_{n\to\infty}\Pr\left\{\sup_{t\geq 0}(W(t)-at)>x\right\}=e^{-2at}.$$

This suggests the approximation

$$Q(x)\approx e^{-2(1-\rho)x}.\qquad(15.1.4)$$

For an exact derivation of a heavy traffic limit theorem for the $GI/G/1$ queue see for example [Asm87]. The waiting time is thus approximately exponentially distributed for large x and heavy traffic. Note also that this distribution has the correct mean value (15.1.2).

A more accurate exponential asymptotic can be obtained by considering the Laplace transform of the complementary distribution function of V. Let

$$F(r)=\int_0^\infty e^{-rx}\Pr\{V>x\}\,dx=\frac{1-V^*(r)}{r}=\frac{1}{r}-\frac{1-\rho}{D(r)},\qquad(15.1.5)$$

where $D(r)=r-\rho(1-e^{-r})$. Our aim is to show that

$$Q(x)\sim C_0 e^{-r_0 x}\qquad(15.1.6)$$

for large x, and to determine r_0 and C_0.

Let us first consider a heuristic argument. Assume that (15.1.6) holds for large x. Fixing a large number X and writing

$$F(r)\sim\int_0^X e^{-rx}Q(x)\,dx+\int_X^\infty e^{-rx}C_0 e^{-r_0 x}\,dx$$

$$=\int_0^X e^{-rx}Q(x)\,dx+\frac{C_0}{r+r_0}e^{-(r+r_0)X}$$

we see that when r approaches $-r_0$ from above, then $F(r)$ grows to infinity as $C_0/(r+r_0)$. Comparing this with (15.1.5), we see that $-r_0$ can be characterized as the (unique) negative zero of $D(r)$. Moreover, by developing $D(r)$

as a Taylor series around $-r_0$ it can be seen that C_0 is determined by the derivative of $D(r)$ at $-r_0$, and we have

$$\rho(e^{r_0} - 1) - r_0 = 0 \qquad (15.1.7)$$

$$C_0 = -\frac{1-\rho}{D'(-r_0)} = \frac{1-\rho}{\rho e^{r_0} - 1}.$$

A proof that (15.1.6) holds with r_0 and C_0 given by (15.1.7) can be obtained applying methods of the theory of complex functions. It is known (see [Doe55], p. 110) that the behaviour of a Laplace inverse at infinity is determined by the poles of the Laplace transform with smallest module. On the other hand, it can be shown, using Rouché's theorem, that $-r_0$ is the only singularity of $F(r)$ in the complex plane. We omit the details. The equation for ρ can also be obtained with the large deviations approach described at the end of Subsection 14.3.3.

In the heavy traffic case, where ρ is close to 1, r_0 is close to 0, and we can approximate in (15.1.7) $e^{r_0} \approx 1 + r_0 + r_0^2/2$. Solving the resulting second order equation gives $r_0 \approx 2(1-\rho)/\rho$, which is close to the expression appearing in (15.1.4).

Table 15.1.1 gives an impression of the accuracy of (15.1.6) with different traffic intensities. The exact values are calculated with (15.1.3), which becomes computationally heavy when the load ρ increases. Note that the approximation is very accurate even for small values of x.

Table 15.1.1: The accuracy of the asymptotic formula (15.1.6).

x	$\rho = 0.1$		$\rho = 0.5$		$\rho = 0.9$	
	(15.1.6)	(15.1.3)	(15.1.6)	(15.1.3)	(15.1.6)	(15.1.3)
1	8.92e-03	5.35e-03	1.89e-01	1.76e-01	7.59e-01	7.54e-01
2	2.40e-04	2.03e-04	5.36e-02	5.30e-02	6.17e-01	6.16e-01
3	6.47e-06	6.30e-06	1.53e-02	1.53e-02	5.01e-01	5.01e-01
4	1.74e-07	1.76e-07	4.34e-03	4.34e-03	4.08e-01	4.08e-01
5	4.69e-09	4.73e-09	1.24e-03	1.24e-03	3.31e-01	3.31e-01
6	1.16e-10	1.26e-10	3.52e-04	3.52e-04	2.69e-01	2.69e-01

15.1.3   The case of a finite buffer

Let us now consider the queueing system $M/D/1/K$ with K queueing positions (not including the server). As before, let the service time be 1, and denote the arrival intensity by λ. (Note that now the server load is strictly smaller than λ because of customer losses.) The number of customers in the system is referred to as the state of the system. The set of possible states is then $\{0, 1, \ldots, K+1\}$. The integral equation method of [Sys86] applies also in this case, resulting in a formula for the virtual waiting time distribution. Here we present another method, providing equations which express the steady state probabilities through the corresponding probabilities in the infinite buffer case.

Let $P_d^\infty(j)$ (respectively $P_d^K(j)$) be the probability that a departing customer leaves exactly j customers in the $M/D/1$ queue (respectively in the $M/D/1/K$ queue). In fact, $P_d^\infty(j)$ is the same as the steady state distribution $P^\infty(j)$ in the infinite buffer case. Now, since the first $K+1$ equations governing the embedded Markov chains for the infinite and finite buffer systems are identical, it can be shown (see [GH85], page 280) that

$$
P_d^K(j) = \frac{P_d^\infty(j)}{\displaystyle\sum_{i=0}^{K} P_d^\infty(i)} = \frac{P^\infty(j)}{\displaystyle\sum_{i=0}^{K} P^\infty(i)} \quad \text{for } j = 0, \ldots, K.
$$

Further, it is known that the entering customer's distribution $P_e^K(j)$ is the same as $P_d^K(j)$. The entering customer's distribution can also be obtained from the arriving customer's distribution $P^K(j)$, which for Poissonian input is the same as the steady state distribution, by conditioning on the event that the arriving customer is not rejected:

$$
P_e^K(j) = \frac{P^K(j)}{1 - P^K(K+1)}. \tag{15.1.8}
$$

One more equation is obtained from the trivial fact that offered traffic equals carried traffic plus rejected traffic:

$$
\lambda = 1 - P^K(0) + \lambda P^K(K+1). \tag{15.1.9}
$$

Now it is easy to deduce from (15.1.8) and (15.1.9) that

$$
P^K(j) = \frac{P_e^K(j)}{\lambda + P_e^K(0)}, \quad j = 0, \ldots, K,
$$

$$
P^K(K+1) = 1 - \frac{1}{\lambda + P_e^K(0)}.
$$

15.1.4   The $Geo^N/D/1$ Queue

We consider the discrete time queueing system $Geo^N/D/1$, where a set of N Bernoulli sources can generate a number of cell arrivals between 0 and N in every time slot, and where at most one departure from the queue may occur per time slot. The number of cells X_k in the system at an arbitrary time slot k (including the currently transmitted cell) evolves according to the following relationship

$$
X_{k+1} = (X_k - 1)^+ + b_k,
$$

where b_k designates the number of arrivals generated in time slot k and the expression $(\cdot)^+$ returns the maximum value between 0 and the argument. We consider then that each of the Bernoulli sources generates a cell in an arbitrary time slot k with probability p and no cell with probability $1 - p$. The number b_k of arrivals at an arbitrary time slot k follows therefore a binomial distribution $\text{Bin}(N, p)$ as follows,

$$\Pr\{b_k = i\} = \binom{N}{i} p^i (1 - p)^{(N-i)}.$$

The offered traffic is clearly $\rho = Np < 1$. A number of results can be obtained by using the generating function approach.

Number of cells in system

The probability generating function (pgf) $X(z)$ of the number of cells in the system equals

$$X(z) = \frac{(1 - \rho)(z - 1)B(z)}{z - B(z)},$$

where $B(z)$ corresponds to the pgf of the number of cell arrivals per slot, $B(z) = (1 - p + pz)^N$. The average number of cells in the system can be obtained by derivation and yields

$$\mathrm{E}\{X\} = \rho + \frac{\rho^2}{2(1 - \rho)}\left(1 - \frac{1}{N}\right).$$

Cell delay

As the service time is constant, expressions for the cell delay (time in system) can also be obtained. The pgf for the cell delay reads (cf. (14.1.5))

$$W(z) = \frac{1}{\rho}(X(z) - (1 - \rho)) = \frac{1 - \rho}{\rho}\frac{z\,(B(z) - 1)}{z - B(z)}.$$

Busy and idle periods

The busy period probability mass function (pmf) for a set of N i.i.d. Bernoulli sources can be computed using Takács result [Tak67]. The busy period B is defined as a contiguous time period delimited by two successive instants at which the system goes from empty state to non-empty state and then returns from non-empty state to empty state. The probability $P_B(k)$ that a busy period lasts k slot times equals that of having $k - j$ more arrivals in the cumulated service time for k cells (i.e., k slot times) and an empty system at the end, given that the busy period started with the arrival of any possible number j of cells between 1 and N. Let $P_B(k, j)$ be the probability that the busy period lasts k slot times given that it started with j cell arrivals in the first time slot. Takács result can be then written as

$$P_B(k,j) = \frac{j}{k} \Pr\{k - j \text{ arrivals in } k \text{ slot times}\}, \quad 1 \leq j \leq N.$$

The probability $P_B(k)$ that a busy period lasts k slots is then

$$P_B(k) = \sum_{j=1}^{N} \Pr\{\text{busy period starts with } j \text{ cells}\} P_B(k,j).$$

Next, for a superposition of Bernoulli sources, the number of arrivals i in the first slot of a busy period is independent of the number of arrivals in previous slots and follows a truncated binomial distribution,

$$\Pr\{b_i = j | b_i > 0\} = \frac{1}{1 - (1-p)^N} \binom{N}{j} p^j (1-p)^{N-j}, \quad 1 \leq j \leq N.$$

We can express $P_B(k,j)$ as follows

$$P_B(k,j) = \frac{j}{k} \binom{kN}{k-j} p^{k-j} (1-p)^{kN-k+j}, \quad 1 \leq j \leq N.$$

With these results, the busy period pmf can be expressed as follows

$$P_B(k) = \frac{1}{k[1 - (1-p)^N]} \sum_{j=1}^{\min(N,k)} j \binom{N}{j} \binom{kN}{k-j} p^k (1-p)^{kN+N-k}, \quad 1 \leq k,$$

which can be simplified to

$$P_B(k) = \frac{p}{k(1-p_0)} \text{bin}(N(k+1), p; k-1), \quad 1 \leq k,$$

where $\text{bin}(N, p; k)$ is the binomial pmf with parameters N and p evaluated at point k,

$$\text{bin}(N, p; k) = \binom{N}{k} p^k (1-p)^{N-k},$$

and p_0 is defined as $p_0 = (1-p)^N$, the probability of no arrivals in an arbitrary time slot. The expectation of the busy period duration is

$$E\{B\} = \frac{p}{(1-p)(1-p_0)}.$$

The probability $P_I(k)$ of having an idle period of k consecutive time slots is equal to that of having no arrivals during $k-1$ slot times and at least one arrival in time slot k. The idle period distribution is therefore geometric and the pmf $P_I(k)$ can be expressed as

$$P_I(k) = p_0^{k-1}(1 - p_0), \quad 1 \leq k.$$

The expectation of the idle period duration is

$$E\{I\} = (1 - p_0)^{-1}.$$

15.2   The $N*D/D/1$ queue

The $N*D/D/1$ system is a basic model for ATM traffic and has received considerable attention in the literature. The input process consists of a superposition of N independent periodic sources with the same period D. We are interested in *ensemble* distributions where the phases of the different sources are chosen at random. The waiting room is assumed to be infinite. Though the assumption of identical periods is a severe limitation for modelling the traffic in a multiservice network, this model, in general, represents an improvement with respect to the simple $M/D/1$ model: the periodic nature of cell emissions (in the cell time scale when burst composition is fixed) is explicitly taken into account. Whereas arrivals in the $M/D/1$ system are completely uncorrelated, successive interarrival times are negatively correlated in the $N*D/D/1$ system. Because of the more orderly behaviour of the arrival process, one can expect queues arising to be shorter than in an $M/D/1$ system with the same mean arrival rate.

An algorithmic solution to the $N*D/D/1$ queueing problem was given by Eckberg [Eck79]. An exact derivation of the number of customers in system and of the virtual waiting time was provided in [DGBP83] and [Gra84]. An closed form solution for the queue length distribution was reported in [RV91] and generalized for the virtual waiting time in [NRSV91]. The same result has been independently obtained by Bhargava et al. [BHH89].

Our presentation here mainly follows [NRSV91]. We first derive the closed form solution for the virtual waiting time distribution. Then we consider an approximation scheme based on the Brownian bridge model which leads to a particularly simple analytic expression valid in the heavy traffic case. The output process from an $N*D/D/1$ system is then studied and the idle and busy period distributionas are derived. Finally, a generalization of the exact solution to the case of on/off modulated streams is presented.

15.2.1   Exact solution

Consider the unfinished work, V_D, in the system at an arbitrary instant which we take to be time D (this device simplifies notation somewhat). We make the following basic observation: if $N < D$, then for any realization of the arrival phases there is necessarily some instant in $[0, D)$ at which the system is empty. Thus, V_D depends only on the arrivals after such an instant and it has the same value as in a system in which the arrivals before time 0 are "switched off". This is illustrated in Fig. 15.2.1 by systems A and B.

Now we can apply the general result (14.2.5) for constant service time systems to the system of type B in which the arrival process consists uniquely of the N arrivals uniformly distributed over $[0, D)$

$$Q(x) = \Pr\{V_D > x\} \tag{15.2.1}$$

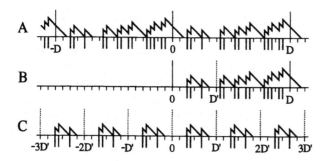

Figure 15.2.1: The evolution of unfinished work in the original $N*D/D/1$ system (A) and modified systems (B and C) for the case $N = 14$ and $D = 16$. The arrival instants of a particular realization are indicated below the time axis.

$$= \sum_{x < n \le N} \Pr\{\nu(D', D) = n\} \cdot \Pr[V_{D'} = 0 \mid \nu(D', D) = n],$$

where $D' = D - n + x$ and $\nu(D', D)$ is the number of arrivals in (D', D). The first probability is binomial:

$$\Pr\{\nu(D', D) = n\} = \binom{N}{n}\left(\frac{n-x}{D}\right)^n\left(1 - \frac{n-x}{D}\right)^{N-n}.$$

To calculate the second, we reverse the above truncation argument: given $\nu(D', D) = n$, $V_{D'}$ of system B depends only on the $N - n$ uniformly distributed arrivals in the interval $[0, D')$. We can now construct system C as shown in Fig. 15.2.1 where the arrivals in $[0, D')$ are extended to a periodic process with period D'. Since the conditions $N < D$ and $\nu(D', D) = n$ imply that $N - n < D'$, system C is always empty at some instant in $[0, D')$ and gives rise to the same value of $V_{D'}$ as system B. As D' is an arbitrary instant with respect to the arrival process of system C, we have $\Pr\{V_{D'} = 0\} = 1 - \rho'$ where $\rho' = (N - n)/D'$ is the corresponding server load. We end up with

$$Q_D^N(x) = \sum_{x < n \le N} \binom{N}{n}\left(\frac{n-x}{D}\right)^n\left(1 - \frac{n-x}{D}\right)^{N-n}\frac{D - N + x}{D - n + x}. \quad (15.2.2)$$

The indices N and D have been added in order to emphasize the dependence on the parameters of the arrival process. As noted in subsection 14.2.2, the complementary distribution of the number in system is given by formula (15.2.2) at integer values of x. It can also be shown that the *virtual* waiting time of (15.2.2) is the same as the *real* waiting time in an $(N + 1) * D/D/1$ system for $N + 1 < D$ [Eck79]. Finally, we note that though (15.2.2) was derived assuming $N < D$, it also gives a solution for the case $N = D$. In this case the queue is indeterminate because the "base level" can be chosen arbitrarily. However, we can remove this ambiguity by requiring that $V_t = 0$ at least at one point in $[0, D)$, i.e., saturation is obtained as a limiting case by approaching it from below. Eq. (15.2.2) corresponds to this limit when $N = D$.

As a matter of fact, we have two different interpretations for expression (15.2.2). On the one hand, it is the *stationary* complementary virtual waiting time distribution of the $N*D/D/1$ system. On the other hand, it represents the complementary ensemble distribution at time D of a *transient* system starting from an empty state at time 0 and receiving N uniformly distributed arrivals in $[0, D)$. The latter interpretation lets itself be extended to the case $N > D$. We readily infer that if $N > D + x$ then necessarily $V_D > x$ and consequently

$$Q_D^N(x) = 1 \qquad \text{for } x < N - D. \tag{15.2.3}$$

For values $x \geq N - D$, the derivation leading to (15.2.2) remains valid. Thus, (15.2.2) together with (15.2.3) give the general distribution of the transient system for any value of N. We choose to *define* $Q_D^N(x)$ by Eqs. (15.2.2) and (15.2.3) with the side remark that for $N \leq D$ it represents the stationary distribution of the $N*D/D/1$ system.

Fig. 15.2.2 shows a sample of numerical results calculated assuming a fixed multiplex load of $\rho = N/D = 0.95$, for a number of sources between 50 and 1000. For comparison, the distribution of an $M/D/1$ queue is also shown. We see that the number of sources is an important determinant of the distribution and the number of required buffer places (approximately determined by the tail probability of the distribution). The Poisson approximation is found to be far too high. For instance, a tail probability $\epsilon = 10^{-10}$ would correspond to some 230 buffer places in an $M/D/1$ system compared to just 84 for a superposition of 1000 deterministic sources. For lower loads, however, the Poisson approximation becomes more accurate.

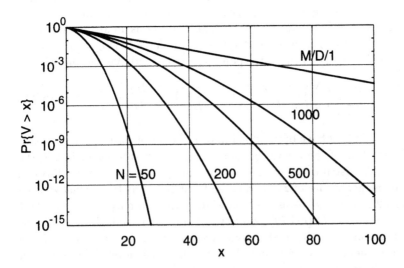

Figure 15.2.2: Distribution $Q_D^N(x)$ for $N*D/D/1$ queues.

15.2.2 Brownian bridge approximation

Even though the exact solution obtained above is rather simple, it is desirable to find a still simpler analytic expression which makes the general behaviour of the virtual waiting time distribution more transparent. Such an expression can be obtained by noting that $\nu(t)/N$ is the empirical distribution function of the uniform distribution on $[0, D]$ (we revert here to the notation of Section 14.2, $\nu(t)$ being the number of arrivals in $[-t, 0)$). We can then approximate the process $\nu(t) - (t/D)N$ by $\sqrt{N} \cdot B(t/D)$ where $B(\cdot)$ is a Brownian bridge, i.e., the standard Brownian motion conditioned on the event $B(1) = 0$ [Don52]. It is well known that

$$\Pr\left\{ \sup_{0 \le u \le 1} (B(u) - au) > x \right\} = \exp\left(- 2x(x + a) \right)$$

(see [SW86a], excercise 2.2.4 and formula (2.2.8)). A heavy traffic limit result can be derived as follows [NRSV91, DRS91]. Let $V^{(N)}$ denote the virtual waiting time in the $N * D/D/1$ system with N sources with period D_N, and assume that $D = D_N$ grows with N in such a way that there is a positive limit

$$\lim_{N \to \infty} \sqrt{N}\left(1 - \frac{N}{D_N}\right) = a.$$

Then

$$\Pr\left\{ \frac{1}{\sqrt{N}} V^{(N)} > x \right\} =$$

$$\Pr\left\{ \sup_{t \in [0,1]} \left[\sqrt{N}\left(\frac{1}{N}\nu^{(N)}(D_N t) - t\right) - \sqrt{N}\frac{D_N}{N}\left(1 - \frac{N}{D_N}\right)t \right] > x \right\}$$

$$\longrightarrow_{N \to \infty} \Pr\left\{ \sup_{t \in [0,1]} (B(t) - at) > x \right\} = e^{-2x(x+a)}.$$

This suggests the approximation

$$Q(x) \approx \exp\left\{ - 2x\left(\frac{x}{N} + 1 - \rho\right) \right\}, \tag{15.2.4}$$

where $\rho = N/D$. For the limiting case $N = D$, we deduce the formulae

$$E\{V\} \approx \sqrt{\frac{\pi N}{8}}; \qquad \text{Var}\{V\} \approx \frac{4 - \pi}{8}N. \tag{15.2.5}$$

It is interesting to compare Eq. (15.2.4) with the corresponding result (15.1.4) for the $M/D/1$ system obtained with the Brownian motion approximation. The exponent of (15.2.4) contains one term linear in x and one term quadratic in x. The linear term is identical to the exponent of (15.1.4).

Thus, for small values of x the queues of these two systems behave in a similar way. For large values of x, however, the quadratic term in (15.2.4) becomes dominant and makes the probability of long queues much smaller. This is in accordance with the fact that the superposition of a large number of periodic sources looks locally Poissonian but, over longer intervals, the fluctuations in the number of arrivals is strongly constrained by the periodicity of the sources.

Table 15.2.1: Accuracy of Brownian Bridge approximation.

x	$\rho = 1.0$		$\rho = 0.8$	
	Exact	Approx.	Exact	Approx.
1	9.73e-01	9.80e-01	5.39e-01	6.53e-01
10	1.27e-01	1.35e-01	1.21e-03	1.50e-03
20	2.78e-04	3.36e-04	1.74e-08	1.52e-08
30	8.86e-09	1.52e-08	1.66e-15	1.04e-15

In Table 15.2.1 we compare exact and asymptotic results for a superposition of streams of period 100 for loads $\rho = 1$ and $\rho = 0.8$. The Brownian bridge approximation is seen to be quite good in these high load cases.

15.2.3   Idle and busy periods

In the previous subsections we have focused on the queue length distribution of a single $N * D/D/1$ system. In real networks where several ATM nodes are connected in tandem and an intermediate node receives its input streams from the outputs of previous stages it is of interest to consider also the output process of an $N * D/D/1$ system. Two of the main characteristics of the output process are its idle and busy period distributions. These have been considered in [DGBP83, PG93b, Pet94, Vir94]. A numerical analysis for a slotted time system with a finite buffer capacity was given in [Hüb91]. Here we give results for the continuous time system with infinite buffer capacity.

First consider the duration I of an idle period. Its complementary distribution $Q_I(x) = \Pr\{I \geq x\}$ (note that equality is allowed by this definition) can be derived by simple probabilistic arguments [Vir94],

$$Q_I(x) = \left(1 - \frac{x}{D}\right)^{N-2} \cdot \left(1 - \frac{x}{D-N+1}\right).$$

This has a positive value for the maximum length of an idle period, $x = D-N$, i.e. this maximum length occurs with a non-zero probability. The average length of an idle period can also be easily derived

$$\mathrm{E}\{I\} = \frac{D}{N} \cdot \frac{D-N}{D-N+1}.$$

Now consider the length B of a busy period. Because of the constant service time B must be an integer (times the service time) and the task is to find the point probabilities $P_B(l) = \Pr\{B = l\}$. We briefly scetch the derivation in [Vir94]. Because of stationarity, the probability is unaltered by conditioning it to a particular start time of the busy period and we can calculate $\Pr[\,B = l \mid \text{busy period starts at time } 0\,]$. In order to characterize such a busy period define four events as follows (see Figure 15.2.3): A) system is empty just before time 0, B) there is an arrival at time 0, C) there are $l - 1$ additional arrivals (uniformly distributed) in the interval $(0, l)$, and D) system is (becomes) empty at time l. With the aid of these events the conditional probability can be written as

$$P_B(l) = \Pr[\,C \cap D \mid A \cap B\,] = \frac{\Pr[C \mid B] \cdot \Pr[A \cap D \mid B \cap C]}{\Pr[A \mid B]}. \qquad (15.2.6)$$

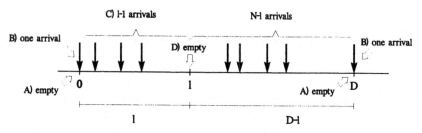

Figure 15.2.3: Events characterizing a busy period

Each of the probabilities in the latter form of (15.2.6) can be determined. First consider the probability $\Pr[A \mid B]$ that the system is empty just before time 0, or, by the periodicity, just before time D, given that there is an arrival at time 0. Note that the conditioning arrival at time 0 does not affect the queue length at the end of the period (for a stable system with $N < D$) but the latter is solely due to the $N - 1$ uniformly distributed arrivals in D. The probability that the system is empty at the end of the period can then be calculated in the same way as for system C in Figure 15.2.1 leading to the "$1 - \rho$" result,

$$\Pr[A \mid B] = 1 - \frac{N - 1}{D}.$$

Second, the probability $\Pr[C \mid B]$ that $l - 1$ additional arrivals occur in the interval $(0, l)$ given that one arrival has occurred at time 0, is given by the binomial distribution

$$\Pr[C \mid B] = \binom{N - 1}{l - 1} \cdot \left(\frac{l}{D}\right)^{l-1} \cdot \left(1 - \frac{l}{D}\right)^{N-l}.$$

Finally, the probability $\Pr[A \cap D \mid B \cap C]$ that the system is empty at time l and just before time 0 (time D) given one arrival at time 0 and $l - 1$ of the remaining $N - 1$ arrivals occurring in the interval $[0, l)$ is

$$\Pr[A \cap D \mid B \cap C] = \left(1 - \frac{N - l}{D - l}\right) \cdot \left(1 - \frac{l - 1}{l}\right) = \frac{D - N}{D - l} \cdot \frac{1}{l}.$$

The first factor represents the empty probability just before time D for a queue solely due to $N - l$ uniformly distributed arrivals in (l, D) and the second factor gives the empty probability at time l for a queue solely due to $l - 1$ uniformly distributed arrivals in $(0, l)$ (cf. [Vir94] for details).

Collecting these results, one obtains the discrete distribution

$$P_B(l) = \binom{N - 1}{l - 1} \cdot \frac{D - N}{D - N + 1} \cdot \frac{l^{l-2} (D - l)^{N-l-1}}{D^{N-2}}, \quad l = 1, \ldots, N$$

with mean

$$\mathrm{E}\{B\} = \frac{D}{D - N + 1}.$$

15.2.4   The modulated $N*D/D/1$ queue

The problems arising from modulated input processes generally belong under the heading "burst scale delay systems" and will be discussed in Chapter 17. However, there is one case which may be appropriately considered in the present context, viz. the modulated $N*D/D/1$ queue with $N \leq D$. When viewed at the burst level, the input rate is always less than or equal to the link capacity and only a cell scale queue can arise.

Each of the N intermittent sources emits cells at constant intervals D during "bursts" and remains inactive during "silences". A typical pattern of cell emissions is illustrated in Fig. 15.2.4. We choose to define the burst as starting at the instant of the first cell emission and ending D time units after the last cell emission. The burst length is then always an integer multiple of D while a silence can have any positive length. The modulating processes are assumed to be independent and stationary but otherwise they can be quite arbitrary and need not be identical for all streams. Thus, the activity probability p_i, defined as the ratio of the mean burst length to the mean burst-silence cycle, may be different for different streams.

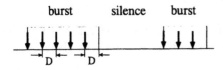

Figure 15.2.4: Bursts of periodic cell emissions in the modulated process.

An important property of the arrival process is that cells from one source can never be closer than D time units apart. Furthermore, since we assumed $N \leq D$, it is clear that the system is empty at some time in any interval of length D. Thus, in order to calculate the virtual waiting time distribution at an arbitrary instant D, we can apply the same method as above: all arrivals before time 0 can be "switched off" without affecting V_D.

The Beneš result (14.2.5) can now be applied to the truncated system. We note that each source can emit at most one cell in the interval $[0, D)$. The emission occurs with probability p_i, the arrival instant being uniformly distributed in the interval. Eq. (14.2.5) can be rewritten by conditioning on the actual number, n, of cells arriving in $[0, D)$:

$$\Pr\{V_D > x\} =$$

$$\sum_{n=0}^{N} \Pr\{\nu(0, D) = n\} \sum_{x < k \le n} \Pr[\nu(D', D) = k \text{ and } V_{D'} = 0 \mid \nu(0, D) = n]$$

with $D' = D - k + x$. The second factor is recognized to be the complementary virtual waiting time distribution, $Q_D^n(r)$, of the $N*D/D/1$ system with $N = n$ sources as given by Eq. (15.2.2). Thus we get

$$\Pr\{V_D > x\} = \sum_{n=0}^{N} P_n \cdot Q_D^n(x), \qquad (15.2.7)$$

with $P_n = \Pr\{\nu(0, D) = n\}$ given by a convolution of the distributions of the N Bernoulli (p_i) variables.

Result (15.2.7), which is exact under the assumptions on the system, states that the virtual waiting time distribution to which the modulated arrival process gives rise is obtained by modulating the virtual waiting time distribution of the basic system. It is a strong case in support of the modulation principle usually introduced heuristically on the basis of a "quasi stationarity" assumption.

Note that the result depends on the modulating processes only through the probabilities P_n and these, in turn, depend solely on the set of activity probabilities $\{p_i\}$ but not on the other details of the modulating processes. For instance, consider the case where the activity probabilities of all the sources are equal $p_i = p$. Then the complementary distribution (15.2.7) is the same as that for the particular realisation of the modulation processes where each source emits cells periodically with period D/p. In this realisation, each burst consists of a single cell, the burst duration is D and the activity probability is clearly p. But this realisation is none other than a $N*D/D/1$ system with N sources and periodicity D/p. Thus we have the simple result

$$\Pr\{V_D > x\} = Q_{D/p}^N(x).$$

15.3   The $\sum D_i/D/1$ queue

In this section we consider a heterogeneous mix of N periodic sources, each characterized by its individual periodicity interval D_i. The load factor $\rho =$

$\sum_i 1/D_i$ is assumed to be less than 1 and the waiting room is assumed to have an infinite capacity. The arrival process is depicted in Fig. 15.3.1. Occasionally, it may be advantageous to consider the traffic mix to be composed of K traffic classes with each of them having a number of sources, N_k, with the same period, D_k. Accordingly, the system is then referred to as the $\sum N_k * D_k/D/1$ system. Obviously, this is just a special case of the $\sum D_i/D/1$ system. An extension to batch arrivals, i.e., the system $\sum N_k D_k^{X_k}/D/1$ queue, is considered in subsection 15.3.2

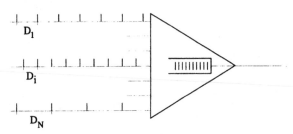

Figure 15.3.1: The $\sum D_i/D/1$ system.

In a restricted case we can immediately write down the exact solution to the $\sum D_i/D/1$ queue based on the results of the previous section on the modulated $N*D/D/1$ queue. Denote by D the shortest of the periods, $D = \min\{D_i\}$. Any of the streams can be considered as a modulated D stream with alternating "bursts" and "silences" of lengths D and D_i-D, respectively, each "burst" consisting solely of one cell. The activity probability of stream i is thus

$$p_i = D/D_i, \qquad i = 1, \ldots, N. \tag{15.3.1}$$

If $N \le D$ the modulated queue result (15.2.7) is directly applicable. So, we only have to calculate the distribution of the number of active "bursts", P_n, by convolving the N Bernoulli (p_i) distributions, and then "modulate" the queue length distribution of the $N*D/D/1$ system with $N = n$ sources by P_n.

The applicability of the exact results discussed above is rather limited. In general, we have to resort to an approximate analysis. The analysis presented in [VR89] turns out to give a very tight upper bound with a reasonably simple algorithm. We derive this bound in the next subsection by a slightly different analysis based on the general approximation (14.2.6).

15.3.1   An approximate analysis - a tight upper bound

It is easy to see that in the specific case of a superposition of N streams we can rewrite (14.2.6) as

$$Q(x) \approx \sum_{n>x} \left\{ \Pr\left\{\nu(n-x) = n\right\} \right. \tag{15.3.2}$$

$$-\sum_{j=1}^{N} \rho_j \cdot \Pr\left[\nu(n-x) = n \mid \text{one arrival at} - (n-x) \text{ from stream } j\right]\Bigg\},$$

where ρ_j is the intensity of the jth stream ($1/D_j$ in the present case). Our task is then to evaluate the probabilities

$$P(x, n) = \Pr\{\nu(n-x) = n\}, \tag{15.3.3}$$

$$P_j(x, n) = \Pr\left[\nu(n-x) = n \mid \text{one arrival at} - (n-x) \text{ from stream } j\right].$$

Let us first focus on $P(x, n)$ and write

$$\nu(n-x) = \sum_{i=1}^{N} \nu_i(n-x), \tag{15.3.4}$$

where $\nu_i(n-x)$ is the number of arrivals in $[-(n-x), 0)$ from stream i. Now, $\nu_i(n-x)$ can be decomposed into a deterministic part and a stochastic part (cf. Figure 15.3.2):

$$\nu_i(n-x) = d_i(n-x) + K_i(n-x), \tag{15.3.5}$$

where

$$d_i(n-x) = \left\lfloor \frac{n-x}{D_i} \right\rfloor,$$

and $K_i(n-x) \in \{0, 1\}$ is a Bernoulli random variable which is 1 with probability

$$p_i(n-x) = \frac{n-x}{D_i} - \left\lfloor \frac{n-x}{D_i} \right\rfloor$$

and zero otherwise. Correspondingly, we have

$$\nu(n-x) = d(n-x) + K(n-x) \tag{15.3.6}$$

with $d(n-x) = \sum_i d_i(n-x)$ and $K(n-x) = \sum_i K_i(n-x)$ standing for the deterministic and stochastic parts, and $P(x, n)$ becomes

$$P(x, n) = \Pr\{K(n-x) = n - d(n-x)\}. \tag{15.3.7}$$

Now, looking at the definition (15.3.3) of $P_j(x, n)$, we see that the condition there excludes stream j from contributing to the stochastic part, i.e., we have $K_i(n-x) \equiv 0$ for $i = j$. Otherwise, the reasoning leading to (15.3.7) remains unaltered and we have

$$P_j(x, n) = \Pr\left\{K^{(j)}(n-x) = n - d(n-x)\right\}, \tag{15.3.8}$$

where

$$K^{(j)}(n-x) = \sum_{i \neq j} K_i(n-x).$$

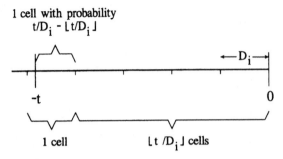

1 cell with probability
$t/D_i - \lfloor t/D_i \rfloor$

Figure 15.3.2: Number of cell arrivals in (-t,0).

The large deviation approximation introduced in subsection 14.3.2 can be used for the estimation of probabilities (15.3.7) and (15.3.8). Focusing again first on $P(n, x)$, i.e., on the distribution of $K(n - x)$, the large deviation approximation (14.3.16) gives (for brevity we suppress the argument $n - x$ everywhere)

$$P(x, n) \approx \frac{\psi(z_n)}{z_n^{n-d}} \cdot \frac{1}{\sqrt{2\pi}\sigma(z_n)}, \qquad (15.3.9)$$

where we have denoted $z = e^\beta$, $z_n = e^{\beta^*}$ and

$$\psi(z) = \mathrm{E}\left\{z^K\right\},$$

and $\sigma(z)$ stands for the standard deviation of K with respect to the transformed distribution. The value of z_n is determined in such a way that the transformed mean $m(z_n)$ equals $n-d$. As K is a sum of independent Bernoulli random variables, we have (cf. equation (14.3.1))

$$\psi(z) = \prod_i (1 - p_i + p_i z),$$

$$m(z) = \sum_i \frac{p_i z}{1 - p_i + p_i z}, \qquad (15.3.10)$$

$$\sigma^2(z) = \sum_i \frac{p_i(1 - p_i)z}{(1 - p_i + p_i z)^2},$$

and, to reiterate, z_n is determined from

$$\sum_i \frac{p_i z_n}{1 - p_i + p_i z_n} = n - d. \qquad (15.3.11)$$

The calculation of $P_j(x, n)$ proceeds along the same lines. We have a result of the form (15.3.9)

$$P_j(x, n) = \frac{\psi^{(j)}(z_n^{(j)})}{(z_n^{(j)})^{n-d}} \cdot \frac{1}{\sqrt{2\pi}\sigma^{(j)}(z_n^{(j)})}, \qquad (15.3.12)$$

where the superscripted quantities now are

$$\psi^{(j)}(z) = \prod_{i \neq j}(1 - p_i + p_i z) = \frac{\psi(z)}{1 - p_j + p_j z},$$

$$m^{(j)}(z) = \sum_{i \neq j} \frac{p_i z}{1 - p_i + p_i z}, \tag{15.3.13}$$

$$\sigma^{(j)2}(z) = \sum_{i \neq j} \frac{p_i(1 - p_i)z}{(1 - p_i + p_i z)^2},$$

and $z_n^{(j)}$ is determined from

$$\sum_{i \neq j} \frac{p_i z_n^{(j)}}{1 - p_i + p_i z_n^{(j)}} = n - d. \tag{15.3.14}$$

Before inserting these results into (15.3.2) it is useful to introduce a further approximation. When the number of streams is large, the omission of one term in the expressions for $m^{(j)}(z)$ and $\sigma^{(j)2}(z)$ in (15.3.13) alters the values of the sums only slightly. Thus, for large systems we have approximately $m^{(j)}(z) \approx m(z)$, $\sigma^{(j)}(z) \approx \sigma(z)$ and, therefore, $z_n^{(j)} \approx z_n$ for all j. The difference between $\psi^{(j)}(z)$ and $\psi(z)$ in (15.3.13) is, however, important. Now, we can write (15.3.2) in the final form

$$Q(x) \approx \sum_{n > x} \frac{\psi(z_n)}{z_n^{n-d}} \cdot \frac{1}{\sqrt{2\pi}\sigma(z_n)} \cdot \left(1 - \sum_{j=1}^{N} \frac{\rho_j}{1 - p_j + p_j z_n}\right). \tag{15.3.15}$$

(Remember that ψ, σ, d, p_j and z_n are all functions of $n - x$ also). This result was derived in [VR89] using a slightly different approach. As in [VR89], we can conjecture that (15.3.15) in fact gives an upper bound for $Q(x)$. This is because the approximation (14.2.6) was based on the local stationarity assumption. However, the arrival intensity ρ' just before $-(n - x)$, conditioned on n arrivals in the subsequent period $[-(n - x), 0)$, is lower than the arrival intensities before $-(n - x)$ which tend to the mean intensity ρ as we get further from $-(n - x)$. Therefore, $1 - \rho'$ gives an upper bound approximation for the probability of empty system at $-(n - x)$.

To get an idea of the accuracy of (15.3.15) we apply it to an $N*D/D/1$ system for which an exact result is known as we saw in the previous section. A comparison is presented in Table 15.3.1. We see that the accuracy is indeed excellent. A numerical example worked out in [VR89] points to an excellent accuracy also in the case of a truly inhomogeneous traffic mix. It is obvious that the accuracy is more than sufficient for any practical dimensioning purposes.

Table 15.3.1: Exact and approximate complementary virtual waiting time distribution of an $N*D/D/1$ system with $N = 80$ and $D = 100$.

	$Q(x)$	
x	Exact (15.2.2)	Approx. (15.3.15)
10	1.21e-03	1.22e-03
15	8.31e-06	8.34e-06
20	1.74e-08	1.74e-08
25	1.04e-11	1.05e-11
30	1.66e-15	1.67e-15

15.3.2   Extension to batch arrivals – the $\sum_k N_k D_k^{X_k}/D/1$ queue

The analysis of the previous section can be extended to the case of batch arrivals with a fixed batch size. This extension is useful, for instance, in analysing queue performance with worst case traffic compatible with a leaky bucket access controller.

Consider K source classes with cells of class k $(1 \le k \le K)$ arriving in constant batches of size b_k separated by a constant batch inter-arrival time of $D_k b_k$. Batch size b_k is assumed here to be an integer. There are N_k sources of class k and the intensity due to all sources of class k is $\rho_k = N_k/D_k$. Let $N = \sum_k N_k$.

Considering the contribution of the different classes, (15.3.2) may be written:

$$Q(x) \approx \sum_{n>x} \left(P(x,n) - \sum_{k=1}^{K} \rho_k P_k(x,n) \right),$$ (15.3.16)

where

$$P(x,n) = \Pr\{\nu(n-x) = n\},$$

$$P_k(x,n) = \Pr[\nu(n-x) = n \mid$$

one arrival at $-(n-x)$ from a stream of type k].

To evaluate the $P(x,n)$, we proceed as above. First note that, if n is not a linear combination of $(b_k)_{1 \le k \le J}$, then $P(x,n) = 0$. Let $\nu_k(n-x)$ be the number of arrivals in the interval from a stream of class k. We can write $\nu_k(n-x) = b_k (d_i(n-x) + K_k(n-x))$, where $d_k(n-x) = \lfloor (n-x)/D_k b_k \rfloor$ is the deterministic part and $K_k(n-x) \in \{0,1\}$ is a Bernouilli random variable which is 1 with probability

$$p_k(n-x) = \frac{n-x}{D_k b_k} - \left\lfloor \frac{n-x}{D_k b_k} \right\rfloor.$$

Let $d(n - x) = \sum_k N_k b_k d_k(n - x)$ represent the deterministic part of $\nu(n - x)$ and $K(n - x)$ the random part. $K(n - x)$ takes values on the set $\{\sum_k N_k b_k k_k\}$ with $0 \leq k_k \leq N_k$. To estimate the distribution of the latter we use the probability change argument (see Section 14.3.2, equation (14.3.16)). This leads to the approximation:

$$P(x, n) = \Pr\left(K(n - x) = n - d(n - x)\right)$$

$$\approx \frac{\psi(\zeta_{\mathbf{n}})}{\zeta_{\mathbf{n}}^{n-d}} \cdot \frac{\mathrm{GCD}(b_k)}{\sqrt{2\pi}\sigma(\zeta_{\mathbf{n}})}$$

where

$$\psi(z) = \prod_{k=1}^{K} (1 - p_k + p_k z^{b_k})^{N_k},$$

$$m(z) = \sum_{k=1}^{K} \frac{N_k p_k b_k z^{b_k}}{(1 - p_k + p_k z^{b_k})},$$

$$\sigma^2(z) = \sum_{k=1}^{K} \frac{N_k p_k b_k^2 (1 - p_k) z^{b_k}}{(1 - p_k + p_k z^{b_k})^2},$$

and $\zeta_{\mathbf{n}}$ is the solution of $m(z) = n - d$. GCD denotes greatest common divisor.

An approximation for $P_k(x, n)$ can be derived as in the previous section to give finally:

$$Q(x) \approx \sum_{\mathbf{n} > x, \mathbf{n} \in \Omega} \frac{\psi(\zeta_{\mathbf{n}})}{\zeta_{\mathbf{n}}^{n-d}} \cdot \frac{\mathrm{GCD}(b_k)}{\sqrt{2\pi}\sigma(\zeta_{\mathbf{n}})} \left(1 - \sum_{k=1}^{K} \frac{p_k}{1 - p_k + p_k \zeta_{\mathbf{n}}^{b_k}}\right) \qquad (15.3.17)$$

where $\Omega = \{\mathbf{n}, \exists (n_1, ..., n_J) \,|\, n = \sum_k n_k b_k\}$.

Line C in Figure (15.3.3) gives the cell loss probability as a function of the buffer size for a balanced mixture of high rate streams ($D_1 = 10$), with batch size $b_1 = 9$ and low rate streams ($D_2 = 400$) with smaller batch size $b_2 = 2$. We observe that the cell loss probability of the mixture is within the envelope given by the cell loss probabilities obtained with homogeneous streams. For small buffers, the cell loss probability is determined by the high rate streams, while for large buffers it is determined by the low rate streams.

The considered model can be used to study the $M^X + D^X/D/1$ queue. A Poisson stream with rate λ and fixed batch size b is considered as the limit of a large number N_1 of streams with low rate $1/D_1$. We then simply make $D_1 \to \infty$ and $N_1 \to \infty$ such thet $N_1/D_1 \to \lambda$ and use (15.3.2). Note that for the Poisson stream $P_1(x, n) = P(x, n)$ and the deterministic part vanishes.

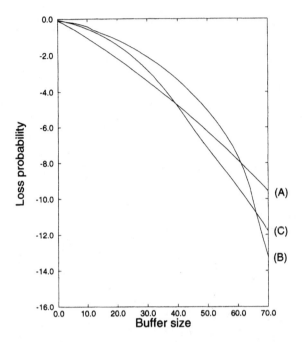

Figure 15.3.3: Loss probability as a function of the buffer size. $\rho = 0.8$. Line (A): homogeneous streams with $D_1 = 400$, $b_1 = 2$, $N_1 = 320$; Line (B): homogeneous streams with $D_1 = 10$, $b_1 = 9$, $N_1 = 8$; Line (C): heterogeneous streams with $D_1 = 10$, $b_1 = 9$, $N_1 = 4$ and $D_2 = 400$, $b_1 = 2$, $N_2 = 160$.

15.4 Multiplexing of a periodic and a batch cell stream

The $M + D/D/1$ queueing model has been applied to study the cell delay variation for a CBR source in an ATM multiplexer. In this section we extend this model and consider a more general case where the input process is the superposition of a general periodic cell stream and a batch arrival background process. This type of model may for instance be used to examine the effect of having different phasing between CBR sources. Another example is the worst case output of a leaky bucket policing device which results in a periodic on/off cell stream. We consider models with both infinite and finite buffers and the main results are the queue length distributions and formulae to calculate cell loss probabilities.

In the analysis the generating function approach is applied. This method usually requires the derivation of some complex roots. It has, however, the advantage that it is applicable both for finite and infinite buffers and therefore the cell loss performance may also be obtained. In addition, asymptotic tail behaviour is usually easy to obtain from the transforms.

We consider a slotted system (with slot length equal to the cell transmission time). The arrival process is the superposition of a periodic stream and a background process. The periodic stream is characterized by a sequence $\{d_k\}$ which is the number of cells arriving from the periodic stream in slot k, and the period T (see Figure 15.4.1), i.e., it is similar to the $N*D$ arrival process but has *fixed phases*. The background process is assumed to be a batch Bernoulli process (B^X), where the batch size in each slot is independent, and with generating function $B(z)$ (e.g. Poisson, Bernoulli or generalized negative binominal).

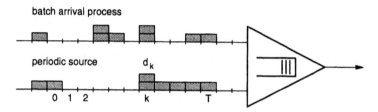

Figure 15.4.1: The queueing model of the ATM multiplexer.

Let $D_k = \sum_{i=1}^{k} d_i$ be the accumulated number of cells arriving from the periodic source up to slot k. To ensure stability, we require $\rho = \rho_P + \rho_B < 1$ where, $\rho_P = D_T/T$ is the mean load from the periodic source and $\rho_B = B'(1)$ is the load from the batch arrival process.

In the analysis of the queueing model above it is useful first to consider the reduced (and deterministic) system where the background stream is switched off. Let

$$U(k, j) = (D_j - D_k) - (j - k - 1)$$

be the excess work entering the reduced system in slots $k + 1, \ldots, j$ $(k < j)$. The queue length (work) at slot k, u_k, may be found (as in section 14.2.1) by taking the supremum of $U(j, k)$ in the backward direction,

$$u_k = \sup_{j < k} U(j, k).$$

Let M be the buffer capacity in the finite buffer case. In the analysis of this case we also apply results for the reduced system served in backward time (reversed order). The queue length (work) at slot k, w_k, for the backward system may be found by taking the supremum of $U(k, j)$ in the forward direction,

$$w_k = \sup_{j > k} U(k, j).$$

If we consider an interval of length T it may be divided into a disjoint union of busy/idle periods for the reduced system denoted B_P and I_P, and also a disjoint union of backward busy/idle periods (for the backward reduced system) denoted BB_P and BI_P (see Figure 15.4.2 where u_k and $M - w_k$ are depicted).

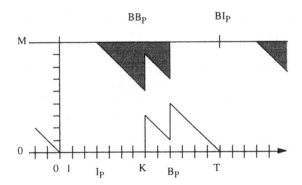

Figure 15.4.2: The busy/idle periods and backward busy/idle periods for the reduced system.

15.4.1 Infinite buffer case

The queue length distribution is found in [Øst94] by using a generating function approach. The complementary distribution of the queue length at the end of slot k, Q_k, is given by the sum

$$Q_k(i) = \Pr\{Q_k > i\} = 1 - \sum_{j=k+1}^{T+k} q_j \Phi(i + U(k,j), j - k), \qquad (15.4.1)$$

where q_j is the probability that the system is empty at the end of slot j (due to the periodicity we obviously have $q_j = q_{j-T}$, $j > T$), and $\Phi(i,j)$ is given by

$$\Phi(i,j) = \frac{1}{i!}\frac{d^i}{dz^i}\left\{\frac{B(z)^{-j}}{1 - z^K B(z)^{-T}}\right\}_{z=0} = \sum_{l=0}^{\lfloor i/K\rfloor} b_{i-Kl}^{-(lT+j)}, \qquad (15.4.2)$$

where b_i^{-k} is the i-th coefficient in the expansion of $B(z)^{-k}$, i.e.

$$b_i^{-k} = \frac{1}{i!}\frac{d^i}{dz^i}\left\{B(z)^{-k}\right\}_{z=0}. \qquad (15.4.3)$$

For Poisson(M), Bernoulli(Geo) and generalized negative binomial (Nb) traffic the coefficients above may be given explicitly (see [Øst95]).

The boundary probabilities q_j can be determined by considering the busy/idle periods of the reduced system and the K $(= T - D_T)$ roots $\{r_1 = 1, r_2, \ldots, r_K\}$ of the complex function $f(z) = z^K - B(z)^T$ inside the unit disk. When $j \in B_P$ then the queue cannot be empty and hence $q_j = 0$. The rest of the boundary probabilities, when $j \in I_P$, are the solution of the following linear system

$$\sum_{j \in I_P} q_j = T(1 - \rho), \qquad (15.4.4)$$

$$\sum_{j \in I_P} q_j \zeta_k^{j-1} r_k^{-D_j} = 0, \qquad k = 2, \ldots, K, \tag{15.4.5}$$

where $\zeta_k = r_k/B(r_k)$.

For some cases of special interest the boundary equations (15.4.4) and (15.4.5) can be solved explicitly. This is the case if the reduced system contains only one single busy/idle period in an interval of length T. We choose the numbering of the axis such that $D_j = 0$; $j = 1, \ldots, K$ (the idle period), and we have $q_j = 0$; $j = K+1, \ldots, T$ (the busy period). In this case the boundary equations (15.4.4) and (15.4.5) will be of Vandermonde type and the polynomial generating the rest of the q_j's may be written as

$$\sum_{j=1}^{K} q_j x^{j-1} = T(1-\rho) \frac{\prod\limits_{k=2}^{K}(x - \zeta_k)}{\prod\limits_{k=2}^{K}(1 - \zeta_k)}. \tag{15.4.6}$$

The coefficients q_j are obtained by the relation between the roots and the coefficients of a polynomial.

Despite the fact that (15.4.1)-(15.4.3) give explicit expressions for the queue length distributions, these formulas are not effecient to perform numerical calculations for large i. This is due to the fact that the function $\Phi(i,j)$ defined by (15.4.2) contains all the roots of $z^K - B(z)^T$ including those with modulus less than unity and hence grows as a power of the inverse of these roots. When i becomes large the expression for $Q_k(i)$ thus becomes numerically unstable.

One way to overcome this problem is to cancel these roots directly by the boundary equations. To do so we also need the roots of $z^K - B(z)^T$ outside the unit circle. Denote these roots by $\{r_{K+1}, r_{K+2}, \ldots\}$ (which in principle can be infinite). We assume that all the roots are distinct. By making partial expansions of the z-transforms (see [Øst94]) we get the complementary queue length distribution as

$$Q_k(i) = \sum_{l=K+1} c_k(r_l) \left(\frac{1}{r_l}\right)^i, \tag{15.4.7}$$

where

$$c_k(z) = -\frac{1}{K - TzB'(z)/B(z)} \sum_{j=k+1}^{T+k} q_j z^{-U(k,j)} B(z)^{k-j}.$$

In the general case the series for $Q_k(i)$ given by (15.4.7) will not be convergent for $i < K$. For numerical calculations it is therefore necessary to use (15.4.1) for small i, (at least for $i < K$) and the partial expansion (15.4.7) for large i ($i \geq K$). The first term in the series for $Q_k(i)$ is the dominant part for large i, and it can be shown that this root, r_{K+1} (with the smallest modulus outside the unit circle), is real.

15.4.2   Finite buffer case

The finite buffer case is analysed by Heiss [Hei92] using an iterative approach to solve the governing equations numerically. However, we take advantage of the analysis for the infinite buffer case. In [Øst94] it is shown that the queue length distribution has the same form, but with slightly different probabilities for an empty system.

We assume that the buffer capacity M is large enough so that the reduced system is without cell loss, i.e., $M > \sup\{u_j\}$. By this assumption the major part of the complementary queue length distribution has the same form as for the infinite buffer case,

$$Q_k^M(i) = 1 - \sum_{j=k+1}^{T+k} q_j^M \Phi(i + U(k, j), j - k), \qquad (15.4.8)$$

valid for $i = 0, 1, \ldots, M - w_k$ (the white area in Figure 15.4.2), and where q_j^M is the probability that the system is empty at the end of slot j.

As for the infinite buffer case we have $q_j^M = 0$ when j lies in a busy period for the reduced system. The rest of the q_j^M's are determined by the fact that (15.4.8) gives the total queue length distribution for the k values in the backward idle periods, and hence

$$\sum_{j=k+1}^{T+k} q_j^M \Phi(M + U(k, j), j - k) = 1 \qquad \text{for } k \in BI_P. \qquad (15.4.9)$$

The K equations (15.4.9) are non-singular and determine the non-zero q_j^M's uniquely. The rest of the $Q_k^M(i)$'s for $i = M - w_k + 1, \ldots, M$ and k laying in a backward busy period may be calculated using the state equations for the system, (see [Øst94]).

Using the cell loss formula (14.1.2) we get

$$P_{loss} = 1 - \frac{1 - \dfrac{1}{T} \displaystyle\sum_{j \in I_P} q_j^M}{\rho}. \qquad (15.4.10)$$

For small buffers (15.4.9) and (15.4.10) are well suited to calculate the coefficients q_j^M. However, to avoid numerical instability for larger buffers we expand (15.4.9) by taking a partial expansion of the function Φ. After a rather lengthy manipulation we end up with a more useful formula for the cell loss probability for large M (see [Øst94] for details),

$$P_{loss} = -\frac{1 - \rho}{\rho} e_1 (I + W)^{-1} W e_1^T, \qquad (15.4.11)$$

where e_1 is the K-dimensional row vector $e_1 = (1, 0, ..., 0)$, and W denotes the $K \times K$ matrix with the elements

$$W(i, j) = \sum_{s=K+1} \left(\frac{r_j}{r_s}\right)^M \frac{KB(r_j) - Tr_jB'(r_j)}{KB(r_s) - Tr_sB'(r_s)} V_i^1(r_s)V_j^2(r_s), \quad (15.4.12)$$

with

$$V_i^1(r_s) = \sum_{k \in I_P} \beta_1(i, k)\zeta_s^{k-1}r_s^{-D_k},$$

$$V_j^2(r_s) = \sum_{k \in BI_P} \beta_2(j, k)\zeta_s^{-k}r_s^{D_k},$$

where β_1 is the inverse of the $K \times K$ matrix $(\zeta_l^{k-1}r_l^{-D_k}); l = 1, ..., K, k \in I_P$, and β_2 is the inverse of the $K \times K$ matrix $(\zeta_l^{-k}r_l^{D_k}); l = 1, ..., K, k \in BI_P$.

For an important subclass of periodic sources explicit expressions for $V_i^1(r_s)$ and $V_j^2(r_s)$ may be obtained. If the reduced system contains only one busy/idle period, and only one backward busy/idle period in an interval of length T, the matrices defined above will be of Vandermonde type. We choose the numbering so that the idle period starts at slot 1, and we let L be the smallest integer in $B_P \cap BI_P$ and set $N = L + K - 1$. We get the following important simplification

$$V_i^1(r_s) = \prod_{K=1, k \neq i}^{K} \frac{\zeta_s - \zeta_k}{\zeta_i - \zeta_k},$$

$$V_j^2(r_s) = \left(\frac{r_j}{r_s}\right)^{-D_N} \left(\frac{\zeta_j}{\zeta_s}\right)^N \prod_{K=1, k \neq j}^{K} \frac{\zeta_s - \zeta_k}{\zeta_j - \zeta_k}.$$

Further simplification is possible if $K = 1$. In this case the periodic source must send exactly $T - 1$ cells in the period T. We get

$$P_{loss} = -\frac{1-\rho}{\rho} \frac{w}{w+1}, \quad (15.4.13)$$

where

$$w = \sum_{s=2} \left(\frac{1}{r_s}\right)^{M-D_L} \left(\frac{1}{\zeta_s}\right)^L \frac{1-\rho}{B(r_s) - Tr_sB'(r_s)}. \quad (15.4.14)$$

The case with only the background stream present, $T = 1$, (and $L = 0$, $D_L = 0$) is also covered by (15.4.13) and (15.4.14).

From (15.4.11) and (15.4.12) it is easy to obtain the large buffer approximation of the cell loss probability. For large M we expand $(I + W)^{-1}$, and by taking only the first term in the expression for W we get the following simple approximate formula for the cell loss probability

$$P_{loss} \approx \alpha \left(\frac{1}{r_{K+1}} \right)^M, \qquad (15.4.15)$$

where

$$\alpha = -\frac{(1-\rho)^2}{\rho} \frac{TV_1^1(r_{K+1})V_1^2(r_{K+1})}{KB(r_{K+1}) - Tr_{K+1}B'(r_{K+1})}.$$

The exact cell loss formula above is numerically applicable for buffers and sources of real interest, i.e., for buffers and sources with a period of several hundred, as well as for cell losses well below the target values for ATM networks. However, the CPU time may become quite long. For dimensioning purposes the simple (and fast) asymptotic formula (15.4.15) will give satisfactory accuracy.

15.4.3 Applications

As a first example we consider how the phasing of CBR sources will affect the cell loss performance. A superposition of CBR sources (with the same period) will result in a variety of different periodic cell streams depending on their mutual phasing. Among all the possible outcomes there are obviously two extreme cases. The worst case occurs when all the sources are fully synchronized resulting in a deterministic batch process, whereas the best case is obtained when the phases are equally spread over one period. All the other ways of merging different CBR sources will, from a performance point of view, fall between these two limits. To study the effect of different phasing we have analysed the two limit cases together with a case we call moderate shaping:

A. Worst case assumption and no shaping; i.e., the streams are fully synchronized.

B. Moderate shaping. The phases of the deterministic sources are shifted one cell place so that the superposition leads to a periodic on/off source.

C. Perfect shaping. The phases of the deterministic sources are spread evenly over the period

We also assume the background stream to be Poisson. We take the period of the CBR sources to be 100. This corresponds to CBR sources of speed 1.5 Mbit/s on a 150 Mbit/s link. The load from CBR sources is taken to be 0.2 and 0.5, respectively, which corresponds to 20 and 50 CBR sources. The buffer capacity is taken to be 100 in the example below. The load on the multiplexer is increased by increasing the load from the background source.

Figure 15.4.3 depicts the cell loss probability for increasing load. The effect of shaping the different CBR sources is visible also for moderate load of the deterministic sources ($= 0.2$). In this case the decrease in cell loss

from case A (worst case) to case C (best case) is almost two decades. For the
case with 50 CBR sources, the decrease from case B to case C is also nearly
two decades. These examples show clearly the negative effect an unfortunate
phasing of CBR sources will have on the cell loss performance in an ATM
network.

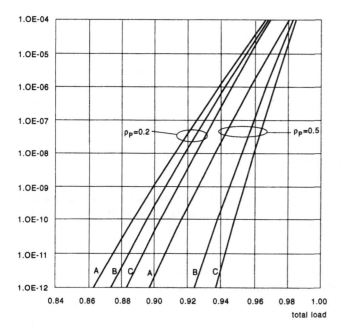

Figure 15.4.3: Cell loss probability as a function of the total load. The buffer
capacity is 100 and the load from the CBR sources is 0.2 and 0.5, respectively.

As a second example we study the effect of multiplexing a periodic on/off
source with a Poisson background stream. The period in this example is also
set to 100 cell times, and we increase the length of the on period (thereby
increasing the load). In Figure 15.4.4 we have used the exact formula to calculate the cell loss probability as a function of the buffer size and background
load, while keeping the total load fixed. From Figure 15.4.4 we see the effect
of increasing the on period. For small buffers the increase results in worse
performance. This is due to the fact that the buffer is not able to absorb
all the cells in an on period (and a part of them is lost). However for larger
buffers this is not the case and it is likely that all the cells in an on period are
buffered without loss. The curves seem to have a common intersection point
at approximately 50. We may say that a multiplexer with small buffers sees
a periodic on/off source as a bursty source while for larger buffers the source
is smooth. This is just what we would expect since on a long time scale the
periodic on/off source has a more regular behaviour than a Poisson stream,
for instance.

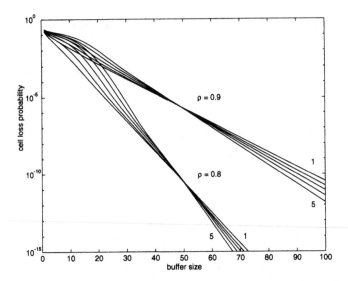

Figure 15.4.4: Cell loss probability as a function of buffer size for different loads of the periodic source. The period is 100 and the total load 0.8 and 0.9. (1) $\rho_P = 0.05$, (2) $\rho_P = 0.10$, (3) $\rho_P = 0.15$, (4) $\rho_P = 0.20$, (5) $\rho_P = 0.25$.

In Figure 15.4.5 we compare the approximate formula for cell loss with exact calculations. The approximation is best when the load from the background stream is high (compared with the load from the periodic source). As the buffer space increases the difference between exact and approximate values decreases rapidly.

15.5 Models for CDV

The objective of this section is to present models which provide a tool for obtaining quantities needed in the characterization of cell delay variation (CDV) in various contexts as described in Chapter 3 of Part I.

First an exact Markovian model to evaluate the CDV introduced on a CBR cell stream in a FIFO multiplex is presented, together with an approximate diffusion model which is computationally much simpler. Secondly, the change of traffic characteristics as a main stream passes a tandem of queues with interfering traffic is modelled by assuming the system to be under heavy load.

15.5.1 CDV in a single FIFO multiplex - the $D + B^X/D/1$ queue

The queueing model applied to describe CDV in a single FIFO multiplexer is a special case of the model presented in section 15.4, i.e., we assume:

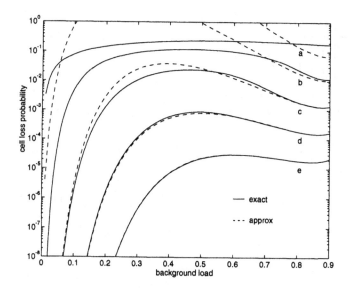

Figure 15.4.5: Exact and approximate cell loss probabilities as a function of the background load for different buffer sizes. The period is 100 and the total load 0.9. (a) $M = 1$, (b) $M = 10$, (c) $M = 20$, (d) $M = 30$, (e) $M = 40$.

- The system is a discrete time system with slot size equal to the transmission time of one cell and an infinite waiting room receiving traffic from two independent cell streams.

- The background stream is batch Bernoulli, i.e., in each timeslot i, arrivals occur in independent and identically distributed batches of size B_i with common distribution $b(k) = \Pr\{B_i = k\}$ and mean $\rho_B = \mathrm{E}\{B_i\}$. The generating function of the distribution $b(k)$ is denoted by $B(z)$.

- The periodic stream generates a single arrival in each period T such that the interarrival time is always equal to T.

- Arrivals of cells are assumed to take place just after slot boundaries while departures are completed just prior to slot boundaries. Multiple arrivals from the background streams are served in random order while CBR arrivals are served first in slots where both a CBR arrival and background arrivals occur. This is referred to as CBR cells having "mini priority" over background cells, but the influence this priority rule has on the results is negligible.

An equivalent approach in continuous time applied to the $M + D/D/1$ queue is presented in [RG92].

With the above assumptions the waiting time of CBR cells entering the multiplexer constitutes a discrete time Markov process W_n with transition

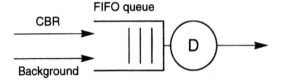

FIFO queue

CBR

D

Background

Figure 15.5.1: A CBR cell stream multiplexed with a background traffic stream

matrix $\mathbf{P}$. Entry (j, k) in $\mathbf{P}$ is defined by $p_{j,k} = \Pr[W_l = k \mid W_{l-1} = j]$ and can be computed as follows:

$$p_{j,k} = P(j, k-1) - P(j, k), \tag{15.5.1}$$

where

$$P(j, k) = \Pr[W_l > k \mid W_{l-1} = j] = \sum_{n \geq 0} P_n(j, k), \tag{15.5.2}$$

with $P_n(j, k) = \Pr[W_l > k \mid W_{l-1} = j$ and n arrivals in $(Tl - T, Tl]] \cdot b^{T*}(n)$. Equation (15.5.2) decomposes $P(j, k)$ according to the number of background arrivals occuring between the two CBR arrivals. $b^{T*}(n)$ is the T-fold convolution of the batch size distribution and, evaluated at n, it represents the probability of n arrivals in $(Tl - T, Tl]$. It can be shown [MB96] that

$$P_n(j, k) = \begin{cases} 0, & \begin{array}{l} j + 1 \geq T \text{ and } n \leq T + k - j - 1 \text{ or} \\ j + 1 < T \text{ and } k \geq n, \end{array} \\ b^{T*}(n), & n > T + k - j - 1, \\ \sum_{s=1}^{n-k} b^{s*}(k + s) \times b^{(T-s)*}(n - k - s) \dfrac{T - n + k}{T - s}, & \\ & j + 1 < T \text{ and } k < n \leq T + k - j - 1 \ . \end{cases} \tag{15.5.3}$$

The stationary waiting time distribution $w_k = \Pr\{W_n = k\}$ can be found from solving the equilibrium system $\mathbf{w} = \mathbf{w}P$ where $\mathbf{w} = (w_0, w_1, ..., w_i, ...)$ and $\mathbf{P} = \{p_{j,k}\}$. In numerical applications, a truncation of the state space is in general necessary.

For the steady state queue length distribution, a closed form expression for the generating function is provided in [BMS93] as well as in the previous section. [BMS93] also provides the generating function of the waiting time for cell $n + 1$ conditioned on the event that the waiting time for cell n is i. From this the entries in the transition matrix can in principle be found. However, this requires not only an inversion but also the determination of $T - 1$ boundary probabilities.

Define $t_n = nT + W_n + 1$. Thus t_n denotes the departure time of cell n. Define *the shifted interdeparture time* of cell n as: $U_n = t_n - nT - t_0$

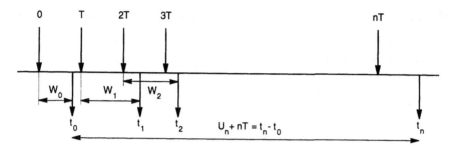

Figure 15.5.2: U_n is the difference between the actual and expected departure time of cell n when only the departure time of cell 0 is known

(see Figure 15.5.2). It is a fundamental random variable from which all CDV characteristics of interest can be derived. Let $f_n(k) = \Pr\{U_n = k\}$ denote the probability distribution of U_n. A straightforward application of the law of total probability yields

$$
f_n(k) = \begin{cases} \displaystyle\sum_{i=0}^{\infty} w_i p_{i,i+k}^{(n)}, & \text{if } k \geq -nT + 1, \\ 0, & \text{otherwise,} \end{cases}
$$

where $p_{j,k}^{(n)}$ denotes entry (j, k) in the nth power of the transition matrix $\mathbf{P} = \{p_{j,k}\}$.

The usual point process characteristics of the CBR departure process can be derived from U_n as illustrated by the examples below.

Interdeparture time distributions

The interdeparture time distribution between cell 0 and n is simply obtained by the translation nT of the shifted interdeparture time distribution U_n i.e.

$$
\Pr\{U_n = k - nT\} = f_n(k - nT).
$$

Index of dispersion for intervals

The index of dispersion for intervals can be obtained as follows:

$$
\frac{n\text{Var}\{t_n - t_0\}}{\text{E}\{t_n - t_0\}^2} = \frac{\text{Var}\{U_n\}}{nT^2} = \frac{\displaystyle\sum_{k=-nT+1}^{\infty} k^2 f_n(k)}{nT^2}
$$

since $E\{U_n\} = 0$.

Number of departures in a window synchronized with a departure

Let $N(t_j, t_j + t)$ denote the number of departures in a window of length t starting just after an arbitrary CBR cell departure at time t_j. The probability distribution is

$$\Pr\{N(t_j, t_j + t) = n\} = \Pr\{U_n + nT \leq t\} - \Pr\{U_{n+1} + (n+1)T \leq t\}$$

$$= \sum_{j \leq -nT+t} f_n(j) - \sum_{j \leq -(n+1)T+t} f_{n+1}(j).$$

Number of departures in an arbitrary window

Based on a basic result in point process theory (see, for example, Section 4.2 in [CL66]) a simpler formula for the distribution of the number of departures in an arbitrary window than that presented in [GR91] can be obtained. Let 0 denote an arbitrary point in time and $N(0, t)$ denote the number of departures in $(0, t]$. Then

$$\Pr\{N(0, t) \geq n\} = \Pr\{Y + X_2 + \ldots + X_n \leq t\}$$

$$= \frac{1}{T} \sum_{u=1}^{t} \sum_{k=1}^{u} (f_{n-1}(k - [n-1]T) - f_n(k - nT)),$$

in which Y denotes the forward recurrence time and X_i denotes the interdeparture time between cell $i - 1$ and cell i. From (15.5.4) we obtain

$$\Pr\{N(0, t) = n\} = \frac{1}{T} \sum_{u=1}^{t} \sum_{k=1}^{u} f_{n-1}(k - (n-1)T) - 2f_n(k - nT)$$

$$+ f_{n+1}(k - (n+1)T).$$

Limit distributions

The Markovian property implies that in the limit $n \to \infty$, $p_{j,k}^{(n)} \to w_k$, independent of j, yielding the limit distribution

$$f_\infty(k) = \sum_{i=0}^{\infty} w_i w_{i+k}$$

and

$$\Pr\{N(t_j, t_j + t) = n\} = \sum_{j=-(n+1)T+t+1}^{-nT+t} f_\infty(j).$$

The exact Markovian model is computationally hard especially when T is large or when multiplex load approaches 1. The limit distributions which are computationally easy to obtain cannot be used, however, since applying them implies neglecting the transient part of the analysis which is an essential part. An approximate model based on modelling the virtual waiting time between CBR arrivals by a Brownian motion, which is simpler from a computational point of view and which captures the transient part, has therefore been developed.

15.5.2 Modeling virtual waiting time by Brownian motion

The idea in this diffusion model is to model the evolution of the virtual waiting time between successive arrivals of the CBR connection with a reflected Brownian motion.

Let $\tilde{V}_t$ denote the waiting time a fictitious observer would experience if he joined the diffusion queue at time t. Assuming that $\tilde{V}_t$ is a Brownian motion with drift m (m assumed smaller than zero in order to ensure a stable queue), variance σ^2 and a reflection at zero, the conditional distribution for $\tilde{V}_t$ is given by (see [Kle75b]):

$$\Pr\left[\tilde{V}_t \leq x \mid \tilde{V}_0 = y\right] = \qquad\qquad (15.5.4)$$

$$\begin{cases} \Phi\left(\dfrac{x - y - mt}{\sigma\sqrt{t}}\right) - e^{2mx/\sigma^2}\Phi\left(\dfrac{-x - y - mt}{\sigma\sqrt{t}}\right), & \text{for } x \geq 0, \\ 0, & \text{for } x < 0, \end{cases}$$

where Φ denotes the standard Gaussian probability distribution. The right hand side of formula (15.5.4) is a distribution function in x for all $y \geq 0$ and all $t > 0$, and it converges to the exponential distribution with mean $(-\sigma^2/m)$, independently of the initial value y, when t tends to infinity.

Arrivals to the queue are modelled in accordance with the diffusion approach, i.e., by an appropriate choice of drift and variance in the diffusion process as shown below. The waiting time W_n for cell n is naturally approximated by the virtual waiting in the diffusion queue at time nT:

$$W_n = \tilde{V}_{nT}.$$

Thereby, we can proceed in the same way as in the exact case with equation (15.5.4) as the basis instead of equations (15.5.1) to (15.5.3).

A basic weakness of the diffusion model is observed in equation (15.5.4): there is a positive probability that the waiting time between time t and time $(t + u)$ decreases more than u (which is of course impossible in the original $D + B^X/D/1$ queue). In order to illustrate this weakness, let t be the arrival time of CBR cell n and let $u = T$. If the virtual waiting time decreases by more than T between t and $t + T$, cell $n + 1$ departs from the queue before cell n. This weakness is inherent to the model and therefore the model should only be used for values of T for which this probability is low, implying that T needs to be large.

The diffusion approximation of the M/D/1 queue as carried out in [Kle75b] (for the more general M/G/1 queue) provides the following values for the drift and variance of the diffusion process: $m = \rho - 1$ and $\sigma^2 = \rho$. However, the server's idle time is not properly taken into account resulting in a steady state waiting time distribution with too large a decay rate, especially for a lightly loaded queue.

The asymptotic behaviour of the waiting time of CBR cells in the slotted $D + B^X/D/1$ queue has been characterized in [BS87]. It is asymptotically geometric with parameter $1/z_\infty$ where z_∞ is the unique real solution of $B^*(z) = z^{1-1/T}$ larger than 1. By appropriately choosing the variance parameter σ^2 in the diffusion process properly, we are thus able to match the decay rate of the stationary waiting time distribution in the diffusion approximation with the decay rate of the exact $D + B^X/D/1$. The diffusion decay rate is $(-2m)/\sigma^2$ and matching then enforces

$$\sigma^2 = -\frac{2m}{\ln z_\infty},$$

where the drift m is set equal to $\rho - 1$ as in the standard diffusion approximation of the $M/G/1$ queue.

Probability distribution for $\tilde{U}_t$

In the diffusion model, the variable corresponding to U_n is defined as $\tilde{U}_t = \tilde{V}_t - \tilde{V}_0$ and

$$\Pr\left\{\tilde{U}_t \leq x\right\} = \int_0^\infty \Pr\left[\tilde{V}_t \leq x + y \mid \tilde{V}_0 = y\right] dP\{\tilde{V}_0 \leq y\}.$$

The following result for the probability distribution function $\tilde{F}_t$ of $\tilde{U}_t$ is obtained in [Bla93]:

$$\tilde{F}_t(x) = \Pr\{\tilde{U}_t \leq x\}$$

$$= \begin{cases} \dfrac{1}{2} + \dfrac{1}{2}\Phi\left(\dfrac{x - mt}{\sigma\sqrt{t}}\right) - \dfrac{1}{2}e^{2mx/\sigma^2}\Phi\left(-\dfrac{x + mt}{\sigma\sqrt{t}}\right), & \text{for } x \geq 0, \\[3mm] \dfrac{1}{2}e^{-2mx/\sigma^2}\Phi\left(\dfrac{x - mt}{\sigma\sqrt{t}}\right) + \dfrac{1}{2}\Phi\left(\dfrac{x + mt}{\sigma\sqrt{t}}\right), & \text{for } x < 0. \end{cases}$$

It is thus seen that $\tilde{F}_t(x) + \tilde{F}_t(-x) = 1$, which implies that the density function for $\tilde{U}_t$ is symmetric.

Number of departures in a window synchronized with a departure

Consider an interval of the form $(t_j, t_j + t]$ starting just after the departure of cell j. Then

$$\Pr\{N(t_j, t_j + t) \geq n\} = \Pr\left\{\tilde{U}_{nT} + nT \leq t\right\} = \tilde{F}_{nT}(t - nT).$$

Number of departures in an arbitrary window

Let 0 denote an arbitrary point in time, and consider the interval $(0, t]$. According to Section 4.2 in [CL66]:

$$\Pr\{N(t_0, t_0 + t) \geq n\} = \Pr\{Y + X_2 + .. + X_n \leq t\}$$

$$= \frac{1}{T}\int_0^t (F_{n-1}(u) - F_n(u))\, du,$$

where Y denotes the forward recurrence time, X_i denotes the interdeparture time between cell $i - 1$ and cell i, and $F_i(u)$ denotes the distribution function of $\sum_{j=1}^{i} X_j$. Clearly:

$$F_n(u) = \Pr\left\{\tilde{U}_{nT} \leq u - nT\right\} = \tilde{F}_{nT}(u - nT).$$

A closed form expression for $F_n(u)$ has been obtained previously, which yields, by integration, the probability distribution of the number of departures in an arbitrary window. The result can be formulated as a closed form formula that does not involve summations (see [Bla93] and [MB96]).

15.5.3 CDV in tandem queues

In this subsection we present a simple model by which it is possible to see how the traffic characteristics of a main cell stream change as it passes through a tandem of queues with interfering traffic. A similar model has been developed in [MSB94a] based on a thorough investigation of the $GI + B^X/D/1$ queue carried out in [MSB94b]. However, the model presented here does not *a priori* assume the departure process of the main stream to be renewal as required in [MSB94a].

Consider M discrete-time queues in series, each with unit (deterministic) service time. A main cell stream enters the system at the first queue and proceeds successively through all tandem queues. Furthermore, in each queue, an interfering cell stream enters the queue and leaves the system immediately after service completion (Figure 15.5.3).

The main cell stream is modelled by a general discrete stochastic process. The sequence $\{X_n^0\}_{n \geq 0}$ denotes the sequence of original interarrival times. For example, if the main stream is CBR, $\Pr\{X_n^0 = T\} = 1$, where $1/T$

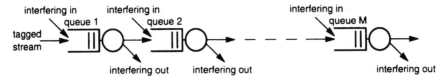

Figure 15.5.3: Main cell stream passing through M tandem queues

denotes the cell rate of the stream. No assumptions are made concerning the independence of the successive interarrival times for the main cell stream. The interfering stream at each queue is modelled as a batch Bernoulli process just as in the single queue case, and $B(z)$ denotes the generating function of the batch size distribution B_i^k at slot i at queue k.

The sequence $\{X_n^k\}_{n\geq 0}$ denotes the interdeparture time of cells of the main stream from the k-th queue. The evolution of the statistical characteristics of $\{X_n^k\}_{n\geq 0}$ when k varies from 0 to M is the main object under investigation in the rest of this subsection. It is assumed that cells from the main stream have priority over the cells of the interfering stream that arrive in the same time slot. The main cell stream output process $\{X_n^k\}_{n\geq 0}$ from the k-th queue can theoretically be obtained as a function of the input process into the k-th queue; this input process is the superposition of the main stream that is characterized by $\{X_n^{k-1}\}_{n\geq 0}$ and of the interfering stream that is characterized by $\{B_n^k\}_{n\geq 0}$.

Slots between two successive departures from queue k of cells of the main stream are due to either:

- departures of cells from the interfering stream, or

- idleness of the server in queue k.

The idleness of the server is difficult to handle, which implies that the derivation of $\{X_n^k\}_{n\geq 0}$ from $\{X_n^{k-1}\}_{n\geq 0}$ and $\{B_n^k\}_{n\geq 0}$ is generally not possible. However, if all queues are overloaded or loaded to 1, successive departures of cells of the main stream can be separated only by slots occupied by departing cells from the interfering streams. In this particular case, the analysis of $\{X_n^k\}_{n\geq 0}$ becomes tractable.

Therefore, it is now assumed that for each n and k, the number of cells in queue k at time n is strictly positive. The interdeparture time between cell n and cell $(n+1)$ of the main stream from the kth queue is then given by:

$$X_n^k = 1 + \sum_{i=1}^{X_n^{k-1}} B_i^k, \qquad (15.5.5)$$

where the B_i^k's are the i.i.d. batch size random variables with generating function $B(z)$. It follows that, for fixed n, the sequence $X_n^0, \ldots, X_n^M$ is a homogeneous discrete-time Markov chain. More specifically, the process

$\{X_n^k, \; k = 0, \ldots, M\}$, as defined in equation (15.5.5), is a branching process with immigration where both the offspring distribution and the immigration distribution have generating function $B(z)$ (see [AN72]). If we introduce $X^k(z)$ as the generating function of X_n^k, one easily obtains:

$$X^k(z) = zX^{k-1}(B(z)). \qquad (15.5.6)$$

Equation (15.5.6) allows in principle the iterative computation of moments for X_n^k.

The general theory for branching processes with immigration [AN72] yields the following asymptotic result: If $E\{B_i^k\} < 1$, then X_n^k converges in distribution to X_n^∞ when k tends to infinity, and the generating function $X(z)$ of X_n^∞ satisfies the functional equation

$$X(z) = zX(B(z)).$$

For numerical investigation, the following iterative scheme is useful:

$$X(z) = \prod_{l=0}^{\infty} B_l(z),$$

with $B_l(z)$ defined by

$$\begin{cases} B_0(z) = z, \\ B_1(z) = B(z), \\ B_{l+1}(z) = B_l(B(z)), \qquad l = 1, 2 \ldots . \end{cases}$$

When the interarrival times of cells of the main stream at the first queue are dependent, successive interdeparture times of cells of the main stream become asymptotically independent when the number of queues grows to infinity. To show this, consider X_n^0 and $X_{n'}^0$ to be two possibly dependent random variables and define

$$(X_n^{k+1}, X_{n'}^{k+1}) = \left(1 + \sum_{i=1}^{X_n^k} B_i^k, 1 + \sum_{i=1}^{X_{n'}^k} \tilde{B}_i^k\right),$$

where $\{B_i^k\}_{i\geq 0, k\geq 0}$ and $\{\tilde{B}_i^k\}_{i\geq 0, k\geq 0}$ are independent of each other, with the same generating function $B(z)$, satisfying $B'(1) < 1$. Then the pair $(X_n^k, X_{n'}^k)$ also converges in distribution to a limit $(X_n^\infty, X_{n'}^\infty)$ when k tends to infinity, where X_n^∞ and $X_{n'}^\infty$ are independent.

Indeed, if X_n^0 and $X_{n'}^0$ are independent, the above result follows immediately from the independence property of $\{B_i^k\}_{i\geq 0, k\geq 0}$ and $\{\tilde{B}_i^k\}_{i\geq 0, k\geq 0}$. Furthermore, since the process $\{(X_n^k, X_{n'}^k)\}_{k\geq 0}$ is an ergodic two-dimensional Markov chain, its limit distribution, when it exists, does not depend on the initial distribution. Hence the above result is still true for any arbitrary distribution of $(X_n^0, X_{n'}^0)$.

Convergence results

Assume now that the main process is an arbitrary renewal process with mean interarrival time $E\{X_n^0\} = T$ and arbitrary variance $\text{Var}\{X_n^0\}$. The arrival rate is then $1/T$. Since all queues are loaded to 1, $E\{B\} = 1 - 1/T$. and from the recursion (15.5.5), we get

$$E\left\{X_n^{k+1}\right\} = 1 + E\{B\} E\left\{X_n^k\right\},$$

which implies that $E\{X_n^k\} = T$ for all k, just as expected. Furthermore (see, for example, exercise XII.6.1 in [Fel68]),

$$\begin{aligned}
\text{Var}\{X_n^{k+1}\} &= E\{X_n^k\}\text{Var}\{B\} + E\{B\}^2\text{Var}\{X_n^k\} \\
&= T\text{Var}\{B\} + (1 - 1/T)^2\text{Var}\left\{X_n^k\right\}.
\end{aligned}$$

Therefore, the sequence $\{\text{Var}\{X_n^k\}\}_{k\geq 0}$ converges monotonically (decreasing or increasing) to $\text{Var}\{X_n^\infty\}$, where

$$\text{Var}\left\{X_n^\infty\right\} = \frac{\text{Var}\{B\}}{(1/T)^2(2 - 1/T)}. \tag{15.5.7}$$

This is an illuminating result. It shows that when a very large number of queues has been passed the squared coefficient of variation is approximately the half of the variance of the interfering batch size distribution, at least if T is large, and *independent* of the original characteristics of the main stream.

As an example, consider a CBR cell stream with $T = 10$ passing 25 queues in series where the interfering traffic in all cases is Poisson traffic. The number B of interfering arrivals in each slot follows a Poisson distribution with $E\{B\} = \text{Var}\{B\} = 0.9$. Table 15.5.1 provides the values for $\text{Var}\{X_n^k\}$ when k increases. As the table shows the squared coefficient of variation is approximately $1/2$ after 25 queues.

Table 15.5.1: Variance of interarrival times of main process with Poisson interfering traffic.

k	0	1	2	4	6	10	25	∞
$\text{Var}\{X_n^k\}$	0	9	16.3	27	34	41.6	47.1	47.4

The above results shows that, if the number of heavy loaded tandem queues grows to infinity, the main process converges to a renewal process (under the assumption of the main stream having priority over the background stream) with the same mean as the original process and a variance (given by equation (15.5.7)) that is a function of the mean of the original process and of the variance of the batch size of the interfering streams.

When the loads on the tandem queues are light to moderate and the traffic volume of the main stream is small, it is possible by a different approach, described in [vdBDRvdW95], to obtain approximate results. The main idea in this approach is to utilize the fact that successive cells of the main stream see independent waiting times when the load is light and the interarrival times between cells of the main stream are large.

Practical considerations

The above model for CDV in tandem queues is burdened with two assumptions which with an overwhelming probability will not be fulfilled in any "real" applications. These are

- The assumption that all queues are loaded to 1.

- The assumption of background traffic constituting a batch Bernoulli process.

The first assumption may be considered to constitute a "worst case" scenario and any outcome from the model therefore constitutes "worst case" figures which, in general, is of practical significance.

The second assumption, in a strict sense, obliges the background stream to be the superposition of a number of Bernoulli streams. However, it is suggested in [MB96] that the CDV introduced by a bursty background process can be modelled by applying a batch Bernoulli distribution with high peakedness like the negative binomial distribution for the background traffic. Such a rough approach is to some extent supported by the investigation carried out in [WB94].

15.6   Priority queues

ATM networks have to accomodate widely different types of information services; for each service, a given set of attributes defines the user oriented quality of service. The main grade of service requirements deal with cell loss rate and transfer delay characteristics. Some services are delay sensitive (e.g., VBR video) and others are loss sensitive (e.g., computer interconnections for parallel computing).

Priorities can be implemented using two types of discriminators. If *implicit priority* is used, cells are discriminated on the basis of the VPI/VCI value; in this case, all the cells of a given connection are allocated the same priority level. If *explicit priority* is used, cells are discriminated on the basis of the value taken by the CLP bit in the cell header

Both *time* and *loss* priorities have been envisaged:

- in time priority, some cells have priority over others concerning the access to *transmission* facilities;

- in loss priority, some cells have priority over others concerning the access to *buffering* facilities.

Due to the high bitrate of the links (150 or 600 Mbit/s), queueing delays inside international or national ATM networks are small compared to propagation delays [Tra89] provided that the buffer sizes in switching elements are moderate (of the order of 100 places). This is however not true if buffer sizes are significantly larger, or if non-FIFO queueing disciplines are implemented. Time priority may thus be needed in order to ensure that connections carrying delay-sensitive services experience only small queueing delays.

Loss priorities may be needed in order to offer different Cell Loss Ratios to different types of cells: in case of congestion, low priority cells are discarded in order to ensure a low Cell Loss Ratio to high priority cells.

In the present section, the issue of offering loss and/or time priorities to cells in ATM networks is addressed. In network elements, buffer space may or may not be statistically shared between classes of traffic. The following two methods have been proposed for which statistical sharing of buffer space can be avoided:

- allocating specific buffer space to *each* connection in *every* buffer along the VC [Mak90].

- using implicit priority in order not to mix different classes of cells in buffers [KHBG91];

The first solution is considered to be very uneconomical since it implies very large buffers if the number of connections sharing the buffer is not small.

The second solution is currently envisaged in order to discriminate between time sensitive connections that share only small buffers and less time sensitive connections that share larger buffers. Absolute time priority is offered to the cells of the small buffers; further levels of priority can be envisaged in both small and large buffers.

In the following, we consider FIFO buffers that are statistically shared by different classes of cells. Fair queueing mechanisms offer an alternative to priority mechanisms that are implemented in FIFO buffers as discussed in Part I Chapter 6.

Much work has been done on modelling time priorities, mostly with Poisson inputs (see for example [Tak64]). More recent references address the modelling of time priority implementation in ATM switching elements: a numerical solution of the $Geo/D/1/N$ queue with several levels of head-of-line (HOL) priority is presented in [PS90]; and some partial results concerning a HOL $MMPP/GI/1$ queue are given in [Fis89].

A few references related to loss priorities handle overload control strategies with only one class of customers [DH86, Li89, Neu85, Tak85]. More recently, loss priorities with several classes of customers have been studied under various assumptions concerning the statistics of input streams and the physical implementation of the selective discard algorithms. The following references present an exact analysis of the models under study allowing the computation of loss probabilities: selective discard algorithms have been studied under the assumption of Poisson input streams in [HG90, KHBG91, GH90, GH91b]; exact analysis of the threshold mechanism has also been carried out for a queue with MMPP (Markov Modulated Poisson Process) input [GC90] and for a slotted discrete time queue with an MMBP (Markov Modulated Bernoulli Processes) input process [LeB91].

15.6.1   Time priority mechanism

We consider here the case where two buffers are accessed by high and low priority cells: a small (respectively large) buffer is accessed by high (respectively low) priority cells. Whenever the small buffer is not empty, high priority cells are transmitted; low priority cells are transmitted only if the small buffer is empty. This is a particular case of the more general one where more than two levels of time priorities are considered.

It is assumed here that the Cell Loss Ratios are small in both buffers. In this particular case, the delay behaviour of the global system is very similar to that of a single queue with infinite buffer that is shared by both input streams.

In the particular case where the input streams are Poisson, the delay characteristics can be approximated by those of customers in an $M/D/1$, non-pre-emptive HOL queue with unlimited capacity. Let V_i be the virtual waiting time of class-i customers in an $M/D/1$ queue with non pre-emptive HOL priority and deterministic service time D. Takács addresses the analysis of time priorities in the $M/G/1$ queue [Tak64]. In the present case, expressions for the Laplace-Stieltjes transform $\tilde{V}_i$ of the distribution of V_i are as follows:

$$\tilde{V}_1(s) = \frac{s(1-\rho) + \lambda_2\left(1 - \exp\left(-sD\right)\right)}{s - \lambda_1\left(1 - \exp\left(-sD\right)\right)}, \tag{15.6.1}$$

$$\tilde{V}_2(s) = (1-\rho)\frac{s + \lambda_1\left(1 - \tilde{B}(s)\right)}{s - \lambda_2\left(1 - \tilde{B}(s)\right)}, \tag{15.6.2}$$

where $\tilde{B}(s)$ is the Laplace-Stieltjes transform of the distribution of a busy period duration B in an $M/G/1$ queue with service time D and arrival rate λ_1. $\tilde{B}(s)$ is therefore solution of the following functional equation:

$$\tilde{B}(s) = \exp\left(-D\left(s - \lambda_1 + \lambda_1\tilde{B}(s)\right)\right)$$

It is convenient to use the above expressions in order to obtain moments for V_i. For example,

$$E\left\{V_1\right\} = \frac{\lambda D}{2\left(1 - \rho_1\right)}, \qquad E\left\{V_2\right\} = \frac{\lambda D}{2\left(1 - \rho_1\right)\left(1 - \rho\right)}.$$

In order to evaluate the jitter experienced by both classes of cells, it is necessary to evaluate the percentiles of the waiting time in the switching elements. Equations (15.6.1) and (15.6.2) allow these computations. The derivation for V_1 is straightforward since closed form expressions are known for the waiting time distribution in an $M/D/1$ queue (see Section 6.1, [Sys86]). The distribution for B is given by the following infinite sum [Rio62]

$$\tilde{B}(s) = \sum_{k=1}^{\infty} \frac{1}{k} e^{-k\rho_1} \frac{(k\rho_1)^{k-1}}{(k-1)!} e^{-kDs}$$

and an appropriate inversion routine (e.g. that described in [PAB88]) can be used to obtain similar results for non-HOL customers.

The above analyses easily extends to the case where more than two levels of time priority are considered, as long as the Cell Loss Ratio for each priority class can be considered as negligible.

15.6.2 Loss priority mechanisms

A simple queue models an output buffer of a network element. Two input streams (representing two classes of cells) share a buffer which is managed in order to guarantee a given cell loss probability level to at least one class of cells. Let λ_i be the arrival rate of class-i customers and let μ be the service rate; $\rho_i = \lambda_i/\mu$ is the load offered by class-i customers and $\rho = \rho_1 + \rho_2$ is the overall load. The following equation relates the loss probabilities for both classes:

$$\rho\pi = \rho_1\pi_1 + \rho_2\pi_2 \tag{15.6.3}$$

where π_i is the loss probability for class-i customers and π is the overall loss probability. This equation does not depend on the statistical nature of the input streams.

Threshold mechanism

The mechanism presented below is termed "Partial Buffer Sharing (PBS)", or "Threshold Mechanism": when the buffer occupancy reaches a given threshold M, arriving class-2 cells are discarded whereas class-1 cells are accepted as long as the buffer (of capacity N) is not full.

The analysis of the threshold mechanism is performed using an embedded Markov chain approach. The main steps of the solution in case of Poisson input streams are briefly described below.

We are interested in the number of cells in the system. The service completion instants are embedded points for the underlying Markov chain. Let $\mathbf{Q} = (q_1, \cdots, q_N)$ be the departing customer's distribution of the number of cells in the system. The transition matrix $\mathbf{T}$ for the Markov chain is derived using the fact that as soon as the number of waiting cells is larger than M,

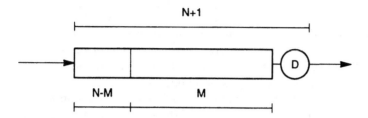

Figure 15.6.1: Threshold mechanism: threshold M, capacity N.

only class-1 cells are allowed to queue. Therefore, the cell arrival rate depends on the number of waiting cells. The vector $\mathbf{Q}$ is obtained by solving $\mathbf{Q} = \mathbf{QT}$.

Let p_k, $k = 0, \cdots, N+1$, be the steady state probability distribution of the number of cells in system. Using the PASTA property [Wol82] the loss probabilities for class-1 and class-2 cells are given by

$$\pi_1 = p_{N+1}, \tag{15.6.4}$$

$$\pi_2 = \sum_{k=M+1}^{N+1} p_k. \tag{15.6.5}$$

Now, the departure rate is equal to the arrival rate of cells which join the queue:

$$\lambda\,(1 - \pi) = \mu\,(1 - p_0).$$

Furthermore, departing customer's and entering customer's distributions are identical [Coo72]. Using the above equation together with equations (15.6.3) and the fact that the entering customer's distribution is obtained by conditioning the arriving customer's distribution on the event {the arriving cell is not rejected}, the steady state distribution is related to the departing customer's distribution as follows (equations (15.6.4) and (15.6.5) yield the value for p_{N+1})

$$p_k = \begin{cases} \dfrac{q_k}{q_0 + \rho} & 0 \le k \le M, \\[3ex] \dfrac{\lambda}{\lambda_1}\dfrac{q_k}{q_0 + \rho} & M+1 \le k \le N, \\[3ex] 1 - \dfrac{1}{q_0 + \rho}\left(1 + \dfrac{\lambda_2}{\lambda_1}\displaystyle\sum_{k=M+1}^{N} q_k\right) & k = N+1. \end{cases} \tag{15.6.6}$$

The above derivation needs to be modified if Markov Modulated processes are considered as inputs to the shared buffer.

In the discrete time queue presented in [LeB91], cell service time can start only at the beginning of a time slot and lasts exactly one slot. Arrivals occur between time slot boundaries and departures occur at time slot boundaries; the input process is a Markov modulated Bernoulli process (MMBP). An efficient numerical method is used to derive the distribution of the state of the system (number of cells in system and phase of the input process) at time slot boundaries which directly yields loss probabilities.

For MMPP input streams, a classical embedded Markov chain approach is used to obtain the distribution of the state of the system (number of cells in system and phase of the input process) at departure time; a regenerative theorem then yields the stationary distribution and the loss probabilities are obtained by using the Markovian nature of the input process (see [GC90]).

Pushout mechanism

The following selective discard mechanism is termed "pushout": a class-1 cell may join the queue, even if the buffer is full as long as some class-2 cells are buffered; the last class-2 cell to have entered the queue is discarded (pushed out) and the class-1 cell is buffered in the last position of the queue (cell sequence integrity is thus maintained).

It is not easy to derive the class-1 loss probability π_1, even in the case of Poisson input streams. We therefore derive π_2 and equation (15.6.3) yields π_1.

A class-2 cell arriving to an empty system cannot be rejected since the pushout mechanism is not pre-emptive (only *buffered* class-2 cells can be rejected). Assume now that a test class-2 cell finds k cells $(C_1, \cdots, C_k)$ in system at the time of its arrival and a remaining service time equal to v. The rejection of the test class-2 cell depends on the number ν_i of class-1 arrivals during the service time of cell C_i for $i = 1, \cdots, k$. More precisely, the probability $\Pr[\text{served} \mid k]$ that a class-2 cell which finds k cells in system is not rejected is given by

$$
\Pr\left[\text{served} \mid k\right] = \Pr \left\{ \begin{array}{l} \nu_1 \leq N - k \\ \vdots \\ \nu_1 + \cdots + \nu_i \leq N - k + i - 1 \\ \vdots \\ \nu_1 + \cdots + \nu_i + \cdots + \nu_k \leq N - 1 \end{array} \right\} \qquad (15.6.7)
$$

The variables ν_i in turn depend only on v and on the length of the service times for $(C_2, \cdots, C_k)$. The above equation is solved recursively for Poisson input streams and general independent service times in [KHBG91]. An alternative direct derivation for the case of deterministic service times is proposed in [HG90].

Mixing time and loss priorities

The need may arise in some cases to offer time priority in a single buffer in order to enhance a loss priority mechanism that is already implemented.

A buffer in which both time and loss priority mechanisms are implemented is studied in [GH90] and [GH91b]. There are 2 classes of cells: class-1 cells have non pre-emptive HOL priority over class-2 cells; the class of pushout cells is either class-1 or class-2.

Note that no improvement concerning loss probabilities can be attained by simultaneously implementing HOL priority and a threshold mechanism. In this case loss probabilities are the same as in a system where only the threshold mechanism is implemented: the rejection of a non-priority customer by the threshold mechanism is independent of the buffered customers' respective classes since the service times are identical for both classes.

The loss probabilities in the HOL+pushout $M/G/1/N$ queue are obtained via regenerative theory in the 2 cases where pushout customers are class-1 or class-2.

The method used in the previous subsection (pushout mechanism) for the derivation of the loss probability of non-pushout cells cannot be used in the present case since the probability that a non-pushout cell is rejected now depends on the number of class-1 (HOL priority) customers found in the system by an arriving customer. It is still possible, using the embedded Markov chain method, to compute the distribution of the numbers of class-1 and class-2 customers left in the system by a departing customer. However, we cannot deduce directly from this distribution the entering customer's distribution : if there are several classes of customers, the entering customer's distribution can differ from the departing customer's distribution.

We therefore resort to the following equation in order to evaluate loss probabilities:

$$\lambda_i \left(1 - \pi_i\right) = \mu P_i, \tag{15.6.8}$$

which relates π_i to P_i, the proportion of time during which class-i customers are served. The computation of P_i itself is performed using semi-regenerative theory which yields a closed-form expression for P_1 and P_2:

$$P_1 = \frac{\rho_1 P(0,0) + \rho \sum_{n_1 > 0} P(n_1, n_2)}{P(0,0) + \rho}, \tag{15.6.9}$$

$$P_2 = \frac{\rho_2 P(0,0) + \rho \sum_{n_1 = 0, n_2 > 0} P(n_1, n_2)}{P(0,0) + \rho}. \tag{15.6.10}$$

where $P(n_1, n_2)$ is the stationary probability distribution of the numbers of customers at departure time.

Now, the departing customer's distribution $P(n_1, n_2)_{0 \leq n_1 \leq N}$ is obtained using the classical Markov chain approach. It turns out that the derivation of $\sum_{n_1 > 0} P(n_1, n_2)$ (which is the parameter of interest) reduces to a 1-dimensional system (an $M/D/1/N$ analysis) if class-1 cells are pushout cells. In this case, for the high priority cells, the use of both time and pushout priorities for the high priority cells in a single buffer is equivalent to the use of two buffers with a single time priority for the high priority cells. In the other case (class-1 cells are non-pushout cells), the system is 2-dimensional and the analysis is more cumbersome.

15.6.3   Numerical results

First of all, we see in Figure 15.6.2 the typical influence of selective discard mechanisms on loss probabilities for both classes of cells. We assume a balanced load of class-1 and class-2 cells and a global load equal to 0.8. Loss probabilities for both classes are displayed versus buffer size in a finite queue in which pushout is implemented. As soon as the overall loss probability is small enough (say 10^{-3}), loss probabilities for both classes differ by more than 2 orders of magnitude. This is a common feature of all selective discard algorithms: as soon as the mechanism performs properly (i.e., provides cell loss rates differing by several orders of magnitude to class-1 and class-2 cells), an approximate value for the loss probability of class-2 cells is given by $\pi_2 \simeq \rho \pi / \rho_2$ (which follows from equation (15.6.3)).

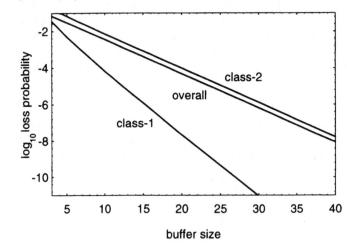

Figure 15.6.2: Loss probabilities for class-1 and class-2 cells versus buffer size in a pushout queue.

In Figure 15.6.3, we compare PBS with pushout: for a given buffer size ($N = 32$) and given offered loads ($\rho_1 = 0.2, \rho = 0.8$) the loss probabilities for both classes of cells are displayed versus threshold value. The loss probabilities for the pushout queue are displayed as a reference. This figure illustrates the influence of the threshold value on the behaviour of the PBS mechanism. Moreover, in the area of interest, the pushout cell loss probability constitutes a lower bound for the loss probability of loss sensitive cells in the PBS queue.

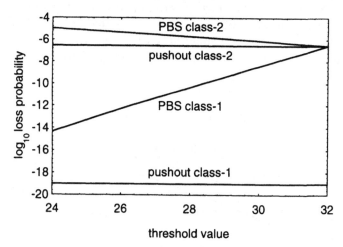

Figure 15.6.3: Loss probabilities for class-1 and class-2 cells versus threshold value in a PBS queue.

Now, consider the problem of buffer dimensioning: buffer dimensioning obtained for $(\pi_1 = 10^{-10}, \pi_2 = 10^{-6})$ with PBS and pushout mechanisms are compared in Table 1 for several values of the total load ρ with $\rho_1 = \rho/5$. It turns out that PBS is clearly not the optimal selective discard mechanism but that it compares well with pushout. It should also be noted that PBS is simpler to implement than HOL or pushout and is therefore a more likely candidate for implementation.

Table 15.6.1: Buffer size obtained with various selective discard mechanisms; $\rho_1 = \rho/5$.

total load	$M/D/1/N$	pushout	PBS
.55	20	12	16
.70	32	19	22
.80	49	29	31
.85	67	38	41

Buffer dimensioning is sensitive to the assumptions made about the intensities of both input streams. If class-2 load is not controlled the loss probability for loss-sensitive cells may increase dramatically. Figure 15.6.4 compares the following three selective discard mechanisms (PBS, pushout,

HOL+pushout with HOL class=pushout class) when the load offered by loss-sensitive traffic is fixed ($\rho_1 = .25$) and the class-2 load varies. The figure plots the loss-sensitive cell loss rate versus overall load. Under the above assumptions, in an $M/D/1/10$ queue with no priority, the loss probability for class-1 cells increases from 10^{-9} to 10^{-2}, the range of variation decreases to 4 orders of magnitude for the pushout and PBS queues and is less than one order of magnitude when HOL and pushout are simultaneously implemented. In the latter case, loss-sensitive traffic is hardly disturbed by transient overloading of the buffers by class-2 cells.

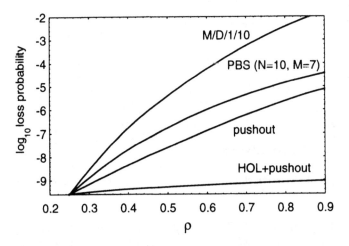

Figure 15.6.4: Loss probabilities for loss-sensitive cells versus total load. Comparison of selective-discarding algorithms.

This problem needs to be taken into account if the traffic offered as class-2 is not controlled. This can be the case for example if the CLR offered to CLP=0 cells is bounded, whereas the CLR offered to the aggregate CLP=0+1 cell stream is unbounded, as can be expected for SBR2 and SBR3 connection types. Indeed, under the assumption that class-2 traffic is not controlled, if PBS is implemented in a buffer, it is necessary to choose the threshold value M such that the buffer space $M - N$ that is dedicated to class-1 traffic is sufficient to ensure the guaranteed CLR to this traffic [GBH91].

16 Burst scale loss systems

At burst level we are interested in phenomena occuring in a time scale typical of an on/off source activity period or video frame duration (burst scale) rather than in the time scale of an inter-cell interval (cell scale). In considering this kind of gross behaviour of the system, we can ignore the discrete nature of the cell arrivals and regard the incoming stream of cells as a *continuous fluid* characterized by its instantaneous arrival (flow) rate λ_t.

We will consider, in particular, congestion phenomena occuring in the burst scale. These arise when the arrival rate λ_t exceeds multiplex capacity and leads to a more or less constantly growing queue as long as the arrival rate excess lasts. There are two basic alternatives to handle burst scale congestion as discussed in Part I Section 4.1:

- burst scale loss (rate envelope multiplexing) - a small buffer is provided to absorb cell scale congestion only with an overflow probability of 10^{-9}, say, given that the arrival rate is less than capacity; the acceptable multiplex load is then such that the arrival rate only exceeds capacity with a very small probability (10^{-9}, say).

- burst scale delay (rate sharing) - a large buffer is provided to absorb burst scale congestion with the buffer overflow probability limited to 10^{-9}, say; the multiplex load can theoretically be as close to 1 as desired;

Burst scale delay systems will be discussed in Chapter 17. In this chapter we consider some properties of burst scale loss systems. The most important of them is related to the efficiency of statistical multiplexing as one of the perceived assets of the ATM concept is its ability to allow better system utilization by means of resource sharing. This is considered in Section 16.1 where our purpose is to assess to what extent the multiplexing gain can actually be realized without burst scale buffering. In Sections 16.2-16.4 we study at some length different aspects of the cell loss process in burst scale loss systems.

16.1 Multiplexing sources

16.1.1 Cell loss probability

The purpose of the call admission procedure is to guarantee some planned quality of service parameter for those calls which have been accepted. With any given QoS objective it is possible to determine the allowed traffic mix, the corresponding system load and other indicators of the efficiency of the multiplexing.

We first consider the case where the QoS is defined in terms of the overall cell loss probability, P_{loss}, by which we mean the long term probability for an arbitrary arriving cell to be lost. This is given by the ratio of the mean loss rate to the mean arrival rate,

$$P_{loss} = \frac{1}{m} \cdot \mathrm{E}\left\{(\lambda_t - c)^+\right\} \tag{16.1.1}$$

where c denotes the link capacity and $m = \mathrm{E}\{\lambda_t\}$ is the mean rate. We also define an instantaneous loss probability p_t,

$$p_t = \frac{(\lambda_t - c)^+}{\lambda_t} = \frac{(\rho_t - 1)^+}{\rho_t} \tag{16.1.2}$$

where $\rho_t = \lambda_t/c$ is the instantaneous load of the system. The loss probability p_t is itself a stochastic process. We can then re-write (16.1.1) as

$$P_{loss} = \frac{1}{m} \cdot \mathrm{E}\left\{p_t \cdot \lambda_t\right\}. \tag{16.1.3}$$

In writing Eqs. (16.1.1) and (16.1.2) we have assumed that when $\lambda_t \leq c$ no losses occur and when $\lambda_t > c$ only the surplus flow is lost. This corresponds to our fluid approximation. If we make a digression back to the cell level, we note that neither of these assumptions is valid unless a (small) buffer is provided to absorb the cell scale fluctuations. In particular, a buffer is needed in an overloaded system in order to validate the use of (16.1.2) as an instantaneous loss probability. In fact, we saw in Section 14.1 that the actual loss probability is given by Eq. (14.1.2), $p_t = (\rho_t - 1 + \phi)/\rho_t$, where ϕ is the empty system probability. The role of the buffer in that case is to function as a dam to ensure continuous server workload (i.e., to ensure that $\phi \approx 0$). Even a very small buffer is sufficient for that purpose; certainly a buffer which is able to eliminate cell losses in the case of non-overload will do the job of the dam for the overload case, too. (If the buffer were large the system would become more like a burst scale delay system).

16.1.2   Superposition of identical sources

Given the process λ_t it is easy to calculate P_{loss} from (16.1.1). Note that only the stationary distribution of λ_t is needed for such a calculation. Therefore, this parameter is much less sensitive to individual source characteristics than any delay based QoS criterion. As an example, we consider a superposition of N identical independent on/off sources. For this offered traffic, in particular, it is sufficient to know the source mean rate m and peak rate h. The burst structure of the cell flow (characterised e.g. by the Intrinsic Burst Tolerance

τ_{IBT}) has no influence on P_{loss}. We denote the "on" probability of a source by $\alpha = m/h$ and the ratio of the link capacity to the peak rate by N_0, i.e., $N_0 = c/h$ is the number of streams the link can carry when capacity is allocated according to the peak rate. The probability p_n of n sources being simultaneously "on" is given by the binomial distribution

$$p_n = \binom{N}{n} \alpha^n (1-\alpha)^{N-n} \tag{16.1.4}$$

and the loss probability reads

$$P_{loss}(N, N_0, \alpha) = \frac{1}{N\alpha} \sum_{n=\lceil N_0 \rceil}^{N} p_n \cdot (n - N_0). \tag{16.1.5}$$

We can now determine an admissible load $N_\epsilon(N_0, \alpha)$ such that

$$P_{loss}(N_\epsilon, N_0, \alpha) \le \epsilon. \tag{16.1.6}$$

(For $\epsilon = 0$ we have $N_\epsilon = N_0$ making our notation consistent.) The *effective bandwidth*, i.e., the bandwidth required by one source is $(N_0/N_\epsilon) \cdot h$. It lies somewhere between the peak rate and the mean rate. The efficiency of statistical multiplexing may be gauged by comparing the effective bandwidth to these bounds. A comparison to the peak rate is represented by the multiplexing gain $G_\epsilon = N_\epsilon/N_0$ whereas a comparison to the mean rate yields the admissible load factor $\rho_\epsilon = N_\epsilon \alpha/N_0 = \alpha \cdot G_\epsilon$. These two measures of the efficiency (which contain the same information) are illustrated in Figs. 16.1.1 a and b where G_ϵ and ρ_ϵ, respectively, are plotted against α for the QoS level $\epsilon = 10^{-9}$. Each curve in the family represents a given value of N_0: from bottom to top, the link capacity corresponds to 10, 15, 30 or 100 "on" sources.

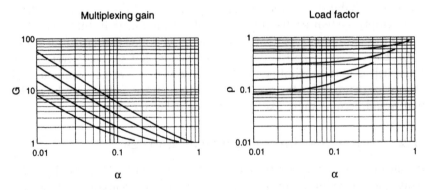

Figure 16.1.1: Multiplexing gain and admissible multiplex load against α for link capacities of 10, 15, 30 and 100 "on" sources, and a loss probability 10^{-9}.

For small values of α the admissible load ρ_ϵ is approximately constant. In fact, in this asymptotic region the binomial distribution (16.1.4) may be approximated by the Poisson distribution which depends on N and α only through their product $N\alpha = \rho N_0$ (the mean number of active sources). Thus we get in place of (16.1.5)

$$P_{loss}(\rho, N_0) \approx \frac{1}{\rho N_0} \sum_{n=\lceil N_0 \rceil}^{\infty} \frac{(\rho N_0)^n}{n!} e^{-\rho N_0}(n - N_0) \qquad (16.1.7)$$

$$\approx \frac{1}{(1-\rho)^2 N_0} \frac{(\rho N_0)^{\lfloor N_0 \rfloor}}{\lfloor N_0 \rfloor!} e^{-\rho N_0},$$

and the QoS criterion $P_{loss}(\rho, N_0) \leq \epsilon$ yields $\rho_\epsilon = \rho_\epsilon(N_0)$ independent of α. The latter approximation is due to [Lin91]. In tabular form, the admissible load factors in this asymptotic region for $\epsilon = 10^{-9}$ are given in Table 16.1.1.

Table 16.1.1: Admissible load factors

N_0	10	30	50	60	75	100	150	300
ρ_ϵ (%)	7	29	41	45	50	55	62	72

For $N_0 = 100$ the admissible load in the asymptotic region is about 55%, while for $N_0 = 10$ the corresponding figure is less than 8%. This suggests that to achieve high link utilization with burst scale loss, the source peak rate should barely exceed one hundredth of multiplex capacity. However, even for higher peak rates the multiplexing gain may be quite appreciable and is the higher the smaller is α. For instance, for $N_0 = 10$ and $\alpha = 0.01$ the gain is about 8; the utilization with statistical multiplexing is much higher than without it, though it falls short of 100%.

Table 16.1.2: Load factor, burstiness and the loss probability during congestion as a function of the safety factor for two different levels of P_{loss}.

$P_{loss} = 10^{-4}$				$P_{loss} = 10^{-9}$			
η	ρ	B	P_{loss}^c(%)	η	ρ	B	P_{loss}^c(%)
2.5	0.925	0.033	1.6	5.0	0.915	0.019	0.3
2.7	0.832	0.075	2.4	5.2	0.776	0.056	0.8
2.9	0.679	0.163	3.6	5.4	0.518	0.172	1.6
3.1	0.480	0.349	4.8	5.6	0.244	0.553	2.3
3.3	0.286	0.755	5.6	5.8	0.085	1.845	2.5

Another case of general interest is a system where a large number of independent streams with a small peak rate are superposed to form a λ_t which is approximately normally distributed with a mean m and standard deviation σ. Then it is again easy to evaluate $P_{loss}(\rho, \sigma)$ which on imposing the QoS criterion leads to $\rho_\epsilon = \rho_\epsilon(\sigma)$. Instead of σ it is more illuminating to use some combination of σ and m, two parameters of choice being the "safety factor" $\eta = (c - m)/\sigma$ and the "burstiness factor" $B = \sigma/m$. The dependence of the admissible ρ_ϵ on these parameters is shown by Table 16.1.2. (The meaning of the last column will be explained in Subsection 16.3).

16.1.3   Superposition of general sources: an equivalent burst approximation

In the case of a general superposition of independent VBR streams, the stationary distribution of λ_t can be expressed as the convolution of the individual stream stationary distributions. Although this convolution can be calculated exactly with fast algorithms, we may prefer to use the following approximations proposed in [Lin91] which prove accurate as long as the streams are not too different (e.g., not a mixture of a small number of very high peak rate streams and many low rate streams).

The approximations are based on the replacement of the considered superposition by an equivalent process of simpler characteristics. The equivalent process is chosen to have the same values for the following parameters:

- mean cell rate, m,

- variance of instantaneous rate, σ^2,

- asymptotic variance of the number of cell arrivals in an interval of length t, vt. (This parameter, however, does not affect the stationary distribution of λ_t.)

These parameters can be derived from measurements of the total stream. Alternatively, they can also be deduced from the corresponding parameters of the individual streams, viz. for independent streams we have $m = \sum_i m_i$, $\sigma^2 = \sum_i \sigma_i^2$ and $v = \sum_i v_i$.

A first equivalent process is a burst process with bursts of rate h and exponentially distributed length of mean b arriving according to a Poisson process of intensity λ. The fitting relations are

$$\lambda = 2m^2/v,$$

$$h = \sigma^2/m,$$

$$b = v/2\sigma^2.$$

The stationary distribution of λ_t is then:

$$\Pr\{\lambda_t = nh\} = \frac{(m^2/\sigma^2)^n}{n!} \exp\{-(m^2/\sigma^2)\}, \quad \text{for } n \geq 0.$$

The loss probability can be calculated from (16.1.7) where we now have $N_0 = c/h = cm/\sigma^2$ and $\rho N_0 = m^2/\sigma^2$.

16.1.4   Large deviation approximation

So far we have used the requirement $P_{loss} \leq \epsilon$ as the QoS criterion. Another criterion, more common in the literature (e.g., [Hui88]), is based on the congestion or saturation probability,

$$P_{sat} = \Pr\{\lambda_t > c\} = \int_{\lambda_t > c} dP, \qquad (16.1.8)$$

requiring $P_{sat} \leq \epsilon$. In Subsection 14.3.2 it was shown how large deviation theory can be used to obtain a very accurate estimate for this type of probability when λ_t is composed of a number of independent streams [Hui88]. The idea was to introduce a shifted (by an exponential factor) probability measure P_β such that $dP_\beta = e^{\beta \lambda_t} dP / \psi(\beta)$ where $\psi(\beta) = E\{e^{\beta \lambda_t}\}$ is the normalizing factor. Then we trivially have for the original probability

$$dP = e^{-\beta \lambda_t} \psi(\beta) dP_\beta. \qquad (16.1.9)$$

Now the shifted distribution can be approximated very accurately by a normal distribution around its own mean $m(\beta) = E_\beta\{\lambda_t\}$. As β is a free parameter we can shift the mean to any place we wish and thus (16.1.9) provides a means to approximate the original probability accurately at any interesting point.

The integral of (16.1.8) gets its main contribution close to the lower limit $\lambda_t = c$. To apply the above method, the mean is therefore shifted to the value c. The corresponding value of β is denoted β^*,

$$m(\beta^*) = c. \qquad (16.1.10)$$

Evaluation of the integral leads to (14.3.14) which is repeated here [Hui88]

$$P_{sat} \approx \frac{1}{\sqrt{2\pi \beta^* \sigma(\beta^*)}} e^{-\beta^* c + \mu(\beta^*)}, \qquad (16.1.11)$$

where $\mu(\beta) = \ln \psi(\beta)$ and $\sigma^2(\beta) = \mu''(\beta)$ is the variance of the shifted distribution.

In passing, we note that the same technique can also be used to calculate

$$P_{loss} = m^{-1} \int_{\lambda_t > c} (\lambda_t - c)\, dP$$

of Eq. (16.1.1). The result is

$$P_{loss} \approx \frac{1}{\sqrt{2\pi} m \beta^{*2} \sigma(\beta^*)} e^{-\beta^* c + \mu(\beta^*)}, \qquad (16.1.12)$$

which differs from (16.1.11) only by the appearance of an extra factor $m\beta^*$ in the denominator. This factor is typically of the order of 100 which means that QoS criteria based on P_{sat} are roughly two orders of magnitude tighter than those based on P_{loss}. The impact of this apparently large difference on the dimensioning of the network elements, however, is relatively small.

Formulae (16.1.11) and (16.1.12) are applicable to any system with a number of independent constituent streams. They are also accurate and relatively easy to calculate. Note that both $m(\beta)$ and $\mu(\beta)$ are additive quantities with respect to the independent constituents, e.g., $m(\beta) = \sum_i m_i(\beta)$. Eq. (16.1.10) then suggests that for any constituent stream i the shifted mean $m_i(\beta^*)$ may be interpreted as the effective bandwidth of that stream. If β^* is relatively insensitive to variations in the composition of the other traffic components, as often seems to be the case, then this notion of effective bandwidth has some invariant meaning, intrinsic to that particular source type.

16.2 Loss distribution among sources

We consider how cell loss is distributed among different sources assuming that losses occur solely due to burst scale overflows [NV91]. When an overflow situation occurs, the local cell loss probability is given by (16.1.2). We assume that this probability does not distinguish between the sources: all the sources have the same local loss probability. The reason why losses are distributed unevenly is that sources send different proportions of their cells during the periods of congestion: for bursty sources the bursts tend to be correlated with high local loss probabilities as the bursts are themselves partially causing the congestion.

We assume now that the total rate λ_t is the sum of the rates $\lambda_t^{(i)}$ of N independent (not necessarily identical) streams,

$$\lambda_t = \sum_{i=0}^{N} \lambda_t^{(i)}.$$

The total and individual streams have mean rates m and m_i, respectively. The individual loss probability $P_{loss}^{(i)}$ of source i is, in analogy with (16.1.3),

$$P_{loss}^{(i)} = \frac{1}{m_i} \, \mathrm{E}\left\{ p_t \lambda_t^{(i)} \right\}.$$

We now derive an approximation for $P_{loss}^{(i)}$. First, since p_t depends on λ_t only, we can write

$$P_{loss}^{(i)} = \frac{1}{m_i} \, \mathrm{E}\left\{ p_t \lambda_t^{(i)} \right\} = \frac{1}{m_i} \, \mathrm{E}\left\{ p_t \mathrm{E}\left[\lambda_t^{(i)} \mid \lambda_t \right] \right\}. \tag{16.2.1}$$

Furthermore, since p_t is non-zero only when $\lambda_t > c$ and the probability that λ_t appreciably exceeds c is very small, we can use inside (16.2.1) the approximation

$$\mathrm{E}\left[\lambda_t^{(i)} \mid \lambda_t \right] \approx \mathrm{E}\left[\lambda_t^{(i)} \mid \lambda_t = c \right].$$

We now apply the probability shift method introduced in Subsection 14.3.2, and denote

$$m(\beta) = E_\beta\{\lambda_t\},$$

$$m_i(\beta) = E_\beta\{\lambda_t^{(i)}\}.$$

It is always true that $E[\lambda_t^{(i)} \mid \lambda_t = c] = E_\beta\left[\lambda_t^{(i)} \mid \lambda_t = c\right]$. If we choose for β a particular value β^* such that $m(\beta^*) = c$, then the condition $\lambda_t = c$ represents a typical case under the shifted distribution and we can replace the conditional expectation by an unconditional one to get

$$E\left[\lambda_t^{(i)} \mid \lambda_t = c\right] \approx E_{\beta^*}\{\lambda_t^{(i)}\} = m_i(\beta^*)$$

which upon substitution in (16.2.1) leads to the simple formula

$$P_{loss}^{(i)} \approx \frac{m_i(\beta^*)}{m_i} \, E\{p_t\}.$$

From (16.1.2), we have approximately $E\{p_t\} \approx E\{(\lambda_t - c)^+\}/c = \rho\, P_{loss}$. This gives the ratio of the individual to overall loss probability

$$\frac{P_{loss}^{(i)}}{P_{loss}} \approx \rho \, \frac{m_i(\beta^*)}{m_i}. \tag{16.2.2}$$

If both the individual source and the total stream have normal distributions, $N(m_i, \sigma_i^2)$ and $N(m, \sigma^2)$, (16.2.2) becomes (see also [Lin91])

$$\frac{P_{loss}^{(i)}}{P_{loss}} \approx \rho + (1 - \rho)\frac{m}{m_i}\frac{\sigma_i^2}{\sigma^2}. \tag{16.2.3}$$

Note that the minimum value of the ratio (16.2.2) is ρ which is obtained when $m_i(\beta^*) = m_i$, i.e., when $\lambda_t^{(i)}$ is constant. The maximum value of $P_{loss}^{(i)}/P_{loss}$ is obtained when the considered stream is the only variable bit rate stream in the whole traffic mix. In the Gaussian case (16.2.3) this means that $\sigma_i^2/\sigma^2 = \max = 1$, whence

$$\frac{P_{loss}^{(i)}}{P_{loss}} = \rho + (1 - \rho)\frac{m}{m_i}.$$

The ratio may be high, i.e., the losses are distributed unevenly among the streams, if m_i/m is small. Such a case where a small stream represents a major part of the rate variations is, however, exceptional. Normally we can assume that $m\sigma_i^2/m_i\sigma^2$ is of the order of one and the loss probabilities experienced by different streams are more or less of the same magnitude.

16.3   Consecutive cell losses

As we are considering congestion occurring at the burst level, the duration of such a congestion is of the order of a typical burst length and can be assumed to be long in comparison with the cell interarrival times. Thus, once a cell is lost the probability that the congestion still prevails when the next cell arrives is high. We can therefore approximate the probability of consecutive cell losses by the overall loss probability of a cell during congestion

$$P^c_{loss} = \Pr\left[\text{cell is lost} \mid \lambda_t > c \text{ upon cell arrival}\right] = P_{loss}/P^*_{sat},$$

where $P^*_{sat} = \Pr\{\lambda_t > c \text{ upon cell arrival}\}$ is the probability that a cell arrives during the congestion period. By considering a long time interval and equating the number of arrivals during the congestion periods to the sum of the number of cells served during these periods and the number of lost cells one sees that $\rho P^*_{sat} = P_{sat} + \rho P_{loss}$, where $P_{sat} = \Pr\{\lambda_t > c\}$ as defined in (16.1.8). Thus we get

$$P^c_{loss} = \frac{\rho P_{loss}}{P_{sat} + \rho P_{loss}}. \tag{16.3.1}$$

With the approximations (16.1.11) and (16.1.12), this becomes $P^c_{loss} \approx 1/(1 + c\beta^*)$ which is easy to evaluate with any given distribution of λ_t. For instance, assuming that λ_t has the normal distribution $N(m, \sigma^2)$ we have $\beta^* = (c - m)/\sigma^2$. The results for maximally loaded systems, according to the loss criterion $P_{loss} = \epsilon$, were given in Table 16.1.2 for the cases $\epsilon = 10^{-4}$ and $\epsilon = 10^{-9}$. In terms of the "burstiness factor" $B = \sigma/m$ and "safety factor" $\eta = (c - m)/\sigma$ used there, we can write $P^c_{loss} = 1/(1 + \eta/\rho B)$. We see that, irrespective of the required GOS level ϵ and the value of the "burstiness factor" B, P^c_{loss} remains relatively constant, around 1%. For instance, if we have $\rho = 0.8$, then $P^c_{loss} \approx 0.7\%$ in the case $P_{loss} = 10^{-9}$ and $P^c_{loss} \approx 2.6\%$ in the case $P_{loss} = 10^{-4}$. The probability for consecutive cell losses is not very high but for some applications it may have to be taken into account.

It may be noted from the results of Table 16.1.2 that for a given value of $P_{loss} = \epsilon$, P^c_{loss} is higher for higher values of the safety factor. This apparently surprising relation only means that once losses occur despite the higher safety factor, they are of more severe nature (i.e., as P_{sat} in (16.3.1) becomes smaller with increasing η then, for a constant P_{loss}, P^c_{loss} must be higher).

A similar "surprising" relation holds also in the case of delay systems: the loss probability during a buffer overflow is greater for larger buffers when P_{loss} is kept constant.

16.4   Cell loss process

The cell loss probability is a global measure of QoS which does not necessarily reflect the impact of congestion on user perceived performance. In this section we investigate the duration of rate overloads and the volume of information lost under certain assumptions concerning the arrival rate process λ_t. Define the duration θ of a congestion period by

$$\theta = \inf_{t>0}\{t, \lambda_t \leq c | \lambda_{0-} \leq c, \lambda_{0+} > c\} \tag{16.4.1}$$

and the volume V of information lost in such a period by

$$V = \int_0^\theta (\lambda_t - c)dt. \tag{16.4.2}$$

(Note that the above definition of θ is also equivalent to $\theta = \sup_{t>0}\{t, \lambda_s > c \,\forall s \in (0,t) \,|\, \lambda_{0-} \leq c, \lambda_{0+} > c\}$ as roughly introduced in [GS95, p.863]). The definitions of random variables θ and V are illustrated in Figure 16.4.1. Note the identity

$$P_{loss}^c \sim \frac{E\{V\}}{cE\{\theta\}}$$

derived from taking expectations in (16.4.2).

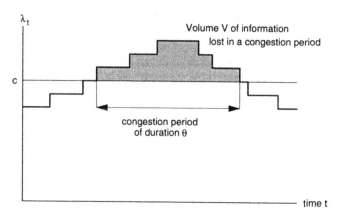

Figure 16.4.1: Instantaneous bit rate created by the superposition of bursts – the individual peak rate is taken as unity.

We consider the distributions of θ and V in the case of a specific traffic model: a large number of sources emit bursts at constant rate such that the burst arrival process is Poisson and the burst durations are exponentially distributed. The burst peak rate and mean duration are taken for units of rate and time, respectively. The arrival rate process is thus represented by the population of an $M/M/\infty$ system and the stationary distribution of λ_t is Poisson with mean m.

16.4.1   Large deviations analysis

We consider the case where the link capacity c and the mean arrival rate m scale together so that the load $\rho = m/c$ remains constant. This corresponds to a large deviations situation where the average of $\{\lambda_t\}$ is far from the congestion domain $(\lambda_t > c)$. In view of the Poisson distribution of λ_t with parameter m, approximation (16.3.1) implies that for fixed load ρ and large c,

$$P^c_{loss} \approx \frac{1}{c\,(1-\rho)}. \tag{16.4.3}$$

We thus deduce that the consecutive cell loss probability P^c_{loss} remains relatively small in the considered large deviations conditions (if $c - m$ is of order 100 Mbit/s while the individual peak bit rate is 1 Mbit/s, P^c_{loss} is about 1% only).

Concerning the congestion period θ and the volume V of work lost during a congestion period, it can be shown [GS95] that their respective mean values are given by

$$E\{\theta\} = \frac{1}{m}\sum_{j>c}\frac{c!}{j!}m^{j-c}, \tag{16.4.4}$$

$$E\{V\} = \frac{1}{m}\sum_{j>c}(j-c)\frac{c!}{j!}m^{j-c}. \tag{16.4.5}$$

Again, for fixed load ρ and large c, these formulae can be estimated by

$$E\{\theta\} \sim \frac{1}{c\,(1-\rho)}, \tag{16.4.6}$$

$$E\{V\} \sim \frac{1}{c\,(1-\rho)^2}. \tag{16.4.7}$$

Note, from estimation (16.4.6), that $c\,E\{\theta\}$ is equal to the mean busy period of an $M/M/1$ queue with unit service rate and arrival rate ρ. It can also be shown that $c\,E\{V\}$ from (16.4.7) is equal to the mean area swept by the queue length process of the same $M/M/1$ queue [GS95]. Estimations (16.4.3), (16.4.6) and (16.4.7) are thus consistent with the heuristic approach proposed by Aldous which claims that the behaviour of the occupancy process associated with the $M/M/\infty$ system around a state c can be approximated for large c by a linearization of the local characteristics of the process [Ald89]. In fact, it is shown in [GS95] that the Aldous heuristic is true for the entire *distributions* of variables θ and V. For fixed ρ and large c, we can state that

- the distribution of variable $c\theta$ converges to that of the busy period β of an $M/M/1$ queue with input rate ρ and unit service rate having density

$$\beta(t) = \frac{e^{-(1+\rho)t}}{t\sqrt{\rho}}I_1(2t\sqrt{\rho}), \quad t>0, \tag{16.4.8}$$

where I_1 is the modified Bessel function of the first kind of order 1;

- the distribution of variable cV converges to that of the area $\mathcal{A}$ swept during a busy period by the occupation process of an $M/M/1$ queue with input rate ρ and unit service rate; the Laplace transform of its density is given by

$$A^*(s) = \frac{1}{\sqrt{\rho}} \frac{J_{\nu(s)+1}(u(s))}{J_{\nu(s)}(u(s))} \qquad (16.4.9)$$

for $s > 0$, with $\nu(s) = (1 + \rho)/s$, $u(s) = 2\sqrt{\rho}/s$ and J_ν denoting the Bessel function of order ν.

Estimates (16.4.6) and (16.4.7) for averages and (16.4.8) and (16.4.9) for probability density functions therefore enable us to derive useful information on the distributions of congestion duration θ and volume V of lost information. The distributions of θ and V can also be derived using classical Markovian techniques. The resulting expressions involve the exponential of a matrix which can be evaluated numerically with arbitrary accuracy using a uniformization technique [GRSS95].

For large c and fixed t (resp. x), it is seen from numerical experiments that the convergence of probability $\Pr\{c\theta > t\}$ (resp. $\Pr\{cV > x\}$) towards $\Pr\{\beta > t\}$ (resp. $\Pr\{\mathcal{A} > x\}$) occurs from below. The limit $\Pr\{\beta > t\}$ (resp. $\Pr\{\mathcal{A} > x\}$) is thus an upper bound for the associated probability $\Pr\{c\theta > t\}$ (resp. $\Pr\{cV > x\}$). When $\rho = 0.8$ for example, this bound is seen to be very tight for c larger than 500 as shown in Figure 16.4.2.

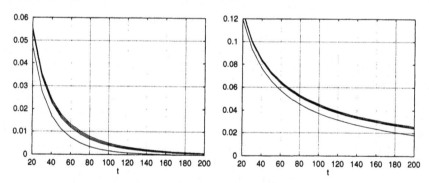

Figure 16.4.2: Distribution of $c\theta$ (upper figure) from top to bottom: $\Pr\{\theta_{M/M/1} > t\}$, $\Pr\{1000\theta > t\}$, $\Pr\{500\theta > t\}$, $\Pr\{100\theta > t\}$ when $\rho = 0.8$. Distribution of cV (lower figure) from top to bottom: $\Pr\{V_{M/M/1} > t\}$, $\Pr\{1000V > t\}$, $\Pr\{500V > t\}$, $\Pr\{100V > t\}$ when $\rho = 0.8$.

16.4.2 Heavy traffic analysis

Alternative expressions can be derived in the heavy traffic regime. Assume link capacity and traffic increase together with $c = m + \eta\sqrt{m}$ where η, the so-called "safety factor", is a positive constant so that $\rho \to 1$. The stationary distribution of the centered process $(\lambda_t - m)/\sqrt{m}$ can be then approximated by a standard Gaussian distribution.

For large m and fixed η, approximation (16.3.1) gives

$$P^c_{loss} \approx \frac{k}{\sqrt{m}}, \qquad (16.4.10)$$

where

$$k = \frac{\int_{x>\eta} (x - \eta) d\Phi(x)}{1 - \Phi(\eta)}$$

and Φ denotes the Gaussian distribution. Equivalent expressions derived for $E\{\theta\}$ and $E\{V\}$ read [GMS96]

$$E\{\theta\} \sim \frac{1}{\sqrt{m}} e^{\eta^2/2} \int_{x>\eta} e^{-x^2/2} dx, \qquad (16.4.11)$$

$$E\{V\} \sim e^{\eta^2/2} \int_{x>\eta} (x - \eta) e^{-x^2/2} dx. \qquad (16.4.12)$$

Note that the mean volume of lost information tends to a finite limit for fixed safety factor η while $E\{\theta\}$ and P^c_{loss} decrease like $1/\sqrt{m}$ (i.e., less rapidly than the large deviation estimates where η is assumed to be large).

It can be easily verified that heavy load estimates (16.4.10), (16.4.11), and (16.4.12) for P^c_{loss}, $E\{\theta\}$, and $E\{V\}$ tend to the respective large deviations estimates (16.4.3), (16.4.6), and (16.4.7) as the safety factor η increases. The former estimates can therefore be employed in either load situation.

Figures 16.4.3 and 16.4.4 compare the large deviations and heavy traffic estimates with simulation results [SGLR92]. The mean input rate is fixed at 240 Mbit/s with an individual burst peak rate of 2 Mbit/s while the link capacity varies between 240 Mbit/s and 310 Mbit/s (thus varying η between 0 and 4.5). The simulation results are obtained for four different burst duration distributions: constant, exponential, hyperexponential with squared coefficient of variation 2 and 10. The results show that expected congestion duration and the expected loss volume are relatively insensitive with respect to the burst length distribution and are accurately predicted in the considered load range ($\rho > 0.8$) by the heavy traffic estimates.

16.4.3 Stationary versus transient congestion probability

Define the transient congestion probability Π_T as the probability that the input rate process λ_t exceeds the capacity c during the interval $[0, T]$,

$$\Pi_T = \Pr\left\{ \max_{0 \leq t \leq T} \lambda_t > c \right\}. \qquad (16.4.13)$$

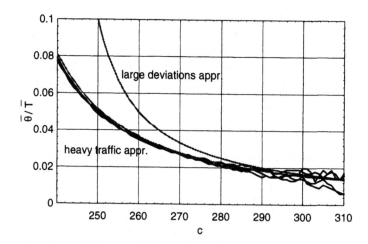

Figure 16.4.3: Ratio $\bar{\theta}/\bar{T}$ vs. link capacity.

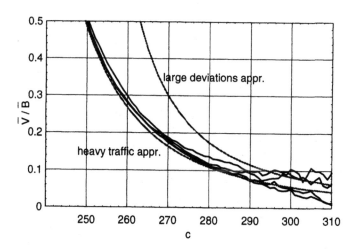

Figure 16.4.4: Ratio $\bar{V}/\bar{B}$ vs. link capacity.

Π_T can be interpreted as the probability of congestion (and consequent cell loss) during the transmission of a given burst of length T on a link of capacity $c+1$. Note that $P_{\text{sat}} = \lim_{T \to 0} \Pi_T$.

Applying the Poisson clumping heuristic of Aldous [Ald89, GS95], the point process $\{\mathcal{N}_t\}$ counting the number of times λ_t passes above c is approximately Poisson with intensity $\nu = P_{\text{sat}}/\mathrm{E}\{\theta\}$. Consequently, we have $\Pi_T = \Pr\{\mathcal{N}_T > 0\} \approx 1 - e^{-\nu T}$ which entails for small ν and fixed T, $\Pi_T \approx \nu T$. It follows that for sufficiently large c the ratio Π_T/P_{sat} can be estimated by

$$\frac{\Pi_T}{P_{\text{sat}}} \sim \frac{T}{\mathrm{E}\{\theta\}}. \qquad (16.4.14)$$

Using now the approximation (16.4.6) for $\mathrm{E}\{\theta\}$ along with (16.4.3), we derive from (16.4.14)

$$\frac{\Pi_T}{P_{\text{sat}}} \sim \frac{T}{P_{loss}^c}. \qquad (16.4.15)$$

Since $\mathrm{E}\{T\} = 1$, we deduce from (16.4.15) that the probability an arbitrary burst suffers cell loss is approximately equal to $P_{\text{sat}}/P_{loss}^c$. This is typically around one hundred times the stationary saturation probability P_{sat} (see Table 16.1.2).

17 Burst scale delay systems

In this chapter we consider several models for burst scale delay systems, i.e. systems where the buffering capacity is sufficient to absorb even large scale traffic variations. This leads to queueing problems where the input process can be regarded as a fluid flow process. (In another terminology, these models are *dams* with constant leak rate. One difference in orientation is that in the theory of dams the main concern is keeping the dam non-empty, whereas we are concerned about overflows. Moreover, constant leak rate is rather exceptional in connection with dams.)

The main object of interest is the distribution of the buffer content. This turns out to be difficult to obtain even in the logically simplest models. In Section 17.1 we consider a special class of fluid flow models where the fluid rate is modulated by an underlying finite state space Markov process. We review the theory of such systems and give several examples. The notion of effective bandwidth of buffered systems is also discussed in this context.

In Section 17.2 the Beneš method presented in Section 14.2.3 is applied to a number of traffic models, to derive exact and approximate expressions for the queue length distribution. The main advantage of the Beneš approach is that its applicability extends beyond the realm of Markovian systems.

Another useful modelling approach is reviewed in Section 17.3, viz. the discrete-time Batch Markovian Arrival Process (D-BMAP) models. These are in fact cell arrival models but because of the underlying Markov chain they are also able to capture burst scale phenomena.

The last two sections are devoted to more specialised topics. Section 17.4 addresses the problems that arise in studying the buffer content distribution of systems with long range dependent input traffic. Finally, in Section 17.5, modelling the leaky bucket performance is considered.

The following notation is used throughout this chapter. The fluid arrival rate at time t is denoted by $\Lambda(t)$ and the channel service rate by c. The system load ρ is thus equal to $E\{\Lambda(t)\}/c$. The amount of fluid (work) that arrives in time interval $(s, t]$ is

$$A(s,t) = \int_s^t \Lambda(u)du. \qquad (17.0.1)$$

and the amount of fluid in the buffer at time t is denoted by $X(t)$.

The behaviour of $X(t)$ is governed by the equation

$$X(t) = X(0) + \int_0^t (\Lambda(s) - c)1_{\{\Lambda(s)>c \text{ or } X(s)>0\}}ds, \quad t \geq 0. \qquad (17.0.2)$$

Note that if $\Lambda(s) > c$, then $X(s) > 0$ except for a set of measure zero, so that the first condition can be left out of the indicator in (17.0.2) without changing the value of the integral. However, $X \equiv 0$ is then also a solution of (17.0.2) so that for a given realisation of $\Lambda(t)$ there are several solutions for $X(t)$.

17.1   The case of Markov driven input

In many important models, $\Lambda(t)$ has the form

$$\Lambda(t) = r(Z(t)), \qquad (17.1.1)$$

where $Z(t)$ is a Markov process and $r(z)$ is a (deterministic) function. Such models are referred to as Markov Modulated Rate Processes (MMRP). In most models considered below, $r(z)$ is one-to-one so that Λ is itself a Markov process and we could also choose $r(z) = z$. The two-dimensional process $(Z(t), X(t))$ is then also a Markov process, which makes it possible to apply Markovian methods in the study of the (non-Markovian) process $X(t)$.

More specifically, we consider the case where $Z(t)$ is a birth-death process. Let the birth and death rate in state $i = 0, 1, \ldots$ be λ_i and μ_i, respectively. Denote by $r(i) = r_i$ the input rate when the state is i. Then the Chapman-Kolmogorov equations for the partial distribution functions $F_i(x) = \Pr\{Z = i, X \leq x\}$, $i = 0, 1, \ldots$ read

$$\mu_{i+1}F_{i+1} + \lambda_{i-1}F_{i-1} - (\lambda_i + \mu_i)F_i - (r_i - c)F_i' = 0. \qquad (17.1.2)$$

(Quantities with negative indices are assumed to be identically zero.) This can be written in matrix notation as

$$\mathbf{DF'} = \mathbf{MF}, \qquad (17.1.3)$$

where $\mathbf{F} = (F_0, F_1, \ldots)^T$, $\mathbf{D} = \mathbf{R} - c\mathbf{I}$ with $\mathbf{R} = \mathbf{diag}(r_0, r_1, \ldots)$, and $\mathbf{M}$, the infinitesimal generator of the modulating Markov chain, is given by the tri-diagonal matrix

$$\mathbf{M} = \begin{pmatrix} -\lambda_0 & \mu_1 & & \\ \lambda_0 & -(\lambda_1 + \mu_1) & \mu_2 & \\ & \lambda_1 & -(\lambda_2 + \mu_2) & \mu_3 \\ & & \ddots & \ddots & \ddots \end{pmatrix}.$$

Assuming that $r_i \neq c$ for all i, we have the formal solution

$$\mathbf{F}(x) = e^{\mathbf{D}^{-1}\mathbf{M}x}\mathbf{F}(0). \qquad (17.1.4)$$

We see that in any sufficiently regular case the asymptotic behaviour of the complementary buffer content distribution is exponential, that is,

$$Q(x) = \Pr\{X > x\} = 1 - \mathbf{1} \cdot \mathbf{F}(0) = 1 - \sum_i F_i(x) \sim \eta e^{-\zeta x}, \qquad (17.1.5)$$

where $-\zeta$ is the eigenvalue with the largest (closest to zero) negative real part, the so called dominant eigenvalue of $\mathbf{D}^{-1}\mathbf{M}$, and $\mathbf{1}$ is a row vector with all components equal to one.

A more explicit solution is obtained in terms of the full set of eigenvalues z_n and eigenvectors $\boldsymbol{\phi}_n$ of $\mathbf{D}^{-1}\mathbf{M}$. One of the eigenvalues is 0 because $\mathbf{M}$ is singular (all columns add up to 0) and the corresponding eigenvector $\boldsymbol{\pi}$ is the stationary probability vector of the modulating Markov process, i.e., $\mathbf{M}\boldsymbol{\pi} = 0$. Let this eigenvalue be z_0 and let the others be indexed in such a way that eigenvalues with negative real part have positive indices while those with positive real part have negative indices. Assume further the following ordering: $0 > \mathbb{R}(z_1) \geq \mathbb{R}(z_2) \geq \ldots$, i.e., z_1 is the dominant eigenvalue. Then the solution, which must be bounded, can be written [EM93]

$$\mathbf{F}(x) = \sum_{n>0} a_n \boldsymbol{\phi}_n e^{z_n x} + \boldsymbol{\pi}.$$

The coefficients a_n for $n > 0$ must be determined by the boundary conditions

$$F_i(0) = 0 \qquad \text{if} \quad r_i > c, \tag{17.1.6}$$

which express the fact that the queue cannot be empty when the input rate is greater than the output rate c and leads to the linear equations

$$\sum_{n>0} a_n \phi_{n,i} = -\pi_i \qquad \forall i: \ r_i > c. \tag{17.1.7}$$

These are in principle sufficient to determine the a_n for $n > 0$, though for a large state space dimension the solution of these equations may be computationally difficult. Given the a_n, we have $Q(x) = -\sum_{n>0} a_n \mathbf{1} \cdot \boldsymbol{\phi}_n e^{z_n x}$ and the coefficient η in (17.1.5) is given by $-a_1 \mathbf{1} \cdot \boldsymbol{\phi}_1$.

17.1.1   Effective bandwidth of a buffered system

We wish to determine the minimal required service rate $C_B(\epsilon)$ such that the probability $P_B = \text{Pr}\{X > B\}$ that the buffer occupancy exceeds some level B is below ϵ. In the sequel, a prescribed value ϵ is implicitly assumed and we suppress the dependence on it.

Our starting point is equation (17.1.5). Note that the matrix $\mathbf{D}$ and therefore also the dominant eigenvalue $-\zeta$ is a function of the service rate c. To emphasize this we write $\zeta = \zeta(c)$. In general, it is difficult to determine the value of the intercept of the asymptote, η. However, using importance sampling simulations ([KWC93], [MR95]) or analytical approximations ([BB93]) estimates for η have been derived.

Let us now assume that $\eta \approx 1$, cf. [GAN91]. Then the required service rate is

$$C_B = \zeta^{-1}(-\log \epsilon / B).$$

The inverse function $C(\cdot) = \zeta^{-1}(\cdot)$ is called *effective bandwidth* function. This term is justified by the important property shown in [EM93] that for a superposition of independent MMRP sources these functions are additive, i.e., if $-\zeta^{(k)}(c)$ is the dominant eigenvalue of the matrix $(\mathbf{R}^{(k)} - c\mathbf{I})^{-1}\mathbf{M}^{(k)}$ of the kth source and $C^{(k)}(\cdot)$ the corresponding inverse function, then the dominant eigenvalue $-\zeta$ of the superposed system satisfies

$$\sum_k C^{(k)}(-\zeta) = c,$$

and assuming again $\eta \approx 1$ the required service rate for given B and ϵ is given by

$$C_B = \sum_k C^{(k)}(-\log \epsilon / B).$$

Alternatively, the effective bandwidth function can be written as (see, e.g., [KWC93]),

$$C(\zeta) = \frac{1}{\zeta} \lim_{t \to \infty} \frac{1}{t} \log \mathrm{E} \left\{ \exp(\zeta A(t)) \right\} \tag{17.1.8}$$

where the additivity with respect to independent sources is obvious.

The validity of the above approach depends on how well η may be approximated by one. This approximation yields quite accurate results for a broad range of parameter values. However, η can be considerably smaller than 1 for sources with positive autocorrelation (bursty sources), implying that the approximation considerably overestimates the required bandwidth, while η can be greater than 1 for sub-bursty sources with an underestimation of the required bandwidth. Refinements to this approximation, particularly for a large number of sources, can be found in [CLW95, BD95].

In the next subsection it is shown that in the case of a high load η indeed converges to 1. In addition, we derive an insightful asymptotic formula for the effective bandwidth C_B, which holds when B becomes large. It is noticed that there are already several studies on the heavy traffic behaviour of Markov fluid queueing models. Knessl and Morrisson [KM91] derive approximations for the loss probability in the A-M-S model (cf. Section 17.1.2). Kobayashi and Ren [KR93] treat an extension to the case of heterogeneous traffic.

Asymptotic results for the case of a large buffer

It is clear that the effective bandwidth C_B decreases to the mean input rate m as the buffer size B tends to ∞. In this section it is shown that asymptotically the decay of C_B to m is hyperbolic, and

$$\lim_{B \to \infty} (C_B - m)B = -(\log \epsilon)\frac{v}{2}, \tag{17.1.9}$$

where ϵ denotes the predefined loss probability and v is the *asymptotic variance* of the arrival process: $v = \lim_{t \to \infty} (\mathrm{E}\{A(t)^2\} - \mathrm{E}\{A(t)\}^2)/t$, with $A(t)$ denoting the total amount of fluid that has arrived up to time $t \geq 0$.

To prove this result we first derive an asymptotic expression for the required level $B = B_c$ to achieve $P_B = \epsilon$, when $c \to m$, which is in some sense the dual problem of finding C_B. From this result for B_c we then deduce (17.1.9).

We start with the asymptotic behaviour of C_B. It should be realized that P_B, a_n, ϕ_n, and z_n depend on c. We therefore write $P_B(c), a_n(c), \phi_n(c)$, and $z_n(c)$. Consider now the following limit

$$\lim_{c \to m} \left(\frac{B_c}{c} - \frac{B_c}{m} \right).$$

Multiplying both sides in $\epsilon = -\sum_{n>0} a_n (1 \cdot \phi_n) \exp(z_n B)$ by $\exp(\zeta B)$ yields

$$B_c = \frac{1}{\zeta(c)} \log \left(\frac{-\sum_{n>0} a_n(c)(1 \cdot \phi_n(c)) e^{z_n(c) B_c} e^{\zeta(c) B_c}}{\epsilon} \right).$$

Now we show that $\zeta(c) \to 0$ as $c \to m$. It is known that $\zeta(c)$ is the unique positive solution to $C(\zeta) = c$, see [EM93]. They also showed that the effective bandwidth function $C(\cdot)$ attains m in 0, and is an increasing function. Letting $c \to m$, we consequently have $\zeta(c) \to 0$. For reasons of continuity, $\phi_1(c) \to \pi$. Also, for $n = 2, \ldots$, $z_n(c) \to z_n(m)$ and $\phi_n(c) \to \phi_n(m)$. From the set of linear equations (17.1.7), it follows that $a_1(c) \to -1$, whereas $a_n(c) \to 0$, for $i = 2, \ldots$. Thus we get

$$\lim_{c \to m} \left(\frac{B_c}{c} - \frac{B_c}{m} \right) = (\log \epsilon) \lim_{c \to m} \left(\frac{1}{m \zeta(c)} - \frac{1}{c \zeta(c)} \right)$$

$$= (\log \epsilon) \lim_{c \to m} \left(\frac{c - m}{mc \, \zeta(c)} \right). \qquad (17.1.10)$$

The right limit in (17.1.10) ('0 over 0') can be evaluated as follows. We recall that $C(\cdot)$ is the inverse of $\zeta(\cdot)$. The effective bandwidth function $C(\cdot)$ of a Markov fluid source is an analytical, increasing function defined on $[0, \infty)$, with $C(0) = m$. Therefore, $\zeta(\cdot)$ is increasing on $[m, \infty)$ and $\zeta(m) = 0$. We write $\zeta(\cdot)$ as a power series in a neighborhood of m: $\zeta(c) = \zeta(m) + (c - m)\zeta'(m) + O((c - m)^2)$. We get that the limit under consideration equals $(\log \epsilon)(m^2 \zeta'(m))^{-1} = (\log \epsilon) m^{-2} C'(0)$.

But using the alternative definition of the effective bandwidth (17.1.8) and inserting the Taylor expansions of log and exp provides us the first order approximation of $C(\cdot)$ in the neighborhood of 0: $C(\zeta) = m + \zeta v/2 + O(\zeta^2)$, where v denotes the asymptotic variance of the arrival process as defined above. Therefore, $C'(0) = v/2$ and we conclude

$$\lim_{c \to m} \left(\frac{B_c}{c} - \frac{B_c}{m} \right) = (\log \epsilon) \frac{v}{2m^2} \qquad (17.1.11)$$

Now it is a matter of algebra to derive the behaviour of C_B in (17.1.9) from the behaviour of B_c.

We saw that under heavy traffic $a_1(c) \to -1$ and $\phi_1(c) \to \pi$, implying that $\eta(c)$ in the asymptotic formula $\eta(c) \exp(-\zeta(c)B)$ tends to 1, so that the [GAN91] results become quite accurate as $c \to m$. But in that region, we found that $C(\zeta)$ can be approximated by the simpler formula $m + \zeta v/2$.

Implications of the asymptotic result

This subsection deals with some reflections on the simple asymptotic formula (17.1.9) for the effective bandwidth C_B. It is easily seen that

$$C_B \sim m - (\log \epsilon) \frac{v}{2B}, \qquad \text{for large } B. \qquad (17.1.12)$$

The right hand side of (17.1.12) has a very intuitive structure: C_B equals the mean input rate plus an additional amount to cope with the variability of the input process. This additional amount is, of course, decreasing in both ϵ and B. Further it is noticed that, in case of a group of independent sources, a simple *additivity* property holds (cf. [GAN91]): due to the additivity of means and variances of independent sources, the effective bandwidth equals the sum of the effective bandwidths of the individual sources.

The general Markov fluid framework allows us to study the case of on/off sources ($r_0 = 0$, $r_1 = r$) with any phase-type distribution for the on and off times. However, for most of these distributions the effective bandwidth function $C(\cdot)$ of a single source cannot be given explicitly. Now notice that the asymptotic formula (17.1.12) for the effective bandwidth only requires the mean input rate m and the derivative of the bandwidth function at zero $C'(0) = v/2$. The latter can be calculated from the implicit relations between ζ and $C(\zeta)$, as found in [Kos86]. For instance, in case of an Erlang(n) on-time (mean $1/\lambda$) and an exponential off-time (mean $1/\mu$), we have

$$(\lambda n + \zeta C(\zeta))^n (\mu + \zeta(C(\zeta) - r)) - (\lambda n)^n \mu = 0,$$

yielding after implicit differentiation

$$v \equiv 2C'(0) = \frac{\lambda \mu r^2}{(\lambda + \mu)^3} \frac{n+1}{n}.$$

In the limit $n \to \infty$, i.e., when the on-times become deterministic, we get $v = (\lambda \mu r^2)/(\lambda + \mu)^3$.

Numerical example

We now consider a numerical example in order to illustrate the asymptotic behaviour of C_B, when B becomes large. Suppose two types of input traffic. The first (second) group of sources consists of 4 exponential on/off sources, with $1/\lambda = 1$ (0.25), $1/\mu = 4$ (4.75), and $r/c = 1$ (4). Hence, the mean input rate $m = 1.6$. The superposition of the two groups of sources yields a single,

25-state Markov fluid source. Table 17.1.1 contains, for increasing values of B, simulation and approximation results for the required capacity C_B to keep P_B below $\epsilon = 10^{-4}$. The approximation results have been obtained by applying the asymptotic formula (17.1.9). The simulation results have been obtained by 'fast simulation' using importance sampling, see e.g. [KWC93] and [MR95].

Table 17.1.1: The accuracy of the asymptotic approximation.

B	$C_B^{(\text{sim})}$	$C_B^{(\text{app})}$	B	$C_B^{(\text{sim})}$	$C_B^{(\text{app})}$
5	5.04	3.87	30	2.05	1.98
10	3.31	2.73	35	1.97	1.92
15	2.63	2.36	40	1.93	1.88
20	2.32	2.17	45	1.89	1.85
25	2.15	2.05	50	1.86	1.83

It is seen that the approximation results $C_B^{(\text{app})}$ indeed approach the simulation results $C_B^{(\text{sim})}$, when B grows. It appears that the asymptotic formula, which has been derived under heavy traffic conditions, seems to be useful in situations with lower (i.e., practically relevant) loads as well.

17.1.2 Examples of Markov driven systems

The Kosten model

In this model, extensively studied in [Kos74], the arrival process consists of a superposition of independent bursts all having the same constant rate h. This can be considered as a superposition of infinitely many on/off sources with an infinitesimal on-probability. The bursts start independently according to a Poisson process with parameter λ and their lengths are exponentially distributed with parameter β. This is a special case of a more general Poisson burst model considered in Section 13.3.3.

The fluid arrival rate is then $\Lambda(t) = Z(t)h$, where $Z(t)$ is the number of bursts going on at time t. $Z(t)$ is a birth-death process with birth rate λ and death rate $Z(t)\beta$. It follows from the equilibrium equations that the stationary distribution of $Z(t)$ is Poisson(λ/β). Thus we have

$$E\{\Lambda\} = \frac{\lambda h}{\beta}, \quad \text{Var}\{\Lambda\} = \frac{\lambda h^2}{\beta}. \tag{17.1.13}$$

For the covariance of the process $\Lambda(t)$ and the variance of $A(0,t)$ we refer to Section 13.3.3.

Let us now turn to the buffer content $X = X(t)$. Its distribution was thoroughly studied by Kosten [Kos74]. It turns out that all components of the matrix form solution (17.1.4) cannot be expressed in closed form, but an algorithmic solution involving numerical integration of (17.1.2) is given in [Kos74]. Kosten also obtains all eigenvalues of $\mathbf{D}^{-1}\mathbf{M}$. The spectrum consists of a numerable set of discrete eigenvalues with a negative lower bound $-\beta/h$ which is also a cumulation point of the spectrum. The largest negative eigenvalue, determining the asymptotic exponential behaviour of $\Pr\{X > x\} \sim \eta \exp(-\zeta x)$, is

$$-\zeta = -\frac{\beta}{h}(1 - \rho). \tag{17.1.14}$$

The Anick-Mitra-Sondhi (A-M-S) model

The model studied by Anick, Mitra and Sondhi [AMS82] only differs from the preceding one by the upward intensities of the birth-death process $Z(t)$. We now consider a fixed number, say N, of identical sources which switch between burst and silence periods like independent two-state Markov processes. Again let the rate within a burst be h, and assume that each burst length has distribution $\text{Exp}(\beta)$ and each silence period has distribution $\text{Exp}(\alpha)$. A superposition of more general on/off sources was considered in Section 13.3.2. Note that with large N and small α the A-M-S model is close to the Kosten model.

In the stationary case, the number of active sources K now has the binomial distribution $Bin(N, (\alpha/(\alpha + \beta)))$, so that the mean and variance of the input rate are

$$\text{E}\{\Lambda\} = \frac{N\alpha h}{\alpha + \beta}, \quad \text{Var}\{\Lambda\} = \frac{N\alpha\beta h^2}{(\alpha + \beta)^2}. \tag{17.1.15}$$

The covariance function of $\Lambda(t)$ and the variance of $A(0,t)$ were given in Section 13.3.2.

Let us consider the queue. The equation system (17.1.2) is now finite, and correspondingly the spectrum is finite consisting of N discrete eigenvalues. Anick et al. obtain an explicit solution using several analytic and matrix-theoretic arguments. The distribution of the buffer content X is a mixture of exponential distributions together with an atom at the origin, i.e.,

$$Q(x) = \Pr\{X > x\} = \sum_{n=1}^{N-\lfloor 1/h \rfloor} b_n e^{z_n x}, \tag{17.1.16}$$

where the z_n's are the negative eigenvalues of $\mathbf{D}^{-1}\mathbf{M}$ in (17.1.3). Reference [AMS82] gives straightforward procedures for calculating the z_n's and the b_n's. The largest negative eigenvalue, which determines the asymptotic behaviour of (17.1.16), turns out to be

$$-\zeta = -\frac{(\alpha + \beta)(1 - \rho)}{h - c/N}. \tag{17.1.17}$$

Note that (17.1.17) converges to (17.1.14) when $N \to \infty$ and $\alpha \to 0$ so that ρ remains unchanged.

A modification of the A-M-S model with finite buffer was analysed by Tucker [Tuc88], and a further generalisation with heterogeneous input traffic has been studied by Kosten [Kos84] and by Jacobsen and Dittmann [JD90]. In the latter models each source can have its individual on/off and rate characteristics α_i, β_i and h_i. The state space of the arrival process is then multidimensional. Remarkably, the general solution (17.1.4) still holds with $\mathbf{F}(x)$, $\mathbf{D}$ and $\mathbf{M}$ having substantially higher dimensions but essentially the same structure. The finiteness of the buffer is taken into account through the special initial conditions that the buffer is never full when the input rate is smaller than server capacity. The eigenvalues of $\mathbf{D}^{-1}\mathbf{M}$ can again be derived from simple equations, whereas the eigenvectors and thus the buffer occupancy distribution have to be calculated numerically. For this finite buffer case all eigenvectors (positive and negative) must be retained in the expansion of $\exp\{\mathbf{D}^{-1}\mathbf{M}\}$.

M/M/1 output driven fluid queue

Another model of the same genre was studied in [VN94]. In this model the fluid queue, which is served at rate c, receives its input from the output of an $M/M/1$ queue such that the input rate to the fluid queue is c_0 whenever the $M/M/1$ queue is non-empty, i.e., during its busy periods, and is zero during the idle periods when the $M/M/1$ queue is empty. This model was motivated in the context of studying an interworking unit which receives traffic from a shared medium network. In the present setting of MMRP models, the system can be summarized as follows: $Z(t)$ is the buffer occupancy in an $M/M/1$ system and $r(z)$ is the non-linear function

$$r(z) = \begin{cases} 0, & \text{if } z = 0; \\ c_0, & \text{if } z > 0. \end{cases}$$

The analysis in [VN94] was based on the study of the eigenvalues and eigenvectors of $\mathbf{D}^{-1}\tilde{\mathbf{M}}$ where $\tilde{\mathbf{M}} = \mathbf{E}\mathbf{M}\mathbf{E}^{-1}$ is the symmetrized form obtained with the similarity transform by $\mathbf{E} = \mathrm{diag}(1, \rho^{-1/2}, \rho^{-2/2}, \ldots)$. There is a marked qualitative difference between the spectrum of this system and those of the Kosten system and the A-M-S system. In the present case the spectrum, depicted in Figure 17.1.1, consists of a continuous part in the interval $[-(1 + \sqrt{\rho})^2, -(1 - \sqrt{\rho})^2]$ and a discrete eigenvalue at 0, and additionally, when $\rho_0 > \kappa$, of another discrete eigenvalue at $-(1 - \kappa)(1 - \rho_0)$. Here ρ and ρ_0 denote the loads of the $M/M/1$ and the fluid queues, respectively, and $\kappa = c/c_0$ is the "slow-down" factor.

After some rather technical steps the analysis leads to the complementary buffer content distribution

$$Q(x) = \frac{\kappa - \rho}{(1 - \kappa)^2} \cdot \left(\int_{-1}^{1} w(u) e^{z(u)x} du + w_1 e^{-(1-\kappa)(1-\rho_0)x} \right), \qquad (17.1.18)$$

Figure 17.1.1: The spectrum for $\rho_0 < \kappa$, $\rho_0 = \kappa$, and $\rho_0 > \kappa$.

where

$$z(u) = 2\sqrt{\rho}\,u - (1+\rho),$$

$$w(u) = \frac{2}{\pi} \frac{(1-u^2)^{1/2}}{1-u^2+(au-b)^2}, \qquad w_1 = 1_{\rho_0 > \kappa} \frac{(\rho_0 - \kappa)(1-\kappa)}{\kappa - \rho},$$

and

$$a = \frac{1+\kappa}{1-\kappa}, \qquad b = \frac{\kappa + \rho}{(1-\kappa)\sqrt{\rho}}.$$

The same result has been rederived by Adan and Resing [AR96] and by Cohen [Coh95]. They both first derive the Laplace transform of the buffer content distribution. Adan and Resing start from the observation that the input process of the fluid queue is an on/off renewal process and cast the problem into an equivalent $M/G/1$ queue problem. Cohen analyses equations (17.1.2) directly by means of a z-transform. His expression for the Laplace transform is particularly simple, $(1 - \rho_0)/(1 - \rho_0 G^*(s))$, where $G^*(s)$ is the well-known Laplace transform of the busy period length in an $M/M/1$ system (see e.g. [Kle75b], equation (5.144)). By making the inverse transform they arrive at (17.1.18) (also correcting the sign error in the formula for w_1 in [VN94]).

A two-level model

Since the solutions of the above models are rather complicated, let us consider a very crude model where the solution of (17.1.2) is readily computable. Assume that the Markov process $Z(t)$ in (17.1.1) has only two states, say 0 and 1, with $r_0 < c < r_1$. Denote the transition intensities of $Z(t)$ by $\lambda = \lambda_0$ and $\mu = \mu_1$. The stationary probabilities of the states 0 and 1 are then, respectively, $\pi_0 = \mu/(\lambda + \mu)$ and $\pi_1 = \lambda/(\lambda + \mu)$.

The stationary first and second order moments of $\Lambda(t)$ are

$$E\{\Lambda\} = r_0 + (r_1 - r_0)\pi_1, \tag{17.1.19}$$

$$\text{Cov}\{\Lambda_u, \Lambda_v\} = (r_1 - r_0)^2 \pi_0 \pi_1 e^{-(\lambda + \mu)|u - v|}.$$

The equations (17.1.2) reduce to

$$(r_0 - c)F_0' = -\lambda F_0 + \mu F_1,$$

$$(r_1 - c)F_1' = \lambda F_0 - \mu F_1.$$

The solution is obtained by standard methods. Using the boundary conditions $F_1(0) = 0$, $F_1(\infty) = \pi_1$ we finally get

$$F_0(x) = \pi_0 - \pi_1 \frac{r_1 - c}{c - r_0} e^{-\zeta x},$$

$$F_1(x) = \pi_1(1 - e^{-\zeta x}), \tag{17.1.20}$$

where

$$\zeta = \frac{(c - r_0)\mu - (r_1 - c)\lambda}{(r_1 - c)(c - r_0)}.$$

The complementary queue length distribution is then

$$Q(x) = \Pr\{X > x\} = \pi_1 \frac{r_1 - r_0}{c - r_0} e^{-\zeta x}. \tag{17.1.21}$$

The two-level model can be used as an approximation for more complicated systems by choosing the parameters so that certain characteristics of the input process match with the true ones. For three conditions we have natural choices: the mean m, the variance σ^2 and the correlation function exponent γ of $\Lambda(t)$ should match (assuming that the true correlation is also exponential). As a fourth fitting relation, we choose to equate also the expected surplus arrival rates $\Sigma = E\{(\Lambda - c)^+\}$ of the two processes. The parameters of the two-level model can then be calculated by the following formulae:

$$r_1 = m + \frac{\sigma^2}{2\Sigma} - \sqrt{\left(\frac{\sigma^2}{2\Sigma}\right)^2 - \sigma^2 - \frac{\sigma^2}{\Sigma}(c - m)},$$

$$r_0 = m - \frac{\sigma^2}{r_1 - m}, \tag{17.1.22}$$

$$\mu = \gamma \frac{r_1 - m}{r_1 - r_0},$$

$$\lambda = \gamma - \mu.$$

The Ornstein-Uhlenbeck model

Here we consider the queueing problem where the fluid input rate is assumed to constitute an Ornstein-Uhlenbeck process introduced in Subsection 13.3.5. We apply the Beneš method presented in Section 14.2.3. The analysis is presented in [SV91]. By (14.2.7) we have for $Q(x) = \Pr\{X > x\}$ the upper bound

$$Q(x) \leq \int_{u>0} du \int_0^1 d\lambda \, (c - \lambda) \frac{\partial^2}{\partial x \partial \lambda} \Pr\{A_u \leq x + u, \Lambda_u \leq \lambda\}.$$

The integral with respect to λ turns out to be expressible in closed form, so that only the outer integral remains to be calculated numerically. Denote, as before, $\sigma^2 = \gamma^2/(2\beta)$, $\eta = (c - m)/\sigma$, and

$$f(\theta) = 2(e^{-\theta} - 1 + \theta),$$

$$g(\theta) = 2\theta - 3 + 4e^{-\theta} - e^{-2\theta},$$

$$h(\theta) = 1 - e^{-\theta},$$

and let Φ be the standard Gaussian distribution function. The solution is then:

$$Q(x) \leq q\left(\frac{\beta x}{\sigma}\right), \tag{17.1.23}$$

with

$$q(y) = \frac{1}{\sqrt{2\pi}} \int_0^\infty \frac{d\theta}{\sqrt{f(\theta)}} \exp\left(-\frac{S(\theta, y)}{2}\right) R(\theta, y),$$

where

$$S(\theta, y) = \frac{(\eta\theta + y)^2}{f(\theta)},$$

$$R(\theta, y) = \sqrt{\frac{g(\theta)}{f(\theta)}}\left(T(\theta, y)\Phi(T(\theta, y)) + \frac{1}{\sqrt{2\pi}}e^{-T^2(\theta, y)/2}\right),$$

$$T(\theta, y) = \sqrt{\frac{f(\theta)}{g(\theta)}}\left(\eta - (\eta\theta + y)\frac{h(\theta)}{f(\theta)}\right).$$

The asymptotic behaviour of the upper bound (17.1.23) can be studied analytically yielding

$$q(y) \sim \frac{e^{-\eta^2}}{\eta\sqrt{2\pi}} \cdot e^{-\eta y},$$

where

$$\eta = \frac{2\beta^2}{\gamma^2}(1 - \rho).$$

A comparison with the exact results in [Sim91] shows that the asymptotic given by (17.1.2) is fairly accurate if the load is not too high (see [SV91]).

17.1.3   Characterization of the output process

In this subsection our purpose is to characterize the output process from a stochastic fluid queue driven by a Markov jump process, i.e., a Markov process with discrete state space. In certain cases, e.g., if the system is loaded by homogeneous and exponential on/off sources (as in the A-M-S [AMS82] model discussed above), it is possible to find another Markov jump process that modulates the output rate. As we will see, the process modulating the output rate is essentially more complex (having usually more dimensions and always an infinite state space) than the process that modulates the input rate. However, under certain assumptions, both modulating processes are quite simple birth-death processes. The results of the subsection appeared first in [Aal94].

Model

Consider the A-M-S model [AMS82] discussed in the previous Subsection. The system is loaded by N identical and exponential on/off sources. Denote by h the rate within a burst of a source and by c the leak rate of the buffer. For the input rate $\Lambda(t)$, we have

$$\Lambda(t) = Z(t)h,$$

where the modulating process $Z(t)$ is a birth-death process with a finite state space $\mathcal{E} = \{0, 1, \ldots, N\}$. The rates $q(i, j)$ for possible state transitions are given in the following table.

i	j	$q(i,j)$
$i < N$	$j = i+1$	$(N-i)\alpha$
$i > 0$	$j = i-1$	$i\beta$

Denote by $R(t)$ the output rate. It is modulated together by $Z(t)$ and the buffer content process $X(t)$ as follows:

$$R(t) = \begin{cases} \min\{Z(t)h, c\} & X(t) = 0, \\ c, & X(t) > 0. \end{cases}$$

Although the pair $(Z(t), X(t))$ is a Markov process, the problem is that it is not a Markov *jump* process.

Assumptions

Assume that $Nh > c$. In the opposite case the buffer would always be empty, and the output rate would equal the input rate. Let

$$M = \max\{m \in \mathcal{E} \mid mh \leq c\}.$$

We split the state space $\mathcal{E}$ into two complementary sets as follows:

$$\mathcal{E}_0 = \{0, 1, \ldots, M\}, \qquad \mathcal{E}_1 = \{M+1, M+2, \ldots, N\}.$$

The assumption made above implies that both sets are non-empty. States in $\mathcal{E}_0$ ($\mathcal{E}_1$) may be called underloaded (overloaded) states, since the buffer content is decreasing (increasing) when $Z(t) \in \mathcal{E}_0$ ($\mathcal{E}_1$). An underloaded (overloaded) interval corresponds to a visit to set $\mathcal{E}_0$ ($\mathcal{E}_1$). Another assumption is that $Mh < c$ (excluding only the case $Mh = c$).

Results

Under the assumptions given above, it was proved in [Aal94] that the output rate $R(t)$ in the A-M-S model is modulated by another Markov jump process $\tilde{Z}(t) = (\tilde{Z}_1(t), \tilde{Z}_2(t), \tilde{Z}_3(t))$ as follows:

$$R(t) = \begin{cases} \tilde{Z}_1(t)h, & \tilde{Z}(t) \in \tilde{\mathcal{E}}_0, \\ c, & \tilde{Z}(t) \in \tilde{\mathcal{E}}_1. \end{cases}$$

The state space $\tilde{\mathcal{E}}$ of $\tilde{Z}(t)$ consists of the following two complementary sets:

$$\tilde{\mathcal{E}}_0 = \mathcal{E}_0 \times \{\Delta\} \times \{0\}, \qquad \tilde{\mathcal{E}}_1 = \mathcal{E}_0 \times \mathcal{E}_1 \times \{1, 2, \ldots\},$$

where Δ denotes an element not included in $\mathcal{E}$. The rates $\tilde{q}(\mathbf{x}, \mathbf{y})$ for possible state transitions are given in the following table.

$\mathbf{x}$		$\mathbf{y}$		$\tilde{q}(\mathbf{x}, \mathbf{y})$
$\mathbf{x} \in \tilde{\mathcal{E}}_0$	$x_1 < M$	$\mathbf{y} \in \tilde{\mathcal{E}}_0$	$\mathbf{y} = \mathbf{x} + \mathbf{e}_1$	$q(x_1, x_1 + 1)$
$\mathbf{x} \in \tilde{\mathcal{E}}_0$	$x_1 > 0$	$\mathbf{y} \in \tilde{\mathcal{E}}_0$	$\mathbf{y} = \mathbf{x} - \mathbf{e}_1$	$q(x_1, x_1 - 1)$
$\mathbf{x} \in \tilde{\mathcal{E}}_0$	$x_1 = M$	$\mathbf{y} \in \tilde{\mathcal{E}}_1$	$\mathbf{y} = (M, M+1, 1)$	$q(x_1, x_1 + 1)$
$\mathbf{x} \in \tilde{\mathcal{E}}_1$	$x_2 = M + 1,$ $x_3 = 1$	$\mathbf{y} \in \tilde{\mathcal{E}}_0$	$y_1 = x_1$	$\dfrac{\tilde{f}(\mathbf{x})}{f(x_2)} q(x_2, x_2 - 1)$
$\mathbf{x} \in \tilde{\mathcal{E}}_1$	$x_1 < M$	$\mathbf{y} \in \tilde{\mathcal{E}}_1$	$\mathbf{y} = \mathbf{x} + \mathbf{e}_1$	$\dfrac{\tilde{f}(\mathbf{x})}{f(x_1)} q(x_1, x_1 + 1)$
$\mathbf{x} \in \tilde{\mathcal{E}}_1$	$x_1 > 0$	$\mathbf{y} \in \tilde{\mathcal{E}}_1$	$\mathbf{y} = \mathbf{x} - \mathbf{e}_1$	$\dfrac{\tilde{f}(\mathbf{x})}{f(x_1)} q(x_1, x_1 - 1)$
$\mathbf{x} \in \tilde{\mathcal{E}}_1$	$x_2 < N$	$\mathbf{y} \in \tilde{\mathcal{E}}_1$	$\mathbf{y} = \mathbf{x} + \mathbf{e}_2$	$\dfrac{\tilde{f}(\mathbf{x})}{f(x_2)} q(x_2, x_2 + 1)$
$\mathbf{x} \in \tilde{\mathcal{E}}_1$	$x_2 > M + 1$	$\mathbf{y} \in \tilde{\mathcal{E}}_1$	$\mathbf{y} = \mathbf{x} - \mathbf{e}_2$	$\dfrac{\tilde{f}(\mathbf{x})}{f(x_2)} q(x_2, x_2 - 1)$
$\mathbf{x} \in \tilde{\mathcal{E}}_1$	$x_1 = M$	$\mathbf{y} \in \tilde{\mathcal{E}}_1$	$\mathbf{y} = \mathbf{x} + \mathbf{e}_3$	$\dfrac{\tilde{f}(\mathbf{x})}{f(x_1)} q(x_1, x_1 + 1)$
$\mathbf{x} \in \tilde{\mathcal{E}}_1$	$x_2 = M + 1,$ $x_3 > 1$	$\mathbf{y} \in \tilde{\mathcal{E}}_1$	$\mathbf{y} = \mathbf{x} - \mathbf{e}_3$	$\dfrac{\tilde{f}(\mathbf{x})}{f(x_2)} q(x_2, x_2 - 1)$

Here $\mathbf{e}_k$ is a unit vector to direction k,

$$f(i) = \begin{cases} c - ih, & i \in \mathcal{E}_0, \\ ih - c, & i \in \mathcal{E}_1. \end{cases}$$

and

$$\tilde{f}(\mathbf{x}) = \frac{1}{\dfrac{1}{f(x_2)} + \dfrac{1}{f(x_1)}} = \frac{(x_2 h - c)(c - x_1 h)}{(x_2 - x_1)h}, \qquad \mathbf{x} \in \tilde{\mathcal{E}}_1.$$

Remarks

The essential point in the proof of the previous result is the observation that the number of overloaded intervals during a non-empty period of the fluid buffer coincides with the number of customers during a busy period in a queue constructed as follows. The amount of excessive fluid that flows in the buffer during an overloaded interval corresponds to the service time of a customer, and the amount of excessive capacity during an underloaded interval corresponds to the interarrival time between two consecutive customers.

Note that, in general, a non-empty period and the following empty period of the fluid buffer are not independent. The length of the empty period depends on the state of the modulating process $\tilde{Z}(t)$ (in fact, on the first component only) at the end of the non-empty period.

We also note that the previous result may be quite easily generalized to stochastic fluid queues driven by other Markov birth-death processes with a finite state space. In fact, the only direct modifications concern state transition rates $q(i, j)$ and the function f.

Special cases

Assume now that the input rate even from a single active source is greater than the leak rate of the buffer, i.e., $h > c$. The modulating process $\tilde{Z}(t)$ is in this case essentially two-dimensional, since $\mathcal{E}_0 = \{0\}$. Moreover, it is easy to see that the output rate is modulated by the sum $\hat{Z}(t) = \tilde{Z}_2(t) + \tilde{Z}_3(t)$ (understanding that $\hat{Z}(t) = 0$ whenever $\tilde{Z}(t) \in \tilde{\mathcal{E}}_0$),

$$
R(t) = \begin{cases} 0, & \hat{Z}(t) = 0, \\ c, & \hat{Z}(t) > 0, \end{cases}
$$

and $\hat{Z}(t)$ is a birth-death process with an infinite state space $\{0, 1, \ldots\}$. The rates $\hat{q}(i, j)$ of possible state transitions are given in the following table.

i	j	$\hat{q}(i, j)$
$i = 0$	$j = i + 1$	$N\alpha$
$i > 0$	$j = i + 1$	$(N - c/h)\alpha$
$i > 0$	$j = i - 1$	$\beta c/h$

Note that the fluid queue behaves just like an on/off source with transmission rate c. Idle periods are exponential with intensity $N\alpha$, and active periods behave as busy periods in an ordinary $M/M/1$ queue with arrival rate $(N - c/h)\alpha$ and service rate $\beta c/h$. Furthermore, non-empty and empty periods of the buffer are mutually independent in this case.

Finally we note that the previous result may be generalized to stochastic fluid queues loaded by homogeneous on/off sources with arbitrarily distributed active periods (idle periods are still assumed to be exponential).

Denote by $F(x)$ the distribution function of an active period of a source. In this case active periods on the output line behave as busy periods in an $M/G/1$ queue with arrival rate $(N - c/h)\alpha$ and service time distribution function $F(xc/h)$.

17.2   Beneš method for fluid queues

In this section we apply the Beneš result for queues with fluid input (cf. Section 14.2.3) to a number of traffic models, deriving exact and approximate expressions for the queue length distribution. We first consider the superposition of a finite number of on/off sources where the lengths of successive on and off periods are independent having general Coxian distributions. A different approach is necessary for the practically important case of a superposition of deterministic on/off sources. Large deviations techniques are used to derive asymptotic approximations for systems where the relevant buffer occupancy levels and the server capacity scale with the number of sources. Finally, we consider the queue length distribution when the offered traffic derives from a set of Poisson sources emitting generally distributed bursts. The case of long range dependent traffic, which is also partially amenable to the Beneš approach, is deferred to Section 17.4.

17.2.1   Superposition of on/off sources

We assume that the channel capacity is c and that the input consists of a superposition of N heterogeneous on/off sources, source i being defined as follows:

- successive burst and silence durations are independent;

- each silence duration has density $a_i(\cdot)$ and mean $1/\alpha_i$;

- each burst duration has density $b_i(\cdot)$ and mean $1/\beta_i$;

- during a burst, work is generated at rate h_i.

Defining $D_i(t) = 1_{\{\text{source } i \text{ is on at } -t\}}$, the Beneš result for fluid queues (14.2.7) may be written:

$$Q(x) = \sum_{\mathbf{d \cdot h} < c} (c - \mathbf{d \cdot h}) \int_0^\infty \frac{\partial}{\partial x} \Pr\{A_t \le x + ct, \mathbf{D}(t) = \mathbf{d}, V_{-t} = 0\}\, dt$$

$$\tag{17.2.1}$$

$$= \sum_{\mathbf{d \cdot h} < c} (c - \mathbf{d \cdot h}) \int_0^\infty \phi_{\mathbf{d}}(t, x + ct) \cdot \Pr[V_{-t} = 0 \mid A_t = x + ct, \mathbf{D}(t) = \mathbf{d}]\, dt$$

$$\tag{17.2.2}$$

where

$$\phi_{\mathbf{d}}(t, w) \;=\; \frac{\partial}{\partial w}\,\mathrm{Pr}\,\{A_t \le w\,,\; D_i(t) = d_i,\; i = 1,\ldots,N\}\,.$$

In the particular case where $h_i \ge c$ for all i (i.e., all sources emit at least at the channel rate), the summation in (17.2.2) disappears since only the idle state $\mathbf{D}(t) = \mathbf{0}$ counts. If we further assume that all source silence intervals have the exponential distribution, the memoryless property allows us to evaluate the conditional probability and we have [BGRS94]:

$$Q(x) \;=\; c\left(1 - \sum_{i=1}^{N} \frac{\alpha_i h_i}{c(\alpha_i + \beta_i)}\right)\prod_{i=1}^{N}\left(1 + \frac{\alpha_i}{\beta_i}\right)\cdot\int_{0}^{\infty}\phi_{\mathbf{0}}(t, x + ct)dt\,.$$

In general it is not possible to evaluate the conditional probability in (17.2.2). However, the bound obtained simply on replacing this term by 1 constitutes a good approximation in many cases, notably when the value of c is large with respect to the h_i or the link load is not too close to 1. Introduce, therefore, the following bound:

$$Q(x) \le q(x) \;=\; \sum_{\mathbf{d}\cdot\mathbf{h}<c} (c - \mathbf{d}\cdot\mathbf{h})\int_{0}^{\infty}\phi_{\mathbf{d}}(t, x + ct)\,dt \qquad (17.2.3)$$

which we write in the form

$$q(x) \;=\; \sum_{\mathbf{d}\cdot\mathbf{h}<c} (c - \mathbf{d}\cdot\mathbf{h})\,\psi_{\mathbf{d}}(x),$$

where $\psi_{\mathbf{d}}(x) = \int_{0}^{\infty}\phi_{\mathbf{d}}(t, x + ct)\,dt$. Taking the Laplace transform of $\psi_{\mathbf{d}}(x)$ and inverting the integrals, we have:

$$\psi_{\mathbf{d}}^{*}(z) \;=\; \int_{0}^{\infty}\int_{ct}^{\infty}\phi_{\mathbf{d}}(t, w)e^{-zw}\,dw \cdot e^{zct}\,dt$$

$$=\; \phi_{\mathbf{d}}^{**}(-cz, z) - \int_{0}^{\infty}\int_{0}^{ct}\phi_{\mathbf{d}}(t, w)e^{z(ct-w)}dw\,dt\,. \qquad (17.2.4)$$

Now, it is shown in [Gui94] that all the singularities of $\psi_{\mathbf{d}}^{*}(z)$ lie in the left hand half plane while the double integral in (17.2.4) is regular in the same region. We can therefore apply the Laplace transform inversion formula to (17.2.4) with an integration contour consisting of a vertical line L strictly to the left of the imaginary axis and to the right of all singularities of $\phi_{\mathbf{d}}^{**}(-cz, z)$. The second term then disappears and we deduce:

$$\psi_{\mathbf{d}}(x) \;=\; \frac{1}{2\pi i}\int_{L}\phi_{\mathbf{d}}^{**}(-cz, z)\,e^{zx}\,dz\,. \qquad (17.2.5)$$

Note that $\phi_d^{**}(-cz, z)$ may have singularities for $\mathbb{R}(z) \geq 0$ so it is not a Laplace transform itself and (17.2.5) is not a Laplace transform inversion formula.

We now turn our attention to the evaluation of $\phi_d(t, w)$. Let $\omega_i(t)$ be the work arriving in $(-t, 0)$ from a given source i, normalized with respect to the burst rate h_i (i.e., a burst of duration 1 produces 1 source i unit of work). Let $f_i(t, .)$ and $g_i(t, .)$ denote the conditional densities:

$$
\begin{cases}
f_i(t, w) & = & \dfrac{\partial}{\partial w} \Pr\left[\omega_i(t) \leq w \mid \text{source } i \text{ off at } -t\right], \\[2mm]
g_i(t, w) & = & \dfrac{\partial}{\partial w} \Pr\left[\omega_i(t) \leq w \mid \text{source } i \text{ on at } -t\right],
\end{cases}
$$

for $0 \leq w \leq t$. As the sources are independent, the function $\phi_d(t, .)$ is the convolution of N individual densities. Its Laplace transform may therefore be written

$$
\phi_d^*(t, z) = \prod_{i=1}^{N} (\eta_i g_i^*(t, h_i z))^{d_i} \cdot ((1 - \eta_i) f_i^*(t, h_i z))^{1-d_i}, \tag{17.2.6}
$$

where $\eta_i = \alpha_i / (\alpha_i + \beta_i)$ is the probability source i is on at $-t$. It is shown in [BGRS94] that the double Laplace transforms of f_i and g_i are given by:

$$
\begin{cases}
f_i^{**}(s, z) & = & \dfrac{\alpha_i z \left(1 - a_i^*(s)\right) \left(b_i^*(z + s) - 1\right)}{s^2(z + s) \left(1 - b_i^*(z + s)a_i^*(s)\right)} + \dfrac{1}{s}, \\[4mm]
g_i^{**}(s, z) & = & \dfrac{\beta_i z \left(1 - a_i^*(s)\right) \left(1 - b_i^*(z + s)\right)}{s(z + s)^2 \left(1 - b_i^*(z + s)a_i^*(s)\right)} + \dfrac{1}{z + s}.
\end{cases} \tag{17.2.7}
$$

To evaluate $\psi_d(x)$ using (17.2.5), we must first invert (17.2.7) with respect to s, form the product (17.2.6) and then take its Laplace transform with respect to t. These operations are feasible, though intricate, when the densities $a_i(\cdot)$ and $b_i(\cdot)$ are Coxian and consequently have rational Laplace transforms. In this case, f_i^* and g_i^* may be written as

$$
\begin{cases}
f_i^*(t, z) & = & \sum_j f_{ij}(z) e^{s_{ij}(z)t}, \\[2mm]
g_i^*(t, z) & = & \sum_j g_{ij}(z) e^{s_{ij}(z)t},
\end{cases} \tag{17.2.8}
$$

where the $s_{ij}(z)$ are the roots of the equation in s:

$$
a_i^*(s) b_i^*(s + z) = 1 \tag{17.2.9}
$$

and $f_{ij}(z)$ and $g_{ij}(z)$ are the corresponding residues of $f_i^{**}(s, z)$ and $g_i^{**}(s, z)$, respectively. Substitution in (17.2.6) yields an expression of the form:

$$
\phi_d^*(t, z) = \sum_k \gamma_k(z) e^{\sigma_k(z)t}, \tag{17.2.10}
$$

where the summation extends over all possible combinations taking one root $s_{ij}(h_i z)$ (i.e., evaluated from (17.2.9) with $h_i z$ in place of z) from each source i and $\sigma_k(z)$ is the sum of the roots in combination k. The coefficient $\gamma_k(z)$ is a function of the residues $f_{ij}(z)$ and $g_{ij}(z)$ and the d_i (see [Gui94] for details but note that the definition of density $b_i(\cdot)$ therein is different to ours).

Taking the Laplace transform of (17.2.10) with respect to t yields:

$$\phi_{\mathbf{d}}^{**}(-cz, z) = \sum_k \frac{-\gamma_k(z)}{\sigma_k(z) + cz}.$$

Finally, the contour integral in (17.2.5) is evaluated using the residue theorem to give

$$\psi_{\mathbf{d}}(x) = \sum_k \sum_l \frac{-\gamma_k(z_{kl})}{\sigma'_k(z_{kl}) + c} e^{z_{kl} x}, \qquad (17.2.11)$$

where the z_{kl} are roots of

$$\sigma_k(z) + cz = 0. \qquad (17.2.12)$$

Relation (17.2.11) substituted in (17.2.3) constitutes a spectral expansion for the Beneš bound $q(x)$. In general the z_{kl} are complex with negative real part. It can be verified that in the case of exponential burst and silence distributions, we have the same set of (real) roots $\{z_{kl}\}$ as that obtained for the queue content survivor function $Q(x)$ using the A-M-S approach (see Section 17.1.2). The above can thus be seen as a generalization of the A-M-S result, albeit for a bound and not the actual distribution.

Asymptotic behaviour

For large buffer sizes or low overflow probabilities, $q(x)$ can be approximated by its asymptote derived by retaining only the term in (17.2.11) corresponding to the root z_{kl} with largest real part. To simplify the presentation we consider only the special case of a superposition of homogeneous sources, dropping therefore the index i and, without loss of generality, setting $h = 1$. Let $s_0(z)$ be the root of equation (17.2.9) of largest real part. The dominant term in (17.2.11) then corresponds to the combination k where each source contributes the root s_0 and we have:

$$\psi_{\mathbf{d}}(x) \sim \frac{-\gamma_0}{N s'_0(\zeta) + c} e^{\zeta x}, \qquad (17.2.13)$$

where ζ is the root of largest negative real part of the equation

$$a^*(-\frac{cz}{N}) b^*(-\frac{cz}{N} + z) = 1 \qquad (17.2.14)$$

and γ_0 is the coefficient corresponding to the combination of N roots $s_0(\zeta)$ in (17.2.10). It may be demonstrated that ζ is real. Equation (17.2.14) derives from (17.2.9) on noting that $\sigma_k(z)$ in (17.2.12) is here equal to $Ns_0(z)$ so that $s_0(z) = -cz/N$.

The derivative s_0' in (17.2.13) is obtained on differentiating (17.2.9). We have

$$s_0'(z) = \frac{-a^*(s)b^{*\prime}(s+z)}{a^{*\prime}(s)b^*(s+z) + a^*(s)b^{*\prime}(s+z)}.$$

The coefficient γ_0 takes the form

$$\gamma_0 = \prod_{i=0}^{N}(\eta g_0)^{d_i}((1-\eta)f_0)^{1-d_i},$$

where f_0 and g_0 are the residues of $f^{**}(s,z)$ and $g^{**}(s,z)$, respectively, evaluated at the point $(s_0(\zeta), \zeta)$.

Finally then, we have the asymptotic expression for $q(x)$:

$$q(x) \sim \sum_{k=1}^{\lfloor c \rfloor} \binom{N}{k}(c-k)(\eta g_0)^k((1-\eta)f_0)^{N-k} \cdot \frac{e^{\zeta x}}{Ns_0'(\zeta)+c}. \qquad (17.2.15)$$

In heavy traffic it is possible to derive a more explicit expression for ζ. Denote by $\rho = N\eta$ the channel load. It can be shown that as $\rho \to 1$ we have $\zeta \to 0$ so that the left hand side of (17.2.14) can be expanded to second order terms to give:

$$\zeta \approx -\frac{2\beta\rho(1-\rho)}{(1-\eta)^2(cv_a^2 + cv_b^2)}, \qquad (17.2.16)$$

where cv_a^2 and cv_b^2 are the squared coefficients of variation of silence and burst durations, respectively. Expression (17.2.16) shows clearly how the queue performance depends significantly on the variability of the silence and burst distributions.

It may further be verified that the coefficient multiplying the exponential term in (17.2.15) tends in heavy traffic to

$$C = \frac{1}{c(1-\rho)}\sum_{k=1}^{\lfloor c \rfloor}(c-k)\left(\frac{\alpha}{\alpha+\beta}\right)^k\left(\frac{\beta}{\alpha+\beta}\right)^{N-k}.$$

Of course, $q(x)$ is not a good approximation for $Q(x)$ in heavy traffic since the ignored conditional probability is then not close to 1. To deduce a *corrective factor* to make $q(x)$ a good approximation even in this regime we reason as follows. The true heavy traffic behaviour of the distribution $Q(x)$ is intuitively $Q(x) \sim e^{\zeta x}$ (the asymptotic slope should be the same as that of the bound, the asymptote is attained rapidly in heavy traffic and $\lim_{\rho \to \infty} Q(0) = 1$). We therefore propose to multiply $q(x)$ in (17.2.15) by $1/C$ to derive the approximation:

$$Q(x) \approx Q_{approx}(x) = \frac{c(1-\rho)}{\displaystyle\sum_{k=1}^{\lfloor c \rfloor}(c-k)\left(\frac{\alpha}{\alpha+\beta}\right)^k\left(\frac{\beta}{\alpha+\beta}\right)^{N-k}} \cdot q(x).$$

Note that $Q_{approx}(x)$ tends to $q(x)$ both for large c and small ρ, i.e., precisely when $q(x)$ is a tight upper bound of $Q(x)$. We conjecture therefore that the approximation is accurate over the range of traffic levels.

The corrective factor can in fact be generalized to the case of heterogeneous traffic. We have:

$$Q(x) \approx Q_{approx}(x) = \frac{c(1-\rho)}{\sum_{\mathbf{d \cdot h} < c}(c - \mathbf{d} \cdot \mathbf{h}) \cdot P(\mathbf{d})} \cdot q(x), \qquad (17.2.17)$$

where $P(\mathbf{d})$ is the stationary probability of arrival state $\mathbf{d}$.

17.2.2   Superposition of periodic sources

Consider now a superposition of N identical independent deterministic on/off sources corresponding, for example, to a worst case traffic assumption. This particular case cannot be handled by the method of the previous section since the Laplace transform of burst and silence distributions are not rational. In a period of unit length, each source is active continuously during d time units; we assume the start of the burst of each source is uniformly distributed within the period, independently of other sources. We assume $d < 1/2$ and, without loss of generality, set $h = 1$.

In applying the Beneš result (14.2.7) to calculate the distribution of the queue length at an arbitrary instant in this case, it is sufficient to consider only the traffic arriving in the preceding unit length period (including the remainder of any bursts begun within d time units of the end of the preceding period). It is then convenient to distinguish "arrival states" according to the position of bursts with respect to the instant $-t$. Let $\{\text{state} = \{k, l\}_t\}$ denote the arrival patterns where exactly l bursts start in the interval $(-t - d, -t)$ (and are therefore active at $-t$) and exactly k bursts start in $(-t, \min(0, 1 - t - d))$ (the min ensuring that the l and k bursts in question are distinct in the case where $1 - d \leq t \leq 1$). We denote by $\pi_{kl}(t)$ the probability of state $\{k, l\}_t$:

$$\pi_{kl}(t) = \begin{cases} \dfrac{N!}{k!l!(N-k-l)!} \, t^k d^l (1-t-d)^{N-k-l}, & \begin{array}{l} 0 \leq t < 1 - d, \\ k + l \leq N, \end{array} \\[2em] \dfrac{N!}{l!(N-l)!} \, (1-d)^k d^l, & \begin{array}{l} 1 - d \leq t \leq 1, \\ k + l = N. \end{array} \end{cases}$$

$$(17.2.18)$$

The Beneš result for the queue length distribution survivor function (cf. (17.2.2)) may then be written:

$$Q(x) = \sum_{l=0}^{\lfloor c \rfloor} \sum_{k=0}^{N-l} (c - l) \int_0^1 dt \, \phi_{kl}(t, x + ct) \qquad (17.2.19)$$

$$\cdot \Pr\left[V_{-t} = 0 \mid A_t = x + ct, \, \text{state} = \{k, l\}_t\right],$$

where

$$\phi_{kl}(t, w) = \frac{\partial}{\partial w} \Pr\{A_t < w, \text{state} = \{k, l\}_t\}.$$

In the case where $c \leq 1$ (i.e., sources emit at least at the channel rate), the summation in l in (17.2.19) disappears (only $l = 0$ counts) and it is possible to calculate exactly the conditional probability. For this particular case, since the system must empty somewhere in the interval $(-1,0)$, in calculating the distribution of V_0, we need to consider only those burst beginning in $(-1,0)$, i.e., we consider a modified system with exactly N bursts whose starting times are uniformly distributed over $(-1,0)$. Let the event $\{\text{state} = \{k, \cdot\}_t\}$ denote the union $\bigcup_{l \geq 0}\{\text{state} = \{k, l\}_t\}$, i.e., an arrival pattern where exactly k bursts arrive in $(-t, 0)$. We can then write:

$$\Pr[V_{-t} = 0 \mid A_t = x + ct, \text{state} = \{k, 0\}_t]$$

$$= \Pr[V_{-t} = 0 \mid \text{state} = \{k, 0\}_t]$$

$$= \Pr[V_{-t} = 0 \mid \text{state} = \{k, \cdot\}_t] \frac{\Pr\{\text{state} = \{k, \cdot\}_t\}}{\Pr\{\text{state} = \{k, 0\}_t\}}.$$

The first equality derives from the fact that the buffer content at $-t$ is determined here solely by the $N - k$ bursts which begin in $(-1, -t)$. The last step follows since the intersections $\{V_{-t} = 0\} \cap \{\text{state} = \{k, \cdot\}_t\}$ and $\{V_{-t} = 0\} \cap \{\text{state} = \{k, 0\}_t\}$ are identical.

To derive an expression for the conditional probability $\Pr[V_{-t} = 0 \mid \text{state} = \{k, \cdot\}_t]$ we reason as in Subsection 15.2.1 for the $N * D/D/1$ queue. The virtual waiting time at $-t$ is determined solely by the $(N - k)$ sources which start in $(-1, -t)$. The same value of V_{-t} would be obtained if the arrival pattern in $(-1, -t)$ were repeated with period $1 - t$. Since $-t$ is an arbitrary instant with respect to this periodic process the probability of an empty queue is simply $1 - \rho'$ where $\rho' = (N - k)d/((1 - t)c)$.

To derive the value of the quotient it is necessary to distinguish the intervals where t is greater than or less than $1 - d$ and to use the expressions in (17.2.18). We finally deduce the following expression for $Q(x)$ on noting that the probability within the integrand is zero for $k < N - (1 - t)c/d$ with $t \in (0, 1 - d)$ and for $k < N$ with $t \in (1 - d, 1)$:

$$Q(x) =$$

$$c \int_0^{1-d} \sum_{k > N - (1-t)c/d}^{N} \phi_{k,0}(t, x + ct) \left(1 - \frac{(N-k)d}{(1-t)c}\right) \left(\frac{1-t}{1-t-d}\right)^{N-k} dt +$$

$$+ c \int_{1-d}^{1} \phi_{N,0}(t, x + ct) \, dt. \qquad (17.2.20)$$

When $c \gg 1$, the bound derived on replacing the conditional probability in (17.2.19) by 1 constitutes a good approximation in many cases of practical interest:

$$Q(x) \le q(x) = \sum_{l=0}^{\lfloor c \rfloor} \sum_{k=0}^{N-l} (c-l) \int_0^1 \phi_{kl}(t, x+ct)\, dt. \qquad (17.2.21)$$

It is possible in this case to evaluate $\phi_{kl}(t, x+ct)$ exactly. Let $f(t, w)$ and $g(t, w)$ denote the conditional densities:

$$f(t, w) = \Pr[\text{source contributes } w \mid \text{burst starts in } (-t, \min(0, 1-t-d))],$$

$$g(t, w) = \Pr[\text{source contributes } w \mid \text{burst starts in } (-t-d, -t)]$$

so that $\phi_{kl}(t, w)$ may be written as

$$\phi_{kl}(t, w) = \pi_{kl}(t)\, f^{*(k)}(t, w) * g^{*(l)}(t, w), \qquad (17.2.22)$$

where $\pi_{kl}(t)$ is given by (17.2.18). The $*$ in (17.2.22) represents convolution and $f^{*(k)}$ is the kth fold convolution of f.

We derive the following expressions on considering the various possibilities for bursts being entirely in the interval $(-t, 0)$, overlapping one edge or overlapping both edges (assuming $1 - d > d$); see Figure 17.2.1:

$$f(t, x) = \begin{cases} \left(1_{\{0 \le x \le t\}}\right)/t & 0 \le t \le d, \\ \left(1_{\{0 \le x \le d\}} + (t-d)\delta_d(x)\right)/t & d \le t \le 1-d, \\ \left(1_{\{t-1+d \le x \le d\}} + (t-d)\delta_d(x)\right)/(1-d) & 1-d \le t \le 1, \end{cases} \qquad (17.2.23)$$

$$g(t, x) = \begin{cases} \left(1_{\{0 \le x \le t\}} + (d-t)\delta_d(x)\right)/d & 0 \le t \le d, \\ \left(1_{\{0 \le x \le d\}}\right)/d & d \le t \le 1-d, \\ \left(1_{\{t-1+d \le x \le d\}} + (t-1+d)\delta_{t-1+d}(x)\right)/d & 1-d \le t \le 1, \end{cases} \qquad (17.2.24)$$

where δ_a denotes the Dirac distribution at point a.

Now let $u_a^{*(n)}(x)$ denote the nth fold convolution of the uniform density on $(0, a)$. We deduce from (17.2.23) and (17.2.24) the following relations:

$$\phi_{kl}(t, w) = \begin{cases} \pi_{kl}(t) \displaystyle\sum_{j=0}^{l} \binom{l}{j} \frac{(d-t)^j t^{l-j}}{d^l}\, u_t^{*(k+l-j)}(w - jt) & 0 \le t \le d, \\[4mm] \pi_{kl}(t) \displaystyle\sum_{i=0}^{k} \binom{k}{i} \frac{(t-d)^i d^{k-i}}{t^k}\, u_d^{*(k+l-i)}(w - id) & d \le t \le 1-d, \\[4mm] \pi_{kl}(t) \displaystyle\sum_{i=0}^{k}\sum_{j=0}^{l} \binom{k}{i} \frac{(t-d)^i (1-t)^{k-i}}{(1-d)^k} \binom{l}{j} \frac{(d-1+t)^j (1-t)^{l-j}}{d^l} \\[2mm] \qquad \cdot u_{1-t}^{*(k+l-i-j)}(w - i(1-t) - (k+l)(t-1+d)) \\[2mm] \hfill 1-d \le t \le 1. \end{cases}$$

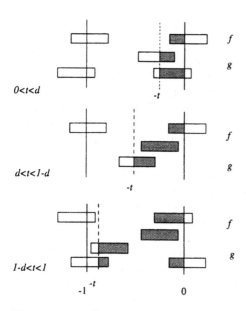

Figure 17.2.1: Illustration of bursts contributing work in $(-t, 0)$.

An expression for the convolutions $u_a^{*(n)}(w)$ is given in [Fel71, vol 2, p27]:

$$u_a^{*(n)}(w) = \frac{1}{a^n (n-1)!} \sum_{i=0}^{n} (-1)^i \binom{n}{i} \{(w - ia)^+\}^{n-1},$$

where $x^+ = \max\{0, x\}$.

The above presentation is based on original work by Garcia, Barceló and Casals [GBC95]. They also derive equivalent expressions in the case of a slotted time system from which numerical results are more readily obtainable. An alternative method for calculating $Q(x)$ exactly in the case $c = 1$ is given by [CGKK94]. In [BGRS94], the function $\phi_{kl}(t, w)$ is evaluated using the approximation (cf. Section 14.3.2, Equation (14.3.16)):

$$\phi_{kl}(t, w) \approx \frac{1}{\sigma\sqrt{2\pi}} \cdot e^{\zeta w} \cdot \phi_{kl}^*(t, \zeta),$$

where $\zeta = \zeta(t, w)$ is a solution to:

$$w = -\frac{1}{\phi_{kl}^*(t, \zeta)} \cdot \frac{\partial \phi_{kl}^*}{\partial z}(t, \zeta) .$$

and

$$\sigma^2 = \frac{1}{\phi_{kl}^*(t, \zeta)} \frac{\partial^2 \phi_{kl}^*}{\partial^2 z}(t, \zeta) - w^2.$$

The Laplace transform $\phi_{kl}^*(t, \zeta)$ of $\phi_{kl}(t, w)$ is readily derived from (17.2.22) with (17.2.23) and (17.2.24). Finally, the integral in (17.2.20) and (17.2.21) is calculated numerically on distinguishing the different ranges of t.

The bound (17.2.21) is a good approximation for the distribution $Q(x)$ except when c is small and when the server load is high. We have verified the approximation by comparison with simulation results in [KB92] for $c = 20$, $\rho = 0.55$ and $c = 10$, $\rho = 0.4$: the only apparent discrepancy is that due to the fluid approximation of the simulated discrete arrival process. Furthermore, for all loads and link capacities c, the bound intuitively tightens as x increases since the probabilities $\Pr[V_{-t} = 0 \mid A_t = x + ct, \text{ state } = \{k,l\}_t]$, neglected in deriving the Beneš bound, then tend to 1 for the considered deterministic arrival process.

We are thus fairly confident in using (17.2.21) for buffer sizing when the required overflow probability is small (10^{-10}). However, when c is small and the load is close to 1, the Beneš bound is not a good estimate of the queue length survivor function for small x. In particular, its integration will not provide an accurate estimate of the mean delay.

An alternative upper bound is provided from results for the $N * D/D/1$ queue (cf. Section 15.2) which corresponds to the limiting process when c tends to zero. Let $Q_D^N(x)$ be the probability that the virtual waiting time exceeds x in an $N * D/D/1$ queue with N sources of period D (cf. relation 15.2.2): The delay of an arbitrary cell in a burst (i.e., a batch when $c \to 0$) of size d is equal to the sum of the delay of the first cell (the virtual waiting time in an $N * D/D/1$ queue with one less customer) and the time necessary to serve cells between the first and the arbitrary cell (uniformly distributed on $(0, d)$). We have, therefore, the limiting survivor function:

$$Q_{c\downarrow 0}(x) = \frac{1}{d} \int_0^d Q_{c/d}^{N-1}(x - u) \, du$$

which constitutes a bound on the survivor function of the considered fluid queue for any link capacity c.

Figure 17.2.2 compares the survivor function bounds $Q_{c\downarrow 0}(x)$ and $q(x)$ for different values of c in the case where $N = 7$ and the multiplex load is 1 (i.e., when $c/d = N$). The units of x in the figure correspond to bursts. To estimate mean delays, it is proposed to numerically integrate the function $\min\{Q_{c\downarrow 0}(x), q(x)\}$. Note that the mean delay when $c = 1$ is known exactly from the equivalent result for an $N * D/D/1$ queue [RD91]. In the case illustrated in Figure 17.2.2, the computed bound for $c = 1$ is only 6% greater than the exact value.

17.2.3   Large deviations equivalents

In [SG95] approximations are derived for the queue length distribution of a fluid multiplexer receiving a superposition of homogeneous unit peak rate sources using large deviation estimates obtained when the link capacity c and the considered buffer levels scale with the number N of multiplexed sources. Let $c = \gamma N$ $(\gamma < 1)$ and consider the probability $Q(N\xi)$ that the queue

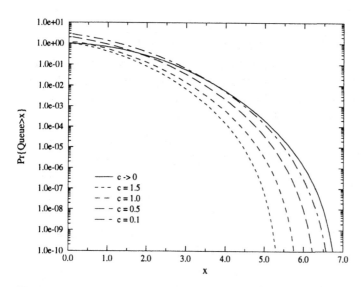

Figure 17.2.2: Survivor function bounds $q(x)$ and $Q_{c\downarrow0}(x)$.

length exceeds $N\xi$. By the Beneš result for fluid queues (14.2.7) we have the bound:

$$Q(N\xi) \leq \tag{17.2.25}$$

$$\leq \sum_{l \leq N\gamma} (N\gamma - l) \int_0^\infty \Pr\left\{ A_t^{(N)} = N(\xi + \gamma t) \text{ and } l \text{ sources on at } - t \right\} dt$$

$$\leq N\gamma \int_0^\infty \Pr\left\{ A_t^{(N)} = N(\xi + \gamma t) \right\} dt,$$

where we write $A_t^{(N)}$ to emphasize dependence on the scaling parameter N.

It is shown in [SG95] that, given a number of technical conditions concerning the burst and silence distributions, the latter integral has the large deviations equivalent (cf. Section 14.3.2):

$$\int_0^\infty \Pr\left\{ A_t^{(N)} = N(\xi + \gamma t) \right\} dt \asymp e^{-NI(\xi)}, \tag{17.2.26}$$

where

$$I(\xi) = \inf_{t>0} J(t, \xi + \gamma t)$$

and $J(t, x)$ is the Cramér-Legendre transform (cf. (14.3.2)) of the probability density $\phi(t, w)$ of the work generated by a single source in the interval $(-t, 0)$:

$$J(t, x) = -\inf_{z \geq 0} \{ zx + \log \phi^*(t, z) \}.$$

We have:

$$\phi^*(t, z) = (1 - \eta)f^*(t, z) + \eta g^*(t, z),$$

where $\eta \; (= \alpha/(\alpha + \beta))$ is the probability a source is on at $-t$ and f^* and g^* are given by (17.2.8) for Coxian burst and silence distributions. For the periodic deterministic sources considered in Section 17.2.2, the corresponding expressions for f^* and g^* are [RBC93]:

$$f^*(t, z) = \begin{cases} (1/\alpha - t) + \dfrac{1}{z}(1 - e^{-zt}), & 0 \le t \le 1/\beta, \\[2ex] (1/\alpha - t) + \dfrac{1}{z}(1 - e^{-z/\beta}) + (t - 1/\beta)e^{-z/\beta}, & 1/\beta \le t \le 1/\alpha, \\[2ex] \dfrac{1}{z}(e^{-z(t-1/\alpha)} - e^{-z/\beta}) + (t - 1/\beta)e^{-z/\beta}, & \\[1ex] & 1/\alpha \le t \le 1/\alpha + 1/\beta, \end{cases}$$

$$g^*(t, z) = \begin{cases} (1/\beta - t)e^{-zt} + \dfrac{1}{z}(1 - e^{-zt}), & 0 \le t \le 1/\beta, \\[2ex] \dfrac{1}{z}(1 - e^{-z/\beta}), & 1/\beta \le t \le 1/\alpha, \\[2ex] \dfrac{1}{z}(e^{-z(t-1/\alpha)} - e^{-z/\beta}) + (t - 1/\beta)e^{-z(t-1/\beta)}, & \\[1ex] & 1/\alpha \le t \le 1/\alpha + 1/\beta. \end{cases}$$

Note that (17.2.26) is a non-standard large deviations result requiring a specific demonstration (cf. [SG95]). By (17.2.26) with bound (17.2.26), we deduce:

$$\limsup_{N \to \infty} \frac{1}{N} \log Q(N\xi) \le -I(\xi). \qquad (17.2.27)$$

We now derive a lower bound starting from (14.2.1),

$$Q(N\xi) \ge \max_{t \ge 0} \Pr\left\{ A_t^{(N)} > N(\xi + \gamma t) \right\}. \qquad (17.2.28)$$

Now, the large deviations equivalent for the probability on the right is given by Chernoff's theorem (cf. Section 14.3.2):

$$\Pr\left\{ A_t^{(N)} > N(\xi + \gamma t) \right\} \asymp e^{-NJ(t, \xi + \gamma t)}$$

and consequently,

$$\max_{t \ge 0} \Pr\left\{ A_t^{(N)} > N(\xi + \gamma t) \right\} \asymp e^{-NI(\xi)}.$$

We deduce from (17.2.28)

$$\liminf_{N \to \infty} \frac{1}{N} \log Q(N\xi) \ge -I(\xi). \qquad (17.2.29)$$

Relations (17.2.27) and (17.2.29) together imply:

$$\lim_{N\to\infty} \frac{1}{N} \log Q(N\xi) = -I(\xi). \qquad (17.2.30)$$

The logarithmic equivalent (17.2.30) is used in [SG95] to derive approximations for $Q(x)$. The simplest approximation relies on the observation that the bound

$$Q(0) \geq \Pr\left\{\Lambda_t^{(N)} > c\right\}$$

is tight, i.e., $Q(0) \approx \Pr\{\Lambda_t^{(N)} > c\}$. An approximation consistent with this is

$$Q(N\xi) \approx \Pr\left\{\Lambda_t^{(N)} > c\right\} e^{-N(I(\xi)-I(0))}. \qquad (17.2.31)$$

It is verified in [SG95] that this approximation is also consistent with the logarithmic equivalent (17.2.30), the two factors $\Pr\{\Lambda_t^{(N)} > c\}$ and $e^{NI(0)}$ cancelling out for large N.

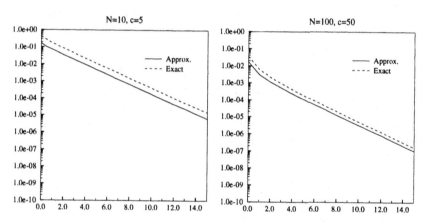

Figure 17.2.3: Exact and large deviations estimates of $Q(x)$ for exponential burst and silence distributions ($N = 10$ and $N = 100$).

We now illustrate the accuracy of large deviations approximation (17.2.31) by first comparing it with either the exact value of probability $Q(N\xi)$ or its upper bound (17.2.26). We take $1/\alpha = 0.6$, $1/\beta = 0.4$ for the mean silence duration and the mean burst length, respectively and we consider either $N = 10$ or $N = 100$ superposed sources with output capacity $c = N/2$. The mean load of the queue is consequently kept constant at the value $N\eta/c = 0.8$. Queue length $x = N\xi$ being measured in average burst lengths, we can then observe the following:

- For exponentially distributed on and off periods (exact results are derived as given in [AMS82]), it may be seen from Figure 17.2.3 that approximation (17.2.31) performs reasonably well as soon as $N \geq 10$.

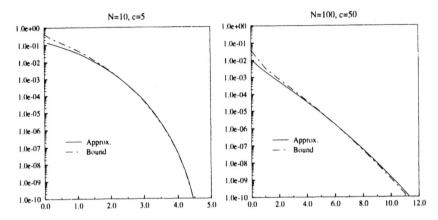

Figure 17.2.4: Exact and large deviations estimates of $Q(x)$ for deterministic burst and silence distributions ($N = 10$ and $N = 100$).

Infimum $I(\xi)$ is almost linear in ξ in this case (except for small values of ξ) and the large deviations approximation gives the correct decreasing rate for increasing ξ.

- For deterministic on and off periods (upper bound (17.2.26) is calculated as in [RBC93]), approximation (17.2.31) gives a close value as soon as $N \geq 10$, cf. Figure 17.2.4. Recall that (17.2.31) is a lower bound for $\xi = 0$ and thus results for $\xi > 0$ are certainly close to the exact values if upper bound (17.2.26) is close to (17.2.31) for small ξ. The variations of rate $I(\cdot)$ are here more pronounced due to the fact that the queue length is bounded within interval $[0, (N - c)/\beta]$. For large values of N, the approximation may even be slightly larger than upper bound (17.2.26). In spite of this small discrepancy, both evaluations are close with identical decreasing rate.

- A trend similar to the above cases can be observed for other Cox type on and off period distributions where approximation (17.2.31) proves reasonably tight as N increases.

17.2.4   Poisson burst arrivals

Consider now the fluid arrival process where constant rate bursts arrive according to a Poisson process, cf. Section 13.3.3. This process corresponds to a superposition of on/off sources in the limit where the number of sources N and the mean silence duration both tend to infinity while their quotient remains finite. Let there be L independent traffic streams where stream i is characterized as follows:

- burst arrivals constitute a Poisson process of intensity ξ_i;

- each burst duration has density $b_i(\cdot)$ and mean $1/\beta_i$;

- during a burst, work is generated at rate h_i.

Let $N_i(t)$ be the number of stream i bursts active at time t. A Beneš relation equivalent to (17.2.2) in this case is:

$$Q(x) = \sum_{\mathbf{n} \mathbf{h} < c} (c - \mathbf{n} \cdot \mathbf{h}) \int_0^\infty dt \, \phi_\mathbf{n}(t, x + ct) \qquad (17.2.32)$$

$$\cdot \Pr\left[V_{-t} = 0 \mid A_t = x + ct, \, \mathbf{N}_t = \mathbf{n}\right],$$

where

$$\phi_\mathbf{n}(t, w) = \frac{\partial}{\partial w} \Pr\{A_t \le w, \, N_i(t) = n_i, \, i = 1, \ldots, L\}. \qquad (17.2.33)$$

If the rates h_i are all greater than c, the summation in (17.2.32) disappears and the conditional probability can be evaluated since V_{-t} is then independent of A_t. We have:

$$Q(x) = c\left(1 - \sum_{i=1}^L \frac{\xi_i h_i}{c\beta_i}\right) \exp\left\{\sum_{i=1}^L \frac{\xi_i}{\beta_i}\right\} \int_0^\infty \phi_0(t, x + ct) \, dt.$$

When the h_i are an order of magnitude lower than c or the system load $(\rho = \sum_{i=1}^L \xi_i h_i / c\beta_i)$ is low, the conditional probability in (17.2.32) is close to 1. The following bound derived on ignoring the conditional probability is then a good approximation:

$$Q(x) \le q(x) = \sum_{\mathbf{n} \cdot \mathbf{h} < c} (c - \mathbf{n} \cdot \mathbf{h}) \int_0^\infty \phi_\mathbf{n}(t, x + ct) \, dt. \qquad (17.2.34)$$

The approach used in Section 17.2.1 for a finite number of sources does not work here because of the complexity of deriving the double Laplace transform $\phi_\mathbf{n}^{**}(s, z)$. It proves possible, however, to estimate $\phi_\mathbf{n}(t, w)$ accurately using the approximation (14.3.16):

$$\phi_\mathbf{n}(t, w) \approx \frac{1}{\sigma\sqrt{2\pi}} \cdot e^{\zeta w} \cdot \phi_\mathbf{n}^*(t, \zeta),$$

where $\zeta = \zeta(t, w)$ is a solution to:

$$w = -\frac{1}{\phi_\mathbf{n}^*(t, \zeta)} \cdot \frac{\partial \phi_\mathbf{n}^*}{\partial z}(t, \zeta)$$

and

$$\sigma^2 = \frac{1}{\phi_\mathbf{n}^*(t, \zeta)} \cdot \frac{\partial^2 \phi_\mathbf{n}^*}{\partial^2 z}(t, \zeta) - w^2.$$

Let $f_i(t, .)$ be the density of the contribution to A_t of a burst of stream i which arrives in $(-t, 0)$ and let $g_i(t, .)$ be the density of the contribution of a rate h_i burst active at $-t$. The transform $\phi_{\mathbf{n}}^*(t, z)$ may be deduced on recognizing that $\phi_{\mathbf{n}}(t, w)$ is here the convolution of n_i densities g_i for $1 \leq i \leq L$ and a compound Poisson distribution accounting for bursts arriving after $-t$:

$$\phi_{\mathbf{n}}^*(t, z) = \prod_{i=1}^{L} \frac{1}{n_i!} \left(\frac{\xi_i}{\mu_i} g_i^*(t, z) \right)^{n_i} \exp\{-\frac{\xi_i}{\mu_i} - \xi_i t(1 - f_i^*(t, z))\} .$$

The densities $f_i(t, .)$ and $g_i(t, .)$ can be written down directly:

$$\begin{cases} f_i(t, h_i w) & = & \dfrac{t - w}{t} b_i(w) + \dfrac{1}{t} \displaystyle\int_w^{\infty} b_i(u) \, du, \\[4mm] g_i(t, h_i w) & = & \tilde{b}_i(w) + \delta_t(w) \displaystyle\int_t^{\infty} \tilde{b}_i(u) \, du, \end{cases} \qquad (17.2.35)$$

for $0 \leq w \leq t$ where $\tilde{b}_i(t) = \beta_i \int_t^{\infty} b_i(u) du$ is the density of the residual burst duration. In the expression for $f_i(t, h_i w)$, $(t - w)/t$ is the probability the burst begins and ends within the interval $(-t, 0)$ and can therefore have a length w (density $b_i(w)$), and the second term accounts for a burst beginning precisely at $-w$ and continuing beyond the end of the interval. The first term in the expression for $g_i(t, h_i w)$ is the density of the residual length of the burst in progress while the second term accounts for the particular case when $w = t$. Note that this infinite source model presents the advantage of giving explicit expressions for densities f_i and g_i in contrast to the finite source model. The necessary Laplace transforms can be derived directly from (17.2.35), see [BGRS94] for details.

Asymptotic behaviour

An explicit asymptotic formula for $q(x)$ can be derived from the above results providing insight into the influence on the performance of particular traffic parameters. The formula derives from applying the Laplace method for evaluating integrals with a large parameter to the integral in t in (17.2.34). The derivation is outlined in [RBC93]. We find:

$$q(x) \sim \left(\sum_{\mathbf{n} \cdot \mathbf{h} < c} (c - \mathbf{n} \cdot \mathbf{h}) \prod_{i=1}^{L} \frac{R_i^{n_i}}{n_i!} e^{-\xi_i/\mu_i} \right) \frac{e^S}{T} e^{z_0 x}, \qquad (17.2.36)$$

where coefficients R_i, S and T are given by

$$R_i = \frac{\xi_i(1 - b_i^*(h_i z_0))}{h_i z_0},$$

$$S = \sum_{i=1}^{L} \xi_i \left[\frac{1 - b_i^*(h_i z_0)}{h_i z_0} + b_i^{*\prime}(h_i z_0) \right],$$

$$T = -\left[c + \sum_{i=1}^{L} \xi_i h_i b_i^{*\prime}(h_i z_0) \right],$$

and z_0 is the largest negative real solution of the equation

$$cz = \sum_{i=1}^{L} \xi_i [1 - b_i^*(h_i z)]. \qquad (17.2.37)$$

For high load it may be verified that z_0 tends to 0. A first order approximation is derived on replacing $b_i^*(h_i z)$ by its Taylor series expansion to order 2 about 0. We find:

$$z_0 \sim -2(1 - \rho) \frac{M_1}{M_2},$$

where $\rho = \sum_i (\xi_i h_i / c\beta_i)$ is the server load and M_1 and M_2 are the first two moments of the work arriving in an arbitrary burst. It is clear from (17.2.37) that z_0 does not depend individually on the burst duration and rate but only on their product. The slope is thus independent of the rate and is identical to that of an M/G/1 queue where the work in a burst is assumed to arrive instantaneously.

In heavy traffic ($\rho \to 1$) the expressions for R_i, S and T simplify and we find:

$$q(x) \sim \sum_{\mathbf{n} \cdot \mathbf{h} < c} (c - \mathbf{n} \cdot \mathbf{h}) \cdot P(\mathbf{n}) \cdot \frac{\exp(-2(1 - \rho) \frac{M_1}{M_2} x)}{c(1 - \rho)},$$

where $P(\mathbf{n}) = \prod_{i=1}^{L} e^{\xi_i / \beta_i} (\xi_i / \beta_i)^{n_i} / n_i!$ is the stationary distribution of the number of bursts of each class in progress. We deduce, as in Section 17.2.1, a *corrective factor* converting the bound $q(x)$ into a good approximation:

$$Q(x) \approx Q_{approx}(x) = \frac{c(1 - \rho)}{\sum_{\mathbf{n} \cdot \mathbf{h} < c} (c - \mathbf{n} \cdot \mathbf{h}) \cdot P(\mathbf{n})} \cdot q(x). \qquad (17.2.38)$$

Note that (17.2.38) reduces to the exact formula (17.2.21) when $c \le h_i$ for $i = 1, \ldots, L$.

Numerical results

Figure 17.2.5 shows the approximate survivor function $Q_{approx}(x)$ for (a) exponential and (b) constant burst lengths. The different curves correspond to different values of the ratio c/h, the mean burst length varying in inverse proportion to conserve the same burst volume. For the lowest burst rate ($c/h = 48$), we have also plotted the asymptote derived from (17.2.36) and

(17.2.38). In the limit $c/h \to 0$, the fluid queue level is exactly the same as the virtual waiting time in an $M/G/1$ queue with appropriate service time distribution.

The asymptotic value of $\log Q(x)/x$ depends only on the distribution of the amount of work in an arbitrary burst and can be deduced from results for $M/G/1$. The accuracy of the asymptote as an approximation to $Q(x)$ for small x improves as the ratio c/h decreases. It is better in the case of constant burst lengths. The intercept $Q(0)$ is largely determined by the stationary input rate distribution.

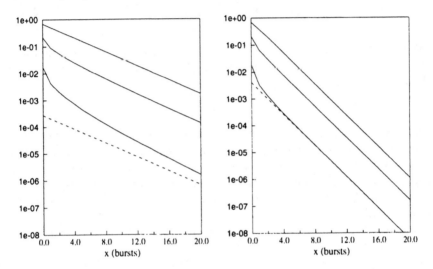

Figure 17.2.5: Queue length survivor function Q_{approx} for Poisson burst arrivals (17.2.38). Left: exponential bursts. Right: constant bursts. Curves from top to bottom: $c/h = 0$, $c/h = 12$, $c/h = 48$, asymptote.

17.3 D-BMAP/D/c models

The class of discrete-time Batch Markovian Arrival Processes (D-BMAP's) encompasses a versatile set of cell arrival models that have been frequently used in the analysis of all kinds of discrete-time queueing models which occur, almost in a natural way, in the performance analysis of slotted systems such as ATM-multiplexers and switching elements. Formally, a D-BMAP can be defined as a two-dimensional discrete-time Markov process $\{A(k), S(k) : k \geq 0\}$ on the state space $\{(i, j) : i \geq 0, 1 \leq j \leq M\}$, where M is the number of states in the irreducible modulating Markov chain [BC92]. The two sets of random variables $\{A(k) : k \geq 0\}$ and $\{S(k) : k \geq 0\}$ represent the number of cell arrivals on a slot-per-slot basis, and the state of the underlying Markov chain respectively. Transitions between subsequent states are governed by the transition matrices $\mathbf{H}_n, n \geq 0$, whose elements $(h_n)_{i,j}$, $1 \leq i,j \leq M$, describe the probability that n cells are generated and that the modulating Markov chain is in state j during a slot, given that it was in state i during the previous slot. The matrix $\mathbf{H} = \sum_{n=0}^{\infty} \mathbf{H}_n$ is the transition matrix of the underlying Markov chain, with stationary probability vector $\bar{\sigma}$ satisfying

$$\bar{\sigma} = \bar{\sigma}\mathbf{H}, \qquad \bar{\sigma}\bar{1} = 1,$$

where $\bar{1}$ is a column vector of appropriate size with all elements equal to 1. The traffic intensity ρ is then given by

$$\rho = \bar{\sigma}\left(\sum_{n=1}^{\infty} n\mathbf{H}_n\right)\bar{1}.$$

17.3.1   Solution methods

Let us consider an infinite-capacity ATM multiplexer model fed by a D-BMAP arrival process. The system equation describing the evolution of the buffer contents $V(k)$ at consecutive time instants then reads

$$V(k + 1) = (V(k) - c)^+ + A(k), \qquad (17.3.1)$$

where c is the maximum number of cell transmissions per slot. Several methods to derive the steady-state distribution of the buffer contents from this equation have been proposed in the literature, some of which are briefly discussed in the following subsections.

The finite buffer queue with a D-BMAP input process has been treated in [Blo93]. Also an efficient numerical method to compute the steady state probability vector and the cell loss probability is given. Furthermore, it is shown that the output process of a D-BMAP/D/1/N queue is a D-MAP (i.e., a D-BMAP with batch size equal to one cell).

Matrix-Geometric solution technique

The two-dimensional Markov chain $\{V(k), S(k) : k \geq 0\}$ can be used to represent the state of the system at time slot k, and its transition matrix $\mathbf{T}$ is given by

$$\mathbf{T} = \begin{pmatrix} \mathbf{H}_0 & \mathbf{H}_1 & \mathbf{H}_2 & \mathbf{H}_3 & \cdots \\ \mathbf{H}_0 & \mathbf{H}_1 & \mathbf{H}_2 & \mathbf{H}_3 & \cdots \\ 0 & \mathbf{H}_0 & \mathbf{H}_1 & \mathbf{H}_2 & \cdots \\ 0 & 0 & \mathbf{H}_0 & \mathbf{H}_1 & \cdots \\ \vdots & \vdots & \vdots & \vdots & \ddots \end{pmatrix}. \qquad (17.3.2)$$

Its stationary distribution $\bar{v} = (\bar{v}(0), \bar{v}(1), \bar{v}(2), \ldots)$ can be computed from

$$\bar{v} = \bar{v}\mathbf{T}, \qquad \bar{v}\bar{1} = 1.$$

The i-th element of $\bar{\mathbf{v}}(j)$ thus equals the steady-state probability

$$\bar{\mathbf{v}}(j)_i = \lim_{k \to \infty} \Pr\left\{V(k) = j, S(k) = i\right\}.$$

The transition matrix in (17.3.2) is of the M/G/1-type, which means that its special structure can be exploited to establish a recursion for the row vectors $\bar{\mathbf{v}}(j)$:

$$\bar{\mathbf{v}}(j) = \left(\bar{\mathbf{v}}(0)\tilde{\mathbf{H}}_j + \sum_{i=1}^{j} \bar{\mathbf{v}}(i)\tilde{\mathbf{H}}_{j+1-i}\right)\left(\mathbf{I} - \tilde{\mathbf{H}}_1\right)^{-1}. \tag{17.3.3}$$

where $\mathbf{I}$ is the identity matrix of appropriate size, and $\tilde{\mathbf{H}}_j$ can be expressed in terms of *the first passage matrix* $\mathbf{G}$ introduced in [Neu89] as

$$\tilde{\mathbf{H}}_j = \sum_{i=j}^{\infty} \mathbf{H}_i \mathbf{G}^{i-j}.$$

The $\mathbf{G}$ matrix, in turn, is the non-negative solution of

$$\mathbf{G} = \sum_{i=0}^{\infty} \mathbf{H}_i \mathbf{G}^i,$$

which can be determined iteratively. Finally, as shown in [Neu89], the boundary probability vector $\bar{\mathbf{v}}(0)$, which still occurs in the right hand side of (17.3.3), can be expressed in terms of $\bar{\mathbf{g}}$, which is the solution of

$$\bar{\mathbf{g}} = \bar{\mathbf{g}}\mathbf{G}, \qquad \bar{\mathbf{g}}\bar{\mathbf{1}} = 1.$$

Functional equation and matrix spectral decomposition approach

In this subsection, we describe a solution technique which is essentially based on a generating-functions technique. The aggregate cell arrival process in the multiplexer buffer is now considered to be a superposition of N homogeneous and independent sources, each of which can be characterized by a probability generating matrix $\mathbf{H}(z)$ defined as

$$\mathbf{H}(z) = \sum_{i=0}^{\infty} z^i \mathbf{H}_i, \tag{17.3.4}$$

and the aggregate process can be represented by the Kronecker product $\mathbf{H}^*(z) = \prod_{i=1}^{N} \otimes \mathbf{H}(z)$.

Let $\bar{\mathbf{m}}$ represent the vector of non-negative integers $\bar{\mathbf{m}} = (m_1, \ldots, m_M)$, and denote by $\{B(i,k) : 1 \le i \le M, k \ge 0\}$ the discrete-time Markov process on the state space $\{\bar{\mathbf{m}} : \sum_{i=1}^{M} m_i = N\}$, where $B(i,k)$ characterizes the number of input links in state i of the modulating Markov chain during slot

$k - 1$. Defining the joint probability generating function of the set of random variables $\{B(1, k), \ldots, B(M, k), V(k)\}$ as

$$P_k(\bar{\mathbf{x}}, z) = \mathrm{E}\left\{ z^{V(k)} \prod_{i=1}^{M} x_i^{B(i,k)} \right\},$$

where $\bar{\mathbf{x}} = (x_1, \ldots, x_M)$, then the following *functional equation* (cf. [Bru88]) for $P_k(\bar{\mathbf{x}}, z)$ can be derived from the system equation (17.3.1):

$$P_{k+1}(\bar{\mathbf{x}}, z) = z^{-c} \left(P_k(\mathbf{H}(z)\bar{\mathbf{x}}, z) + R_k(\mathbf{H}(z)\bar{\mathbf{x}}, z) \right), \qquad (17.3.5)$$

where

$$R_k(\bar{\mathbf{x}}, z) = \mathrm{E}\left[(z^c - z^{V(k)}) \prod_{i=1}^{M} x_i^{B(i,k)} \mid V(k) < c \right] \Pr\{V(k) < c\}.$$

Then, assuming that a steady state is reached after a sufficiently long period of time, the following expression for the steady-state generating function $V(z)$ of the buffer contents can be established by considering repetitive substitutions for the arguments in both sides of (17.3.5) [SX95]:

$$V(z) = \lim_{k \to \infty} \mathrm{E}\left\{ z^{V(k)} \right\} \qquad (17.3.6)$$

$$= \sum_{j=0}^{c-1} (z^c - z^j) \sum_{\bar{\mathbf{m}}} \frac{\prod_{i=1}^{M} \lambda_i(z)^{m_i}}{z^c - \prod_{i=1}^{M} \lambda_i(z)^{m_i}} \sum_{\bar{\mathbf{n}}} F_{\bar{\mathbf{n}}\bar{\mathbf{m}}}(z) p(\bar{\mathbf{n}}, j),$$

where the steady-state probabilities $p(\bar{\mathbf{n}}, j)$ are defined as

$$p(\bar{\mathbf{n}}, j) = \lim_{k \to \infty} \Pr\{B(1, k) = n_1, \ldots, B(M, k) = n_M, V(k) = j\}.$$

The set $\{\lambda_i(z) : 1 \leq i \leq M\}$ contains the eigenvalues of $\mathbf{H}(z)$, the probability generating matrix given by (17.3.4). In addition, $F_{\bar{\mathbf{n}}\bar{\mathbf{m}}}(z)$ is a known function of the right column eigenvectors of $\mathbf{H}(z)$ and, in general, can be obtained recursively. The boundary probabilities $p(\bar{\mathbf{n}}, j)$ can be calculated by using the fact that $V(z)$ is analytic inside the unit disk of the complex z-plane, implying that the zeros in this area of the denominators in the right hand side of (17.3.6) must make the corresponding numerators zero as well; this leads to a set of linear equations for the $p(\bar{\mathbf{n}}, j)$'s which has a unique solution. From expression (17.3.6) for $V(z)$, all kinds of performance measures related to the buffer contents, such as its mean value, variance and tail distribution can be obtained by performing the appropriate operations on this generating function.

The above expression for $V(z)$ is related to the result presented in [Li91], although from a numerical point of view, it presents some simplification in the sense that the calculation of the eigenvectors of the Kronecker product $\mathbf{H}^*(z)$ is not required here since all involved quantities are expressed in terms of the eigenvalues and eigenvectors of $\mathbf{H}(z)$.

17.3.2   Performance measures

One of the major difficulties in obtaining results concerning the buffer performance when applying the solution methods discussed in the previous section, is calculating the unknown boundary probabilities, since this requires numerical techniques for handling matrices of a potentially large size. It was illustrated in [SX95] that, at least for the case $c = 1$, accurate approximations for the boundary probabilities $p(\bar{\mathbf{n}}, j), 0 \leq j \leq c - 1$ can be found which, at the same time, also avoid the calculation of the $F_{\bar{\mathbf{n}}\bar{\mathbf{m}}}(z)$'s which occur in the right hand side of (17.3.6). In [AMR93], for a so-called Markov Modulated Number of Arrivals (MMNA) cell arrival process (which is a special case of the D-BMAP process), an alternative technique for calculating the boundary probability vector is proposed, based on the numerical evaluation of a number of contour integrals. In [Slo94b], for a cell arrival model which is a superposition of N homogeneous on/off sources with fixed spacing between cells generated during an on period, a hybrid technique for obtaining the mean and variance of the buffer contents is used, in the sense that expressions for the latter quantities are derived from the associated functional equation of the queue, whereas the corresponding boundary probability vector is calculated from the first passage matrix $\mathbf{G}$. For the same model, results were also obtained for the busy and idle period distributions [Slo94a]. Based on the matrix-geometric technique, expressions for the variances of the buffer occupancy and the delay were derived in [GN93] assuming a superposition of N homogeneous on/off sources with geometrically distributed off periods and on periods with a PH-distribution.

As results for the complete distribution of the buffer contents are often limited to relatively small systems [Slo95a], buffer dimensioning is often based on the tail distribution of the buffer contents which, for reasons of computational complexity, is often approximated by a geometric form. For a variety of cell arrival models in a single-server multiplexer buffer, a technique was developed in [XB94] and [XB95] for deriving approximate results concerning this geometric-tail approximation from the associated functional equation, by considering those values of $\bar{\mathbf{x}}$ for which the first argument of the $P(\cdot, z)$ functions in both sides of (17.3.5) become equal.

17.4   Queueing with long range dependent traffic

17.4.1   Long range dependence in simple queues

The property of long range dependence is linked to the convergence of the series of autocorrelation coefficients, as seen in Subsection 13.4.1. As we shall see here, it appears quite naturally in the GI/G/1 queue, where several processes may be considered: the queue size, the waiting time and the sojourn time processes. Clearly these three processes are intimately related, and one expects that they have very similar characteristics; however, the picture is still not complete.

Let us consider a GI/G/1 queue, characterized by the sequence $\{T_n\}$ of interarrival intervals and the sequence $\{S_n\}$ of service durations ($n = 0, \pm 1, \ldots$). These are independent random variables with common distribution functions $F(\cdot)$ and $G(\cdot)$, respectively. We define the sequence $\{W_n : n = 0, \pm 1, \ldots\}$ of waiting times. If $E\{U_n\} = E\{S_n - T_n\} < 0$, then it is well known that a stationary distribution exists for the waiting times.

Daley [Dal68] considers the autocorrelation function of the waiting times for the stationary GI/G/1 queue. This is defined as $\{\rho_n : n = 0, 1, \ldots\}$ with

$$\rho_n = E\left\{(W_k - \mu)(W_{k+n} - \mu)\right\}/\sigma^2,$$

for all n and k, where $\mu = E\{W_n\}$ and $\sigma^2 = E\{W_n^2\} - \mu^2$, provided these quantities are finite.

Kiefer and Wolfowitz [KW56] have shown that a sufficient condition for $E\{W^m\}$ to be finite is that $E\{S^{m+1}\}$ should be finite, and that this condition is actually necessary when $E\{T\}$ is finite. Thus, assuming that the interarrival intervals have a finite mean, and that $E\{S\} < E\{T\}$, we see that the autocorrelation sequence $\{\rho_n : n = 0, 1, \ldots\}$ is well defined if and only if the service time distribution has finite moments of order 3 at least.

It is firstly shown in [Dal68] that

$$\rho_n - \rho_{n+1} = \frac{1}{\sigma^2} \int_0^\infty H_n(x)\mathrm{d}x,$$

where the functions $H_n(x) = EW_0 1_{\{W_n > x\}} - EW_0 1_{\{W_{n+1} > x\}}$ satisfy

$$H_{n+1}(x) = \int_{-\infty}^x H_n(x - y)\mathrm{d}U(y),$$

where $U(\cdot)$ is the distribution function of the difference $S_n - T_n$, for all n. It is easy to see that the function $H_0(\cdot)$ is positive, so that the correlation coefficients $\{\rho_n\}$ are seen to decrease monotonically to zero.

Secondly, defining the sequence $Y_n = U_1 + U_2 + \cdots U_n$, Daley shows that

$$\rho_n = \Pr\{M_n + Z > 0\},$$

where $M_n = \min(0, Y_1, \ldots Y_n)$ and Z is a random variable with distribution function

$$\Pr\{Z \leq z\} = \frac{1}{\sigma^2} \int_0^z \int_{0-}^y (\mu - x)\,\mathrm{d}W(x)\,\mathrm{d}y,$$

so that one may write

$$\sum_{n \geq 0} \rho_n = \sum_{n \geq 0} \Pr\{M_n + Z > 0\} = E\{H(Z)\},$$

where the function $H(z) = \sum_{n \geq 0} \Pr\{M_n + z > 0\}$ is the solution of the renewal equation

$$H(z) = \begin{cases} V(z) + \int_{-\infty}^{z} H(z - y)\mathrm{d}V(y), & \text{for } z \geq 0 \\ 0, & \text{for } z < 0, \end{cases}$$

with $V(z) = \Pr\{U_n > -z\}$. This entails that $H(z)$ is asymptotically linear in z, as $z \to \infty$, so that the series $\sum_{n \geq 0} \rho_n$ converges if and only if $E\{Z\}$ is finite. Since

$$E\{Z\} = \frac{1}{2\sigma^2}(E\{W^3\} - E\{W\}E\{W^2\}),$$

one concludes that the series $\sum_{n \geq 0} \rho_n$ converges if the stationary waiting time distribution has a finite third moment.

In summary, assuming that the expected interarrival time is finite, we have that

- if $E\{S^3\} = \infty$, then the second moment of the distribution of W does not exist, and the autocorrelation function is not well defined;

- if $E\{S^3\} < \infty = E\{S^4\}$, then the autocorrelation function exists, but the series diverges and the property of long range dependence holds;

- if $E\{S^4\} < \infty$, then the series converges and we have short range dependence.

Instead of the actual customer waiting time, Ott [Ott77] considers the *virtual* waiting time $W(t)$ at time t. In the case of a stationary M/G/1 queue, for which the service time distribution has a finite third moment, he shows that $\int_0^\infty R(t)\mathrm{d}t$ converges if and only if the service time distribution has a finite fourth moment, where $R(t)$ is the autocovariance function $R(t) = \text{Cov}\{W(\tau), W(t + \tau)\}$, for all τ.

Pakes [Pak71a] examines the queue size process. For the stationary M/G/1 queue, he defines X_n as the queue size after the nth departure, and shows that the series

$$\sum_{n \geq 0} \text{Cov}\{X_0, X_n\}$$

converges if and only if the service time distribution has a finite fourth moment. He provides an expression for the series when it converges. In [Pak71b], the same author extends Daley's analysis to the autocorrelation function of the *sojourn* times for the GI/G/1 queue and shows that the autocorrelation series converges under the same conditions.

17.4.2 A buffer with fractional Brownian traffic as input

The queueing behaviour of fractional Brownian traffic (see Subsection 13.4.4) is not very well understood – only an approximate formula for the queue length distribution is known. Assume that traffic defined by (13.4.12), with parameters m, a and H, is offered to a link with capacity $c > m$ that has an unlimited buffer in front of it. The stationary buffer occupancy process X_t is defined by Reich's formula:

$$X_t = \sup_{s \le t} \left(A(s,t) - c(t-s) \right), \quad t \in (-\infty, \infty), \tag{17.4.1}$$

where $A(s,t) = m(t-s) + \sqrt{ma}(Z_t - Z_s)$ and Z_t is a normalized fractional Brownian motion with Hurst parameter H. As usually, we also denote $A_t = A(0,t)$. The stationarity of X_t follows from the stationarity of the increments of A_t, and its finiteness from the ergodicity of Z_t. Note also that X_t is always non-negative, although the arrival process also has negative increments.

The stationary distribution of X_t obeys a scaling law which is easily deduced from the self-similarity of Z_t [Nor94]. Assume that either the size of the physical buffer or a delay constraint dictates that the probability that the amount of work in system exceeds a certain level x must be less than ϵ, so that the equation

$$\epsilon = \Pr\{X > x\} \tag{17.4.2}$$

holds at the maximal allowed load. The equation (17.4.2) can be written as a relation between the design parameters x (buffer requirement) and c (link capacity) and the traffic parameters m, a and H at the critical boundary:

$$(c-m)(ma)^{-1/(2H)} x^{(1-H)/H} = f^{-1}(\epsilon), \tag{17.4.3}$$

where the function

$$f(y) = \Pr\left\{ \sup_{t \ge 0} (Z_t - yt) > 1 \right\}$$

depends on H but not on m, a, c or x. Equation (17.4.3) can written both as a buffer dimensioning formula:

$$x = f^{-1}(\epsilon) a^{1/(2(1-H))} c^{(2H-1)/(2(1-H))} \frac{\rho^{1/(2(1-H))}}{(1-\rho)^{H/(1-H)}}, \tag{17.4.4}$$

and as a bandwidth allocation rule:

$$c = m + f^{-1}(\epsilon) a^{1/(2H)} x^{-(1-H)/H} m^{1/(2H)}. \tag{17.4.5}$$

From (17.4.4) it is seen that when H is high, a substantial increase in utilization requires a tremendous amount of additional storage space. The important message of (17.4.5) is that for $H > 1/2$, the link requirement c increases slower than linearly in m so that a multiplexing gain is obtained by using links with higher capacity. In the Brownian case $H = 1/2$, c appears in the equations only within the ratio m/c, and no multiplexing gain is obtained in the model.

No explicit formula for the distribution of the fractional Brownian storage seems to be known. Norros [Nor94] approached the distribution of X_t through the lower bound

$$\Pr\{X_t > x\} \geq \sup_{u \geq t} \Pr\{A_u > cu + x\}, \qquad (17.4.6)$$

where the maximum at the right hand side is obtained at $t = Hx/((1 - H)(c - m))$. Explicitly, we have

$$\Pr\{X_t > x\} \geq \overline{\Phi}\left(\frac{(c - m)^H x^{1-H}}{\kappa(H)\sqrt{am}}\right), \qquad (17.4.7)$$

where $\kappa(H) = H^H(1 - H)^{1-H}$ and $\overline{\Phi}(y) = \Pr\{Z_1 > y\}$ is the residual distribution function of the standard Gaussian distribution.

In accordance with the principle that "rare events occur only in the most probable way", (17.4.7) is in fact accurate in the logarithmic sense of the theory of large deviations. This was shown by Duffield and O'Connell [DO96] as a special case of their general approach to queues with long range dependent input. Let us apply the simplified approach of Subsection 14.3.3 to finding the asymptotic distribution of the maximum of the discrete time process $Z_n - n$ (it is easy to understand but requires some technical arguments to prove that the continuous time makes no difference as regards asymptotic behaviour). The free energy functions $\mu_n(\beta)$ have the expression (see Table 14.3.1)

$$\mu_n(\beta) = \frac{1}{2}\beta^2 n^{2H}.$$

With the choice of the scaling function $f(n) = n^\gamma$, where $\gamma = 2 - 2H$, the limit of condition (14.3.3) is trivial and we obtain $\mu^*(x) = x^2/2$. An elementary minimization gives

$$\inf_{y>0} \frac{\mu^*(y + 1)}{y^\gamma} = \frac{1}{2}\left[H^H(1 - H)^{1-H}\right]^{-2} =_{\text{def}} \theta,$$

and the lower bound result (14.3.17) reads

$$\liminf_{x\to\infty} \frac{1}{x^{2-2H}} \ln \Pr\left\{\sup_{n\geq 0}(Z_n - n) > x\right\} \geq -\theta.$$

In this case, it is even easy to verify the upper bound condition (14.3.18). Indeed, we have

$$\sum_{n=1}^{\infty} e^{-\mu_n^*(x+an)-\theta f(x)} = \sum_{n=1}^{\infty} \exp\left(-x^\gamma\left(\frac{(1 + \frac{n}{x})^2}{2(\frac{n}{x})^{2H}} - \theta\right)\right)$$

$$\approx x\int_0^{\infty} \exp\left(-x^\gamma\left(\frac{(1 + u)^2}{2u^{2H}} - \theta\right)\right) du \leq x$$

for x sufficiently large, and (14.3.18) follows.

Thus, the distribution of X_t can be approximated by the Weibull distribution

$$\Pr\{X_t > x\} \sim \exp\left(-\frac{(c-m)^{2H}}{2\kappa(H)^2 am}x^{2-2H}\right). \qquad (17.4.8)$$

Simulation studies show that the approximation (17.4.8) is in fact an upper bound which is roughly as far from the true value as the lower bound (17.4.7). Note that in the Brownian case $H = 1/2$, the right hand side of (17.4.8) reduces to an exponential distribution which in fact gives $\Pr\{X_t > x\}$ *exactly*.

Solving (17.4.8) for c we see that $\Pr\{X_t > x\} = \epsilon$ is achieved approximately when

$$c = m + \left(\kappa(H)\sqrt{-2\ln\epsilon}\right)^{1/H} a^{1/(2H)} x^{-(1-H)/H} m^{1/(2H)}. \qquad (17.4.9)$$

By the exact scaling relation (17.4.5), the only approximate part of formula (17.4.9) is the coefficient in front of the powers of a, x and m.

A different approximation for the queue length distribution was derived in [NSVV95] using the Beneš approach (see also Subsection 14.2.4). The result turned out to be numerically almost identical to the Weibull distribution presented above.

If the traffic is a superposition of fractional Brownian traffic streams with different parameters, the stream with the highest H value dominates in the large deviations asymptotics. Knowing this is, however, often insufficient for getting a useful estimate for the loss probability of a finite buffer. The lower bound (17.4.6) is still easy to calculate numerically and can be used as a rough approximation.

17.4.3   Fluid storage with long range dependent on/off sources

In this subsection, we consider a set of N identical long range dependent on/off sources with peak rate fixed to unity, as already introduced in Subsection 13.4.2, which offer traffic to an unlimited buffer with constant leak rate c. It seems that no closed form expressions, even good approximate ones, are known for the buffer occupancy distribution when N is small. However, asymptotic results for large N and heavy load exist [BRSV95] and are summarized in this subsection.

It is easy to see that for fixed N, the tail of the queue length distribution is qualitatively different in the two cases where long range dependence is caused by a heavy-tailed off-period distribution only and where the on-period distribution is heavy-tailed (or both are). Indeed, if neither distribution is heavy-tailed, the queue length distribution has an exponential tail, and if the silences are made longer, the queueing becomes somewhat lighter but does not change qualitatively. On the other hand, assume that the burst length distribution is of the form $\Pr\{U > t\} = L(t)t^{-\beta}$, where $\beta \in (1, 2)$ and $L(t)$ is slowly varying. Then the distribution of the past lifetime U_0 of an ongoing burst in a stationary process has the form $\Pr\{U_0 > t\} = \tilde{L}(t)t^{-\beta+1}$, and for any $n > c$ we have for the queue length X_t the trivial lower bound

$$\Pr\left[X_t > x \mid n \text{ sources on at time } t\right] \geq$$

$$\geq \Pr\left\{U_0 > \frac{x}{n-c}\right\}^n \sim \left(\frac{x}{n-c}\right)^{-(\beta+1)n},$$

which shows that the $\Pr\{X_t > x\}$ goes to zero as some negative power of x. When a large number of sources is superposed, however, the traffic approaches in both cases a long range dependent Gaussian process and the ultimate tail behaviour of the queue length distribution turns out to be Weibullian, which is "worse" than an exponential tail but "better" than a power tail. Thus, we have the interesting insight that the aggregation of sources makes the traffic "worse behaving" if the long range dependence comes from the silence periods only, but "better behaving" if the on-period distribution is heavy-tailed. This phenomenon has been discussed and illustrated by simulations by Pruthi [Pru95].

After this heuristic introduction, let us now turn to the proven asymptotic results. For each N, we assume that the service capacity of the system is

$$c = c(N) = Nm + \gamma\sqrt{N}, \qquad (17.4.10)$$

where $m \in (0, 1)$ denotes the mean input rate of a single source and γ is some positive constant. For increasing N, the system load $\rho = Nm/c$ of the system consequently tends to 1, thus implying heavy traffic conditions.

As in Subsection 13.4.5, let $\Lambda_t \in \{0, 1\}$ and A_t denote the instantaneous bit rate of a single source at time t and the workload entering the system due to this single source over any interval $[0, t)$, respectively. In general conditions, the probability $Q^{(N)}(x)$ that the queue content of the storage system exceeds a level x can be bounded by

$$L^{(N)}(x) \leq Q^{(N)}(x) \leq U^{(N)}(x), \qquad (17.4.11)$$

where the lower bound $L^{(N)}(x)$ and the upper bound $U^{(N)}(x)$ are essentially determined by the distribution of the workload A_t for each positive t and are derived from Subsection 14.2.1 and Subsection 17.2.1, Eq. (17.2.3), respectively. Now, letting the number N tend to infinity such that the heavy traffic condition (17.4.10) is satisfied and for an occupation level scaled as $x = \xi\sqrt{N}$ (with fixed ξ), these bounds are shown to have a finite limit defined by [BRSV95]

$$L(\xi) = \sup_{t>0} \overline{\Phi}\left(\frac{\xi + \gamma t}{\sqrt{v(t)}}\right), \qquad (17.4.12)$$

where $v(t) = \mathrm{Var}\{A_t\}$ and $\overline{\Phi}$ is the complementary distribution function of the normal distribution, and

$$U(\xi) = \tag{17.4.13}$$

$$\int_{-\infty}^{\gamma} (\gamma - \eta) \frac{e^{-\eta^2/2\sigma^2}}{\sigma 2\pi} \int_0^{+\infty} \exp\left(-\frac{(\xi + \gamma t - \eta M(t))^2}{2(v(t) - \sigma^2 M(t)^2)}\right) \frac{dt}{\sqrt{v(t) - \sigma^2 M(t)^2}} \, d\eta,$$

where $\sigma^2 = m(1 - m)$ and $M(t) = G_1(t) - F_1(t)$ with

$$F_1(t) = \mathrm{E}[A_t \mid \Lambda_0 = 0], \qquad G_1(t) = \mathrm{E}[A_t \mid \Lambda_0 = 1],$$

respectively. Note that these limit bounds are determined via the functions $M(\cdot)$ and $v(\cdot)$ only. Applying the convergence of the global workload process $\{A_t^{(N)}\}$ to the limiting Gaussian process $\{Y_t\}$ introduced in Subsection 13.4.5, Eq.(13.4.13), it is also proved that for all fixed ξ, probability $Q^{(N)}(\xi\sqrt{N})$ also has a finite limit $Q(\xi)$ when $N \to +\infty$, namely

$$Q(\xi) = \Pr\left\{\sup_{t>0}(Y_t - \gamma t) > \xi\right\}. \tag{17.4.14}$$

In view of (17.4.11) applied to $x = \xi\sqrt{N}$ and the existence of the limits $L(\xi)$ and $U(\xi)$, this limit $Q(\xi)$ therefore satisfies

$$L(\xi) \le Q(\xi) \le U(\xi). \tag{17.4.15}$$

We now wish to introduce long range dependence into the source model and evaluate $M(t)$ and $v(t)$, and hence $L(\xi)$ and $U(\xi)$, explicitly for such a case. As discussed in Subsection 13.4.2, assuming that either the silence or activity periods (or both) have a heavy-tailed distribution with respective exponent α and β entails that the variance $v(t)$ grows as a power t^{2H} for large t and with some $H \in (1/2, 1)$. As shown in [BRSV95], the quantity $M(t)$ also behaves as a fractional power of t for large t. Such a behaviour of the functions $M(\cdot)$ and $v(\cdot)$ eventually entails that, for large ξ, the limit bounds $L(\xi)$ in (17.4.12) and $U(\xi)$ in (17.4.13) take the form

$$L(\xi) \sim \frac{\exp(-R\xi^{2(1-H)})}{\sqrt{2\pi}R\xi^{(1-H)}}, \tag{17.4.16}$$

$$U(\xi) \sim \frac{\sigma}{\gamma\sqrt{2\pi}}\sqrt{\frac{H}{1-H}} \exp(-R\xi^{2(1-H)}),$$

where $H = (3 - \min\{\alpha, \beta\})/2$, $R = \gamma^{2H}/2\Sigma^2\kappa(H)^2$, $\kappa(H) = H^H(1-H)^{1-H}$ and, with the notation of Subsection 13.4.2,

$$\Sigma^2 = \frac{2h_b\lambda(1-m)^3}{(\beta-1)(2-\beta)(3-\beta)}$$

if $\beta < \alpha$ or

$$\Sigma^2 = \frac{2h_a\mu m^3}{(\alpha - 1)(2 - \alpha)(3 - \alpha)}$$

if $\alpha < \beta$, or the sum of these in case of equality. The application of (17.4.16) with (17.4.15) therefore implies that $Q(\xi)$ is of order $\exp(-R\xi^{2(1-H)})$ for large ξ. The heavy traffic analysis of the storage system fed by a large number number of on/off sources with long range dependence characteristics thus reveals that the queue length distribution exhibits Weibullian tail behaviour for large buffer content.

For the sake of illustration, we depict the above evaluations in some numerical examples. Consider $N = 100$ on/off sources having activity periods distributed as Pareto random variables. We take the mean burst volume as unity ($\mu = 1$). The mean activity probability $m = \lambda/(\lambda + \mu)$ is set to 0.1 ($\lambda = 1/9$) and the multiplexer load $Nm/c = (1 + \gamma/v\sqrt{N})^{-1}$ to 0.9, implying $c = 11.1..$ ($\gamma = 0.11..$). We assume that 100 sources is sufficient for a useful application of our asymptotic results. Figure 17.4.1 shows a log plot of the complementary distribution function (interpreted as the loss probability against buffer size) measured in units of mean burst volume for a Hurst parameter of 0.7. Curve (b) corresponds to the exact limiting upper bound U given by (17.4.13), and (c) to the exact lower bound L given by (17.4.12). Curves (a) and (d) correspond to the approximation of these two bounds given in (17.4.16). We observe that the buffer size required for a loss probability of 10^{-9} corresponds to around 1000 times the mean burst size.

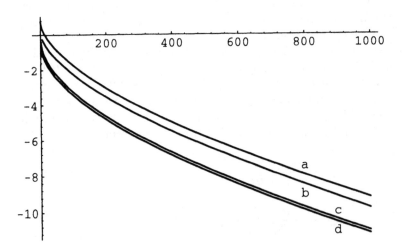

Figure 17.4.1: Loss probability as a function of the buffer size ($H = 0.7$).

Now let T be a large parameter and constant Σ be given as in (17.4.16). As discussed in Subsection 13.4.5, the "stretched" Gaussian process $\{Y_{Tu}\}/\Sigma T^H$ converges for increasing T towards a FBM $\{Z_u\}$ with Hurst parameter H. For fixed δ and y, set $t = Tu$, $\gamma = \delta/T^{1-H}$ and $\xi = yT^H$ in (17.4.14). The "reduced" capacity γ tends to 0 with increasing T; the limit with respect to T thus corresponds to a "super-heavy" load situation. Letting T tend to infinity in (17.4.14), probability $Q(\xi) = Q(yT^H)$ is seen to have a finite limit

$$q(y) = \Pr\left\{\sup_{u>0}(\Sigma Z_u - \delta u) > y\right\}. \tag{17.4.17}$$

This equation simply states that the limit of the (rescaled) distribution Q for large T corresponds to the stationary distribution q of the content of a fluid queue fed by an FBM process with output rate δ. The application of (17.4.15) for $\xi = yT^H$ also entails in turn that

$$\ell(y) \le q(y) \le w(y), \tag{17.4.18}$$

where

$$\ell(y) = \lim_{T\to+\infty} L(yT^H) = \sup_{u>0}\overline{\Phi}\left(\frac{y+\delta u}{\Sigma u^H}\right) \tag{17.4.19}$$

and $w(y) = \lim_{T\to+\infty} U(yT^H)$. While the limit $\ell(y)$ is finite, it is derived [BRSV95] from (17.4.13) that the upper bound $U(yT^H)$ diverges with T, so that $w(y) = +\infty$. To regularize the upper bound U, we can apply the heuristic corrective factor introduced in Section 17.2.1 for improving the estimate $U^{(N)}(x)$ by writing

$$Q^{(N)}(x) \approx \mathcal{F}_N \cdot U^{(N)}(x), \tag{17.4.20}$$

where

$$\mathcal{F}_N = \frac{c - Nm}{\sum_{n<c}(c-n)\Pr\{\Lambda_0^{(N)} = n\}}.$$

The factor $\mathcal{F}_N$ can be understood as reintroducing the empty queue probability which is ignored in the definition of the upper bound $U^{(N)}(x)$ and which is not negligible in the case of heavy load. It is easily derived [BRSV95] that the right-hand side of (17.4.20) has a finite limit $\mathcal{F} \cdot U(\xi)$ for increasing N. The approximation $\hat{q}(y) = \lim_{T\to+\infty} \mathcal{F} \cdot U(yT^H)$ can therefore be proposed for $q(y)$, yielding

$$\hat{q}(y) = \delta \int_0^{+\infty} \Pr\{\Sigma Z_u = y + \delta u\}\,du, \tag{17.4.21}$$

where $\Pr\{Z_u = y\}$ denotes the value at y of the density of Z_u. The following estimate of this integral for large y then exhibits a Weibullian behaviour, namely

$$\hat{q}(y) \sim \sqrt{\frac{H}{1-H}}e^{-ry^{2(1-H)}}, \tag{17.4.22}$$

where $r = \delta^{2H}/2\Sigma^2\kappa(H)^2$. Note that this result gives the exact queue distribution in the case when $H = 1/2$. Estimate (17.4.22) is also in line with the asymptotic evaluation of the lower limit $\ell(y)$ for large y as derived from (17.4.19),

$$\ell(y) \sim \frac{1}{\sqrt{2\pi r}y^{1-H}}e^{-ry^{2(1-H)}}$$

Figure 17.4.2 represents the upper and lower bounds after the operation of time-stretching with $T = 10^5$ and $H = 0.7$. Curve (a) corresponds to the approximation (17.4.21), and (b) to its estimate (17.4.22) for large y. Curve (c) is the lower bound (17.4.19).

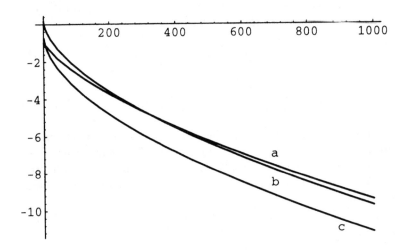

Figure 17.4.2: Loss probability after time-stretching with $T = 10^5$ as a function of the buffer size ($H = 0.7$).

17.5   Leaky Bucket Performance

In view of its importance as a traffic control device, a number of specific queueing models have been developed to analyse the performance of a leaky bucket. In Section 17.5.1 we outline a discrete time model of a leaky bucket policing the Sustainable Cell Rate (SCR) of an on/off source and an MPEG video source. Section 17.5.2 presents an analysis of the same system in continuous time where, however, nonconforming cells are delayed and not discarded. In both cases it is assumed that the cell stream has been shaped to a given Peak Cell Rate (PCR) although the influence of this shaping is not studied. In Section 17.5.3 a dual shaper (two leaky buckets in series) is analysed for input traffic modelled as a renewal process. This analysis also addresses the influence of the PCR shaper on performance.

17.5.1 Leaky Bucket policing

A method for the performance analysis of the continuous state leaky bucket when used for SCR policing was proposed in [RT94b]. The algorithm is applicable for on/off sources with independent and general distributions of on and off periods. The model assumes discrete time, with the slot as the time unit. On periods are assumed to begin always with the first cell arrival of the burst and during the on period cells arrive with a fixed interarrival time equal to d_{cell} slots. On and off periods have distributions $a_{on}(i)$ and $a_{off}(i)$ respectively. The continuous state leaky bucket has leak rate $1/T$ and burst tolerance τ.

Let $Z(t)$ denote the bucket content at time t. $Z(t)$ can also be seen as the expected arrival time of the next cell. A cell arriving at time t will be rejected if $Z(t) > \tau$. The algorithm is based on the relations that determine the bucket content at time t. Let $z_{off,n,k}(i)$ denote the distribution of $Z(t)$ at the k-th slot in the n-th off state, and let $z^-_{on,n,k}(i)$ (respectively $z^+_{on,n,k}(i)$) denote the distribution of $Z(t)$ just before (respectively after) the beginning of the k-th slot in the n-th on state.

For k a multiple of d_{cell} we have

$$z^+_{on,n,k}(i) = z^-_{on,n,k}(i) \qquad \text{for } i = 0, \cdots, T + \tau.$$

If k is not a multiple of d_{cell} and if $\tau \geq T$ (which will be the typical case) then

$$z^+_{on,n,k}(i) = \begin{cases} 0 & : \ 0 \leq i < T, \\ z^-_{on,n,k}(i - T) & : \ T \leq i \leq \tau, \\ z^-_{on,n,k}(i - T) + z^-_{on,n,k}(i) & : \ \tau < i \leq T + \tau. \end{cases} \qquad (17.5.1)$$

The bucket content decreases by one each slot, therefore

$$z_{off,n,k+1}(i) = \begin{cases} z_{off,n,k}(0) + z_{off,n,k}(1) & : \ i = 0, \\ z_{off,n,k}(i + 1) & : \ 0 < i < T, \\ 0 & : \ T, \end{cases} \qquad (17.5.2)$$

and

$$z^-_{on,n,k}(i) = \begin{cases} z^+_{on,n,k}(0) + z^+_{on,n,k}(1) & : \ i = 0, \\ z^+_{on,n,k}(i + 1) & : \ 0 < i < T, \\ 0 & : \ T. \end{cases} \qquad (17.5.3)$$

The system state distribution just before the switching instant to the $(n+1)$-th off (respectively on) state is given by:

$$z_{\text{off},n+1,0}(i) = \sum_{k=1}^{\infty} a_{\text{on}}(k) \cdot z_{\text{on},n,k}^{-}(i) \qquad \text{for } i = 0, \cdots, T+\tau, \qquad (17.5.4)$$

and

$$z_{\text{on},n,0}^{-}(i) = \sum_{k=1}^{\infty} a_{\text{off}}(k) \cdot z_{\text{off},n,k}(i) \qquad \text{for } i = 0, \cdots, T+\tau. \qquad (17.5.5)$$

Using equations 17.5.1 to 17.5.5 iteratively, the system state distributions in equilibrium $z_{\text{on},k}^{-}(i)$ $(k = 0, \cdots, \infty)$ can be derived by

$$z_{\text{on},n,k}^{-}(i) = \lim_{n \to \infty} z_{\text{on},n,k}^{-}(i).$$

Now, from $z_{\text{on},k}^{-}(i)$ we can easily compute the probabilities $p(k)$ that a cell arriving at the k-th slot in an on state is rejected:

$$p(k) = \sum_{i=\tau+1}^{T+\tau} z_{\text{on},n,k}^{-}(i).$$

Finally, the cell rejection probability can be obtained from $p(k)$ with the following equation:

$$p_r = \frac{\sum_{k=1}^{\infty} \delta(k) \cdot F_{A_{\text{on}}}^{c}(k) \cdot p(k)}{\sum_{k=1}^{\infty} \delta(k) \cdot a_{\text{on}}^{c}(k)},$$

where $\delta(k)$ is 1 if $k \bmod d_{cell} = 0$ and 0 otherwise, and $a_{\text{on}}^{c}(k)$ is the cumulative probability distribution of the on state duration.

As an example, Figure 17.5.1 shows the cell rejection probability of the leaky bucket versus the burst tolerance τ_s for three different distributions of on and off state durations. In all cases the cell interarrival time in the on state is $d_{cell} = 5$ and the minimum duration of an off state is 4 slots. The mean durations of on and off periods are 50 slots and the maximum burst size of the on states was truncated to 100 slots.

For each distribution, Figure 17.5.1 shows curves for $T_s = 10$, which corresponds to the mean cell rate, and $T_s=9$. It can be observed that the choice of the burst tolerance τ_s to achieve a desired cell rejection probability is strongly dependent on the distribution of the state lengths.

The same approach can be used to obtain the cell rejection probability for MPEG video sources [RR94]. MPEG sources emit GOPs (groups of frames) of size G frames. During each frame cells arrive back to back at the link speed followed by a silence to complete the frame duration D. The number of cells in successive frames are assumed independent and with distributions $a_n(k)$

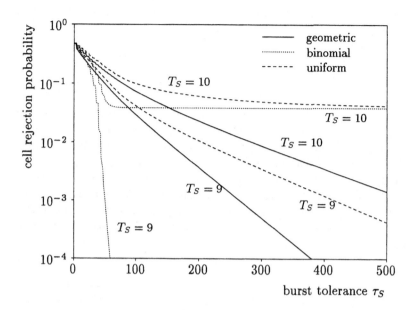

Figure 17.5.1: Cell rejection probability for different on/off distributions.

for the n-th frame of the GOP. In this model, $z_{n,k}^+(i)$ (respectively $z_{n,k}^-(i)$) will denote the distributions of the bucket content just before (respectively after) the k-th slot of the n-th frame of the GOP. Assuming $T \leq \tau$ the equations used to obtain $z_{n,k}^-$ by iteration are

$$
z_{n,k}^+(i) = \begin{cases} 0 & : \ 0 \leq i < T, \\ z_{n,k}^-(i-T) & : \ T \leq i \leq \tau, \\ z_{n,k}^-(i-T) + z_{n,k}^-(i) & : \ \tau < i \leq T + \tau, \end{cases}
$$

and

$$
z_{n+1,k}^-(i) = \sum_{k=1}^{D} a_n(k) \cdot z_{n,k}^*(i)
$$

with

$$
z_{n,k}^*(i) = \begin{cases} \sum_{j=0}^{D-k} z_{n,k}^-(j) & : \ i = 0, \\ z_{n,k}^-(i + (D-k)) & : \ 0 < i \leq T + \tau - (D-k), \\ 0 & : \ T + \tau - (D-k) < i \leq T + \tau. \end{cases}
$$

Finally, the cell rejection probability in a frame of type n is given by the following equation:

$$
P_n = \frac{\sum_{k=1}^{D-1} p_n(k) \cdot a_n^c(k)}{\sum_{k=0}^{D-1} a_n^c(k)},
$$

where $p_n(k)$ is the probability of observing a nonconforming cell at slot k of a frame of type n, and is given by:

$$p_n(k) = \sum_{i=\tau+1}^{T+\tau} z_{n,k}^-(i).$$

Figure 17.5.2 shows the cell rejection probability versus different values of burst tolerance τ_s, computed using the above method compared with simulation results. The distributions $a_n(k)$ were obtained from measurements of a real video source. The figure corresponds to a SCR with $T_s=10$ and a link capacity (peak cell rate) of 34 Mb/s. In this case the analytical results are very accurate, however if τ_s is close to the mean interarrival time, i.e., T_s is greater than two frame durations, the analytical results underestimate the cell rejection probability because the model does not take into account long-range dependence within the video sequence.

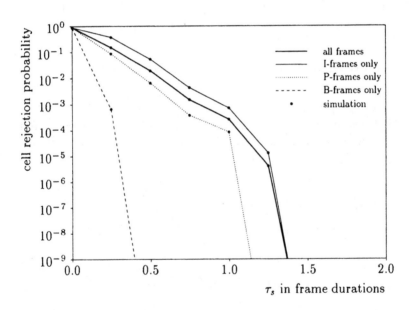

Figure 17.5.2: Cell rejection analysis for MPEG video sources.

17.5.2 Leaky Bucket shaping

Leaky bucket shaping consists in delaying non-conforming cells until the earliest time instant at which the leaky bucket can accept them. Thus, in a leaky bucket with leak rate r and bucket size M (in units of work), arriving cells which find the bucket content smaller than M are directly admitted to the network; otherwise they are queued with FIFO discipline and admitted to the network at rate r.

The leaky bucket shaper can be modelled as a fictitious FIFO queue with deterministic service rate r and a threshold M. All arriving cells are "stored" in this queue on arrival. However, cells are actually admitted to the network directly if the queue length is less than M. Only when the fictitious queue length is greater than M are cells actually delayed (see Figure 17.5.3). The performance of the leaky bucket shaper can be analyzed using the Beneš approach for fluid on/off sources, as described in Section 17.2.1.

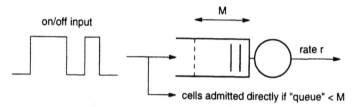

Figure 17.5.3: Leaky bucket as fictitious queue.

In the model, successive silence and burst lengths are assumed to be independent with probability densities $a(\cdot)$ and $b(\cdot)$, respectively. Source peak rate is 1, mean silence length is $1/\alpha$ and mean burst length is $1/\beta$. Let $Q(x)$ represent the probability that the content of the fictitious queue described above exceeds x.

The results of Section 17.2.1 (see also [RBC93]) allow $Q(x)$ to be expressed in terms of the poles with negative real part z_k and corresponding residues f_k of the function $f^{**}(-rz, z)$, where:

$$f^{**}(s,z) = \frac{\alpha z(1 - a^*(s))(b^*(z+s)-1)}{s^2(z+s)(1-b^*(z+s)a*(s))} + \frac{1}{s} \tag{17.5.6}$$

is the double Laplace transform of $f(t, w)$ which is the probability density of the work arriving in an interval of length t given that the source is silent at the start of the interval. $Q(x)$ is then given by the following expression which is exact for exponentially distributed silences and a good approximation otherwise:

$$Q(x) \approx r(1-\rho) \sum_k f_k e^{z_k x}, \tag{17.5.7}$$

where $\rho = \alpha/r(\alpha + \beta)$ is the leaky bucket load.

The performance of the shaper may be assessed through the time necessary for a burst to completely enter the network. This depends on the content of the leaky bucket when the burst leading edge arrives and on the length of the burst itself. Let T_u be the delay of the last cell of a burst of length u. This is the additional transfer time introduced by the leaky bucket access control. If a burst begins at time 0 and ends at time u, the queue length V_u

at u is $V_0 + u - ru$. Let {start in $(0, \Delta t)$} denote the event that the source begins to emit in an infinitesimal interval starting at time 0. This event has probability:

$$\Pr\{\text{start in } (0, \Delta t)\} = \frac{\alpha \beta}{\alpha + \beta} \Delta t.$$

Now,

$$\Pr\{T_u > t\} = \Pr\{V_0 + u(1-r) > rt + M \,|\, \text{start in } (0, \Delta t)\}$$

$$= Q_+(rt - u(1-r) + M).1_{\{rt - u(1-r) + M \geq 0\}} +$$

$$1_{\{rt - u(1-r) + M < 0\}}, \qquad (17.5.8)$$

where Q_+ is defined by:

$$Q_+(x) = \Pr\{V_0 > x \,|\, \text{start in } (0, \Delta t)\},$$

$$= \Pr\{V_0 > x \text{ and start in } (0, \Delta t)\} \cdot \frac{\alpha + \beta}{\alpha \beta \, \Delta t}.$$

Q_+ has a definition very close to that of Q and an expression similar to (17.5.7) can be derived by a parallel analysis. Writing

$$f_+(t, x)\Delta t = \frac{\partial}{\partial x} \Pr[\, w(t) \leq x \text{ and start in } (0, \Delta t) \,|\, \text{source silent at } -t\,],$$

we have,

$$Q_+(x) \approx r(1 - \rho) \sum_k e^{z + kx} f_{+k} \cdot \frac{\alpha + \beta}{\alpha \beta},$$

where the z_{+k} are the poles of negative real part of $z \to f_+^{**}(-rz, z)$ and f_{+k} are their residues. By reasoning as in Section 17.2.1, we find:

$$f_+^{**}(s, z) = \frac{\alpha(1 - a^*(s))}{s(1 - b^*(z + s)a^*(s))} \qquad (17.5.9)$$

so that the poles z_{+k} and z_k are identical.

The expected delay of a burst of duration u, $\overline{T_u}$ is given by integrating expression (17.5.8) for $\Pr\{T_u > \tau\}$:

$$\overline{T_u} = \begin{cases} (1 - \rho) \sum_k \frac{f_{+k}}{-z_k} e^{z_k(M - u(1-r))} \cdot \frac{\alpha + \beta}{\alpha \beta} & \text{if } u < \frac{M}{1 - r}, \\[3mm] \frac{u(1-r) - M}{r} + (1 - \rho) \sum_k \frac{f_{+k}}{-z_k} \cdot \frac{\alpha + \beta}{\alpha \beta} & \text{otherwise.} \end{cases}$$

Given the transit time $\overline{T_u}$, we can deduce the effective throughput of the leaky bucket as a function of the burst size:

$$\theta_u = \frac{u}{u + \overline{T_u}}.$$

This function is used in Subsection 5.3.2 of Part I to investigate the performance of the shaper as a function of parameters r and M.

17.5.3   Analysis for renewal processes

For renewal arrival processes some other performance measures of the leaky bucket shaper can be evaluated. In [Rit95a] the analysis of a dual spacer is considered by an iterative method. The dual spacer, consists of two shapers: one for shaping to a given PCR, $1/T$, and the other for shaping to a given SCR, $1/T_s$ (with a certain BT). For PCR shaping no CDV tolerance is allowed, $\tau = 0$, i.e. cells are delayed if they arrive closer than T slots. For SCR shaping a burst tolerance τ_s is allowed, thus, a cell is delayed if the leaky bucket content at cell arrival is greater than τ_s. Cells can be delayed due to either nonconformance to the PCR or nonconformance to the SCR and BT. A cell is delayed until it is considered conforming with respect to both leaky buckets: PCR (with CDVT equal to zero) and SCR (with a certain BT). Also, a maximum delay W is considered: if a cell needs to be delayed more than W to be conforming it is rejected.

Let P_n^- and P_n^+ be the queue content of the PCR shaper just before and just after the time instant at which the n-th cell arrives respectively; and let S_n^- and S_n^+ be the queue content of the SCR shaper just before and after the arrival of the n-th cell respectively. Let $a_n(k)$ be the interarrival time distribution. Define $u_n^-(i,j) = \Pr\{P_n^- = i \wedge S_n^- = j\}$ and $u_n^+(i,j) = \Pr\{P_n^+ = i \wedge S_n^+ = j.\}$. We can express $u_{n+1}^-(\cdot,\cdot)$ in terms of $u_n^+(\cdot,\cdot)$:

$$u_{n+1}^-(i,j) = \sum_{k=-\infty}^{0} u_n^+(i-k, j-k) \cdot a(-k).$$

To apply the iteration method now we need to compute $u_n^+(\cdot,\cdot)$ from $u_n^-(\cdot,\cdot)$. Define $u_n^*(i,j) = \pi_*(u_n^-(i,j))$, with the operator $\pi_*(\cdot)$ defined as:

$$\pi_*(u_n^-(i,j)) = \begin{cases} 0 & : \ i < 0 \wedge j \leq \tau_s + W, \\ 0 & : \ 0 \leq i \leq W \wedge j < 0, \\ \sum_{i'=-\infty}^{0} u_n^-(i',j) & : \ i = 0 \wedge 0 < j \leq \tau_s + W, \\ \sum_{i'=-\infty}^{0} \sum_{j'=-\infty}^{0} u_n^-(i',j') & : \ i = 0 \wedge j = 0, \\ \sum_{j'=-\infty}^{0} u_n^-(i,j') & : \ 0 < i \leq W \wedge j = 0, \\ u_n^-(i,j) & : \ \text{otherwise.} \end{cases}$$

Thus, u_n^+ can be expressed in terms of u_n^*

$$u_n^+(i,j) = \begin{cases} 0 & : \ (i,j) \in \mathcal{A}_1, \\ u_n^*(i-T, j-T_s) & : \ (i,j) \in \mathcal{A}_2, \\ u_n^*(i-T, j-T_s) + u_n^*(i,j) & : \ (i,j) \in \mathcal{A}_3, \\ u_n^*(i,j) & : \ \text{otherwise.} \end{cases}$$

In the case where $W > T$ and $\tau_s + W \geq T_s$, the regions $\mathcal{A}_i$ are given by

$$\mathcal{A}_1 = \{(i,j) \mid (-\infty \leq i \leq W \wedge -\infty \leq j < T_s) \vee$$

$$(-\infty \le i < T \wedge -\infty \le j \le \tau_s + W)\},$$

$$\mathcal{A}_2 = \{(i,j) \mid (T \le i \le W \wedge T_s \le j \le \tau_s + W)\},$$

$$\mathcal{A}_3 = \{(i,j) \mid (W < i \le T + W \wedge T_s \le j \le \tau_s + W) \vee$$

$$(T \le i \le T + W \wedge \tau_s + W < j \le T_s + \tau_s + W)\}.$$

With the equations presented above the limiting distribution $u^-(i,j)$ can be obtained iteratively:

$$u^-(i,j) = \lim_{n \to \infty} u_n^-(i,j).$$

From $u^-(i,j)$ different performance measures can be obtained:

Cell rejection probability. Cells that should be delayed more than W are rejected. The cell rejection probability, p_r, can be found as

$$p_r = 1 - \sum_{i=-\infty}^{W} \sum_{j=-\infty}^{\tau_s+W} u^-(i,j).$$

Inter departure time distribution. The probability $d(k)$ of observing a time interval of k slots between two consecutive departures is given by:

$$d(k) = \frac{1}{1 - p_r} \cdot \sum_{(i,j)\in \mathcal{B}_k} u^-(i,j).$$

The set $\mathcal{B}_k$ contains those states (i,j) where the departure of the previous not rejected cell has occurred or will occur k slots before the departure of the cell which is currently arriving. The sets $\mathcal{B}_T$ and $\mathcal{B}_{T_s}$ are given by:

$$\mathcal{B}_T = \{(i,j)|(0 < i \le W \wedge -\infty < j \le i+\tau_s) \vee (\min\{T-i, T_s+\tau_s-j\} = T)\},$$

$$\mathcal{B}_{T_s} = \{(i,j) \mid (0 < i \le W \wedge i+\tau_s < j \le \tau_s + W) \vee$$

$$(-\infty < i \le 0 \wedge \tau_s < j \le \tau_s + W) \vee$$

$$(\min\{T - i, T_s + \tau_s - j\} = T_s)\}.$$

For all other values of k, $\mathcal{B}_k$ contains the following states

$$\mathcal{B}_k = \{(i,j) \mid (\min\{T-i, T_s+\tau_s-j\} = k)\}.$$

Delay distribution. The distribution, $w(k)$, of the delay introduced by the shaper is given by

$$w(k) = \frac{1}{1 - p_r} \cdot \sum_{(i,j)\in \mathcal{C}_k} u^*(i,j),$$

where C_k is the set containing those states (i,j) where an arriving cell is delayed k slots. For $0 \le k \le W$, C_k is given by:

$$C_k = \{(i,j)|(i = k \wedge i > j - \tau_s) \vee (j = k + \tau_s \wedge i \le j - \tau_s)\}.$$

As an example, Figure 17.5.4 presents the delay distribution when the interarrival time follows a negative binomial distribution with mean $E_A = 10$ slots and coefficient of variation $c_A = 2$. The spacer parameters are $T = 5$, $T_s = 8$ and τ_s varies from 0 to ∞. A maximum delay of $W = 200$ slots is tolerated. It can be observed that as τ_s decreases the effect of the SCR shaping increases and, thus, the delay increases. With $\tau_s = \infty$ delay is only due to PCR shaping.

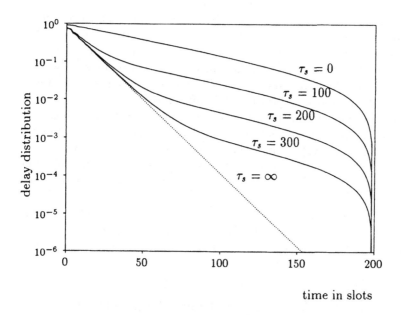

Figure 17.5.4: Delay distribution in the dual spacer.

18 Multi-rate models

Multi-rate models are widely used to analyze CAC mechanisms in ATM networks. One of the problems with these models is how to determine the call blocking probabilities of the different traffic streams. It is well-known that for Poisson arrivals in the absence of controls the distribution of the busy circuits has a product form, which is insensitive to the holding time ratios, from which the call blocking probabilities can be calculated. However, faster methods are often needed and these are discussed in section 18.1 along with some behavior peculiar to multi-slot systems.

In the absence of controls, large bandwidth calls can experience a much higher blocking probability than low bandwidth calls. Ways of controlling this are discussed in section 18.2, which particularly concentrates on the use of trunk reservation as a protection mechanism. The section looks at ways of calculating call blocking probabilities when trunk reservation is used. A general rule to equalize blocking for calls with different bandwidth requirements is given and the influence of the call mean holding time ratio is studied.

Since a CAC mechanism has to be defined with respect to a certain GOS for accepted calls, the evaluation of the blocking behavior on the burst or the cell level is also of interest. Section 18.3 examines call and burst blocking for ATM networks. A model which allows the consideration of call and burst blocking simultaneously is presented.

The problem of calculating end-to-end blocking in multi-rate networks becomes even more complicated than in the single link case. Section 18.4 examines some preliminary issues concerning call blocking in multi-rate networks.

18.1   Call blocking in multi-rate systems

This section focuses on the evaluation of the call blocking probabilities in multi-rate networks in the absence of controls. Generally, a multi-rate system, such as an ATM network, behaves like a multi-slot circuit-switched system if an equivalent bandwidth is allocated for accepted VBR calls. To take into account the given fact that calls may differ notably in their individual resource requirements and holding times, we consider the single link traffic model presented in Figure 18.1.1 for the analyses presented in this section.

Calls belonging to N different traffic classes have access to a transmission link with a fixed capacity of c Mbps using a complete sharing admission policy. The arrival process for class i calls is assumed to be a Poisson process with a rate λ_i and the holding time of class i calls follows a general distribution function with mean $1/\mu_i$. During the life-time of a call of class i, a constant rate denoted by c_i is allocated and released immediately after the call has completed. This rate may either be the rate required by CBR calls or an equivalent bandwidth in case of VBR calls.

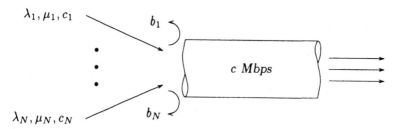

Figure 18.1.1: Basic link model for the complete sharing policy.

The analysis of this model and extensions of it is quite well understood in the existing literature. In Enomoto *et al.* [EM73], a product form solution for the state probabilities is developed for an infinite and a finite number of traffic sources. Aein [Aei78] extended these results by showing that the product form solution is valid for all *coordinate convex* admission policies. It is obvious, that the numerical evaluation of these product form solutions becomes intractable for a large number of traffic classes. Therefore, a simpler recursive solution was developed by Kaufman [Kau81] and Roberts [Rob81], who also have shown that their solution provides exact results for general holding time distributions. A more general arrival process, the *Bernoulli-Poisson-Pascal* process, was considered by Delbrouck [Del83] and van de Vlag [vdVA94].

In Barberis [BB82] the *intermediate sharing* policy is investigated as a generalization of complete and partial sharing and in [Kau81] it was stated that the restriction on *coordinate convex* admission policies is not necessary for the product form solution to exist. Johnson [Joh85] has shown that complete sharing performs better from the resource requirement point of view than partial sharing.

A generalization of partial sharing where a call needs capacity of more than one resource was considered in [PC90]. Furthermore, the recursion formula was generalized by Dziong and Roberts [DR87] to investigate a network where a call requires a variable number of slots on each trunk group of its transmission path. A general method for computing call, time and traffic congestion for multi-rate links, based on a convolution approach, is presented in [Ive87, Ive91]. In [Kau92] it is assumed that blocked calls can retry immediately to be admitted to the system with a smaller capacity requirement. To fulfill their total rate requirement, the product of the rate and the holding time is kept constant.

In the following, two algorithms to determine the exact call blocking probabilities for the traffic model described above are outlined. Additionally, we present a rough approximation of the call blocking probabilities, which is of low complexity. A final subsection compares the complexity and the accuracy of various approximation methods for traffic streams, which are characterized by a peakedness factor.

18.1.1   Exact algorithms

Product form solution

The first algorithm we present computes the state probabilities using the product form solution described in [EM73]. Therefore, the system state is defined by the number of accepted calls of each traffic class, i.e. $(n_1, \ldots, n_N)$, which results in a multi-dimensional state space having as many dimensions as the number of traffic classes. The set of allowed states S is determined by the capacity constraint

$$S = \{(n_1, \ldots, n_N) \mid \sum_{i=1}^{N} n_i c_i \leq c\} \quad . \tag{18.1.1}$$

To derive the state probabilities $p(n_1, \ldots, n_N)$, the following equation is used

$$p(n_1, \ldots, n_N) = \prod_{i=1}^{N} \frac{(\lambda_i/\mu_i)^{n_i}}{n_i!} \cdot G(S)^{-1} \quad , \tag{18.1.2}$$

where the normalizing constant $G(S)$ is given by the sum of the unnormalized probabilities over the allowed states S. Note that they only depend on the offered load $a_i = \lambda_i/\mu_i$ and not on the arrival and service rates individually. The probability b_i, that an arriving class-i call is blocked, is obtained by

$$b_i = \sum_{(n_1, \ldots, n_N) \in S_i} p(n_1, \ldots, n_N) \quad , \tag{18.1.3}$$

with the sets of blocking states S_i defined as

$$S_i = \{(n_1, \ldots, n_N) \mid c - c_i < \sum_{j=1}^{N} n_j c_j \leq c\} \quad . \tag{18.1.4}$$

For theoretical developments the following form of writing b_i may be of interest. By defining the sum over the blocking states S_i in an analogous fashion, we write b_i as

$$b_i = \frac{G(S_i)}{G(S)} \quad . \tag{18.1.5}$$

Noting further that

$$\frac{\partial}{\partial a_i} G(S) = G(S) - G(S_i) \tag{18.1.6}$$

and substituting $G(S_i)$ from this equation into equation (18.1.5), we get

$$b_i = 1 - \frac{\partial}{\partial a_i} \log G(S) \quad .$$

(18.1.7)

Thus, all blocking probabilities can be derived from $G(S)$. Immediately, the reciprocity relation

$$\frac{\partial b_i}{\partial a_j} = \frac{\partial b_j}{\partial a_i}$$

(18.1.8)

follows from equation (18.1.7), cf. [Vir88].

Recursive solution

The recursive solution of the model is based on a mapping of the multi-dimensional state space into a one dimensional state space in accordance with a proper bandwidth discretization (cf. [Kau81, Rob81]). Therefore, a basic bandwidth unit Δc is defined by

$$\Delta c = gcd(c_i) \quad \text{for} \quad 1 \leq i \leq N,$$

(18.1.9)

where $gcd(\cdot)$ denotes the greatest common divisor. In broadband network environments this basic bandwidth unit could be, for example, 64 *kbps* or 2.048 *Mbps*. The maximum number of available basic bandwidth units is denoted by $M = \lfloor c/\Delta c \rfloor$ and the number of basic bandwidth units required per class-i call is expressed by $m_i = \lfloor c_i/\Delta c \rfloor$. Contrary to the product form solution, the system states are now defined by the number m of occupied basic bandwidth units.

The unnormalized state probabilities $\tilde{p}(m)$ are derived by the following recursive algorithm

$$\tilde{p}(m) = \begin{cases} 1 & : \quad m = 0, \\ 0 & : \quad m < 0, \\ (1/m) \sum\limits_{i=1}^{N} \tilde{p}(m - m_i) \cdot m_i \cdot (\lambda_i/\mu_i) & : \quad 0 < m \leq M. \end{cases}$$

(18.1.10)

After normalization, we arrive at the state probabilities $p(m)$

$$p(m) = \tilde{p}(m) \cdot \left(\sum_{m=0}^{M} \tilde{p}(m) \right)^{-1} \quad ,$$

(18.1.11)

and the blocking probability b_i for class-i calls is obtained by

$$b_i = \sum_{m=M-m_i+1}^{M} p(m) \quad .$$

(18.1.12)

Note that the recursive solution is also of an exact nature regarding the call blocking probabilities.

To illustrate the results derived in the last subsections, we make use of an example parameter set considering $N = 4$ traffic classes. According to [TGH93], the following parameters are chosen. The transmission speed is set to 150 $Mbps$ and the bandwidth requirements of the calls are $c_1 = 2$ $Mbps$, $c_2 = 4$ $Mbps$, $c_3 = 10$ $Mbps$ and $c_4 = 20$ $Mbps$, respectively. Calls from each traffic class are assumed to offer the same load.

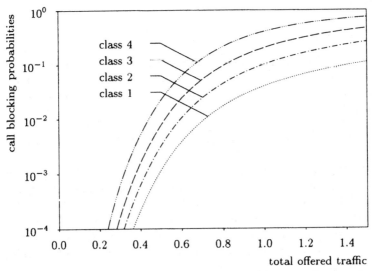

Figure 18.1.2: Call blocking probabilities in a multi-rate environment.

Figure 18.1.2 shows the call blocking probabilities as functions of the offered load. Obviously, calls having higher bandwidth demands experience higher blocking than calls which require less bandwidth, since the blocking state sets for lower bandwidth calls are always subsets of those, which belong to calls having higher bandwidth demands. Thus, a kind of unfairness is observed due to the preference given to low bandwidth calls. This unfairness also appears if we look at the link utilization shared among the traffic classes (cf. Figure 18.1.3). Remember that the offered load is the same for all call classes.

In Figure 18.1.2, the call blocking probabilities increase monotonically with the load. Generally, they can be non-monotonic functions of the offered load. This behavior, however, strongly depends on the composition of the traffic mixture. An illustration of this effect is depicted in Figure 18.1.4.

We observe an interference between the two curves and the GOS for class-1 calls is improved at some particular ranges even though the offered load increases. This behavior was also noticed in [Joh85] and [Vir88]. The explanation for this perhaps surprising effect is the following. If the offered load

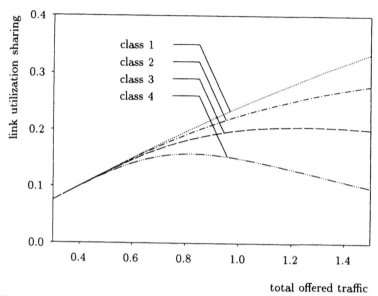

Figure 18.1.3: Link utilization sharing in a multi-rate environment.

increases, class-1 calls can *take over* an *imaginary block* of the link capacity with a size equal to the bandwidth demand of a class-2 call. At this point, the blocking probability of class-2 calls increases whereas the class-1 calls experience a better GOS, since this *imaginary block* can be used by them *exclusively*. If the offered load further increases, the blocking probability for class-1 calls grows rapidly until another *imaginary block* is *taken over*. Thus, oscillations of the blocking probability for class-1 calls occur, which become stronger if the class-2 bandwidth approaches the link capacity.

18.1.2   Simple approximations

In the last section, two algorithms for computing the exact call blocking probabilities have been outlined. Thus, it might look unnecessary to consider the prospect of an approximation. A first-shot approximation could, however, give a clearer view of the importance of certain key parameters, which can be useful in network planning. This section briefly discusses such an approximation, which is derived in [Lin94]. It provides reasonably accurate results in certain ranges of the parameter space.

The superposition of N traffic classes offered to a link can be characterized by its mean m and variance v

$$m = \sum_{i=1}^{N} \frac{\lambda_i}{\mu_i} c_i \qquad v = \sum_{i=1}^{N} \frac{\lambda_i}{\mu_i} c_i^2 \quad . \qquad (18.1.13)$$

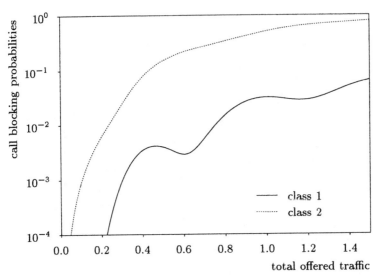

Figure 18.1.4: Oscillation of call blocking in multi-rate environments.

For the definition of an equivalent model, the mean bandwidth $\bar{c}$ of a traffic mixture is determined by $\bar{c} = v/m$. Following the derivation in [Lin94], the average call blocking probability $\bar{b}^*$ and the highest individual call blocking probability $b^*_{\max}$ can be estimated by

$$\bar{b}^* = \mathrm{E}\left(\frac{c - \bar{c} + c_{\min}}{\bar{c}}, \frac{m}{\bar{c}}\right) \tag{18.1.14}$$

and

$$b^*_{\max} = \bar{b}^* \cdot \frac{c_{\max}}{\bar{c}} \left(\frac{c}{m}\right)^{(c_{\max} - \bar{c})/2\bar{c}} \quad . \tag{18.1.15}$$

The function $\mathrm{E}(\cdot, \cdot)$ denotes Erlang's loss formula and the parameters $c_{\min}$ and $c_{\max}$ represent the smallest and the largest value of the bandwidth requirements c_i respectively. Defining

$$\epsilon = (c/m)^{1/2\bar{c}} - 1 \approx (c/m - 1)/2\bar{c} \quad , \tag{18.1.16}$$

the individual call blocking probabilities b^*_i can be estimated by

$$\frac{b^*_i}{\bar{b}^*} = \frac{c_i(1 + c_i\epsilon)}{\bar{c} + \epsilon \sum_{i=1}^{N} \frac{\lambda_i}{\mu_i} c_i^3 / m} \quad . \tag{18.1.17}$$

If the offered load is rather close to 1.0, i.e. ϵ is rather small, we can simplify the approximation by

$$b^*_i / \bar{b}^* \approx c_i / \bar{c} \quad . \tag{18.1.18}$$

Since all the individual call blocking probabilities are approximated, it is important that they add up to $\bar{b}^*$, i.e.

$$\sum_{i=1}^{N} \frac{\lambda_i}{\mu_i} c_i \cdot b_i^* = m \cdot \bar{b}^* \quad . \tag{18.1.19}$$

This holds for the equations (18.1.17) and (18.1.18).

The accuracy of these approximations is checked for a wide variety of traffic mixtures in [Lin94]. Results have shown that the relative error is in the magnitude of typically 5%-10%. Of course, approximation formulae of this type will have unacceptable errors at least in some areas of the parameter space which are not typical, so limitations are required. The mixtures chosen in [Lin94] show, however, that the results are rather good in large regions, covering the essential cases for the intended application.

18.1.3   Approximations for peaked traffic sources

If we look at multi-rate networks implementing an alternative routing algorithm, a product form solution to compute the call blocking probabilities no longer exists. Furthermore, exact evaluation methods having a reasonable computation effort can not be found due to the complexity of the system. A solution to this problem are decomposition methods with two-moment traffic approximations, which have been successfully applied in single-rate environments. Compared to a one moment traffic approximation, the consideration of two moments takes into account the substantial effect of peakedness on the blocking probabilities.

The success of these methods clearly depends on the availability of efficient evaluation algorithms for multi-rate traffic on a single link, which include two-moment traffic approximations. In this subsection, we compare the complexity and the accuracy of various approximation methods for traffic streams which are characterized by a peakedness factor. We use the traffic model presented in Figure 18.1.1, however describing traffic class i additionally by the mean intensity m_i and the peakedness h_i of the traffic streams generated by calls of this class.

The approximation methods are based upon the first two moments of the traffic streams, i.e. the mean intensities m_i and the variances v_i. The latter determine the peakedness $h_i = v_i/m_i$ of the traffic streams. For their description, we arranged them according to the basic formulae used, which also determine the computational complexity of the methods.

Methods based on Erlang's loss formula

Basically, these methods convert the offered multi-rate traffic into one composite traffic stream. The call blocking probability of the composite stream is evaluated applying Erlang's loss formula and used together with a transformation to arrive at the call blocking probabilities of the individual traffic classes. The methods described in the following result from different ways of transforming the composite blocking probability into the individual ones. Since Erlang's formula is applied on a composite stream and its complexity is proportional to the link capacity c, the computational effort is best described by $O(c + N)$.

A first method is based on the observation that the blocking probability will be higher for classes with a higher peakedness factor and a higher rate. Therefore, we compute the call blocking probabilities of the individual classes by multiplying that of the composite stream with a factor $h_i \cdot c_i$, weighted by the average of all of these factors. Our second method is an adjusted method. The transformation from the first method is replaced by a technique developed in [Ell75], which substitutes the weight factor by the peakedness of the composite traffic stream. A third method is obtained by using another refinement for the computation of the weight factor, as suggested in [Roo85].

Methods based on Kaufman's and Roberts' recursive solution

These methods are based on the computation of the link occupancy distribution described in section 18.1.1. The computational complexity of these methods is $O(c \cdot N)$. Following methods have been examined.

The first method approximates the peaked traffic streams with appropriate Poisson traffic. This corresponds to a one-moment matching. By multiplying the time congestion computed by means of this method with the peakedness of the corresponding traffic class, we obtain a second solution. A third method is based on a technique developed in [SHW83], which adds traffic streams, one for each traffic class, so that the composite traffic in each class has a peakedness equal to 1. Thus, we add traffic with mean $v_i - m_i$ and variance 0. At the same time, the capacity of the link is increased by $(v_i - m_i) \cdot c_i$. For the resulting system, the time congestion can be evaluated using the recursive solution. The call blocking probabilities of the original streams can be found by an appropriate transformation.

Based on the assumption that the dependencies of c_i and h_i can be decomposed, a fourth method is obtained. This approximation uses two correction terms for the blocking probability of a composite stream. The first one corrects the error made by assuming Poisson traffic and applying the recursive solution, and the second term corrects the error made by measuring the peakedness by assuming single-rate traffic and applying a transformation [Lin83].

Methods based on Delbrouck's solution

These methods use another expression for the line occupancy distribution. This distribution was first introduced in [Del83] and has been proven to be equivalent to a process of batch arrivals with stochastic batch sizes [vDP93]. The distribution is determined by the first and second moment and the rates of the multi-rate traffic streams. Time congestion is calculated by summation over the line occupation distribution and the computational complexity of Delbrouck's formula is $O(c^2)$ if $N \ll c$. The different methods apply different transformations from time congestion to call congestion.

The first two methods are proposed by Delbrouck for renewal processes and non-renewal processes. They make use of the average number of simultaneously active calls of a certain class. Another two methods combine the time-congestion computed using Delbrouck's solution with the transformations suggested in [Ell75] and [Roo85]. In [Rob83], it is proposed to estimate the call blocking probability by the use of the time congestion formula of Delbrouck with an adjusted mean, which leads to a fifth method.

Methods based on product form solutions

With the assumption of Poisson arrivals, the system state distribution described by the number of active calls of each traffic class satisfies a product form, cf. section 18.1.1. The computational complexity of the product form solution is proportional to the summation over all possible system states, i.e. $O(c^N)$.

This method is a generalization of a method published in [Fre80]. It transforms a peaked traffic stream into a Poisson stream by dividing the mean by the peakedness factor and the capacity by the same factor. This relation is very similar to Erlang's loss formula for multi-rate single class traffic, where the capacity is divided by the rate. The generalization for multi-rate traffic is achieved by dividing the mean by the peakedness of the individual traffic classes and multiplying the rate with the peakedness. Due to the division by the peakedness factor, the rates become non-integer, and therefore the evaluation of the call blocking probabilities must be done by a product form solution. This method is also known as Hayward's approximation.

Comparison of the approximation accuracy

The most accurate methods are those based on Delbrouck's solution followed by those employing the recursive solution and finally by those based on Erlang's loss formula, which have an accuracy comparable to the product form based method. The fourth method based on the recursive solution shows a good absolute accuracy but the relative error is high. It can be improved by applying the transformations proposed in [Ell75] and [Roo85]. The transformation suggested in [Roo85] also improves the absolute error of the approximation methods while that of [Ell75] worsens the absolute accuracy especially for high blocking probabilities. Methods making use of the latter transformation give in some cases unrealistic estimates (larger than 1.0), resulting in a poor absolute accuracy. The Erlang based methods do not explain the non-monotonic behavior of the blocking probabilities. Therefore, the relative accuracy is limited. The combination of Delbrouck's method with the transformation proposed in [Roo85] seems to be the most accurate approximation method considered.

18.2 Trunk reservation and forming of call blocking

With the model presented in section 18.1, calls with higher bandwidth demands experience in general higher call blocking probabilities than calls which require less bandwidth. Therefore, a more sophisticated CAC algorithm is necessary to provide a fair access to the link for all traffic classes. This section deals with an access regulation strategy called the *trunk reservation* mechanism, which is one among a number of other access control mechanisms considered for multi-rate systems. For a comparison of such strategies, which includes e.g. *partial sharing* and *class limitation*, see [KW88]. In this section we exclusively focus on trunk reservation, since it is one of the simplest and most effective methods to adjust call blocking.

The aim of trunk reservation is to influence performance parameters such as the call blocking probabilities. To provide such a capability, thresholds θ_i are assigned to every traffic class i. According to this, a call of class i is blocked if $c_r < c - \theta_i$, and accepted otherwise. The capacity c_r denotes the unused bandwidth of the transmission link upon the arrival of a call. In other words, trunk reservation is a mechanism of bandwidth pre-reservation for traffic classes with higher bandwidth requirements, which leads to a reduction of call acceptance for call classes with lower bandwidth demands. The basic system environment is shown in Figure 18.2.1.

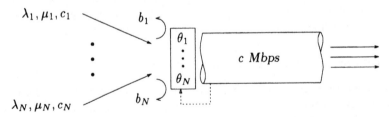

Figure 18.2.1: Basic link model for the trunk reservation operation mode.

The usage of trunk reservation in multi-rate systems is often discussed in the literature. In Roberts [Rob83], an approximate recursive solution for the state probabilities is proposed, and it is stated how to choose the thresholds θ_i to minimize the maximum call blocking probability. An approximation of the blocking probability for two traffic classes characterized by a peakedness factor is derived in [Lin83]. In [GL89], trunk reservation was investigated by an approximate iterative algorithm for blocking probability evaluation. A general cost function for GOS management was introduced and an iterative algorithm for the minimization of this cost function was proposed.

Out of the number N of traffic classes we may want to influence the GOS in such a way, that the blocking probabilities of certain classes should be equalized. This can be achieved by choosing appropriate values for the trunk reservation thresholds. In the following, we present a simple rule for equalizing call blocking probabilities of arbitrary traffic classes. Subsequently, two approximation methods for the call blocking probabilities are outlined. The first one is based on the recursive solution presented in section 18.1.1 and proves to be quite accurate. To obtain a second approximate algorithm, the simple solution mentioned in section 18.1.2 is extended. It provides less accurate results but the complexity remains quite low, even for models with a large number of traffic classes. Finally, a heavy traffic approximation is presented and the influence of the holding time characteristics are studied.

18.2.1   Rule for equalizing call blocking

The adjustment of the call blocking probabilities by the use of trunk reservation is achieved by setting the threshold parameters θ_i to appropriate values. In [TGH93], the following simple and general rule for the balancing of call blocking probabilities of arbitrary traffic classes was proposed, which is more general than that suggested in [Rob83].

> For any subset of traffic classes $\omega \subseteq \{1, \ldots, N\}$, the corresponding call blocking probabilities b_i ($i \in \omega$) are equal, if all thresholds θ_i for $i \in \omega$ are set to $c - \max\{c_k \mid k \in \omega\}$.

As we will discuss later, a positive side effect occurring if trunk reservation is employed for blocking equalization, is the enforcement of a fairer link utilization. In general, the link utilization by class-i calls will be proportional to the offered load by this class, if the blocking is equalized.

18.2.2   Accurate approximation

Considering trunk reservation for access regulation we are leaving the area of state spaces which can be solved by a product form solution. In principle, at least for smaller state spaces, the state equation systems can be formulated and solved using iterative algorithms. This possibility is, however, numerically intractable for realistic parameter sets having large state spaces. This motivates the necessity of an approximate solution.

The recursive technique discussed in the following was first proposed in [Rob83] and is based on a mapping of the multi-dimensional state space into a one-dimensional state space, as already done in section 18.1.1. By employing trunk reservation, a call of class i is accepted only if at least m_i basic bandwidth units are available upon arrival and not more than θ_i Mbps are occupied. If calls of a class j should not be subject to trunk reservation, θ_j is set to $c - c_j$. This setting is straightforward, because class-j calls are blocked anyway if more than $c - c_j$ *Mbps* of the transmission link are occupied.

The unnormalized state probabilities can be approximated using the following recursion algorithm [Joh85, Rob83] which is similar to that described in section 18.1.1.

$$\tilde{p}^*(m) = \begin{cases} 1 & : \quad m = 0 \\ 0 & : \quad m < 0 \\ 1/m \sum_{i=1}^{N} \tilde{p}^*(m - m_i) \cdot m_i(m) \cdot (\lambda_i/\mu_i) & : \quad 0 < m \leq M \end{cases} ,$$

$$(18.2.1)$$

where $m_i(m)$ is defined by

$$m_i(m) = \begin{cases} m_i & : \quad m \cdot \Delta c \leq \theta_i \\ 0 & : \quad m \cdot \Delta c > \theta_i \end{cases} . \qquad (18.2.2)$$

To emphasize the approximation, we denote the state probabilities by $p^*(m)$ and the resulting blocking probabilities by b_i^*. The difference between this recursion and that presented in section 18.1.1 are the missing transitions for arrivals which are blocked due to the trunk reservation mechanism instead of capacity constraints. After normalization we arrive at the approximate state probabilities $p^*(m)$

$$p^*(m) = \tilde{p}^*(m) \cdot \left(\sum_{m=0}^{M} \tilde{p}^*(m) \right)^{-1} , \qquad (18.2.3)$$

and the blocking probability b_i^* for calls of class i can be calculated by

$$b_i^* = \sum_{m=\min\{M-m_1, \theta_i/\Delta c\}+1}^{M} p^*(m) . \qquad (18.2.4)$$

In this approximation approach, the results depend only on the offered traffic $a_i = \lambda_i/\mu_i$ but not on the arrival and service rates themselves. However, this independence is not assured if trunk reservation is employed and therefore the approximation accuracy should be checked. The following numerical results are taken from [TGH93] and show two main effects:

- the approximation accuracy of the recursive algorithm and

- the efficiency of the trunk reservation mechanism and accordingly, the blocking probability and GOS management issues.

The first example, depicted in Figure 18.2.2, deals with two traffic classes with bandwidth requirements of $c_1 = 2$ *Mbps* and $c_2 = 20$ *Mbps*, offering the same load to a transmission link of capacity $c = 150$ *Mbps*. When trunk reservation is not employed, the blocking probability of class-2 calls is clearly higher than that of class-1 calls. For the trunk reservation mode, the thresholds are chosen according to the rule stated in Section 18.2.1 ($\theta_1 = \theta_2 = 130$ *Mbps*), and therefore the blocking is equalized.

First, the simulation results show that the accuracy of the recursive approximation is good over the whole range of traffic intensities. We have observed this level of accuracy also for a large number of other parameter sets. In general, it can be stated that the approximate solution can be considered as sufficiently accurate. The differences between the exact blocking probabilities b_i and the approximate ones b_i^* are negligible for practical purposes.

Second, we observe that the blocking probability b_1^* increases much more than b_2^* decreases. This effect becomes more distinctive if the ratio between the bandwidth demands is larger.

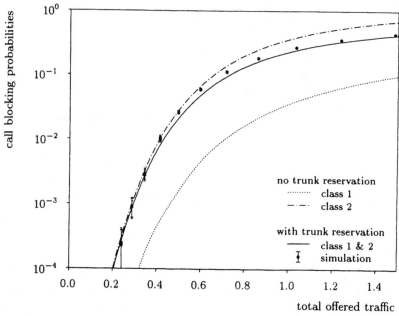

Figure 18.2.2: Approximation accuracy of the recursive solution.

Results for a more complex example are presented in the figures 18.2.3 and 18.2.4 respectively. The transmission link capacity is $c = 150$ *Mbps* and we consider four traffic classes with bandwidth demands $c_1 = 2$ *Mbps*, $c_2 = 5$ *Mbps*, $c_3 = 10$ *Mbps*, and $c_4 = 20$ *Mbps*. Calls of each traffic class are assumed to offer the same load and we want to equalize the blocking probability for class-1 and class-3 calls. Figure 18.2.3 shows, that a trunk reservation threshold of $\theta_1 = 140$ *Mbps* not only equalizes the blocking probabilities of class-1 and class-3 calls, but also decreases the blocking of class-2 and class-4 calls.

Without trunk reservation, the utilization of the transmission link by class-1 calls is higher than that by class-3 calls. By employing trunk reservation with a threshold $\theta_1 = 140$ *Mbps*, both are equalized, cf. Figure 18.2.4. The transmission link utilization by class-2 and class-4 calls is increased as the blocking decreases. The price to be paid for that equalization is, that the total link utilization decreases.

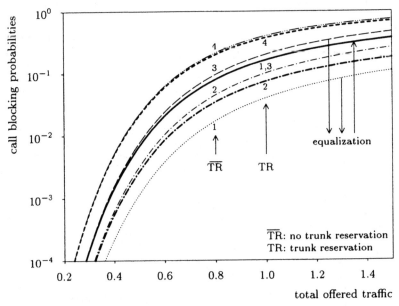

Figure 18.2.3: Blocking probability equalization in a multi-rate system.

18.2.3   Simple approximations

After presenting very accurate approximations for the call blocking probabilities, we focus on a simpler method which is in most cases less accurate. The method was suggested by Lindberger [Lin94] and is based on the results derived in section 18.1.2. By assuming that the blocking probabilities of all classes are equalized, we have a common blocking probability b_0^*. In general, the common blocking probability is dependent on the ratio of the mean holding times of the different classes of calls. However, it would be very complicated to consider also the relationships between all the mean holding times when the number of traffic classes is large. Thus, we assume that the mean holding times are equal and the result is meant to cover cases when they are in the same order of magnitude.

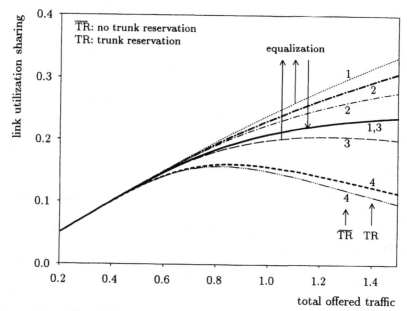

Figure 18.2.4: Link utilization sharing in a multi-rate system.

This assumption allows us to describe the traffic classes by the same two characteristics m and v as in the case without trunk reservation. Following the approach in [Lin94], we obtain an estimate for the ratio between b_0^* and $\bar{b}^*$ by

$$\frac{b_0^*}{\bar{b}^*} = (c/m)^{(c_{\max}-\bar{c})/\bar{c}} \quad . \tag{18.2.5}$$

If b_0^* is expressed in terms of Erlang's loss formula, we get

$$b_0^* = \mathrm{E}\left(\frac{c - c_{\max} + c_{\min}}{\bar{c}}, \frac{m}{\bar{c}}\right) \quad . \tag{18.2.6}$$

The results show a relative error of typically 5%-10% which is still reasonably accurate to get a first impression of the system behavior. However, for parameter sets with very different holding times, the error may be significantly higher.

18.2.4   Heavy traffic approximation

In this section, we discuss the analysis with special emphasis on its practical implementation for solving real world examples that arise in the study of multi-rate networks. The extension of this approach to networks is possible by means of the reduced load approximation (see Whitt [Whi85]). The assumption of heavy traffic holds in this section, but there is good reason to expect results that are also accurate when the resource is near to critical loading.

Formally we study the system depicted in Figure 18.2.1. In order to ensure irreducibility of all the stochastic processes involved, we assume, without loss of generality, that the capacities c_i have a greatest common divisor equal to 1.

By studying the stochastic process which merely describes the free capacity of the system at any time, good approximations for the call blocking probabilities may be obtained. These approximations are asymptotically exact under the limiting scheme described in the following. We consider some numerical aspects of the determination of these approximations and give numerical examples.

Let $N(t) = (n_i(t), i \in I)$, where $n_i(t)$ is the number of calls of type i in progress at time t. Then $N(\cdot)$ is a Markov process with state space S. Let $\pi = (\pi(n), n \in S)$ denote its equilibrium distribution. For each time t, define also $M(t)$ as the free capacity available at time t which takes values in the state space $M = \{0, 1, \ldots, c\}$. The equilibrium distribution of this process is denoted by $\bar{\pi} = (\bar{\pi}(m), m \in M)$, and the set containing states $n \in S$ corresponding to a free capacity of m by $S(m)$. Note that a knowledge of the one-dimensional distribution $\bar{\pi}$ is sufficient for the determination of blocking probabilities. We now show how $\bar{\pi}$ may be determined, at least to a good approximation, without the very much more difficult evaluation of the higher-dimensional distribution π.

Define $r_i = c - \theta_i$ and $\bar{\nu}_i(m)$ as the expected value of $n_i(t)$ under the equilibrium distribution π, conditioned on the event $N(t) \in S(m)$. Define also $\bar{\nu} = (\bar{\nu}_i(m), m \in M, i \in I)$. Then $(\bar{\nu}, \bar{\pi})$ satisfies the system of equations (18.2.7) to (18.2.10) in (ν^*, π^*), where $\nu^* = (\nu_i^*(m), m \in M)$ and $\pi^* = (\pi^*(m), m \in M)$.

$$\pi^*(m) \left[\sum_{i:r_i+c_i \le m} \lambda_i + \sum_{i:m+c_i \in M} \mu_i \nu_i^*(m) \right] = \qquad (18.2.7)$$

$$\sum_{\substack{i:r_i \le m \\ m+c_i \in M}} \pi^*(m+c_i)\lambda_i + \sum_{i:c_i \le m} \pi^*(m-c_i)\mu_i \nu_i^*(m-c_i) \qquad m \in M \quad,$$

$$\sum_{m \in M} \pi^*(m) = 1 \quad, \qquad (18.2.8)$$

$$\sum_{i \in I} c_i \nu_i^*(m) + m = c \qquad m \in M \quad, \qquad (18.2.9)$$

$$g_i(\nu^*, \pi^*) = 0 \qquad i \in I \quad, \qquad (18.2.10)$$

where, for each i the function g_i is defined by

$$g_i(\nu^*, \pi^*) = \lambda_i \sum_{m \geq c_i + r_i} \pi^*(m) - \mu_i \sum_{m \in \mathcal{M}} \pi^*(m) \nu_i^*(m) \quad . \qquad (18.2.11)$$

Note that the equations (18.2.7) and (18.2.8) may be regarded as the equations determining the unique equilibrium distribution π^* of a Markov process on $\mathcal{M}$ with transition rates given, for each $i \in \mathcal{I}$, by

$$m \to \begin{cases} m - c_j & \text{at rate} \quad \lambda_i \, \delta(m \geq r_i + c_i) \\ m + c_j & \text{at rate} \quad \mu_i \nu_i^*(m) \, \delta(m + c_i \in \mathcal{M}) \end{cases} \quad . \qquad (18.2.12)$$

Thus, as usual, any one of the equations (18.2.7) and any one of the equations (18.2.10) may be omitted. In general, the equations (18.2.7) and (18.2.10) are insufficient to determine $(\bar{\nu}, \bar{\pi})$. We may however use them to determine $(\bar{\nu}, \bar{\pi})$ approximately by making appropriate assumptions, for each i, about the dependence of $\bar{\nu}_i(m)$ on m.

To motivate our approximation scheme, we first consider some asymptotic theory. Bean et al. [BGZ95] show that, when c is large, and under a heavy traffic condition the process $N(\cdot)/c$ evolves approximately as a deterministic dynamical system and that the process $M(\cdot)/c$ eventually remains close to 0. The fixed points of this dynamical system are in one-to-one correspondence with the solutions of the system of equations (18.2.7), (18.2.8) and (18.2.10) modified by replacing $\mathcal{M}$ throughout by $\mathbb{N}_+$ and, for each $i \in \mathcal{I}$, by replacing $\nu_i^*(m)$ by a positive constant ν_i^* independent of m, where we additionally require in place of the equations (18.2.9)

$$\sum_{i \in \mathcal{I}} c_i \nu_i^* = c \quad . \qquad (18.2.13)$$

For each solution (ν^*, π^*) of this modified system of equations, the corresponding fixed point of the dynamical system is given by ν^*/c.

Now suppose that the dynamical system $N(\cdot)/c$ possesses a unique fixed point to which all of its trajectories converge. Bean et al. [BGZ95] show that the component π^* of the corresponding unique solution (ν^*, π^*) of the above modified system of equations then approximates the equilibrium distribution $\bar{\pi}$ of the process $M(\cdot)$.

While the above approximation to $\bar{\pi}$ becomes exact under the limiting scheme described, it is nevertheless too crude for most practical applications. We therefore seek to improve it by retaining the state space $\mathcal{M}$ and, for each i, approximating $\bar{\nu}_i(m)$ as a linear function of m. Although this linear approximation is less than ideal from a theoretical viewpoint, it nevertheless appears to work well in practice. We also expect that it will continue to work reasonably well even when the heavy traffic condition is not satisfied, particularly if the resource is instead at or near to critical loading. Recalling that $\bar{\nu}_i(m)$ must be always positive, we thus replace the equations (18.2.9) by

$$\nu_i^*(m) = a_i(c - m) \qquad i \in \mathcal{I},\ m \in \mathcal{M} \ , \qquad (18.2.14)$$

where $a_i \geq 0$ for each $i \in \mathcal{I}$ and

$$\sum_{i \in \mathcal{I}} c_i a_i = 1 \ . \qquad (18.2.15)$$

The equilibrium distribution $\bar{\pi}$ is then approximated by the component π^* of the solution $(\boldsymbol{a}, \pi^*)$, where $\boldsymbol{a} = (a_i, i \in \mathcal{I})$, provided that this solution is unique. Note that, after eliminating $\boldsymbol{\nu}^*$ and the redundant equations, the system of equations consists of $|\mathcal{M}| + |\mathcal{I}|$ equations in $|\mathcal{M}| + |\mathcal{I}|$ unknowns. Multiple solutions, were they to occur, would correspond to quasi-equilibrium distributions of the process $\boldsymbol{M}(\cdot)$.

In the following, we present a numerical example to show the accuracy of our approach. We consider a single resource and two types of offered traffic. Call type I has an effective bandwidth of 0.04 *Mbps* and a mean holding time of 180 s. For call type II, we assumed an effective bandwidth of 0.50 *Mbps* and a mean holding time of 1200 s. Figure 18.2.5 presents the analytic results and compares them to simulations of the system, for which the relevant 99% confidence intervals are displayed. The call blocking probabilities are plotted as functions of $s = r_1 - r_2$, while holding at least one of r_1 or r_2 equal to zero.

Figure 18.2.5 shows accurate estimates of the blocking probabilities for both types of calls over a wide range of values for the trunk reservation parameters. In this example, the link is critically loaded. It shows the typical behavior of the blocking probabilities at such a link. The relationship between the blocking probabilities of the call types is governed by the relative values of $c_1 + r_1$ and $c_2 + r_2$. So, for example, when $c_1 + r_1 = c_2 + r_2$, the blocking probabilities are equal for both call types. The blocking probability for the calls of type I is almost zero when the trunk reservation parameters are such that s is large and negative and as s increases, the blocking probability decreases for calls of type II and increases for calls of type I. Note that when there is no trunk reservation, i.e. $s = 0$, the calls with the higher equivalent bandwidth have a higher blocking probability than those of the other call type.

Further results have shown that the algorithm described here can be used in a wide range of practical examples to give very accurate estimates for the blocking probabilities in multi-rate networks.

18.2.5   Influence of the holding time characteristics

Since in the case of trunk reservation the call blocking probabilities depend on the mean holding times and the holding time distribution type itself, the next two subsections deal with the role of these two quantities. First, we

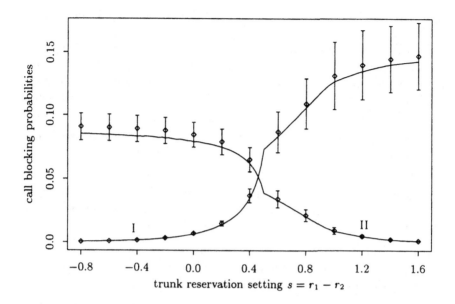

Figure 18.2.5: Analytical and simulation results with varying trunk reservation settings.

focus on the influence on the blocking probabilities if the mean holding times of the traffic classes differ in orders of magnitude. Afterwards, we show that the trunk reservation mechanism is quite robust against the holding time distribution type.

Influence of the mean holding time

In the case of trunk reservation, no product form solution exists due to the fact that one-way transitions are present in the state space which destroy reversibility. Although in principle, the call blocking probabilities could be calculated by solving the multidimensional birth-death equations, this becomes intractable as the number of call types and the link capacity grow. Consequently approximations are sought. In the first part of this section, two approaches have been presented. However, these methods do not cater for different mean holding times.

The influence of the call holding time has been investigated in a number of papers, see e.g. [AM94, BGZ94, Hin90, LH92]. The outcome of these studies can be summarized as follows.

In principle, we observe a sensitivity of the call blocking probabilities of different traffic types on changes in the mean holding time ratio, even when they are equalized. This effect is, however, hardly recognizable for typical traffic scenarios. It becomes more apparent if the system is lightly loaded and if the mean holding time of higher bandwidth calls is significantly larger than that of the lower bandwidth calls.

Even though the result for a mixture of two traffic classes can, in cases where the mean holding time ratio is very great, differ significantly from the case with similar mean holding times, a simple method which does not include mean holding time relations is applicable for practical dimensioning for the following reasons:

- Trunk reservation is only recommended to be used in mixtures where the bandwidth ratio between calls is relatively small, i.e. of the order of 10.

- Mean holding time ratios as large as, for example, 100 are very difficult to combine with the assumption of constant arrival intensities for all types of calls during common busy hour-like periods.

- The result will always be more extreme in cases with only two types of bandwidths. The methods are usually meant to be used in mixtures with a greater number of traffic classes.

- Methods depending upon all of the mean holding time relations in a mixture with many classes of bandwidths are difficult to use in the practical dimensioning and optimization of networks anyway.

All in all, trunk reservation is still the simplest and most effective method for the equalization of call blocking.

Influence of the holding time distribution type

As pointed out in section 18.1.1, for Poisson arrivals the call blocking probabilities depend only on the mean holding time and not on the holding time distribution type itself, if trunk reservation is not employed. However, in a system with trunk reservation the holding time distribution type may have an influence on the blocking probabilities.

Using simulation studies, the authors in [TGH93] have observed that the results for the system with trunk reservation are almost insensitive to the holding time distribution type, as can be seen in table 18.2.1. The blocking probabilities of the two call classes are presented for a number of coefficients of variation c_{T_H} of the holding time distribution, which vary from 0 (deterministic), 1/3 (Erlang-9), 1 (Poisson), to 4 (hyper-exponential). It is easy to recognize that the blocking probabilities of the two call classes are almost identical, no matter what kind of holding time distribution type is used. For practical purposes, the blocking probabilities can be regarded as equal for these cases too. Furthermore, the equalized blocking probabilities for the four considered holding time distribution types are also very close to each other. Thus, the results derived with the approximation methods presented in this section are quite accurate also for models with general holding time distributions.

Table 18.2.1: Illustration of the robustness against the holding time distribution type.

load	$c_{T_{H_1}} = c_{T_{H_2}} = 0$		$c_{T_{H_1}} = c_{T_{H_2}} = 1/3$		$c_{T_{H_1}} = c_{T_{H_2}} = 1$		$c_{T_{H_1}} = c_{T_{H_2}} = 4$	
	class 1	class 2	class 1	class 2	class 1	class 2	class 1	class 2
0.3	1.28E-3	1.34E-3	1.28E-3	1.32E-3	1.27E-3	1.29E-3	1.27E-3	1.28E-3
0.4	8.10E-3	8.06E-3	8.05E-3	8.05E-3	7.90E-3	7.96E-3	7.80E-3	7.74E-3
0.5	2.68E-2	2.69E-2	2.68E-2	2.68E-2	2.66E-2	2.65E-2	2.56E-2	2.57E-2
0.6	5.97E-2	5.95E-2	5.95E-2	5.95E-2	5.94E-2	5.92E-2	5.73E-2	5.73E-2
0.7	1.05E-1	1.05E-1	1.05E-1	1.05E-1	1.04E-1	1.04E-1	1.00E-1	1.00E-1
0.8	1.56E-1	1.56E-1	1.56E-1	1.56E-1	1.55E-1	1.55E-1	1.51E-1	1.51E-1
0.9	2.08E-1	2.09E-1	2.08E-1	2.09E-1	2.07E-1	2.07E-1	2.03E-1	2.03E-1
1.0	2.58E-1	2.58E-1	2.58E-1	2.58E-1	2.57E-1	2.57E-1	2.53E-1	2.53E-1
1.1	3.04E-1	3.05E-1	3.04E-1	3.04E-1	3.03E-1	3.03E-1	2.99E-1	2.99E-1
1.2	3.47E-1	3.47E-1	3.47E-1	3.47E-1	3.45E-1	3.45E-1	3.44E-1	3.44E-1

18.3   Call and burst blocking in ATM systems

In this section, we present an analytical approach to investigate the blocking behavior at the call and the burst level simultaneously. Therefore, the analysis of the traffic model presented in [HR93] is outlined. The model is similar to that used in section 18.1, i.e. we consider a transmission link with a fixed capacity c and an input traffic which consists of the superposition of the traffic of N call classes. Some classes contain CBR and others VBR traffic. The CBR calls are considered only at the call level whereas the VBR calls are looked at the call and at the burst level. CBR calls of class i have a constant rate c_i during their holding time. VBR calls of class j are assumed to behave like on/off-sources. They generate a constant rate c_j in times of activity (on-state) and nothing during times of silence (off-state). The distributions of the length of the on- and off-state periods is negative exponential with a mean $1/\beta_j$ and $1/\gamma_j$ respectively.

If a VBR call is accepted, it may start with an active or a silent phase. Calls which start in the on- and off-state arrive with rates $\lambda_{j_{on}}$ and $\lambda_{j_{off}}$ respectively. At the arrival of a VBR call of class j a capacity $\tilde{c}_j$ is allocated at the call level and released if the call has completed. This capacity could be the mean rate of the source or an equivalent bandwidth. The basic system environment is shown in Figure 18.3.1.

The connection acceptance control and blocking behavior is basically different for CBR and VBR calls:

(i) CBR calls, class-i say, are accepted when their required bandwidth c_i is available at the call and at the burst level. Thus, the sum of the bandwidths of already accepted CBR calls plus the equivalent bandwidths $\tilde{c}_j$ of already accepted VBR calls must also be less than or equal to $c - c_i$,

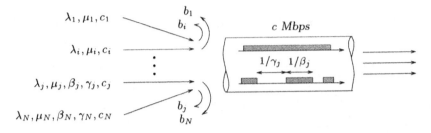

Figure 18.3.1: Basic link model for CBR and VBR call classes.

and the sum of the bandwidths of already accepted CBR calls plus the sum of the peak bandwidths of currently active VBR calls must be less than or equal to $c - c_i$.

(ii) VBR calls, e.g. class-j calls, are accepted when their equivalent bandwidth $\tilde{c}_j$ is available at the call level, i.e. when the sum of the bandwidths of already accepted CBR calls plus the sum of the equivalent bandwidths of already accepted VBR calls is less than or equal to $c - \tilde{c}_j$.

Calls of class i or j are blocked at the call level with probability b_i^c and b_j^c respectively, and do not influence the system any longer. VBR calls can also be blocked at the burst level with probability b_j^b, since the sum of the bandwidths from the actual number of VBR calls being in the on-state together with the bandwidths from the accepted CBR calls can exceed the link capacity. If burst blocking occurs, the affected burst is not transmitted and the VBR source remains in the off-state.

The following presents analytical approaches to derive approximate state probabilities and related performance measures such as blocking probabilities and link utilization of our model. The accuracy of the approximation is investigated, too.

To illustrate the structure of the state space for a system with only VBR sources, we show in Figure 18.3.2, which is taken from [HR93], a simple example state space considering only one VBR traffic class.

The system states $(\tilde{n}_1, n_1, \ldots, \tilde{n}_N, n_N)$ are defined by the number $\tilde{n}_i$ of accepted calls of class-i and by the number n_i of class-i calls, which are currently in the on-state. Thus, the multi-dimensional state space has twice as many dimensions as VBR traffic classes. In our example, a transmission link with $c = 12$ $Mbps$ serves one class of VBR sources with a bandwidth demand of $c_1 = 4$ $Mbps$ while being in the on-state. The arrival rate is λ_1 and the mean holding time is $1/\mu_1$. The ratio of mean active and silent times is assumed to be $3\beta_1 = \gamma_1$. An equivalent bandwidth of $\tilde{c}_1 = 3$ $Mbps$, which is equal to the mean rate, is used for CAC.

The structure of the state space contains an exception between states (3,3) and (4,3). Being in state (3,3), an arriving call which wants to start

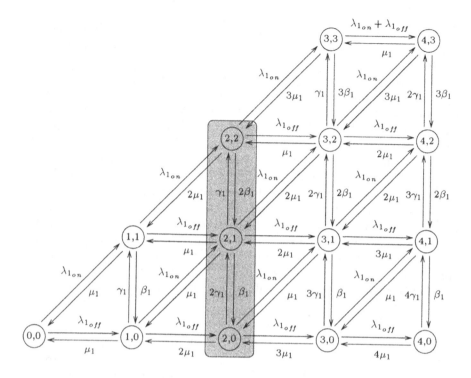

Figure 18.3.2: Example state space for one VBR traffic class.

in the on-state is accepted at the call level but blocked at the burst level. Therefore, it has to start in the off-state, causing a change of the rate $\lambda_{1_{off}}$ to the rate $\lambda_{1_{on}} + \lambda_{1_{off}}$.

For state spaces of this type, a product form or recursive solution does not exist. In principle, an exact computation, at least for smaller state spaces, can be done by solving the system of state equations. For realistic parameter sets which lead to larger state spaces, this possibility is, however, numerically intractable.

If we know the state probabilities $p(\tilde{n}_1, n_1, \ldots, \tilde{n}_N, n_N)$, the call and burst blocking probabilities can be derived as

$$b_i^c = \sum_{(\tilde{n}_1, n_1, \ldots, \tilde{n}_N, n_N) \in S_i^c} p(\tilde{n}_1, n_1, \ldots, \tilde{n}_N, n_N) \quad , \qquad (18.3.1)$$

and

$$b_i^b = \frac{\sum\limits_{\vec{n} \in S_i^b} p(\vec{n}) \cdot \gamma_i \cdot (\tilde{n}_i - n_i) + \sum\limits_{\vec{n} \in S_i^b/S_i^c} p(\vec{n}) \cdot \lambda_{i_{ON}}}{\sum\limits_{\vec{n} \in S} p(\vec{n}) \cdot \gamma_i \cdot (\tilde{n}_i - n_i) + \sum\limits_{\vec{n} \in S/S_i^c} p(\vec{n}) \cdot \lambda_{i_{ON}}} \quad , \qquad (18.3.2)$$

if S contains all existing states, and the sets S_i^c and S_i^b those, where call and burst blocking occurs for class-i calls respectively. Thus, an estimation of the state probabilities would be helpful. To obtain an approximation for them, we follow the approach presented in [HR93].

Looking at the model, we distinguish two different processes. The first one is given by the arrival and the departure of calls and the second one is the switching between the on- and off-states. State transitions of the first process are independent of the present state of the second process, if we only deal with VBR sources. Because of this independence, we can calculate the exact state probabilities for macro states, which represent the number of accepted calls of each class, by the product form solution described in section 18.1.1. In Figure 18.3.2, the macro state for two accepted calls is shaded.

Knowing the macro state probability, the probability distribution for the states within an macro state can be determined. This is done by assuming that there is always enough bandwidth available at the burst level. By doing this, we get an approximate solution for the state probabilities by evaluating the Markov chain in each macro state.

The unnormalized state probabilities of the macro states $\tilde{Q}(\tilde{n}_1, \ldots, \tilde{n}_N)$ are given by

$$
\tilde{Q}(\tilde{n}_1, \ldots, \tilde{n}_N) = \begin{cases} \prod_{i=1}^{N} \frac{(\lambda_i/\mu_i)^{\tilde{n}_i}}{\tilde{n}_i!} & : \quad \sum_{i=1}^{N} \tilde{n}_i \tilde{c}_i \leq c \\ 0 & : \quad \sum_{i=1}^{N} \tilde{n}_i \tilde{c}_i > c \end{cases} , \qquad (18.3.3)
$$

and the local state probabilities $\tilde{q}(\tilde{n}_1, n_1, \ldots, \tilde{n}_N, n_N)$ within the macro state $(\tilde{n}_1, \ldots, \tilde{n}_N)$ by

$$
\tilde{q}(\tilde{n}_1, n_1 \ldots, \tilde{n}_N, n_N) = \begin{cases} \prod_{i=1}^{N} \frac{\tilde{n}_i!}{(\tilde{n}_i - n_i)!} \cdot \frac{(\gamma_i/\beta_i)^{n_i}}{n_i!} & : \quad \sum_{i=1}^{N} n_i c_i \leq c \\ 0 & : \quad \sum_{i=1}^{N} n_i c_i > c \end{cases} .
$$
$$(18.3.4)$$

After normalization, we obtain the macro state probabilities $Q(\tilde{n}_1, \ldots, \tilde{n}_N)$ and the local state probabilities $q(\tilde{n}_1, n_1, \ldots, \tilde{n}_N, n_N)$. Finally, the approximate state probabilities $p^*(\tilde{n}_1, n_1, \ldots, \tilde{n}_N, n_N)$ are computed by multiplying the local state probabilities with the probabilities of the corresponding macro states

$$
p^*(\tilde{n}_1, n_1, \ldots, \tilde{n}_N, n_N) = q(\tilde{n}_1, n_1, \ldots, \tilde{n}_N, n_N) \cdot Q(\tilde{n}_1, \ldots, \tilde{n}_N) . \quad (18.3.5)
$$

Using these state probabilities, an approximate solution for the blocking probabilities and the link utilization can be obtained. Note, that the call blocking probabilities are exact due to the accurate calculation of the macro state probabilities. With a similar approach we obtain approximate state probabilities for systems with CBR and VBR sources.

To compare the approximate solutions with exact values, we make use of a numerical example presented in [HR93]. It deals with one VBR traffic class with $c_1 = 20$ *Mbps* and a transmission link capacity equal to $c = 150$ *Mbps*. The mean durations of the on- and off-states are assumed to be equal, i.e. $\beta_1 = \gamma_1$, and the equivalent bandwidth is chosen as the mean bandwidth $\tilde{c}_1 = 10$ *Mbps*. For the computation of the exact solution the complete system of state equations was solved by an iterative algorithm. The resulting blocking probabilities are depicted in Figure 18.3.3.

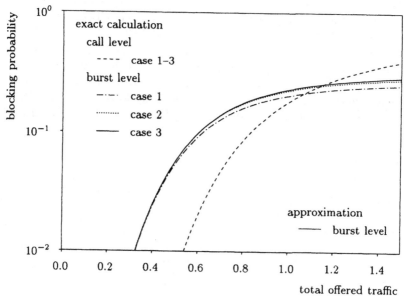

Figure 18.3.3: Accuracy of the approximate blocking probabilities.

For the exact solution, the burst blocking probabilities depend on the ratio of the rates β_1 and μ_1, which represents the mean number of active phases within a call duration. This is not the case for the approximation. Therefore, Figure 18.3.3 shows the call and burst blocking probabilities as functions of the offered load for three different ratios β_1/μ_1. The ratios are chosen as 1 (case 1), 5 (case 2), and 50 (case 3). Note that $\beta_1/\mu_1 = 1$ indicates that the VBR sources behave similar to CBR sources, since there is only one active phase per call in equilibrium.

Due to the exact calculation of the call blocking by our algorithm, we exclusively focus on the burst blocking with this example. We observe a slight difference between the curves of the exact solution and the curve computed with the approximate solution, which is the same for all three cases. The distance between the curves decreases as the ratio β_1/μ_1 increases. This effect occurs due to the fact that most state changes take place inside the macro states if the ratio is large and therefore the exact probability distribution within the macro states is closer to the one determined approximately. The ratio $\beta_1/\mu_1 = 1$ can be seen as a worst case of the approximation, as mentioned above. For realistic parameter sets the ratio β_1/μ_1 is of the order of 10^2 or higher and thus, the approximation provides values which are very close to the exact ones. Furthermore, they constitute upper bounds.

Next, numerical results for a model with CBR and VBR input traffic are presented. In contrast to the model with only VBR sources, the results for the call blocking probabilities are now also approximative, since the acceptance of arriving CBR calls is not independent of the utilization at the burst level. In the figures 18.3.4 and 18.3.5 the influence of the choice of the equivalent bandwidth on the blocking behavior is studied for one CBR and one VBR traffic class. These figures are also taken from [HR93].

The authors considered a transmission link capacity of $c = 600$ *Mbps* with CBR calls requiring $c_1 = 2$ *Mbps* and VBR calls having a peak rate of $c_2 = 20$ *Mbps*. The mean durations of the on- and off-states are assumed to be equal and both traffic classes offer the same load. In Figure 18.3.4, the equivalent bandwidth is chosen equal to the mean rate of class-2 calls, i.e. $\tilde{c}_2 = 10$ *Mbps*.

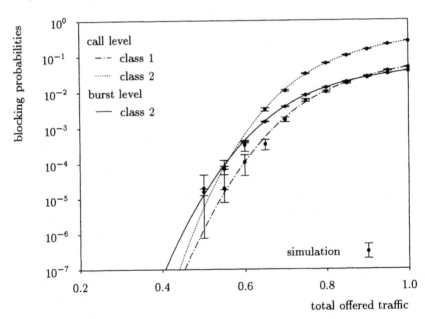

Figure 18.3.4: Blocking probabilities for an equivalent bandwidth equal to the mean.

As discovered in section 18.1, calls with higher bandwidth requirements experience higher blocking probabilities if only CBR calls are considered. Looking at mixes of CBR and VBR traffic, it turns out that CBR calls of

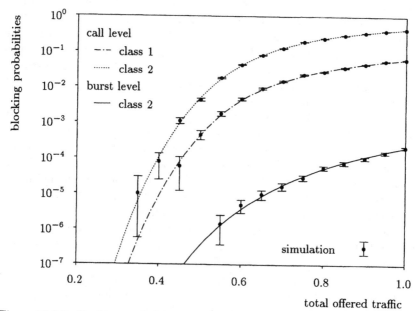

Figure 18.3.5: Blocking probabilities for an increased equivalent bandwidth.

lower bit rate requirements than VBR calls can be blocked at the call level with a higher probability. This surprising effect clearly depends on the total offered load and occurs due to sparse capacity available on the burst level. Moreover, Figure 18.3.4 shows that the burst blocking probability of the VBR calls is unacceptably high.

To reduce the burst blocking, a higher equivalent bandwidth of $\tilde{c}_2 = 17.5$ *Mbps* is assigned to the VBR calls. From Figure 18.3.5 we observe that increasing the equivalent bandwidth leads to considerably lower burst blocking probabilities for VBR calls, but on the other hand to higher call blocking for both classes. Thus, it can be concluded that the burst blocking probabilities are strongly dependent on the estimation of the equivalent bandwidth for the VBR calls. In both cases, the approximate solution delivers results which are quite close to those obtained by simulation. A large number of further examples have shown the same degree of accuracy.

18.4   Extensions to multi-rate networks

From the conceptual point of view, the problem of calculating end-to-end blocking in a multi-rate network without alternative routing can be handled in a nice and simple way, since the product form solution in section 18.1.1 still applies, see for example [Kel91b]. Therefore, we have to consider one factor for each link on each route. The only difference with respect to a completely shared link is that the set of allowed states is now limited by several linear capacity constraints, but still the set is coordinate convex.

However, from the practical point of view, the problem of calculating end-to-end blocking probabilities when analyzing multi-rate networks becomes even more complicated, because the evaluation using the product form solution becomes numerical intractable. Another problem is the use of trunk reservation in such a network.

Ideally, methods such as reduced load approximations or fixed point models are used. These models should be built upon a suitably efficient and accurate algorithm to calculate the blocking probabilities for different call types on a single link. In a single-rate circuit-switched network, the appropriate fixed point model uses Erlang's loss formula, or an extended version of it, for the calculation of single link blocking. A number of authors [LH92] have also used Erlang's loss formula for approximate methods in the multi-rate case, but these do not allow for unequal mean holding times when trunk reservation is used. Let us assume that, for the time being, only fixed routing is used as even this case has not yet been sufficiently analyzed when trunk reservation is used. Fixed point methods based on one or two moment models have been reported for networks without use of trunk reservation and allowing for unequal mean holding times [CR93, PKW89].

In this section, two fixed point models are presented to calculate end-to-end blocking in multi-rate networks without alternative routing. The first one is based on Kelly's fixed point model [Kel89], which has been extended to the multi-rate case. We allow VBR calls to have different equivalent bandwidths on different links of their path. The second fixed point model is based on a more accurate calculation of single link blocking probabilities using the algorithm presented in section 18.1.1. What we have also allowed for in the latter model is the calculation of an approximate mean holding time for a typical channel on a link in the network, which forms an integral part of the approximation technique. The idea used could equally apply to any fixed point method which uses a fast model to calculate blocking on a single link when trunk reservation is used and mean holding times are not explicitly modeled.

18.4.1 Fixed point model without trunk reservation

First consider the following notation which is used throughout this section:

λ_{kr} Poisson arrival rate for call type k on route r,

μ_{kr} mean holding time for call type k on route r,

c_{kjr} equivalent bandwidth of call type k on link j of route r,

θ_{kj} trunk reservation threshold associated with call type k on link j,

c_j total capacity on link j,

b_{kj} blocking probability seen by call type k on link j,

β_{kr} approximation for the loss probability of call type k on route r,

μ_j approximate mean holding time for a typical channel on link j.

It is very important to assume that links block independently [Kel89]. Then, the blocking probabilities as a solution to the fixed point equations (18.4.1) and (18.4.2) are asymptotically correct as the link capacity and the offered traffic are increased together, cf. [Kel91b].

The link blocking probabilities are calculated using Erlang's loss formula

$$b_j = \mathrm{E}(\sigma_j, c_j) \quad , \tag{18.4.1}$$

where $\mathrm{E}(.)$ represents Erlang's loss function. The *reduced load* or *effective offered traffic* on link j after the thinning effect in the network is denoted by σ_j, which is calculated by

$$\sigma_j = (1 - b_j)^{-1} \sum_k \sum_{r,j} \lambda_{kr} \mu_{kr} c_{kjr} \prod_i (1 - b_i)^{c_{kjr}} \quad . \tag{18.4.2}$$

The stream blocking for type k calls on route r is then approximated by

$$\beta_{kr} = 1 - \prod_j (1 - b_j)^{c_{kjr}} \quad . \tag{18.4.3}$$

Note that the term c_{kjr} has been used as in ATM networks calls may require different bandwidths on different links of their route.

18.4.2   Fixed point model with trunk reservation

The fixed point model presented now is based on the single link solution of Kaufman [Kau81] and Roberts [Rob81]. It is a special case of the model described in [PLK90]. The fixed point equations (18.4.4) to (18.4.7) are solved by repeated substitution until convergence is reached. The end-to-end blocking probabilities for call type k on route r are then given by β_{kr}. The calculation of b_{kj} is done using the recursive method of Section 18.1.1.

$$b_{kj} = E_k(\sigma_{1j}, \ldots, \sigma_{Nj}, a_{1j}, \ldots, a_{Nj}, \theta_{1j}, \ldots, \theta_{Nj}, c_j) \quad , \tag{18.4.4}$$

$$\sigma_{kj} = \mu_j(1 - b_{kj})^{-1}\sum_r c_{kjr}\lambda_{kr}(1 - \beta_{kr}) \quad , \qquad (18.4.5)$$

$$1 - \beta_{kr} = \prod_j(1 - b_{kj}) \quad , \qquad (18.4.6)$$

$$\mu_j = \frac{\sum_r\sum_k c_{kjr}\lambda_{kr}(1 - \beta_{kr})\mu_{kr}}{\sum_r\sum_k c_{kjr}\lambda_{kr}(1 - \beta_{kr})} \quad , \qquad (18.4.7)$$

where σ_{kj} is the *reduced load* or *effective offered traffic* of call type k on link j after the thinning effect in the network. The parameter μ_j is the approximate mean holding time for a typical channel on link j and $E_k(.)$ is the solution for the single link model.

The idea behind the approximation is that the stream of rate λ_{kr} is thinned by a factor $1 - b_{ki}$ at link i before being offered to link j. If this thinning could be assumed to be independent both from link to link and over all routes containing link j, then the type-k traffic offered to link j would be Poisson with a rate given by σ_{kj}. The approximation for the proportion of lost calls of type k requesting route r is β_{kr}.

18.4.3   Numerical results

First, we consider the case without trunk reservation and compare the results obtained using our solution with those presented in [CR93] for a star network depicted in Figure 18.4.1.

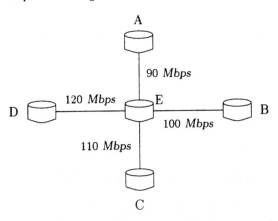

Figure 18.4.1: A multi-rate star network.

In [CR93], the blocking probabilities are calculated using three different methods: Kelly's fixed point model, a Knapsack fixed point model and a Pascal fixed point model. It is concluded that Kelly's approximation gives the least accurate results especially at light and moderate loads what is also known from [Kel91b].

Our results based on Kelly's fixed point model match those in [CR93]. Kelly's fixed point model is used without normalizing the different bandwidth requirements as Erlang's loss formula can only cope with homogeneous traffic characteristics. The approximation improves as the network load increases but still underestimates the blocking probabilities sometimes considerably. It is fairly poor under light and moderate traffic profiles. In [CR93], it is shown that the Knapsack and Pascal approximation methods are significantly more accurate, with the Pascal approximation having the same computational complexity as Kelly's.

To check the approximation accuracy for the solution taking into account trunk reservation, we use the ring network depicted in Figure 18.4.2 and compare the results with values obtained by simulation.

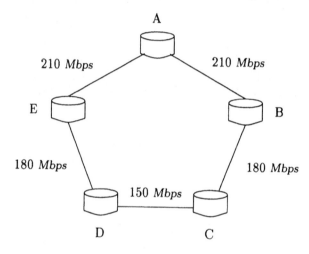

Figure 18.4.2: A multi-rate ring network.

Two call types with bandwidth demands $c_1 = 1$ Mbps and $c_2 = 2$ Mbps are modeled and trunk reservation was used to equalize the blocking probabilities. The offered traffic is such that the network is critically loaded on all links. Details are summarized in table 18.4.1. The first observation is, that the results from the fixed point model agree well in the majority of cases with the simulation results if the mean holding times are in the same order of magnitude as considered for table 18.4.1. Other configurations have shown that our model tends to underestimate the blocking probabilities slightly if the mean holding time of call type 1 is by orders of magnitude shorter than that of call type 2, and vice versa. This is due to the sensitivity of the end-to-end blocking probabilities to the mean holding time.

Concluding we can say that fixed point models based on algorithms such as the recursive solution presented in section 18.1.1 are viable for small or medium networks without alternative routing. In order to cater for alternative routing, fixed point models based on two moments have been suggested

Table 18.4.1: Heavy traffic with trunk reservation.

route	offered traffic		simulation 95% confidence intervals		fixed point model	
	λ_1	λ_2	β_1	β_2	β_1	β_2
A B	60	2	(17.8, 21.6)	(17.5, 24.2)	20.5	20.5
A B C	30	2	(28.5, 32.9)	(29.5, 34.4)	34.0	34.0
B C D	30	2	(34.3, 39.9)	(35.4, 41.0)	36.7	36.7
C D	30	1	(25.0, 29.6)	(22.6, 29.5)	23.7	23.7
D E	30	2	(22.5, 26.5)	(20.0, 25.3)	22.2	22.2
D E A	30	2	(36.3, 40.4)	(36.7, 43.8)	37.3	37.3
E A	60	2	(19.1, 21.9)	(18.1, 23.1)	19.4	19.4

e.g. in [PKW89]. When trunk reservation is present and the mean holding time ratio of the calls is quite large, a more appropriate fixed point model should be used. This will be, however, not the typical case as discussed in section 18.2.5.

Bibliography

[Aal94]   S. Aalto. Output from an A-M-S type fluid queue. In *14th Int. Teletraffic Congress Proceedings*, volume 1a, pages 421–430. North Holland-Elsevier Science Publishers, June 1994. (pp 469, 470)

[ACS93]   M. Ammar, S. Cheung, and C. Scoglio. Routing multipoint connections using virtual paths in an ATM network. In *IEEE Infocom 93 Proceedings*, pages 98–105, 1993. (p 271)

[Aei78]   J. M. Aein. A multi-user-class, blocked-calls-cleared, demand access model. *IEEE Trans. Comm.*, 26(3):378–385, 1978. (p 514)

[AJN95]   A. T. Andersen, A. Jensen, and B. Friis Nielsen. Modelling and performance study of packet-traffic with self-similar characteristics over several timescales with Markovian Arrival Processes (MAP). In I. Norros and J. Virtamo, editors, *Proceedings of the 12th Nordic Teletraffic Seminar, NTS12, VTT symposium 154*, pages 269–283. VTT, Finland, 1995. (p 348)

[AK91]   B. W. Abeysundara and A. E. Kamal. High-speed local area networks and their performance: A survey. *ACM Computing Surveys*, 23(2):221–264, September 1991. (p 213)

[Ald89]   D. Aldous. *Probability approximations via the Poisson clumping heuristic*. Springer-Verlag, 1989. (pp 451, 455)

[AM94]   M. Azmoodeh and R. N. Macfadyen. Multi-rate call congestion: Fixed point models and trunk reservation. In *Proceedings of IEEE U.K. Teletraffic Symposium*, Cambridge, March 1994. (p 532)

[AMO93]   R. K. Ahuja, T. L. Magnanti, and J. B. Orlin. *Network Flows: Theory, Algorithms and Applications*. Prentice-Hall, 1993. (p 252)

[AMP94]   J. Andrade and M. J. Martinez-Pascua. Use of the IDC to characterize LAN traffic. In *2nd IFIP Workshop on Performance Modelling and Evaluation of ATM Networks*, Bradford, July 1994. (pp 309, 310, 311)

[AMR93]   J. Andrade and E. Martín-Rebollo. Method for the perfor-
            mance evaluation of an ATM queue with correlated input.
            Technical Report Internal Document, Telefonica Investiga-
            cion y Desarrollo, Madrid, 1993.   (p 493)

[AMS82]   D. Anick, D. Mitra, and M. M. Sondhi. Stochastic theory of
            a data handling system with multiple sources. *Bell System
            Tech. J.*, 61(8):1871–1894, 1982.   (pp XIX, 345, 464, 468,
            469, 484)

[AN72]   K. B. Athreya and P. E. Ney. *Branching Processes.* Springer-
            Verlag, 1972.   (p 428)

[AN96]   A.T. Andersen and B.F. Nielsen. An application of super-
            positions of two-state Markovian sources to the modelling
            of self-similar behaviour. Technical report, Technical Uni-
            versity of Denmark, DK - 2800, Lyngby, 1996.   (pp 345,
            347, 348)

[And94]   J. Andrade. ATM source traffic descriptor based on the
            peak, mean and second moment of the cell rate. In *14th
            Int. Teletraffic Congress Proceedings*, volume 1a, pages 223–
            232. North Holland-Elsevier Science Publishers, June 1994.
            (p 158)

[And95a]   A. T. Andersen. *Modelling of Packet Traffic with Matrix
            Analytic Methods.* Ph.D dissertation, Institute of Mathe-
            matical Modelling, Technical University of Denmark, 1995.
            IMM-PHD-1995-18.   (pp 343, 345, 348)

[And95b]   J. Andrade. Characterization and control of the burst
            structure of ATM traffic. Technical Report Internal Doc-
            ument, Telefonica Investigacion y Desarrollo, Madrid, 1995.
            (p 161)

[AR96]   I. Adan and J. Resing. A simple analysis of a fluid queue
            driven by an $M/M/1$ queue. *Queueing Systems*, 22:171–174,
            1996.   (p 466)

[Arv94]   Å. Arvidsson. Management of reconfigurable virtual path
            networks. In *14th Int. Teletraffic Congress Proceedings*.
            North Holland-Elsevier Science Publishers, June 1994.
            (pp 288, 291, 292, 293)

[Asm87]   S. Asmussen. *Applied Probability and Queues.* John Wiley
            & Sons, 1987.   (p 392)

[ASV95]   J. Andrade, M. F. Sanchez-Canabate, and M. Villen-Altamirano. ATM cell traffic description based on queueing performance. In *International Teletraffic Seminar*, Bangkok, November 1995.   (p 313)

[AT91]   AOTC (Australian Telecom). A generic flow control protocol for B-ISDN. Technical Report Contribution No. D.1660, CCITT Study Group XVIII, December 1991.   (p 205)

[AT92]   AOTC (Australian Telecom). A controlled point-to-point generic flow control protocol. Technical Report Contribution No. D.2197, CCITT Study Group XVIII, Geneva, June 1992.   (p 205)

[AV94]   J. D. Angelopoulos and I. S. Venieris. A distributed FIFO spacer/multiplexer for access to tree APONs. Submitted for publication, 1994.   (p 220)

[AV95]   S. Aalto and J. T. Virtamo. Real-time estimation of call arrival intensities. In *Proceedings of the seminar ATM hot topics on traffic and performance: from RACE to ACTS*, Milano, June 1995.   (pp 349, 353)

[BB82]   G. Barberis and R. Brignolo. Capacity allocation in a DAMA satellite system. *IEEE Trans. Comm.*, 30(7):1750–1757, 1982.   (p 514)

[BB93]   A. Baiocchi and N. Bléfari-Melazzi. An error-controlled approximate analysis of a stochastic fluid flow model applied to an ATM multiplexer with heterogeneous on-off sources. *IEEE/ACM Trans. Networking*, 1:628–637, 1993.   (p 459)

[BBM93]   A. Baiocchi and N. Bléfari-Melazzi. Steady-State Analysis of the MMPP/G/1/K Queue. *IEEE Trans. Comm.*, 41(4):531–534, April 1993.   (p 338)

[BC92]   C. Blondia and O. Casals. Statistical multiplexing of VBR sources: A matrix-analytic approach. *Performance Evaluation*, 16(5):5–20, 1992.   (pp 28, 30, 489)

[BC95]   H.-W. Braun and K. C. Claffy. Web traffic characterization: an assessment of the impact of caching documents from NCSA's web server. *Comput. Networks ISDN Systems*, 28:37–52, 1995.   (p 12)

[BCT91]   M. Butto, E. Cavallero, and A. Tonietti. Effectiveness of the leaky bucket policing mechanism in ATM networks. *IEEE J. Selected Areas in Comm.*, 9, April 1991.   (p 60)

[BD95]   D. D. Botvich and N. G. Duffield. Large deviations, the shape of the loss curve, and economies of scale in large multiplexers. *Queueing Systems*, 20:293–320, 1995.   (pp 167, 460)

[BDMS94]   C. Bruni, P. D'Andrea, U. Mocci, and C. Scoglio. Optimal capacity management of virtual paths in an ATM network. In *IEEE Globecom 94 Proceedings*, 1994.   (p 283)

[Bea93]   N. G. Bean. *Statistical Multiplexing in Broadband Communication Networks*. PhD thesis, University of Cambridge, 1993.   (pp 137, 170)

[Bea94]   N. G. Bean. Robust connection acceptance control for ATM networks with incomplete source information. *Ann. Oper. Res.*, 48:357–379, April 1994.   (p 138)

[Bel70]   R. Bellman. *Introduction to Matrix Analysis*. McGraw-Hill, 2nd edition, 1970.   (p 345)

[Ben63]   V. E. Beneš. *General Stochastic Processes in the Theory of Queues*. Addison-Wesley, 1963.   (p 374)

[Ber92]   J. Beran. Statistical methods for data with long-range dependence. *Statistical Science*, 7(4):404–427, 1992.   (p 361)

[Ber94]   J. Beran. *Statistics for Long-Memory Processes*, volume 61 of *Monographs on Statistics and Applied Probability*. Chapman & Hall, 1994.   (pp 25, 325, 354, 358, 362, 363)

[BF95]   F. Bonomi and K. W. Fendick. The rate based flow control framework for the Available Bit Rate ATM service. *IEEE Network*, March 1995.   (p 58)

[BG93]   G.-J. Böcker and R. Gobbi. A techno-economic comparison of the R2024 BAF system and other possible optical access systems. In *Proceedings of the RACE Open Workshop on Broadband Access*, June 1993. Nijmegen.   (p 215)

[BG96]   C. Blondia and F. Geerts. The correlation structure of the output of an ATM multiplexer. Submitted for publication, 1996.   (pp 316, 318)

[BGRS94]   B. Bensaou, J. Guibert, J. W. Roberts, and A. Simonian. Performance of an ATM multiplexer queue in the fluid approximation using the Beneš approach. *Ann. Oper. Res.*, 49:137–160, 1994.   (pp 383, 473, 474, 480, 487)

[BGZ94]   N. G. Bean, R. J. Gibbens, and S. Zachary. The perfor-
            mance of single resource loss systems in multiservice net-
            works. In *14th Int. Teletraffic Congress Proceedings*, vol-
            ume 1a, pages 13–21. North Holland-Elsevier Science Pub-
            lishers, June 1994.   (p 532)

[BGZ95]   N. G. Bean, R. J. Gibbens, and S. Zachary. Asymptotic
            analysis of single resource loss systems in heavy traffic with
            applications to integrated networks. *Adv. in Appl. Probab.*,
            27:273–292, March 1995.   (p 530)

[BHH89]   A. Bhargava, P Humblet, and M. Hluchyj. Queueing anal-
            ysis of continuous bit stream transport in packet networks.
            In *IEEE Globecom 89 Proceedings*, October 1989.   (p 397)

[BHJ48]   E. Brockmeyer, H. L. Halstrom, and A. Jensen. *The Life
            and Works of A. K. Erlang.* Academy of Technical Sciences,
            1948. Copenhagen.   (p 138)

[BJ76]   G. E. P. Box and G. M. Jenkins. *Time Series Analysis.*
            Holden-Day, 1976.   (p 338)

[BK93]   H. Bruneel and B. G. Kim. *Discrete-time models for com-
            munication systems including ATM.* Kluwer Academic Pub-
            lishers Group, Boston, 1993.   (p 217)

[Bla92]   S. Blaabjerg. Estimating the effect of cell delay variation
            by an application of the heavy limit theorem (COST 242
            TD(92)017). Technical report, Technical University of Den-
            mark, 1992.   (p 68)

[Bla93]   S. Blaabjerg. Cell delay variation in a FIFO queue: A dif-
            fusion approach. In *High Speed Networks*, 1993.   (pp 425,
            426)

[Blo89]   C. Blondia. The N/G/1 Finite Capacity Queue. *Commun.
            Statist.-Stochastic Models*, 5(2):273–294, 1989.   (p 338)

[Blo93]   C. Blondia. A discrete-time batch Markovian arrival process
            as B-ISDN traffic model. *Belgian Journal of Operations Re-
            search, Statistics and Computer Science*, 32(3,4):3–23, 1993.
            (pp 316, 490)

[BMPS95]   C. Bruni, U. Mocci, P. Pannunzi, and C. Scoglio. Efficient
            capacity assignment for ATM virtual paths. In *ATM Hot
            Topics on Traffic and Performance: from RACE to ACTS*,
            1995.   (p 283)

[BMS93]   C. Bisdikian, W. Matragi, and K. Sohraby. A study of the jitter in ATM multiplexers. In *High Speed Networks*, 1993. (p 421)

[Bor76]   A. A. Borovkov. *Stochastic Processes in Queueing Theory*. Springer-Verlag, 1976.   (p 374)

[BR60]   R. R. Bahadur and R. R. Rao. On deviations of the sample mean. *Ann. Math. Statist.*, pages 1015–1027, 1960.   (p 382)

[BRS95]   F. Brichet, J. W. Roberts, and A. Simonian. Performance of an ATM multiplexer fed by leaky bucket controlled sources. In *IFIP Workshop an ATM Traffic Management*, Paris, December 1995.   (p 154)

[BRSV95]   F. Brichet, J. Roberts, A. Simonian, and D. Veitch. Heavy traffic analysis of a storage model with long range dependent on/off sources. Submitted for publication, 1995.   (pp 328, 333, 498, 499, 500, 502)

[Bru88]   H. Bruneel. Queueing behavior of statistical multiplexers with correlated inputs. *IEEE Trans. Comm.*, 36, 1988. (p 492)

[BS87]   P. Brown and A. Simonian. Perturbation of a periodic flow in a synchronous server. In *Performance '87, Brussels*, 1987. (p 425)

[BS90]   J.-C. Bolot and A. U. Shankar. Dynamical behavior of rate-based flow control mechanisms. *Computer Communication Review*, pages 35–49, April 1990.   (p 97)

[BSTW95]   Jan Beran, Robert Sherman, Murad S. Taqqu, and Walter Willinger. Variable-bit-rate video traffic and long-range dependence. *IEEE Trans. Comm.*, 43(2/3/4):1566–1579, 1995.   (pp 15, 28)

[BT 91]   BT and NTT. Combined proposal for GFC. Technical Report Contribution No. D.1860, CCITT Study Group XVIII, Melbourne, December 1991. Attachment to NTT, "Proposal for Enhanced GFC Protocol".   (p 205)

[BT92a]   P. E. Boyer and D. P. Tranchier. A reservation principle with applications to the ATM traffic control. *Comput. Networks ISDN Systems*, 24:321–334, 1992.   (pp 57, 88)

[BT92b]   BT. BT/NTT protocol proposal for GFC. Technical Report Contribution No. D.1958, CCITT Study Group XVIII, Geneva, June 1992.   (p 205)

[BT93]   D. Bertsimas and J. Tsitsiklis. Simulated annealing. *Statistical Science*, 8(1):10–15, 1993.   (p 261)

[Buc90]   J. A. Bucklew. *Large Deviation Techniques In Decision, Simulation, And Estimation*. John Wiley & Sons, 1990. (pp 379, 384)

[Bur90]   J. L. Burgin. Broadband ISDN resource management. *Comput. Networks ISDN Systems*, 20:323–331, 1990.   (p 271)

[Cam94]   P. Camarda.   Bandwidth management for ATM-based LANs. In *19Th Conference on Local Computer Networks*, October 1994.   (p 207)

[CB96]   Mark E. Crovella and Azer Bestavros.   Self-similarity in World Wide Web traffic: Evidence and possible causes. In *Proceedings of the 1996 ACM SIGMETRICS International Conference on Measurement and Modeling of Computer Systems*, May 1996.   (p 12)

[CBC95]   Carlos A. Cunha, Azer Bestavros, and Mark E. Crovella. Characteristics of WWW client-based traces. Technical Report TR-95-010, Boston University Department of Computer Science, April 1995.   (p 12)

[CBP95]   O. Crochat, J.-Y. Le Boudec, and T. Przygienda. A path selection method in ATM using Pre-Computation. Technical Report 95/128, Ecole Polytechnique Fédérale de Lausanne, May 1995.   (pp 288, 289)

[CCC⁺91]   J. Choi, M. Choi, T. Chung, Y. Shin, and K. Kim. Overall design requirements of broadband ATM network. In *IEEE Infocom 91 Proceedings*, April 1991.   (p 267)

[CFZ93a]   I. Chlamtac, A. Faragó, and T. Zhang. How to establish and utilize virtual paths in ATM networks. In *IEEE International Conference on Communications (ICC'93)*, pages 1368–1372, May 1993.   (p 272)

[CFZ93b]   I. Chlamtac, A. Faragó, and T. Zhang. Optimizing the system of virtual paths in ATM network architecture. Internal report, Technical University of Budapest, 1993.   (pp 271, 272)

[CFZ94]   I. Chlamtac, A. Faragó, and T. Zhang.   Optimizing the system of virtual paths. *IEEE/ACM Trans. Networking*, 2(6):581–587, 1994.   (p 272)

[CGB93]   O. Casals, J. García, and C. Blondia. A medium access control protocol for an ATM access network. In *Proceedings of 5th International Conference on Data Communication Systems and Their Performance*, October 1993. Raleigh, North Carolina.   (p 219)

[CGKK94]   I. Cidon, R. Guérin, I. Kessler, and A. Khamisy. Analysis of a statistical multiplexer with generalized periodic sources. In S. Fdida, editor, *High Performance Networking '94*, pages 325–339. IFIP, 1994.   (p 480)

[Cha91]   H. J. Chao. A novel architecture for queue management in the ATM network. *IEEE J. Selected Areas in Comm.*, 9(7):1110–1118, September 1991.   (p 182)

[Çin75]   E. Çinlar. *Introduction To Stochastic Processes*. Prentice-Hall, 1975.   (p 318)

[CL66]   D. R. Cox and P. A. W. Lewis. *The Statistical Analysis of Series of Events*. Methuen, 1966.   (pp 423, 426)

[CLK95]   Simon Crosby, Ian Leslie, and Peter Key. Cell delay variation and burst expansion in ATM networks: results from a practical study using fairisle, 1995. Submitted for publication.   (p 83)

[CLW94]   G. L. Choudhury, D. M. Lucantoni, and W. Whitt. On the effectiveness of effective bandwidths for admission control in ATM networks. In *14th Int. Teletraffic Congress Proceedings*, pages 411–420. North Holland-Elsevier Science Publishers, 1994.   (p 147)

[CLW95]   G. L. Choudhury, D. M. Lucantoni, and W. Whitt. Squeezing the most out of ATM. Preprint, 1995.   (p 460)

[CM88]   Y-H. Choi and M. Malek. A fault tolerant systolic sorter. *IEEE Trans. on Computers*, 29(5):621–624, May 1988.   (p 182)

[CMST92]   C. Cavallero, U. Mocci, C. Scoglio, and A. Tonietti. Optimisation of virtual path / virtual circuit management in ATM networks. In *Networks 92, Kobe*, May 1992.   (pp 260, 263, 272)

[Coh95]   J. W. Cohen. Private communication, 1995.   (p 466)

[COMM95]   D. Cavendish, Y. Oie, M. Murata, and H. Miyahara. Proportional rate-based congestion control under long propagation delay. *International Journal of Communication Systems*, 8:79–89, 1995.   (p 97)

[Coo72]   R. B. Cooper. *Introduction to Queueing Theory*. MacMillan, 1972.   (p 434)

[Cou77]   P. J. Courtois. *Decomposability*. ACM Monograph Series, 1977.   (p 338)

[Cox84]   D. R. Cox. Long-range dependence: A review. In H. A. David and H. T. David, editors, *Statistics: An Appraisal*, pages 55–74. The Iowa State University Press, Ames, Iowa, 1984.   (p 325)

[CR93]   S. Chung and K. W. Ross. Reduced load approximations for multirate loss networks. *IEEE Trans. Comm.*, 41(8):1222–1231, 1993.   (pp 541, 543, 544)

[Cru91]   R. L. Cruz. A calculus for network delay, part 1: network elements in isolation. *IEEE Trans. on Information Theory*, 37:114–31, 1991.   (p 90)

[CSF82]   I. P. Cornfeld, Ya. G. Sinai, and S. V. Fomin. *Ergodic Theory*. Springer-Verlag, Berlin, 1982.   (p 330)

[CSZ92]   D. D. Clark, S. Shenker, and L. Zhang. Supporting real-time applications in an integrated services packet network: Architecture and mechanism. In *ACM SIGCOM 92 Proceedings*, 1992.   (pp 51, 61)

[CW96]   C. Courcoubetis and R. Weber. Buffer overflow asymptotics for a switch handling many traffic sources. *J. Appl. Probab.*, 33, 1996.   (p 167)

[Dah89]   R. Dahlhaus. Efficient parameter estimation for self-similar processes. *Ann. Statist.*, 17:1749–1766, 1989.   (p 361)

[Dal68]   D. J. Daley. The serial correlation coefficients of waiting times in a stationary single server queue. *J. Austral. Math. Soc.*, 8:683–699, 1968.   (p 494)

[DDH94]   B. T. Doshi, S. Dravida, and P. Harshavardhana. Performance and roles of bandwidth and buffer reservation schemes in high speed networks. In *14th Int. Teletraffic Congress Proceedings*, volume 1a, pages 23–34. North Holland-Elsevier Science Publishers, June 1994.   (p 88)

[DeG86]   M. H. DeGroot. *Probability and Statistics*. Addison-Wesley, second edition, 1986.   (p 135)

556   Bibliography

[Del83]   L. E. N. Delbrouck. On the steady-state distribution in a service facility carrying mixtures of traffic with different peakedness factors and capacity requirements. *IEEE Trans. Comm.*, 31(11):1209–1211, 1983.   (pp 514, 521)

[DGBP83]   A. Dupuis, A. Gravey, P. Boyer, and J.-M. Pitie. The output process of the single server queue with periodic arrival process and deterministic service time. In *Modelling and Performance Evaluation Methodology, Lecture Notes in Control and Information Sciences no 60*. Springer-Verlag, 1983. (pp 397, 401)

[DH86]   B. T. Doshi and H. Heffes. Overload performance of several processor queueing disciplines for the M/M/1 queue. *IEEE Trans. Comm.*, 34(6), June 1986.   (p 431)

[DH90]   J. R. Davin and A. T. Heybey. A simulation study of fair queueing and policy enforcement. *ACM Comp. Comm. Rev.*, 20(5), October 1990.   (p 173)

[DH95]   A. Dupuis and G. Hébuterne. Dimensionning the continuous state leaky bucket for geometric arrivals. In S. Fdida and R. O. Onvural, editors, *IFIP WG6.3 Data Communications and their Performance*, Istanbul, Turkey, 1995. Chapman & Hall.   (p 69)

[DKS89]   A. Demers, S. Keshav, and S. Shenker. Analysis and simulation of a fair queueing algorithm. In *ACM SIGCOM 89 Proceedings*, 1989.   (pp 173, 174)

[DLO+94]   N. G. Duffield, J. T. Lewis, N. O'Connell, R. Russell, and F. Toomey. Statistical issues raised by the Bellcore data. In *11th UK Teletraffic Symposium*, 1994.   (p 17)

[DO96]   N. G. Duffield and N. O'Connell. Large deviations and overflow probabilities for the general single-server queue, with applications. *Mathematical Proceedings of the Cambridge Philosophical Society*, 1996.   (pp 167, 308, 385, 497)

[Doe55]   G. Doetsch. *Handbuch der Laplace Transformation*, volume II. Birkhäuser, Basel, 1955.   (p 393)

[Don52]   M. D. Donsker. Justification and extension of Doob's heuristic approach to Kolmogorov-Smirnov theorems. *Ann. Math. Statist.*, 23:277–281, 1952.   (p 400)

[Dos94]   B. T. Doshi. Deterministic rule based traffic descriptors for broadband ISDN: Worst case behavior and connection acceptance control. In *14th Int. Teletraffic Congress Proceedings*, volume 1a, pages 591–600. North Holland-Elsevier Science Publishers, June 1994.   (pp 53, 151)

[DR87]   Z. Dziong and J. W. Roberts. Congestion probabilities in a circuit-switched integrated services network. *Performance Evaluation*, 7:267–284, 1987.   (p 514)

[DRS91]   L. G. Dron, G. Ramamurthi, and B. Sengupta. Delay analysis of an ATM switch for continuous-bit-rate traffic. In *13th Int. Teletraffic Congress Proceedings*, pages 33–38. North Holland-Elsevier Science Publishers, 1991.   (p 400)

[Eck79]   A. E. Eckberg. The single server queue with periodic arrival process and deterministic service time. *IEEE Trans. Comm.*, 27:556–562, March 1979.   (pp 397, 398)

[EGHS96]   V. Elek, Z. Gal, P. L. Huong, and C. Szabo. ATM LAN network design. *Journal on Telecommunications*, Vol. XLVII., January-February 1996.   (p 211)

[Ell75]   A. Elldin. Basic traffic theory. In *ITU Seminar on Traffic Engineering and Network Planning*, pages 11–133, New Delhi, 1975.   (pp 521, 522)

[Ell85]   R. S. Ellis. *Entropy, Large Deviations and Statistical Mechanics*. Springer-Verlag, New York, 1985.   (p 379)

[EM73]   O. Enomoto and H. Miyamoto. An analysis of mixtures of multiple bandwidth traffic on time division in switching networks. In *7th Int. Teletraffic Congress Proceedings*, pages 635.1–8, 1973.   (pp 514, 515)

[EM93]   A. I. Elwalid and D. Mitra. Effective bandwith of general Markovian traffic sources and admission control of high speed networks. *IEEE/ACM Trans. Networking*, 1, June 1993.   (pp 146, 167, 459, 460, 461)

[EMW95]   A. Elwalid, D. Mitra, and R. H. Wentworth. A new approach to allocating buffers and bandwidth to heterogeneous regulated traffic in an ATM node. *IEEE J. Selected Areas in Comm.*, 13(6):1115–1128, August 1995.   (pp 115, 130, 131, 132)

[Ens94]   Jürgen Enssle. Modelling and statistical multiplexing of VBR MPEG compressed video in ATM networks. In *Proceedings of the 4th Open Workshop on High Speed Networks, Brest*, September 1994.   (p 25)

558   Bibliography

[ENW94]   A. Erramilli, O. Narayan, and W. Willinger. Experimental queueing analysis with long-range dependent packet traffic. Technical report, Bellcore, Morristown, 1994.   (p 348)

[ES90]   A. Erramilli and R. P. Singh. Application of deterministic chaotic maps to characterize broadband traffic. In *7th ITC Specialist Seminar, Morristown*, Morristown, USA, 1990. (p 329)

[ET86]   E. Eberlein and M. S. Taqqu, editors. *Dependence in Probability and Statistics*, volume 11 of *Progress in Prob. and Stat.* Birkhäuser, Boston, 1986.   (p 327)

[FBA$^+$95]   A. Faragó, S. Blaabjerg, L. Ast, G. Gordos, and T. Henk. A new degree of freedom in ATM network dimensioning: optimizing the logical configuration. *IEEE J. Selected Areas in Comm.*, 13(7):1199–1206, 1995.   (pp 293, 294)

[FBH$^+$94a] A. Faragó, S. Blaabjerg, W. Holender, T. Henk, L. Ast, A. Szentesi, and Z. Ziaja. Optimal partitioning of ATM networks into virtual subnetworks. Internal report, Technical University of Budapest, 1994.   (pp 278, 279)

[FBH$^+$94b] A. Faragó, S. Blaabjerg, W. Holender, T. Henk, A. Szentesi, and Z. Ziaja. Resource separation - an efficient tool for optimizing ATM network configuration. In *Networks 94, Budapest*, pages 83–88, September 1994.   (pp 278, 279)

[FBHA94]   A. Faragó, S. Blaabjerg, T. Henk, and L. Ast. A new degree of freedom in ATM network dimensioning: optimizing the logical configuration. Internal report, Technical University of Budapest, 1994.   (p 293)

[Fel68]   W. Feller. *An Introduction to Probability Theory and its Applications*, volume 1. John Wiley & Sons, New York, 1968.   (p 429)

[Fel71]   W. Feller. *An Introduction to Probability Theory and its Applications*, volume 2. John Wiley & Sons, New York, 1971.   (pp 144, 325, 480)

[Fil89]   J. Filipiak. Structured systems analysis methodology for design of an ATM network architecture. *IEEE J. Selected Areas in Comm.*, 7(8), October 1989.   (p 271)

[Fis89]   W. Fischer. Waiting times in priority systems with general arrivals. Technical report, INRS Technical Report 89-34, 1989.   (p 431)

[FL91]   H. J. Fowler and W. E. Leland. Local area network traffic characteristics, with implications for broadband network congestion management. *IEEE J. Selected Areas in Comm.*, 9(7):1139–1149, September 1991.   (pp 162, 311, 313)

[FM94]   Victor S. Frost and Benjamin Melamed. Traffic modeling for telecommunication networks. *IEEE Communications Magazine*, pages 70–81, March 1994.   (p 28)

[For96]   ATM Forum. Traffic management specification version 4.0. Technical Report Contribution 95-0013R10, ATM Forum, February 1996.   (pp 4, 48, 55, 95, 96)

[Fre80]   A. A. Fredericks. Congestion in blocking systems - A simple approximation technique. *Bell System Tech. J.*, 59(6):805–828, 1980.   (p 522)

[FRW92]   K. W. Fendick, M. A. Rodrigues, and A. Weiss. Analysis of a rate-based control strategy with delayed feedback. *ACM SIGCOMM Computer Communication Review*, pages 136–148, 1992.   (p 97)

[FT86]   R. Fox and M. S. Taqqu. Large-sample properties of parameter estimates for strongly dependent stationary Gaussian time series. *Ann. of Statistics*, 14(2):517–532, 1986. (p 361)

[FW95]   V. J. Friesen and J. W. Wong. Migrating to ATM in the local area. In *6Th IEEE Workshop on Local and Metropolitan Area Networks*, March 1995.   (p 208)

[GA94]   Y. Gong and I. F. Akyildiz. Dynamic traffic control using feedback and traffic prediction in ATM networks. In *IEEE Infocom 94 Proceedings*, pages 91–98, Toronto, 1994. (p 97)

[GAN91]   R. Guérin, H. Ahmadi, and M. Naghshineh. Equivalent capacity and its application to bandwidth allocation in high speed networks. *IEEE J. Selected Areas in Comm.*, 9:968–981, 1991.   (pp 146, 288, 459, 462)

[GB93]   A. Gravey and P. Boyer. Cell Delay Variation specification in ATM networks. In Pujolle Perros and Takahashi, editors, *Modelling and Performance Evaluation of ATM Technology*. North Holland-Elsevier Science Publishers, 1993.   (p 63)

[GBC95]   J. Garcia, J. M. Barceló, and O. Casals. An exact model for the multiplexing of worst case traffic sources. In *Sixth*

Int. Conf. on Data Comm. Syst. and their Performance, Istanbul, October 1995. IFIP.   (p 480)

[GBH91]   A. Gravey, P. Boyer, and G. Hébuterne. Tagging versus policing in ATM networks. In *IEEE Globecom 91 Proceedings*, December 1991.   (p 439)

[GC90]   J. García and O. Casals. Priorities in ATM networks. In *NATO Advanced Research Workshop*, June 1990.   (pp 432, 435)

[GH85]   D. Gross and C. M. Harris. *Fundamentals of Queueing Theory.* John Wiley & Sons, 1985.   (p 394)

[GH90]   A. Gravey and G. Hébuterne. Analysis of a priority queue with delay and/or loss sensitive customers. In *7th ITC Specialist Seminar, Morristown*, October 1990.   (pp 432, 436)

[GH91a]   R. J. Gibbens and P. J. Hunt. Effective bandwidths for the multi-type UAS channel. *Queueing Systems*, 9:17–28, 1991.   (pp 146, 167)

[GH91b]   A. Gravey and G. Hébuterne. Mixing time and loss priorities in a single server queue. In *13th Int. Teletraffic Congress Proceedings*, pages 147–152. North Holland-Elsevier Science Publishers, June 1991.   (pp 432, 436)

[GH94]   A. Gravey and G. Hébuterne. Cell conformance and cell loss ratio commitments in ATM networks. In *ICC'94*, New-Orleans, USA, May 1994.   (p 50)

[Gib96]   R. J. Gibbens. Traffic characterisation and effective bandwidths for broadband network traces. In F. P. Kelly, S. Zachary, and I. B. Ziedins, editors, *Stochastic networks: Theory and Applications*, Royal Statistical Society Lecture Note Series 4, pages 169–179. Oxford University Press, 1996. ISBN 0-19-852399-8.   (p 165)

[GK81]   D. A. Garbin and J. E. Knepley. Marginal cost routing in nonhierarchical networks. In *International Communications Conference*, 1981.   (p 260)

[GK91]   R. J. Gibbens and F. P. Kelly. ATM connection acceptance control, stage 2. Technical report, Lyndewode Research Ltd, January 1991. Report prepared for British Telecommunications plc by Lyndewode Research Ltd.   (p 135)

[GK94]   Rhodri Griffiths and Peter Key. Adaptive call admission control in ATM networks. In *14th Int. Teletraffic Congress Proceedings*, volume 1b, pages 1089–98. North Holland-Elsevier Science Publishers, June 1994.   (pp 135, 136)

[GKK95]   R. J. Gibbens, F. P. Kelly, and P. B. Key. A decision-theoretic approach to call admission control in ATM networks. *IEEE J. Selected Areas in Comm.*, 13(6):1101–1114, August 1995.   (pp 137, 140, 143, 170)

[GKW91]   G. Gopal, C. Kim, and A. Weinrib. Algorithms for reconfigurable networks. In *13th Int. Teletraffic Congress Proceedings*. North Holland-Elsevier Science Publishers, June 1991. (pp 288, 289)

[GL89]   A. Gersht and K. J. Lee. Virtual-circuit load control in fast packet-switched broadband networks. In *IEEE Globecom 89 Proceedings*, pages 214–220, 1989.   (p 523)

[GMP89]   M. Gerla, J. A. Suruagy Monteiro, and R. Pazos. Topology design and bandwidth allocation in ATM nets. *IEEE J. Selected Areas in Comm.*, 7(8), 1989.   (p 297)

[GMS96]   F. Guillemin, R. Mazumdar, and A. Simonian. On heavy traffic approximations for transient characteristics of $M/M/\infty$ queues. *J. Appl. Probab.*, 1996. To appear. (p 453)

[GN93]   J. García and M. F. Neuts. The second moment of the delay in a queue of data packets. Technical Report COST242 TD(93)019, University of Catalonia, 1993.   (p 493)

[GN96]   G. Gripenberg and I. Norros. On the prediction of fractional Brownian motion. *J. Appl. Probab.*, June 1996. To appear. (pp 367, 368)

[Gol94]   S. J. Golestani. A self-clocked fair queueing scheme for broadband applications. In *IEEE Infocom 94 Proceedings*, July 1994.   (pp 173, 181)

[Gol95]   S. J. Golestani. Network delay analysis of a class of fair queueing algorithms. *IEEE J. Selected Areas in Comm.*, 13(6):1057–1070, August 1995.   (p 178)

[Gon94]   Kevin L. Gong. *Berkeley MPEG-1 Video Encoder, User's Guide*. University of California, Berkeley, Computer Science Division-EECS, March 1994.   (p 20)

[GR91]   F. Guillemin and J. W. Roberts. Jitter and bandwidth en-
            forcement. In *IEEE Globecom 91 Proceedings*, December
            1991.   (p 423)

[Gra84]   A. Gravey. Temps d'attente et nombre clients dans une file
            $nD/D/1$. *Annales de l'Institut H. Poincaré - Probabilités
            et Statistiques*, 20(1):53–73, 1984.   (p 397)

[Gra94]   A. Gravey. Cell conformance and Quality of Service guar-
            antees in ATM networks. In *14th Int. Teletraffic Congress
            Proceedings*, pages 1395–1404. North Holland-Elsevier Sci-
            ence Publishers, June 1994.   (p 50)

[GRM95]   F. Guillemin, C. Rosenberg, and J. Mignault. On character-
            izing an ATM source via the Sustainable Cell Rate Traffic
            Descriptor. In *IEEE Infocom 95 Proceedings*, 1995.   (p 49)

[GRSS95]   F. Guillemin, G. Rubino, B. Sericola, and A. Simonian.
            Transient analysis of statistical multiplexing of data con-
            nections on an ATM link. Submitted for publication, 1995.
            (p 452)

[GS95]   F. Guillemin and A. Simonian. Transient characteristics of
            an $M/M/\infty$ system. *Adv. in Appl. Probab.*, 27:862–888,
            1995.   (pp 450, 451, 455)

[Gui94]   J. Guibert. Overflow probability upper bound in fluid
            queues with on-off sources. *J. Appl. Probab.*, 31(3), De-
            cember 1994.   (pp 473, 475)

[Gus91]   R. Gusella. Characterizing the variability of arrival pro-
            cesses with indexes of dispersion. *IEEE J. Selected Areas in
            Comm.*, 9(2), February 1991.   (p 310)

[GW94]   Mark W. Garrett and Walter Willinger. Analysis, modeling
            and generation of self-similar VBR video traffic. In *ACM
            SIGCOM 94 Proceedings*, pages 269–280, 1994.   (pp 11, 21,
            25)

[Haa96]   H. Haapasalo. ATM traffic measurements. Master's thesis,
            Tampere University of Technology, Department of Electrical
            Engineering, 1996. In Finnish.   (p 363)

[HC78]   S. T. Huang and S. Cambanis. Stochastic and multiple
            Wiener integrals for Gaussian processes. *Ann. of Probab.*,
            6:585–614, 1978.   (p 367)

[HCM90]   E. L. Hahne, A. K. Choudhury, and N. Maxemchuk. Improving the fairness of distributed-queue-dual-bus networks. In *IEEE Infocom 90 Proceedings*, 1990.   (p 224)

[HDLK95]   C. Huang, M. Devetsikiotis, I. Lambadaris, and A. R. Kaye. Modeling and simulation of self-similar variable bit rate compressed video: A unified approach. In *ACM SIGCOM 95 Proceedings*, 1995.   (p 363)

[Hei92]   H. Heiss. Impact of jitter on peak cell rate policing with a leaky bucket. In *R1022 Workshop on Traffic and Performance Aspects in IBCN*, Aveiro, Portugal, January 1992. Session 5, paper 1.   (p 415)

[HG90]   G. Hébuterne and A. Gravey. A space priority queueing mechanism for multiplexing ATM channels. *Comput. Networks ISDN Systems*, 20(1–5), December 1990.   (pp 432, 435)

[Hin90]   S. J. Hindmarch. The effect of mean holding times on determining the trunk reservation parameter - A markov decision approach. *BT Internal Paper RTM RT32/22/90*, May 1990.   (p 532)

[HK95]   J. Y. Hui and E. Karasan. A thermodynamic theory of broadband networks with application to dynamic routing. *IEEE J. Selected Areas in Comm.*, 1(6), August 1995.   (p 308)

[HK96]   A. Hung and G. Kesidis. Bandwidth scheduling for wide-area ATM networks using virtual finishing times. *IEEE/ACM Trans. Networking*, 4(1):49–54, February 1996.   (p 175)

[HL86]   H. Heffes and D. M. Lucantoni. A Markov modulated characterization of packetized voice and data traffic and related statistical multiplexer performance. *IEEE J. Selected Areas in Comm.*, 4(6):856–868, September 1986.   (p 310)

[Hos84]   J. R. M. Hosking. Modeling persistence in hydrological time series using fractional differencing. *Water Resources Research*, 20(12):1898–1908, 1984.   (p 337)

[HR93]   F. Hübner and M. Ritter. Blocking in multi-service broadband systems with CBR and VBR input traffic. In *7th ITG/GI Conference*, pages 212–225, Aachen, September 1993.   (pp 534, 535, 537, 538, 539)

[HR95]   M. Hamdi and J. W. Roberts. QoS guarantees for shaped bit rate video connections in broadband networks. In *Mm-Net'95*, 1995. Aizu-Wakamatsu, Japan.   (pp 43, 45)

[HR96]   M. Hamdi and J. W. Roberts. Burstiness bounds based multiplexing schemes for VBR video connections in the B-ISDN. In *International Zurich Seminar on Digital Communications*, February 1996.   (p 46)

[HTL92]   Daniel P. Heyman, Ali Tabatabai, and T. V. Lakshman. Statistical analysis and simulation study of video teleconference traffic in ATM networks. *IEEE Transactions on Circuits and Systems for Video Technology*, 2(1):49–59, March 1992.   (p 28)

[Hüb90]   F. Hübner. Analysis of a finite-capacity asynchronous multiplexer with deterministic traffic sources. In *7th ITC Specialist Seminar, Morristown*, October 1990.   (p 372)

[Hüb91]   F. Hübner. Discrete-time analysis of the output process of an ATM multiplexer with periodic input. Technical Report Report No. 36, Würzburg University, 1991.   (p 401)

[Hui88]   J. Y. Hui. Resource allocation for broadband networks. *IEEE J. Selected Areas in Comm.*, 6(9):1598–1608, 1988. (pp 138, 167, 303, 383, 446)

[Hui90]   J. Y. Hui. *Switching and Traffic Theory for Integrated Broadband Networks*. Kluwer Academic Publishers Group, 1990.   (pp 167, 383)

[HW93]   D Hughes and K. Wajda. Comparison of virtual path bandwidth assignment and routing methods. In *Cracow Workshop on Network Management*, May 1993.   (p 271)

[Ide89]   I. Ide. Superposition of interrupted Poisson processes and its application to packetized voice multiplexers. In *12th Int. Teletraffic Congress Proceedings*, pages 1399–1405. North Holland-Elsevier Science Publishers, April 1989.   (p 345)

[ISO93a]   Draft International Standard: ISO/IEC DIS 13818-2. *Generic Coding of Moving Pictures and Associated Audio Part 2, Video*, 1993.   (p 19)

[ISO93b]   International Standard: ISO/IEC IS 11172-2. *Coding of Moving Pictures and Associated Audio for Digital Storage Media up to 1.5 Mbit/s Part 2, Video*, 1993.   (p 19)

[ITU92]   ATM layer specification for B-ISDN, 1992. ITU Recommendation I.361, Geneva.   (p 271)

[ITU96a]   B-ISDN ATM layer cell transfer performance, May 1996. ITU Recommendation I.356, Geneva.   (p 228)

[ITU96b]   Traffic control and congestion control in B-ISDN, May 1996. ITU Recommendation I.371, Geneva.   (pp 3, 4, 47, 48, 51, 54, 55, 59, 78, 95, 116, 118)

[Ive87]   V. B. Iversen. The exact evaluation of multi-service loss systems with access control. In *7th Nordic Teletraffic Seminar*, Lund, August 1987.   (p 514)

[Ive91]   V. B. Iversen. Multi-dimensional service systems for evaluation of future digital communication systems. In *6th Australian Teletraffic Research Seminar*, Wollongong, November 1991.   (p 514)

[Jac88]   V. Jacobson. Congestion avoidance and control. In *ACM SIGCOM 88 Proceedings*, 1988.   (p 18)

[JD90]   S. B. Jacobsen and L. Dittmann. A fluid flow queueing model for heterogeneous on/off traffic. Traffic and performance aspects in IBCN, R1012 Broadband Local Network Technology, RACE Workshop, Munich, July 1990.   (p 465)

[Jen95]   A. Jensen. Modelling of self-similar behaviour with the Markovian Arrival Process, MAP (In Danish), 1995. IMM-EKS-1995-7 (Masters Thesis).   (p 345)

[Joh85]   S. A. Johnson. A performance analysis of integrated communication systems. *British Telecom Technical Journal*, 3(4), October 1985.   (pp 514, 517, 525)

[JS87]   J. Jacod and A. N. Shiryaev. *Limit Theorems for Stochastic Processes*. Springer-Verlag, 1987.   (p 391)

[Kau81]   J. S. Kaufman. Blocking in a shared resource environment. *IEEE Trans. Comm.*, 29(10):1474–1481, 1981.   (pp 514, 516, 542)

[Kau92]   J. S. Kaufman. Blocking with retrials in a completely shared resource environment. *Performance Evaluation*, 15:99–113, 1992.   (p 514)

[KB92]   K. Kvols and S. Blaabjerg. Bounds and approximations for the periodic on/off queue with application to ATM traffic control. In *IEEE Infocom 92 Proceedings*, pages 487–494, 1992.   (p 481)

[Kel89]   F. P. Kelly. Fixed point models of loss networks. *J. Austral. Math. Soc.*, 31:204–218, 1989.   (pp 294, 541, 542)

[Kel90]   F. P. Kelly. Routing and capacity allocation in networks with trunk reservation. *Math. of Oper. Res.*, pages 771–793, 1990.   (p 294)

[Kel91a]   F. P. Kelly. Effective bandwidths at multi-class queues. *Queueing Systems*, 9:5–15, 1991.   (pp 126, 138, 167)

[Kel91b]   F. P. Kelly. Loss networks. *Ann. of Appl. Probab.*, 1:319–78, 1991.   (pp 279, 540, 542, 543)

[Kel94a]   F. P. Kelly. Mathematical models of multiservice networks. In D M Titterington, editor, *Complex Stochastic Systems and Engineering*. The Institute of Mathematics and its Applications, Oxford University Press, 1994.   (p 69)

[Kel94b]   F. P. Kelly. On tariffs, policing and admission control of multiservice networks. *Oper. Res. Lett.*, 15:1–9, 1994. (pp 164, 168)

[Kel94c]   F. P. Kelly. Tariffs and effective bandwidths in multiservice networks. In Jacques Labetoulle and James W. Roberts, editors, *14th Int. Teletraffic Congress Proceedings*, volume 1a, pages 401–410. North Holland-Elsevier Science Publishers, June 1994.   (pp 126, 164, 169)

[Kel96a]   F. P. Kelly. Charging and accounting for bursty connections. In L. W. McKnight and J. P. Bailey, editors, *Internet Economics*. MIT Press, 1996.   (p 171)

[Kel96b]   F. P. Kelly. Modelling communication networks, present and future. *Philos. Trans. Roy. Soc. London*, 354:437–463, 1996.   (p 164)

[Kel96c]   F. P. Kelly. Notes on effective bandwidths. In F. P. Kelly, S. Zacahry, and I. B. Ziedins, editors, *Stochastic networks: Theory and Applications*, Royal Statistical Society Lecture Note Series 4, pages 141–168. Oxford University Press, 1996. ISBN 0-19-852399-8.   (pp 164, 168)

[Key95]   P. B. Key. Admission control problems in telecommunications. In D. M. Titterington, editor, *Complex Stochastic Systems and Engineering*, pages 235–250. The Institute of Mathematics and its Applications, Oxford University Press, 1995.   (p 93)

[KHBG91]   H. Kröner, G. Hébuterne, P. Boyer, and A. Gravey. Priority management in ATM switching nodes. *IEEE J. Selected Areas in Comm.*, 9(3), April 1991.   (pp 431, 432, 435)

[Kin70]   J. F. C. Kingman. Inequalities in the theory of queues. *J. Roy. Statist. Soc.*, 32:102–110, 1970.   (p 69)

[KK94]   F. P. Kelly and P. B. Key. Dimensioning playout buffers from an ATM network. In *11th UK Teletraffic Symposium*, IEE, Savoy Place, London, WC2R OBL, UK, 1994. The Institution of Electrical Engineers.   (pp 69, 73, 80)

[Kle75a]   L. Kleinrock. *Queueing Systems*, volume 2, Computer applications. John Wiley & Sons, 1975.   (pp 68, 69)

[Kle75b]   L. Kleinrock. *Queueing Systems*, volume 1, Theory. John Wiley & Sons, 1975.   (pp 371, 390, 424, 425, 466)

[KM91]   C. Knessl and J. A. Morrison. Heavy-traffic analysis of a data-handling system with many sources. *SIAM J. Appl. Math.*, 51(1):187–213, February 1991.   (pp 322, 460)

[KM94]   K. P. Kontovasilis and N. Mitrou. Bursty traffic modeling and efficient analysis algorithms via fluid-flow models for ATM IBCN. *Ann. Oper. Res.*, 49:279–323, 1994.   (p 373)

[KM95]   H. T. Kung and R. Morris. Credit-based flow control for ATM networks. *IEEE Network Magazine*, pages 40–48, March 1995.   (p 95)

[Kol40]   A. N. Kolmogorov. Wienersche Spiralen und einige andere interessante Kurven im Hilbertschen Raum. *C.R. (Doklady) Acad. Sci. USSR (N.S.)*, 26:115–118, 1940.   (p 330)

[KOMM95]   K. Kawahara, Y. Oie, M. Murata, and H. Miyahara. Performance analysis of reactive congestion control for ATM networks. *IEEE J. Selected Areas in Comm.*, 13(4):651–661, 1995.   (p 97)

[Kos74]   L. Kosten. Stochastic theory of a multi-entry buffer (I). *Delft Progress Report series F*, 1:10–18, 1974.   (pp 463, 464)

[Kos84]   L. Kosten. Stochastic theory of data handling systems with groups of multiple sources. In *Proceedings of the International Symposium on Performance of Computer Communications Systems*, Zurich, March 1984. IFIP.   (p 465)

568   Bibliography

[Kos86]   L. Kosten. Liquid models for a type of information buffer problems. *Delft Progress Report series F*, 11:71–86, 1986. (p 462)

[KR93]   H. Kobayashi and Q. Ren. A diffusion approximation analysis of an ATM statistical multiplexer with multiple types of traffic, Part I: equilibrium state solutions. In *Proceedings of the 1993 IEEE International Conference on Communications*, volume 2, pages 1047–1053, 1993.   (p 460)

[KS67]   J. G. Kemeny and J. L. Snell. *Finite Markov Chains*. Van Nostrand, New York, 1967.   (p 317)

[KSY84]   J. F. Kurose, M. Schwartz, and Y. Yemini. Multiple-access protocols and time-constrained communication. *Comput. Serv.*, 16:43–70, 1984.   (p 218)

[Kur96]   T. Kurtz. Workload input models with long range dependence. In F. P. Kelly, S. Zachary, and I. B. Ziedins, editors, *Stochastic networks: Theory and Applications*, Royal Statistical Society Lecture Note Series 4, pages 119–140. Oxford University Press, 1996. ISBN 0-19-852399-8.   (pp 322, 333, 335)

[KV94]   M. Koivula and J. Virtamo. VP optimizaton by simulated annealing. In J. Harju, editor, *Proceedings of the Third Summer School on Telecommunications, Vol. II*, pages 87–95. Lappeenranta University of Technology, August 1994. (p 262)

[KW56]   J. Kiefer and J. Wolfowitz. On the characteristics of the general queueing process, with applications to random walks. *Ann. Math. Statist.*, 27:147–161, 1956.   (p 494)

[KW88]   R. Kleinewillinghöfer-Kopp and E. Wollner. Comparison of access control strategies for ISDN-traffic on common trunk groups. In *12th Int. Teletraffic Congress Proceedings*, pages 5.4A.2.1–7. North Holland-Elsevier Science Publishers, 1988.   (pp 267, 523)

[KWC93]   G. Kesidis, J. Walrand, and C-S. Chang. Effective bandwith for multiclass Markov fluids and other ATM sources. *IEEE/ACM Trans. Networking*, 1(4):424–428, August 1993. (pp 146, 167, 459, 460, 463)

[Lam62]   J. W. Lamperti. Semi-stable stochastic processes. *Trans. Amer. Math. Soc.*, 104:62–78, 1962.   (p 326)

[LeB91]   J. Y. Le Boudec. An efficient solution method for Markov models of ATM links with loss priorities. *IEEE J. Selected Areas in Comm.*, 9(3), April 1991.   (pp 432, 434)

[LeG91]   Didier Le Gall. MPEG: A video compression standard for multimedia applications. *Comm. ACM*, 34(4):46–58, April 1991.   (p 19)

[Lel89]   W. E. Leland.   Window-based congestion management in broadband ATM networks: The performance of three access-control policies. In *IEEE Globecom 89 Proceedings*, pages 1794–1800, 1989.   (p 148)

[LH92]   J.-F. P. Labourdette and G. W. Hart. Blocking probabilities in multitraffic loss systems: intensitivity, asymptotic behavior and approximations. *IEEE Trans. Comm.*, 40(8):1355–1366, 1992.   (pp 279, 532, 541)

[Li89]   S. Q. Li. Overload control in a finite message storage buffer. *IEEE Trans. Comm.*, 37(12), December 1989.   (p 431)

[Li91]   S. Q. Li. A general solution technique for discrete queueing analysis of multimedia traffic on ATM. *IEEE Trans. Comm.*, 39, 1991.   (p 492)

[Lin68]   B. W. Lindgren. *Statistical Theory*. MacMillan, New York, 1968.   (p 145)

[Lin83]   K. Lindberger.   Simple approximations of overflow system quantites for additional demands in the optimization. In *10th Int. Teletraffic Congress Proceedings*, pages 5.3–3, 1983.   (pp 257, 521, 523)

[Lin91]   K. Lindberger. Analytical models for the traffical problems with statistical multiplexing in ATM networks. In *13th Int. Teletraffic Congress Proceedings*, pages 807–813. North Holland-Elsevier Science Publishers, 1991.   (pp 127, 128, 336, 444, 445, 448)

[Lin94]   K. Lindberger. Dimensioning and design methods for integrated ATM networks. In *14th Int. Teletraffic Congress Proceedings*, volume 1b, pages 897–906. North Holland-Elsevier Science Publishers, 1994.   (pp 129, 256, 257, 518, 519, 520, 527, 528)

[LMN90]   D. M. Lucantoni, K. S. Meier-Hellstern, and M. F. Neuts. A single-server queue with server vacations and a class of non-renewal arrival processes. *Adv. in Appl. Probab.*, 22:676–705, 1990.   (p 343)

[LPTB93]   J. Lubacz, M. Pióro, A. Tomaszewski, and D. Bursztynowski. A framework for network design and management. Internal report, Institute of Telecommunications, Warsaw University of Technology, 1993.   (p 235)

[LR93]   G. Latouche and V. Ramaswami. A logarithmic reduction algorithm for quasi-birth-death processes. *J. Appl. Probab.*, 30:650–674, 1993.   (p 338)

[LT94]   J. Lubacz and A. Tomaszewski. A lower bound based approach to the design of meshed networks with protected transmission paths. In *14th Int. Teletraffic Congress Proceedings*, pages 1445–1453. North Holland-Elsevier Science Publishers, 1994.   (p 251)

[LT95a]   K. Lindberger and S-E. Tidblom. Weighted fair queueing, a method to control integrated heterogeneous traffic streams having different GOS demands. In *Twelfth Nordic Teletraffic Seminar*. VTT - Technical Research Centre of Finland, 1995.   (p 184)

[LT95b]   J. Lubacz and A. Tomaszewski. Generic network design procedures. Internal report, Institute of Telecommunications, Warsaw University of Technology, 1995.   (pp 248, 250, 251)

[LTB94]   J. Lubacz, A. Tomaszewski, and D. Bursztynowski. A generic approach to network design - virtual network design. Internal report, Institute of Telecommunications, Warsaw University of Technology, 1994.   (p 251)

[LTWW94]   W. E. Leland, M. S. Taqqu, W. Willinger, and D. V. Wilson. On the self-similar nature of Ethernet traffic (extended version). *IEEE/ACM Trans. Networking*, 2(1):1–15, February 1994.   (pp 15, 165, 340, 354, 369)

[Luc91]   D. M. Lucantoni. New results on the single server queue with a Batch Markovian Arrival Process. *Comm. Statist. Stochastic Models*, 7(1):1–46, 1991.   (pp 338, 343)

[LWTW94]   W. E. Leland, W. Willinger, M. S. Taqqu, and D. V. Wilson. Statistical analysis and stochastic modeling of self-similar datatraffic. In Jacques Labetoulle and James W. Roberts, editors, *14th Int. Teletraffic Congress Proceedings*, volume 1a, pages 319–328. North Holland-Elsevier Science Publishers, June 1994.   (pp 25, 165)

[LY93]   F. Y. S. Lin and J. R. Yee. A real-time distributed routing and admission control algorithm for ATM networks. In *IEEE Infocom 93 Proceedings*, 1993.   (p 296)

[Mak90]   B. A. Makrucki. A study of source traffic management and buffer allocation in ATM networks. In *7th ITC Specialist Seminar, Morristown*, October 1990.   (p 431)

[Man69]   B. B. Mandelbrot. Long-run linearity, locally Gaussian processes, H-spectra, and infinite variances. *International Economic Review*, 10:82–113, 1969.   (p 327)

[MAS+88]   Basil Maglaris, Dimitris Anastassiou, Prodip Sen, Gunnar Karlsson, and John D. Robbins. Performance models of statistical multiplexing in packet video communications. *IEEE Trans. Comm.*, 36(7):834–844, July 1988.   (p 28)

[MB95]   R. Mauger and S. Brueckheimer. The role of ATM in 64 kbit/s switching and transmission networks. In *Int. Switching Symposium, ISS 95*, 1995.   (p 8)

[MB96]   S. Molnar and S. Blaabjerg.   Cell delay variation in a fifo multiplexer. submitted to European Transactions on Telecommunications and Related Technologies, 1996. (pp 421, 426, 430)

[MDJ+95]   M. Ajmone Marsan, T. V. Do, L. Jereb, R. Lo Cigno, R. Pasquali, and A. Tonietti. Some simulation results on the performance of traffics shaping algorithms in ATM networks. In D. D. Kouvatsos, editor, *Performance Modelling and Evaluation of ATM Networks*, London, 1995. IFIP, Chapman & Hall.   (pp 296, 297)

[MGZ89]   Neri Merhav, Michael Gutman, and Jacob Ziv. On the estimation of the order of a Markov chain and universal data compression. *IEEE Trans. on Information Theory*, 35(5):1014–1019, September 1989.   (p 42)

[MHWYH91] K. Meier-Hellstern, P. E. Wirth, Y.-L. Yan, and D. A. Hoeflin. Traffic models for ISDN data users: Office automation application. In *13th Int. Teletraffic Congress Proceedings*, pages 167–172. North Holland-Elsevier Science Publishers, 1991.   (p 15)

[MMR96]   D. Mitra, J. A. Morrison, and K. G. Ramakrishnan. ATM network design and optimization: a multirate loss network framework. In *IEEE Infocom 96 Proceedings*, March 1996. (p 295)

[MMST94]   M. Menozzi, U. Mocci, C. Scoglio, and A. Tonietti. Traffic integration and virtual path optimisation in ATM networks. In *Networks 94, Budapest*, September 1994.   (p 263)

[MMV95a]   J. K. MacKie-Mason and H. R. Varian. Pricing congestible network resources. *IEEE J. Selected Areas in Comm.*, 13(7):1141–1149, September 1995.   (p 14)

[MMV95b]   J. K. MacKie-Mason and H. R. Varian. Some FAQs about usage-based pricing. *Comput. Networks ISDN Systems*, 28:257–265, 1995.   (p 14)

[MN68]   B. B. Mandelbrot and J. W. Van Ness. Fractional Brownian motions, fractional noises and applications. *SIAM Review*, 10:422–437, 1968.   (p 330)

[Moc94]   Y. Mochida. Technologies for local-access fibering. *IEEE Comm. Mag.*, pages 64–73, February 1994.   (p 214)

[MPS95]   U. Mocci, P. Pannunzi, and C. Scoglio. Adaptive capacity management of virtual path networks. In *IEEE Globecom 96 Proceedings*, 1995.   (pp 282, 284)

[MR95]   M. Mandjes and A. Ridder.   Finding the conjugate of Markov fluid processes. *Probability in the Engineering and Informational Sciences*, 9, 1995. To appear.   (pp 459, 463)

[MS91]   A. Mukherjee and J. C. Strikwerda. Analysis of dynamic congestion control protocols – A fokker-plank approximation. *ACM SIGCOMM Computer Communication Review*, pages 159–169, 1991.   (p 97)

[MSB94a]   W. Matragi, K. Sohraby, and C. Bisdikian. Jitter calculus in ATM networks: Multiple node case. In *IEEE Infocom 94 Proceedings*, 1994.   (p 426)

[MSB94b]   W. Matragi, K. Sohraby, and C. Bisdikian. Jitter calculus in ATM networks: Single node case. In *IEEE Infocom 94 Proceedings*, 1994.   (p 426)

[MST96]   U. Mocci, C. Scoglio, and A. Tonietti. Robust design of virtual path networks in ATM multiservice environments. In *Networks 96, Sidney*, November 1996.   (p 274)

[MT79]   Benoit B. Mandelbrot and Murad S. Taqqu. Robust R/S analysis of long-run serial correlation. In *42nd Session ISI, Vol. XLVIII, Book 2*, pages 69–99, 1979.   (p 25)

[Nag87]   J. Nagle. On packet switches with infinite storage. *IEEE Trans. Comm.*, 35:435–438, 1987.   (p 173)

[Neu85]   M. Neuts. A queueing model for a storage buffer in which the arrival rate is controlled by a switch with a random delay. *Performance Evaluation*, 5, 1985.   (p 431)

[Neu89]   M. Neuts. *Structured Stochastic Matrices Of M/G/1 Type And Their Applications*. Marcel Dekker Inc, New York and Basel, 1989.   (pp 343, 491)

[New94]   P. Newman. Traffic management for ATM local area networks. *IEEE Communications Magazine*, pages 44–50, August 1994.   (p 95)

[Nie90]   G. Niestegge. The leaky bucket policing method in the ATM network. *Int. Journ. of Dig. and Anal. Comm. Syst.*, 3(2), June 1990.   (p 67)

[Nor94]   I. Norros. A storage model with self-similar input. *Queueing Systems*, 16:387–396, 1994.   (pp 496, 497)

[Nor95]   I. Norros. On the use of fractional Brownian motion in the theory of connectionless networks. *IEEE J. Selected Areas in Comm.*, 13(6):953–962, August 1995.   (p 332)

[NRSV91]   I. Norros, J. W. Roberts, A. Simonian, and J. Virtamo. The superposition of variable bitrate sources in ATM multiplexers. *IEEE J. Selected Areas in Comm.*, 9(3):378–387, April 1991.   (pp 397, 400)

[NSVV95]   I. Norros, A. Simonian, D. Veitch, and J. Virtamo. A Beneš formula for a buffer with fractional Brownian input. In *9th ITC Specialists Seminar '95: Teletraffic Modelling and Measurement*, 1995.   (pp 379, 498)

[NV91]   I. Norros and J. T. Virtamo. Who loses cells in case of burst scale congestion? In *13th Int. Teletraffic Congress Proceedings*. North Holland-Elsevier Science Publishers, 1991.   (p 447)

[OGV95]   A. Ortega, M. W. Garrett, and M. Vetterli. Rate constraints for video transmission over ATM networks based on joint source/network criteria. *Annals des Télécommunications*, 50(7-8):603–616, 1995.   (p 45)

[OMS+95]   H. Ohsaki, M. Murata, H. Suzuki, C. Ikeda, and H. Miyahara. Rate-based congestion control for ATM networks. *ACM SIGCOMM Computer Communication Review*, 1995.   (p 97)

[Øst94]   O. Østerbø. The effect of shaping or phasing of individual VCs in ATM networks. In *14th Int. Teletraffic Congress Proceedings*, volume 1a, pages 461–470. North Holland-Elsevier Science Publishers, 1994.   (pp 413, 414, 415)

[Øst95]   O. Østerbø. Some important queueing models for ATM. *Telektronikk*, 93(2/3):208–219, 1995.   (p 413)

[Ott77]   T. J. Ott. The covariance function of the virtual waiting-time process in an M/G/1 queue. *Adv. in Appl. Probab.*, 9:158–168, 1977.   (p 495)

[OY95]   E. Oki and N. Yamanaka. An optimum logical ATM network design method guaranteeing multimedia QoS requirements. In *IEEE Globecom 95 Proceedings*, 1995.   (p 274)

[PAB88]   L. Platzman, J. Ammons, and J. Bartoldi. A simple and efficient algorithm to compute tail probabilities from transforms. *Oper. Res.*, 36(1), 1988.   (p 433)

[Pak71a]   A. G. Pakes.   The correlation coefficients of the queue lengths of some stationary single server queues. *J. Austral. Math. Soc.*, 13:35–46, 1971.   (p 495)

[Pak71b]   A. G. Pakes. The serial correlation coefficients of waiting times in the stationary GI/M/1 queue. *Ann. Math. Statist.*, 42:1727–1734, 1971.   (p 495)

[Pan96]   F. Panken.   A TDMA based access control scheme for an ATM passive optical tree network. In L. Mason and A. Casaga, editors, *Broadband Communications, Proceedings of the International IFIP-IEEE Conference on Broadband Communications, Canada*, pages 321–332. Chapman & Hall, April 1996.   (pp 221, 223)

[PC90]   E. Pinsky and A. Conway. Performance analysis of sharing policies for broadband networks. In *7th ITC Specialist Seminar, Morristown*, page 11.4, 1990.   (p 514)

[PElZ93]   P. Pancha and M. El Zarki. Bandwidth requirements of variable bit rate MPEG sources in ATM networks. In Pujolle Perros and Takahashi, editors, *Modelling and Performance Evaluation of ATM Technology*, pages 5.2.1–25. North Holland-Elsevier Science Publishers, 1993.   (pp 28, 30)

[Pet94]   J. Petersen. The idle time distribution of a system $\Sigma D/D/1$. *IEEE Trans. Comm.*, 42:854–856, 1994.   (p 401)

[PF95]   V. Paxson and S. Floyd.   Wide-area traffic: The failure of Poisson modeling. *IEEE/ACM Trans. Networking*, 3(3):226–244, 1995.   (pp 15, 328)

[PG93a]   A. Parekh and R. Gallager. A generalized processor sharing approach to flow control in integrated services networks: The single node case. *IEEE/ACM Trans. Networking*, 1(3):344–357, June 1993.   (pp 91, 173, 174, 175, 181)

[PG93b]   J. Petersen and T. Gillen. Multiplexers $\Sigma D/D/1$ with service in cyclic order and service in the order of arrival. *Communication Networks*, 4:455–463, 1993.   (p 401)

[PG94]   A. Parekh and R. Gallager. A generalized processor sharing
            approach to flow control in integrated services networks:
            The multiple node case. *IEEE/ACM Trans. Networking*,
            2(2):137–150, April 1994.   (pp 177, 178)

[PG95a]   M. Pióro and P. Gajowniczek. Design of virtual paths by
            simulated allocation. In 1st *IEEE International Workshop
            on Broadband Switching Systems*, April 1995.   (pp 289, 291)

[PG95b]   M. Pióro and P. Gajowniczek. Stochastic allocation of virtual paths to ATM links. In D. Kouvatsos, editor, *Performance Modelling and Evaluation of ATM Networks*. Chapman & Hall, 1995.   (pp 288, 289, 291)

[PJ95]   A. Pfening and L. Jereb. Comparison of ATM routing algorithms. Internal report, Technical University of Budapest,
            1995.   (pp 296, 297)

[PKW89]   M. Pióro, U. Körner, and B. Wallström. Design methods
            and routing control in integrated services networks with alternative routing. In *12th Int. Teletraffic Congress Proceedings*. North Holland-Elsevier Science Publishers, 1989.
            (pp 541, 545)

[PLK90]   M. Pióro, J. Lubacz, and U. Körner. Traffic engineering
            problems in multiservice circuit switched networks. *Comput.
            Networks ISDN Systems*, 20:127–126, 1990.   (p 542)

[Pra67]   C. W. Pratt. The concept of marginal overflow in alternate
            routing. In *5th Int. Teletraffic Congress Proceedings*, June
            1967.   (p 260)

[Pru95]   P. Pruthi. *An Application of Chaotic Maps to Packet Traffic
            Modeling*. PhD thesis, Royal Institute of Technology, Dept
            of Teleinformatics, 1995.   (pp 329, 499)

[PS90]   J. M. Pitts and J. A. Schormans. Analysis of ATM switch
            model with time priorities. *Electronic Letters*, 26(15), July
            1990.   (p 431)

[Rat91]   E. P. Rathgeb. Modeling and performance comparison of policing mechanisms for ATM networks. *IEEE J. Selected Areas in Comm.*, 9(3), April 1991.   (pp 51, 60)

[RB95]   S. Robert and J.-Y. Le Boudec. Properties of a new class of models designed for self-similar traffic. In *IFIP WG 6.3*, Paris, France, December 1995.   (p 340)

[RB96]   S. Robert and J.-Y. Le Boudec. Can self-similar traffic be modelled by Markovian processes? In *1996 International Zurich Seminar on Digital Communications, IZS'96*, ETH-Zentrum, Zurich, Switzerland, February 1996.   (p 340)

[RBC93]   J. W. Roberts, B. Bensaou, and Y. Canetti. A traffic control framework for high speed data transmission. In H. Perros, G. Pujolle, and Y. Takahashi, editors, *IFIP Transactions C-15: Modelling and Performance Evaluation of ATM Technology*. North Holland-Elsevier Science Publishers, January 1993.   (pp 149, 180, 185, 483, 485, 487, 508)

[RBS95]   J. W. Roberts, P. Boyer, and M. Servel. A real time sorter with application to ATM traffic control. In *Int. Switching Symposium, ISS 95*, 1995.   (p 182)

[RC94]   R. Rooholamini and V. Cherkassky. Moving ATM closer to multimedia applications. In *19Th Conference on Local Communication Networks*, Minneapolis, Minnesota, October 1994.   (p 208)

[RD91]   G. Ramamurthy and R. S. Dighe. Distributed source control: A network access control for integrated broadband packet networks. *IEEE J. Selected Areas in Comm.*, 9(7):990–1002, September 1991.   (p 481)

[RF94]   Oliver Rose and Michael R. Frater. A comparison of models for VBR video traffic sources in B-ISDN. In *IFIP Transactions C-24: Broadband Communications, II*, pages 275–287. North Holland-Elsevier Science Publishers, 1994. URL:http://www-info3.informatik.uni-wuerzburg.de/HTML/publications.html.   (p 28)

[RF95]   Oliver Rose and Michael R. Frater. Impact of MPEG video traffic on an ATM multiplexer. In Ramon Puigjaner, editor, *High Performance Networking VI*, pages 157–168. Chapman & Hall, 1995. URL:http://www-info3.informatik.uni-wuerzburg.de/HTML/publications.html.   (pp 29, 30)

[RG92]   J. W. Roberts and F. Guillemin. Jitter in ATM networks
              and its impact on peak rate enforcement. *Performance Eval-
              uation*, 16:35–68, 1992.   (p 420)

[RH92]   A. R. Reibman and B. G. Haskell. Constraints on variable
              bit rate video for ATM networks. *IEEE Transactions On
              Circuits and Systems for Video Technology*, 2(4):361–372,
              December 1992.   (p 45)

[RH94]   C. Rosenberg and G. Hébuterne. Dimensioning traffic con-
              trol devices in an ATM network. In *Proc. IFIP workshop*,
              March 1994.   (p 49)

[Rio62]   J. Riordan. *Stochastic Service Systems*. John Wiley & Sons,
              1962.   (p 433)

[Rit95a]   M. Ritter. Performance analysis of the dual cell spacer in
              ATM systems. In *IFIP 6th International Conference on
              High Performance Networking*, September 1995.   (p 510)

[Rit95b]   M. Ritter. Steady-state analysis of the rate-based conges-
              tion control mechanism for ABR services in ATM networks.
              *University of Würzburg, Institute of Computer Science, Re-
              search Report Series*, 114, 1995.   (pp 97, 100)

[Rit96]   M. Ritter. Network buffer requirements of the rate-based
              control mechanism for ABR-services. In *IEEE Infocom 96
              Proceedings*, 1996.   (p 104)

[RMR94]   Daniel Reininger, Benjamin Melamed, and Dipankar Ray-
              chaudhuri. Variable bit rate MPEG video: Characteris-
              tics, modeling and multiplexing. In *14th Int. Teletraffic
              Congress Proceedings*, volume 1a, pages 295–306. North
              Holland-Elsevier Science Publishers, June 1994.   (p 28)

[Rob81]   J. W. Roberts. A service system with heterogeneous user
              requirements - application to multi-service telecommuni-
              cations systems. In G. Pujolle, editor, *Performance of
              Data Communication Systems and their Applications*, pages
              423–431. North Holland-Elsevier Science Publishers, 1981.
              (pp 514, 516, 542)

[Rob83]   J. W. Roberts. Teletraffic models for the telecom 1 inte-
              grated services network. In *10th Int. Teletraffic Congress
              Proceedings*, page 1.1.2, 1983.   (pp 522, 523, 524, 525)

[Rob92a]   J. W. Roberts, editor. *COST224 Performance evaluation and design of multiservice networks.* Commission of the European Communities, October 1992. Final Report.   (pp V, 127, 129, 271, 303, 336, 376)

[Rob92b]   J. W. Roberts. Traffic control in the B–ISDN. In *ITC Specialists Seminar*, Cracow, March 1992.   (p 390)

[Rob94]   J. W. Roberts. Virtual Spacing for flexible traffic control. *Int. J. of Communications Systems*, 7:307–318, 1994.   (p 173)

[Rog95]   L. C. G. Rogers. Arbitrage with fractional Brownian motion. To appear in Mathematical Finance, 1995.   (p 367)

[Roo85]   A. H. Roosma. On estimating call congestion and time congestion. Technical Report Internal Document, Notitie 85 INF/130, PTT Research, Leidschendam, Netherlands, 1985.   (pp 521, 522, 523)

[Ros74]   S. M. Ross. Bounds on the delay distribution in GI/G/1 queues. *J. Appl. Probab.*, 11:417–421, 1974.   (p 69)

[Ros95]   Oliver Rose. Statistical properties of MPEG video traffic and their impact on traffic modeling in ATM systems. In *20th Annual Conference on Local Computer Networks, Minneapolis, MN*, October 1995. URL:http://www-info3.informatik.uni-wuerzburg.de/HTML/publications.html.   (pp 30, 165)

[RR94]   O. Rose and M. Ritter. MPEG-video sources in ATM-systems — A new approach for the dimensioning of policing functions. In *IFIP 3rd International Conference on Local and Metropolitan Communication Systems*, December 1994.   (p 505)

[RS92]   G. Ramamurthy and B. Sengupta. Modelling and analysis of a variable bit rate video multiplexer. In *IEEE Infocom 92 Proceedings*, pages 6C.1.1–11, 1992.   (p 28)

[RT94a]   M. Ritter and P. Tran-Gia, editors. *Multi-Rate Models for Dimensioning and Performance Evaluation of ATM Networks, COST224 Interim Report.* Commission of the European Communities, 1994.   (p 127)

[RT94b]   M. Ritter and P. Tran-Gia. Performance analysis of cell rate monitoring mechanisms in ATM systems. In *IFIP 3rd International Conference on Local and Metropolitan Communication Systems*, Kyoto, December 1994.   (p 504)

[RV91]   J. W. Roberts and J. T. Virtamo. The superposition of periodic cell arrival streams in an ATM multiplexer. *IEEE Trans. Comm.*, 39(2):298–303, February 1991.   (p 397)

[SA61]   H. A. Simon and A. Ando. Aggregation of variables in dynamic systems. *Econometrica*, 29, 1961.   (p 338)

[SA95]   R. A. Spanke and J. M. Adrian. ATM composite cell switching for D20 digital switches. In *Int. Switching Symposium, ISS 95*, 1995.   (p 8)

[SB93]   B. Steyaert and H. Bruneel. A general relationship between buffer occupancy and delay in discrete-time multiserver queueing models, applicable in ATM networks. In *IEEE Infocom 93 Proceedings*, March 1993.   (p 373)

[SCK93]   C. Szabo, I. Chlamtac, and Z. Kovacs. ATM-like transmission of isochronous traffic in FDDI and DQDB. In *Proc. 6Th IEEE Workshop on LANs and MANs*, San Diego, CA, October 1993.   (p 201)

[SG95]   A. Simonian and J. Guibert. Large deviation approximations for fluid queues fed by a large number of on/off sources. *IEEE J. Selected Areas in Comm.*, 13(6), August 1995. (pp 167, 308, 383, 481, 482, 483, 484)

[SGLR92]   A. Simonian, F. Guillemin, J. R. Louvion, and L. Romoeuf. Transient analysis of statistical multiplexing on an ATM link: In *IEEE Globecom 92 Proceedings*, 1992.   (p 453)

[She95]   S. Shenker. Fundamental design issues for the future Internet. *IEEE J. Selected Areas in Comm.*, 13(7):1176–1188, 1995.   (p 61)

[SHW83]   B. Sanders, W. H. Haemers, and R. Wilcke. Simple approximation techniques for congestion functions for smooth and peaked traffic. In *10th Int. Teletraffic Congress Proceedings*, page 4.4b.1, 1983.   (p 521)

[Sim62]   H. A. Simon. The architecture of complexity. *Proc. Amer. Phil. Soc.*, 106:467–482, 1962.   (p 338)

[Sim69]   H. A. Simon. The sciences of the artificial. *MIT Press*, Cambridge, Massachussetts, 1969.   (p 338)

[Sim91]   A. Simonian. Stationary analysis of a fluid queue with input rate varying as an Ornstein–Uhlenbeck process. *SIAM J. Appl. Math.*, June 1991.   (p 468)

580   Bibliography

[Slo94a]   R. Slosiar. Busy and idle periods at an ATM multiplexer output resulting from the superposition of homogeneous ON/OFF sources. In *14th Int. Teletraffic Congress Proceedings*, volume 1a, pages 431–440. North Holland-Elsevier Science Publishers, June 1994.   (p 493)

[Slo94b]   R. Slosiar. Moments of the queue occupancy in an atm multiplexer loaded with ON/OFF sources. In *Proceedings of the Singapore International Conference on Communication Systems (ICCS)*, volume 2, pages 754–759, November 1994. Singapore.   (p 493)

[Slo95a]   R. Slosiar. Exact results for an ATM multiplexer with infinite queue loaded with batch Markovian arrivals. In D. Kouvatsos, editor, *Performance Modelling and Evaluation of ATM Networks*, volume 1, pages 297–322. Chapman & Hall, 1995.   (p 493)

[Slo95b]   R. Slosiar. *Performance Analysis Methods of ATM-Based Broadband Access Networks Using Stochastic Traffic Models.* PhD thesis, EPFL, Lausanne, Switzerland, 1995.   (p 217)

[Sny75]   D. L. Snyder. *Random Point Processes.* John Wiley & Sons, New York, 1975.   (p 352)

[SS95]   R. Schlittgen and B. H. J. Streitberg. *Zeitreihenanalyse.* R. Oldenbourg Verlag, 6 edition, 1995. (In German).   (pp 338, 361)

[SSD93]   Paul Skelly, Mischa Schwarz, and Sudhir Dixit. A histogram-based model for video traffic behavior in an ATM multiplexer. *IEEE/ACM Trans. Networking*, 1(4):446–459, August 1993.   (p 30)

[ST94]   G. Samorodnitsky and M. Taqqu. *Stable Non-Gaussian Random Processes.* Chapman & Hall, 1994.   (pp 327, 357)

[Sta93]   J. D. Angelopoulos I. S. Venieris G. I. Stassinopoulos. Dynamic bandwidth distribution for APONs with emphasis on low cell jitter. In *Proceedings of 11th Annual Conference EFOC & N*, pages 86–90, June 1993. The Hague.   (p 220)

[SV91]   A. Simonian and J. T. Virtamo. Transient and stationary distributions for fluid queues and input processes with a density. *SIAM J. Appl. Math.*, 51, December 1991.   (pp 322, 377, 467, 468)

[SW86a]   G. R. Shorack and J. A. Wellner. *Empirical Processes with Application to Statistics*. John Wiley & Sons, 1986. (pp 392, 400)

[SW86b]   K. Sriram and W. Whitt. Characterizing superposition arrival processes in packet multiplexers for voice and data. *IEEE J. Selected Areas in Comm.*, 4(6):833–846, September 1986. (pp 229, 317)

[SW95]   A. Shwartz and A. Weiss. *Large Deviations for Performance Analysis*. Chapman & Hall, 1995. (p 379)

[SX95]   B. Steyaert and Y. Xiong. Buffer requirements in ATM-related queueing models with bursty traffic: an alternative approach. In *Proceedings of the Sixth IFIP WG6.3 Conference on Performance of Computer Networks, Istanbul*, October 1995. (pp 492, 493)

[Sys86]   R. Syski. *Introduction to Congestion Theory in Telephone Systems*. North Holland-Elsevier Science Publishers, 1986. 2nd. edition. (pp 390, 391, 393, 433)

[Tak64]   L. Takács. Priority queues. *Oper. Res.*, 12:63–74, 1964. (pp 431, 432)

[Tak67]   L. Takács. *Combinatorial Methods in the Theory of Stochastic Processes*. John Wiley & Sons, 1967. (p 395)

[Tak85]   H. Takagi. Analysis of a finite-capacity M/G/1 queue with a resume level. *Performance Evaluation*, 5, 1985. (p 431)

[TGH93]   P. Tran-Gia and F. Hübner. An analysis of trunk reservation and grade of service balancing mechanisms in multiservice broadband networks. In *IFIP Workshop TC6, Modeling and Performance Evaluation of ATM Technology*, page 2.1, La Martinique, 1993. (pp 517, 524, 525, 533)

[TKS88]   I. Tokizawa, T. Kanada, and K. Sato. A new transport architecture based on asynchronous transfer model technique. In *ISSLS*, 1988. (p 271)

[TL86]   M. S. Taqqu and J. B. Levy. *Using renewal processes to generate long-range dependence and high variablity*, volume 11 of *Progress in Prob. and Stat. (Dependence in Probability and Statistics)*, pages 77–89. Birkhäuser, Boston, 1986. (pp 333, 334, 335)

[Tra89]   D. Tranchier. Evaluation of jitter and end-to-end delay in ATM networks. Technical report, CNET, December 1989. in French.   (p 431)

[Tuc88]   R. C. Tucker. Accurate method for analysis of a packet speech multiplexer with limited delay. *IEEE Trans. Comm.*, 36(4), April 1988.   (p 465)

[TVC+94]   T. Toniatti, L. Verri, O. Casals, J. García, C. Blondia, J. Angelopoulos, and I. Venieris. Performance of shared medium access protocols for ATM traffic concentration. *European Transactions on Telecommunications*, 5(2), March 1994.   (p 226)

[VAD96]   J. Virtamo, S. Aalto, and D. Down. Window based estimation of the intensity of a doubly stochastic Poisson process. In L. Mason and A. Casaga, editors, *Broadband Communications, Proceedings of the International IFIP-IEEE Conference on Broadband Communications, Canada*, pages 294–305. Chapman & Hall, April 1996.   (pp 284, 351, 352, 353)

[Van71]   H. L. Van Trees. *Detection, Estimation, and Modulation Theory: Part III.* John Wiley & Sons, New York, 1971.   (p 352)

[VB95]   B. Vinck and H. Bruneel. A note on system contents and cell delay in FIFO ATM-buffers. In *Proceedings of the Third Workshop on Performance Modelling and Evaluation of ATM Networks, Bradford*, July 1995.   (p 372)

[vdBDRvdW95] J. L. van den Berg, E. B. Diks, J. A. C. Resing, and J. van der Wal. The change of traffic characteristics in ATM networks 2. Memorandum COSOR 95-09, Eindhoven University of Technology, 1995.   (p 430)

[vdBR92]   J. L. van den Berg and J. A. C. Resing. The change of traffic characteristics in ATM networks (COST 242 TD(92)040). Technical report, KPN Research Netherlands, 1992.   (pp 73, 79)

[vDP93]   E. A. van Doorn and F. J. M. Panken. Blocking probabilities in a loss system with arrivals in geometrically distributed batches and heterogeneous service requirements. *IEEE/ACM Trans. Networking*, 1(6):664–667, December 1993.   (p 522) [vdVA94]   H. A. B. van de Vlag and A. Awater. Exact computation of time and call blocking probabilities in multi-traffic circuit-switched networks. In *IEEE Infocom 94 Proceedings*, pages 56–65, 1994.   (p 514)

[Vei92]   D. Veitch. Novel models of broadband traffic. In *Proc. 7th Australian Teletraffic Research Seminar*, Murray River, Australia, 1992.   (p 328)

[Vir88]   J. T. Virtamo. Reciprocity of blocking probabilities in multiservice loss systems. *IEEE Trans. Comm.*, 36(10):1174–1175, 1988.   (pp 516, 517)

[Vir94]   J. T. Virtamo. Idle and busy period distributions of an infinite capacity $N * D/D/1$ queue. In *14th Int. Teletraffic Congress Proceedings*, volume 1a, pages 453–459. North Holland-Elsevier Science Publishers, 1994.   (pp 401, 402, 403)

[Vir95]   J. T. Virtamo. Numerical evaluation of the distribution of unfinished work in an $M/D/1$ system. *Electronics Letters*, 31(7):531–532, March 1995.   (p 391)

[VN94]   J. T. Virtamo and I. Norros. Fluid queue driven by an $M/M/1$ queue. *Queueing Systems*, 16:373–386, 1994.   (pp 465, 466)

[VR89]   J. T. Virtamo and J. W. Roberts. Evaluating buffer requirements in an ATM multiplexer. In *IEEE Globecom 89 Proceedings*, 1989.   (pp 383, 405, 408)

[WA92]   S. M. Walters and M. Ahmed. Broadband virtual private networks and their evolution. In *Int. Switching Symposium, ISS 92*, October 1992.   (p 186)

[Wag87]   S. S. Wagner. Optical amplifier applications in fiber optic loacal area networks. *IEEE Trans. Comm.*, COM-35:419–426, April 1987.   (p 214)

[WB94]   S. Wittevrongel and H. Bruneel. Queue length and delay for statistical multiplexers with variable-length messages. In *IEEE Globecom 94 Proceedings*, pages 1080–1084, November 1994.   (p 430)

[Whi53]   P. Whittle. Estimation and information in stationary time series. *Arkiv för Matematik*, 2:423–434, 1953.   (p 361)

[Whi85]   W. Whitt. Blocking when service is required from several facilities simultaneously. *AT&T Bell Labs. Tech. J.*, 64:1807–1856, 1985.   (p 528)

[Wol82]   R. W. Wolff. Poisson arrivals see time averages. *Oper. Res.*, 30(2), April 1982.   (p 434)

[Won90]   A. Wong. Queueing analysis for ATM switching of continuous bit oriented traffic — a recursive computation method. In *IEEE Globecom 90 Proceedings*, December 1990. (p 372)

[WTLW95]   W. Willinger, M. S. Taqqu, W. E. Leland, and D. V. Wilson. Self-similarity in high-speed packet traffic: Analysis and modeling of Ethernet traffic measurements. *Statistical Science*, 10(1):67–85, 1995.   (pp 165, 361)

[WTSW95]   W. Willinger, M. S. Taqqu, R. Sherman, and D. V. Wilson. Self-similarity through high-variability: Statistical analysis of Ethernet LAN traffic at the source level. In *ACM SIGCOM 95 Proceedings*, pages 100–113, Cambridge, MA, USA, 1995.   (pp 16, 165, 333)

[XB94]   Y. Xiong and H. Bruneel. On the asymptotic behavior of discrete-time single-server queueing systems with general Markov modulated arrival processes. In *14th Int. Teletraffic Congress Proceedings*, pages 179–189. North Holland-Elsevier Science Publishers, June 1994.   (p 493)

[XB95]   Y. Xiong and H. Bruneel. A simple approach to obtain tight upper bounds for the asymptotic queueing behavior of statistical multiplexers with heterogeneous traffic. *Performance Evaluation*, 22, 1995.   (p 493)

[YH91]   N. Yin and M. G. Hluchyj. A dynamic rate control mechanism for source coded traffic in a fast packet network. *IEEE J. Selected Areas in Comm.*, 9(7):1003–1012, 1991.   (p 97)

[YH94]   N. Yin and M. G. Hluchyj. On closed-loop rate control for ATM cell relay networks. In *IEEE Globecom 94 Proceedings*, pages 99–108, 1994.   (p 97)

[YKH95]   A. Yamashita, R. Kawamura, and H. Hadama. Dynamic VP rearrangement in an ATM network. In *IEEE Globecom 95 Proceedings*, 1995.   (p 274)

[Zha90]   L. Zhang. VirtualClock: A new traffic control algorithm for packet switching networks. In *ACM SIGCOM 90 Proceedings*, 1990.   (pp 173, 175)

Lecture Notes in Computer Science

For information about Vols. 1–1083

please contact your bookseller or Springer-Verlag